DYNAMIC METEOROLOGY

S. PANCHEV

Department of Meteorology and Geophysics,
University of Sofia, Bulgaria

Dynamic Meteorology

D. Reidel Publishing Company

A MEMBER OF THE KLUWER ACADEMIC PUBLISHERS GROUP

Dordrecht / Boston / Lancaster

Library of Congress Cataloging in Publication Data

CIP

Panchev, S. (Stoĭcho), 1933–
Dynamic meteorology.

(Environmental fluid mechanics)
Translation of: Dinamichna meteorologiîa.
Bibliography: p.
Includes index.
1. Dynamic meteorology. I. Title. II. Series.
QC880.P2913 1985 551.5 84-27698
ISBN 90-277-1744-3

Translated by S. Panchev, K. H. Nenkov, and L. A. Kostadinov

This translation has been made and expanded from the original Bulgarian edition of
ДИАМИЧНА МЕТЕОРОЛОГИЯ (*Dynamic Meteorology*)
published in 1981 by the Faculty of Physics, Sofia University
English Edition published by D. Reidel Publishing Company,
P.O. Box 17, 3300 AA Dordrecht, Holland

Sold and distributed in the U.S.A. and Canada
by Kluwer Academic Publishers,
190 Old Derby Street, Hingham, MA 02043, U.S.A.

Sold and distributed in Albania, Bulgaria, Cuba, Czechoslovakia, German Democratic Republic, Hungary, Korean People's Democratic Republic, Mongolia, People's Republic of China, Poland, Rumania, the U.S.S.R., Vietnam, and Yugoslavia by JUSAUTOR, 11 Slaveikov Square, Sofia, Bulgaria.

In all other countries, sold and distributed by
Kluwer Academic Publishers Group,
P.O. Box 322, 3300 AH Dordrecht, Holland

Printed in The Netherlands

Table of Contents

PART I: THE DYNAMICS OF AN IDEAL (WITHOUT FRICTION) ATMOSPHERE

Preface

1. ABOUT THE DISCIPLINE 'DYNAMIC METEOROLOGY'

The name 'dynamic meteorology' is traditional for designating a university course as well as the scientific branch of meteorology as a whole. While there is no need to abandon this name, it needs contemporary treatment and specifications in its definition. A synonym for it could be 'dynamics (more precisely, hydrodynamics or fluid dynamics) of the atmosphere'. It suggests the relationship of this discipline to general hydrodynamics and applied mathematics and its pronounced theoretical nature.

Besides the atmosphere, however, our planet has another (liquid) envelope – the hydrosphere (world's ocean), which also concerns ocean dynamics and, therefore, it is necessary to define, from a unified standpoint, the subject and aims of the disciplines dealing with the dynamics of the processes which take place in both fluid spheres. Such a unified standpoint offers the so-called *geophysical fluid dynamics*. During the past few years this description is encountered quite often in scientific literature concerning the Earth as a planet. Obviously, a scientific branch or a science is created whose subject is our planet and the investigation methods are borrowed from classical fluid dynamics and applied mathematics, including the most recent numerical methods. As can be seen from its very suitable name, it is the dynamics of quite definite geophysical fluids (atmosphere, ocean and even the liquid inside of the Earth) and not of some abstract (often perfect) fluids, as in classical hydrodynamics. What then, distinguishes it from the latter? These differences in total could be accepted as a definition of geophysical fluid dynamics and dynamic meteorology respectively. As we take it, the most important of them are the following:

(a) The characteristic velocities U of motion of the above-mentioned geophysical fluids are considerably lower than the speed of sound a_s (Mach number $\mathrm{Ma} = U/a_s \ll 1$). Therefore, a number of phenomena of nearly-sonic and supersonic hydrodynamics are not presented here.

(b) The motions take place on the rotating Earth which has a spherical (or more complex) shape. This gives rise to additional forces, the most important of them being the Coriolis force which is significant for the motions in the atmosphere and ocean.

(c) The motions have always turbulent nature (with Reynolds numbers $\mathrm{Re} \gtrsim \mathrm{Re}_{cr} \sim 10^3$) and are multi-scale. This means that turbulent fluctuations have a broad frequency and spatial spectrum – periods from fractions of a second to millennia (practically infinity) and wavelengths from millimeters to 30–40 thousands of kilometers, restricted by the size of our planet.

(d) The geophysical fluids are always inhomogeneous (thermally or density stratified) and this leads to the existence of a favoured direction (the vertical) in the field of gravity.

(e) The large-scale motions of the geophysical fluids are quasi-two-dimensional in the sense that the height (depth) of the atmosphere (ocean) is less than their horizontal scale by several orders of magnitude. This feature determines a series of specific peculiarities and regularities of those motions.

Thus, the science which by means of classical physics (at least for the present!) and with the extensive use of applied mathematics and modern numerical methods, studies the motions of geophysical fluids with the above-listed features, is called geophysical fluid dynamics. Obviously, it is composite science, and dynamic meteorology is the part of it which deals with investigating atmospheric motions. The other component is ocean dynamics. However, motions in the atmosphere and the world's ocean are not independent but interrelated and interdeterminate. The atmosphere and hydrosphere interact through the whole spectrum of motion scales (micro-, meso- and macroscale interaction). Therefore, the above division of geophysical fluid dynamics is conventional, or more precisely, it is an historical heredity which is still too strongly emphasized in the presentation of these subjects in universities. Indeed, while in scientific studies the necessity of a joint investigation of the atmosphere and ocean at the present stage is more or less realized and is put into practice (complex expeditions on national and international scale, laboratories, departments, institutes, etc.), such is not the case with the teaching. The geophysical fluid dynamics in a rigorous logical formulation and presentation still appears to practicing meteorologists, oceanologists and, to a much greater extent, geographers as rather abstract while at the same time 'modern' theoretical physicists and 'pure' mathematicians consider it to be a rather applied science and rarely give it any attention. Courses in geophysical fluid dynamics are still rare in universities. More often, separate courses in the dynamics of the atmosphere and the ocean are presented. The situation in Bulgaria is the same.

2. ABOUT THE PRESENT TEXTBOOK

For almost 30 years, a course in dynamic meteorology for physicists in the field of meteorology has been presented at the University of Sofia, and during the past 20 years, the author, a member of the Faculty of Physics, has been engaged in writing and perfecting this textbook.

The first (Bulgarian) edition was published in 1968 in the form of lecture notes and served as a much-needed stop-gap, since such a book in Bulgarian had not previously existed. It was quite natural that that edition should reflect the curricula of the time, the author's experience and, to a considerable extent, the cliché in the presentation of the subject from existing literature. But since then, the curricula have undergone two transformations, mainly for the sake of improvement and updating. New courses were introduced and related textbooks were written in Bulgarian (e.g., in physical oceanography and synoptic meteorology, etc.). All this stimulated the author to prepare a new edition (practically a new textbook) which was published in 1981 by the Sofia University Press.

The English version of the textbook is based on the 1981 Bulgarian edition. The translation was made by the author himself with the aid of his former students K. Nenkov and L. Kostadinov, and further improvements were made during the translation, mainly by adding new material.

The world literature is relatively poor in textbooks on dynamic meteorology and existing ones can be counted on the fingers of one's hands. Some of the most recent are textbooks in the Russian [35], English [11], [27], and Serbian [45] languages. The reader, however, will be easily convinced that [35] is much too rich in time-consuming mathematical derivations and graphical (theoretical or measured) data that makes it difficult to understand; that [11] has a strong theoretical (mathematical) trend; while [27] is synoptically biased and [45] is more a textbook on numerical weather prediction than on dynamic meteorology.

Undoubtedly, it is very difficult to write a textbook for a discipline with such dynamics in its development as dynamic meteorology, which gives rise to new disciplines (e.g., numerical methods of weather prediction), while tending in itself to 'integrate' with ocean dynamics. Probably every lecturer in this field has his own outlook on this and the author is no exception. Nevertheless, we have asked for and have taken into account the many valuable recommendations proposed by Bulgarian and foreign university colleagues, including the reviewers of the 1981 edition.

Besides the obligatory informative and reference material used in the University syllabus, the textbook also includes many additional problems, which are often of increasing difficulty and which follow the most recent original and review articles that could activate the individual work of assiduous students. In order to provide some convenience in this respect, the content of each section is given in several subtitles. Test problems are also included at the end of most chapters to stimulate the individual work of the students.

Some readers may get the impression that this textbook is poorly illustrated, but in defence, we would point out that we have tried to avoid figures with sophisticated and precise drawings, more appropriate for monographs and handbooks, while giving preference to model drawings and diagrams which help the visualization of a given problem.

As far as the references are concerned, they are divided into two sections. The first is devoted to cited textbooks and monographs (they are referenced in the text by numbers), while the second covers cited papers which have been published in scientific journals (referenced in the text by author and year of publication). Sometimes, as is quite normal, names of authors are often met in the text (usually relating to classic findings) without a corresponding literature source being cited and, likewise, the opposite, a problem is presented without being related to a certain author. To do this, some subjective decisions and omissions seem to be unavoidable, for which we apologize, although the author could not resist the temptation of including some of his own results.

The book begins with an introductory chapter on the kinematics of atmospheric motions (questions that can be found in general courses in meteorology or fluid mechanics). The actual text is divided into two parts. The first, consisting of Chapters 2–7, deals with the dynamics of the atmosphere without friction, conditionally called an 'ideal' atmosphere, as opposed to the 'real' (with friction) atmosphere, which is the subject of the second part. Of course, none of the models considered is completely adequate for the *real* atmosphere.

It is realized that this textbook will only fit in with the syllabus of dynamic meteorology at Sofia University. Nevertheless, it is hoped that not only English-speaking students but also university teachers and specialists in this field will find interesting things from a methodological and scientific point of view.

The author would appreciate any comments, suggestions and criticisms, since we believe that this book is not free of inevitable omissions and mistakes.

Acknowledgements

First of all I am very much indebted to my former students K. Nenkov and L. Kostadinov for their help in the translation and to Mr. Richard Freeman of the D. Reidel Publishing Company for his heavy editorial work. The following publishers, organizations and individuals granted me permission to reproduce previously-published figures and pictures: McGraw Hill, New York, Academic Press, New York, 'Nauka' Publishing House, Moscow, The Royal Aeronautical Society, London, Vaisala Co., Helsinki, American Meteorological Society; G. K. Batchelor, J. R. Holton, F. Mesinger, P. Morel, N. Phillips, G. W. Platzman, R. S. Scorer, Ph.D. Thompson, F. Wipperman. To all of them I am most thankful.

Mrs R. Mitzeva and Miss K. Baykusheva produced all the diagrams and the photographic work was undertaken by S. Evtimov. Mrs N. Todorova, a specialist in English, contributed to the improvement of the original translation and typed the manuscript. I am most thankful for all this technical assistance.

Finally, the whole project of translation and publication of the book would have been impossible without the devoted effort of Dr. D. J. Larner from the Acquisition Department of the D. Reidel Publishing Company.

October, 1984
Sofia

S. PANCHEV

List of Commonly-Used Symbols

ROMAN LETTERS

a_s	sound speed
$A(\mu)$, $B(\mu)$, $C(\mu)$	universal resistance-law functions
b	mean kinetic energy
$B(\tau)$, $B(r)$	correlation functions
c	wind velocity magnitude; wave phase velocity
$C_d(C_{dg})$	drag coefficient (geostrophic)
D	divergence
$D(\tau)$, $D(r)$	structure functions
E	energy; spectral density
E	Ekman number
f	Coriolis parameter; unspecified meteorological element
F	external force
g	gravity acceleration
h	surface-layer thickness
h_g	planetary boundary-layer thickness
$H(x, y)$	terrain height
$\mathbf{i}, \mathbf{j}, \mathbf{k}$	unit vectors
$i = \sqrt{-1}$	imaginary unit
J_n, J_0, K_0	Bessel functions
k	wavenumber
$\mathbf{k}$	wave vector
$K(K_H)$	eddy viscosity coefficient (horizontal)
Ki	Kibel number
ℓ	mixing length
L	length scale
m, m_α, m_T	static stability parameters
mb	millibars
p	pressure
p_{00} (= 1000 mb)	standard pressure
P_n^m	spherical functions
q	vertical heat flux; specific humidity
Q	heating (cooling) rate
r	distance; relative humidity

R	gas constant
R	Rayleigh number
$R(\tau)$, $R(r)$	correlation coefficients
Re (Re_λ)	Reynolds number
Ro	Rossby number
s	entropy
S	surface
t, t_d	time, dew-point temperature
T	absolute temperature; time scale; wave period
Tr	rate of energy transformation
$\mathbf{u}$	two-dimensional (wind) velocity vector
$\mathbf{u}_g$ ($\mathbf{u}_T$)	geostrophic (thermal) wind
U, V, W,	velocity scales (sometimes V stands for volume)
$\mathbf{v}$	three-dimensional velocity vector
v_*	friction velocity
w, w_p, w_σ, w_θ, . . .	vertical velocity
$\mathbf{x}$	radius vector of a point
z_0	roughness parameter

GREEK LETTERS

α	reciprocal density ($1/\rho$); angle
α_a	circulation index
β	Rossby parameter ($\mathrm{d}f/\mathrm{d}y$); convective parameter ($g/\bar{\theta}$)
$\gamma(\gamma_a)$	temperature lapse rate (adiabatic)
Γ	gamma function; circulation of velocity
Γ_A	convective temperature lapse rate
δ_{ij}	Kroneker symbol
Δ	specific water content
ϵ	dissipation rate
ϵ_{ijk}	permutation symbol
ζ	vertical coordinate in Monin–Obukhov similarity theory (= z/L)
θ	potential temperature
(θ, λ)	spherical angular coordinates
κ	molecular diffusivity; Von Kármán's constant
λ	wavelength
λ_0	internal (Kolmogorov) scale of turbulence
Λ	external scale of turbulence
μ	static stability parameter for the planetary boundary layer
ν	kinematic viscosity; frequency
ξ, η, ζ	vorticity components
π	mathematical number ($\approx$3.14); also $\pi = p/\rho$
ρ	density
σ	pressure coordinate; Prandtl number
σ_f^2	variance of the quantity f
σ_{ik}	molecular stress tensor

τ	time interval; horizontal stress
τ_{ik}	Reynolds stress tensor
Φ	geopotential
ψ	stream function
ω	angular velocity of the Earth's rotation; wave frequency
ω_{BV}	Brunt–Väisälä frequency
Ω	vorticity vector

SPECIAL SYMBOLS AND ACRONYMS

∇	del operator
∇^2	Laplace's operator (Laplacian)
$\mathcal{E}$	heat influx
$O(a)$	order of magnitude of a
$J(a, b)$	Jacobian
Σ	summation symbol
Diff( )	diffusion term
a_x, a_y, a_z, a_t	partial derivatives with respect to x, y, z, t
$\times$	vector product
SL	surface layer
PBL	planetary boundary layer
GAC	general atmospheric circulation
ITCZ	inter-tropical convergence zone

CHAPTER 1

Introduction to Dynamic Meteorology (Kinematics of the Atmospheric Motions)

1. METHODS FOR THE DESCRIPTION OF CONTINUOUS MEDIA

A system of material points is called a *continuous medium* if these points entirely fill some volume of space. In this case the substance, as well as the physical characteristics of its state or motion, have a continuous distribution. The Earth's atmosphere, precisely the lower part, including the troposphere and the stratosphere, represents a continuous medium. Therefore, in the problems of dynamic meteorology we can ignore the discrete molecular structure of the atmosphere and apply the general methods of fluid dynamics for its description.

a. Lagrange's Method

In the Lagrangian description, the variations of the properties connected with a given fluid particle are considered during its motion through the flow field, starting from some initial instant $t = t_0$. The term 'fluid particle' stands for a volume (parcel) of a fluid (liquid or gas) with linear scales which are much greater when compared with the intermolecular distances, so that we can speak about the macroscopic velocity of the particles in the sense of the mechanics of continuous media. At the same time, these scales are supposed to be so small that the velocity and other characteristics inside the particle are considered to be constant and the particle itself moves 'as a whole' during the considered interval of time. In other words, this individual particle is identified with a material point of the fluid and its motion is studied. Passing from one particle to another, one can describe the motion of the whole system.

The function

$$\mathbf{x}(t) = \mathbf{x}(t \mid \mathbf{x}_0, t_0) \tag{1.1}$$

serves as a basic Lagrangian characteristic of the motion. With respect to a fixed coordinate system K_0 at $t > t_0$, Equation (1.1) defines the coordinates $(x_1, x_2, x_3) = \mathbf{x}$ of a given particle which has a radius vector $\mathbf{x}_0$ at $t = t_0$, that is $\mathbf{x}(t_0) = \mathbf{x}_0$ (Figure 1.1). Therefore, $\mathbf{x}_0$ serves as an identification parameter for the particles. At fixed $\mathbf{x}_0$ and variable t, Equation (1.1) gives the law of motion (the trajectory) of the particle. Obviously $\mathbf{x}(t)$ is a continuous function of t, or otherwise one would come to physically-absurd conclusions. But in contrast, discontinuity of function (1.1) with respect to $\mathbf{x}_0$ is principally allowed.

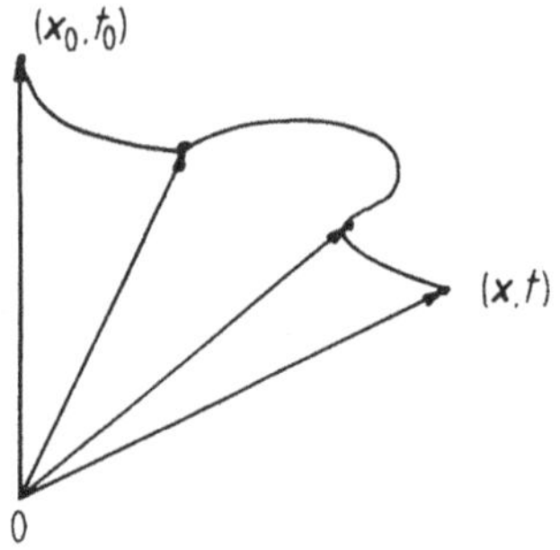

Fig. 1.1. Lagrangian coordinates and the trajectory of a particle.

Differentiating (1.1), one obtains the Lagrangian velocity of the particles

$$\mathbf{v}_L(t \mid \mathbf{x}_0, t_0) = \frac{\mathrm{d}}{\mathrm{d}t}\mathbf{x}(t \mid \mathbf{x}_0, t_0) \tag{1.2}$$

and after one more differentiation, the Lagrangian acceleration

$$\mathbf{a}(t \mid \mathbf{x}_0, t_0) = \frac{\mathrm{d}}{\mathrm{d}t}\mathbf{v}_L = \frac{\mathrm{d}^2}{\mathrm{d}t}\mathbf{x}(t \mid \mathbf{x}_0, t_0). \tag{1.3}$$

Here $\mathbf{x}_0$ is most frequently interpreted as the particle's radius vector at the initial instant. Obviously $\mathbf{x}_0$ cannot depend on t and, consequently, every other set of three (in space) or two (on a plane) conservative characteristics of the particle during its motion can be considered as Lagrangian variables instead of $\mathbf{x}_0$, e.g., the potential temperature θ and the specific humidity a in the case of adiabatic processes and others. In the same way, instead of $\mathbf{x}$ in (1.1) some other characteristics of the particle could be considered, for example, temperature T, and density ρ.

b. Euler's Method

The Lagrangian method has too limited a usage in the problems of dynamic meteorology although, for the time being in many cases it suits the nature of a given problem. Much greater usage has been found for Euler's method. In the Eulerian description, the variation of some property with respect to a fixed co-ordinate system is considered – a fixed geometrical point $\mathbf{x}$ from a space filled with fluid (air) is observed and variations in time, and from point to point, of the various characteristics of the fluid are studied. They are caused by fluid particles passing through these points at a given instant t. Letting $\mathbf{x}$ in (1.1) be their radius vector and solving for $\mathbf{x}_0$, we find $\mathbf{x}_0 = \mathbf{x}_0(\mathbf{x}, t)$, where t is the parameter.

Unlike (1.2), the velocity at the point of observation is denoted by $\mathbf{v}$ (Eulerian velocity). On the basis of the above interpretation, one can write the following relationship between the Lagrangian and Eulerian velocities

$$\mathbf{v}_L(t \mid \mathbf{x}_0, t_0) = \mathbf{v}[\mathbf{x}(t \mid \mathbf{x}_0, t_0), t] \tag{1.4}$$

and also

$$F(t \mid \mathbf{x}_0, t_0) = F[\mathbf{x}(t \mid \mathbf{x}_0, t_0), t] \tag{1.5}$$

for some scalar quantity F. These formulae allow a transition from Lagrangian to Eulerian variables to be carried out, in which

$$\mathbf{v} = \mathbf{v}(\mathbf{x}, t), \qquad F = F(\mathbf{x}, t). \tag{1.6}$$

Again, with respect to time t, functions (1.6), giving the spacetime distribution (i.e., the field) of **v** and F are continuous, but with respect to **x** they may also be discontinuous.

c. Types of Derivatives

Let us differentiate (1.4) following the rule for composed functions. Bearing in mind (1.2) and, once again (1.4), we successively obtain

$$\frac{\mathrm{d}}{\mathrm{d}t}\mathbf{v}_L \equiv \frac{\mathrm{d}}{\mathrm{d}t}\mathbf{v} = \frac{\partial \mathbf{v}}{\partial t} + \frac{\partial \mathbf{v}}{\partial x}\frac{\mathrm{d}x}{\mathrm{d}t} + \frac{\partial \mathbf{v}}{\partial y}\frac{\mathrm{d}y}{\mathrm{d}t} + \frac{\partial \mathbf{v}}{\partial z}\frac{\mathrm{d}z}{\mathrm{d}t} = \frac{\partial \mathbf{v}}{\partial t} + (\mathbf{v}\cdot\nabla)\mathbf{v} \tag{1.7}$$

where

$$\mathbf{v} = \mathbf{i}u + \mathbf{j}v + \mathbf{k}w, \qquad \nabla = \mathbf{i}\frac{\partial}{\partial x} + \mathbf{j}\frac{\partial}{\partial y} + \mathbf{k}\frac{\partial}{\partial z} \tag{1.8}$$

and **i**, **j**, **k** are unit vectors along the three coordinate axes respectively, ∇ is the nabla (del) operator, and $(\mathbf{v}\cdot\nabla)$ stands for a scalar product.

Similarly, from (1.5) one obtains

$$\frac{\mathrm{d}F}{\mathrm{d}t} = \frac{\partial F}{\partial t} + (\mathbf{v}\cdot\nabla F), \tag{1.9}$$

where $F \equiv p, T, \rho$ etc., are scalar quantities.

Expressions (1.7) and (1.8) represent an exceptionally important relationship between the various derivatives of **v** and F. Thus, the left-hand sides, that is $\mathrm{d}\mathbf{v}/\mathrm{d}t$ and $\mathrm{d}F/\mathrm{d}t$, according to (1.4) and (1.5), represent the changes of **v** and F referring to a given particle during its motion. They are called *full* (*total*) or *individual*, also *material* derivatives. But for meteorological applications it is important to know the variations of **v** and F in time in a fixed point **x**. Obviously they are given by the derivatives $\partial\mathbf{v}/\partial t$ and $\partial F/\partial t$ called *local*.

The expression $(\mathbf{v}\cdot\nabla)(F, \mathbf{v})$ is called a *convective* derivative. But in meteorology this term is used only for $w(\partial/\partial z)(F, \mathbf{v})$ while $(\mathbf{u}\cdot\nabla)(F, \mathbf{v})$, where $\mathbf{u} = \mathbf{i}u + \mathbf{j}v$ is the horizontal velocity, and is called an *advective* derivative.

We introduce abbreviated notations for the derivatives which we shall use for brevity when there is no danger of misunderstanding:

$$\frac{\mathrm{d}F}{\mathrm{d}t} = \dot{F}, \qquad \frac{\partial F}{\partial t} = F_t, \qquad \frac{\partial F}{\partial s} = F_s, \qquad s = x, y, z, \tag{1.10}$$

where F may also stand for a vector. Then formula (1.9), when $F \equiv T$ (temperature), can be written in the form

$$T_t = \dot{T} - (uT_x + vT_y) - wT_z = T_t' + T_t'' + T_t'''. \tag{1.11}$$

We see that the local variation of T (the temperature tendency T_t) is caused by three factors: T_t' – warming (cooling) of the air particle during its motion (thermal transformation); T_t'' – advection of warm (cool) air from one place to another; T_t''' – vertical convection. If T_t', T_t'' and T_t''' could be determined with enough accuracy from the observations, then expression (1.11) would have a prognostic value. Indeed, if $T(t_0)$ and $T_t(t_0)$ are known, we can write $T(t_0 + \Delta t) \approx T(t_0) + \Delta t T_t(t_0)$, where Δt is a small enough time interval. Unfortunately, standard meteorological data allow only T_t'' to be easily estimated. That is why other prognostic equations must be looked for, which will not have a formal mathematical character like (1.11), but will be based on fundamental physical laws and acceptable hypotheses. The derivation of such equations, not only for the temperature but also for the other meteorological elements, is one of the basic tasks of dynamic meteorology. Many chapters and sections in the following exposition will be devoted to this.

2. KINEMATIC CHARACTERISTICS OF THE PRESSURE FIELD

a. Pressure Systems

The atmospheric pressure is one of the most important meteorological elements. It is enough to remember that its irregular space distribution yields the origination of air motions.

In the general case, $p = p(x, y, z, t)$ and the equation

$$p(x, y, z, t) = \text{const} \quad \text{or} \quad z = z(x, y, t) \tag{2.1}$$

describes a surface in the space called an *isobaric surface*, upon which the pressure remains constant. In the real atmosphere this surface has a complex topography which changes in time. At fixed z, Equation (2.1) describes a line on the plane z = const called an *isobar*. Consequently, the isobars are cross-lines of the isobaric surfaces with a given horizontal plane. Their equation would be

$$p(x, y, t) = \text{const} \quad \text{or} \quad y = y(x, t). \tag{2.2}$$

Giving different values to 'const' in (2.2) (usually 3 or 5 mb), we obtain a family of isobars with a definite configuration. From the great variety, six typical cases are outlined and are shown in an idealized form in Figure 1.2, borrowed from [56]. They are called *pressure* or *baric systems (formations)*. We shall examine some of their kinematic characteristics. Along with this we shall consider the quantity $p_t(x, y, t) = \partial p/\partial t$ characterizing the pressure tendency in the point (x, y). Evidently

$$p_t(x, y, t) = \text{const} \tag{2.3}$$

represents an equation of *isalobars* – lines connecting points on the plane with equal values of p_t.

Let us now choose the axis $0x$ normally to the axes of the trough and ridge (Figure 1.2a). Then along these axes

$$p_x = 0, \quad p_{xx} \begin{cases} > 0 & \text{for trough,} \\ < 0 & \text{for ridge.} \end{cases} \tag{2.4}$$

In other words, the trough is a 'min' and the ridge is a 'max' in the pressure field. In the case of a front, (Figure 1.2c), some complications appear.

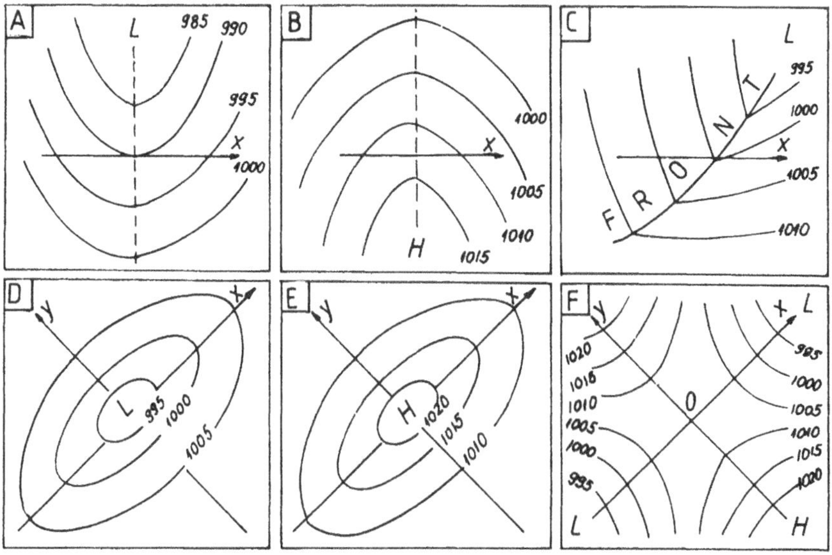

Fig. 1.2. Idealized types of baric systems (after [56]).

The isobars and the isalobars are also called *characteristic curves*. All the points lying on them are *ordinary* points of the pressure field. There may exist, however, special points not belonging to any of the isolines. They are called *isolated* special points of the pressure field and can be centers of low or high pressure – *baric centers*.

If we choose the coordinate axes along the axes of maximum symmetry (Figures 1.2d, e), then in the cyclonic center

$$p_x = p_y = 0; \qquad p_{xx}, p_{yy} > 0, \qquad \nabla^2 p > 0, \tag{2.5}$$

while in the anticyclonic one

$$p_x = p_y = 0. \qquad p_{xx}, p_{yy} < 0, \qquad \nabla^2 p < 0, \tag{2.6}$$

where $\nabla^2 p = p_{xx} + p_{yy}$. At circular isobars $p_{xx} = p_{yy}$.

Figure 1.2f shows the case of col-pressure distribution. Obviously, along the corresponding axes

$$p_x = p_y = 0, \qquad p_{xx} < 0, \qquad p_{yy} > 0. \tag{2.7}$$

At rectangular col $p_{xx} = -p_{yy} < 0$. The point 0 is called the *neutral* or *hyperbolic* point.

b. Pressure-Features Movement

In a pressure field there exist some more complex characteristic curve-lines than isobars and isalobars. Such are the curve-lines connecting points with equal values of ∇p. For instance, the axes of the ridges and troughs which are geometrical places of points with a maximum curvature of the isobars. The characteristic curves and special points in the pressure field move, thus causing weather changes. The success of a given synoptic forecast depends, to a high degree, on how correctly the future places of the pressure centers, ridges, troughs, etc., have been predicted. On purely kinematic considerations, it is not difficult to derive formulae for the speed of this movement and to use them to make a formal extrapolation of their future position. We shall illustrate this possibility in the case of a pressure center (cyclone or anticyclone).

Let $\mathbf{u}_c$ denotes the speed of movement of the pressure center. Obviously

$$\frac{d\nabla p}{dt} = \left(\frac{\partial}{\partial t} + \mathbf{u}_c \cdot \nabla\right) \nabla p = 0 \tag{2.8}$$

is the condition for its conservation as a center during the motion. It is equivalent to two scalar equations

$$p_{xt} + u_c p_{xx} + v_c p_{xy} = 0, \qquad p_{yt} + u_c p_{xy} + v_c p_{yy} = 0. \tag{2.9}$$

Under the condition of $D = p_{xx}p_{yy} - p_{xy}^2 \neq 0$, the solution of (2.9) is

$$u_c = \frac{p_{yt}p_{xy} - p_{xt}p_{yy}}{D}, \qquad v_c = \frac{p_{xt}p_{xy} - p_{yt}p_{xx}}{D} \tag{2.10}$$

provided that p_{xx}, p_{yy} and p_{xy} are not simultaneously equal to zero (simple center – the most frequently observed in practice). These formulae simplify if $p_{xy} = 0$ and the latter is possible if, around the simple center, the isobars represent second-order central curves (ellipses or hyperboles)

$$p(x, y, t) = p_0 + \tfrac{1}{2}\overset{\circ}{p}_{xx}x^2 + \tfrac{1}{2}\overset{\circ}{p}_{yy}y^2 \tag{2.11}$$

whose principal axes coincide with the coordinate axes. Then

$$u_c = -\frac{p_{xt}}{p_{xx}}, \qquad v_c = -\frac{p_{yt}}{p_{yy}} \tag{2.12}$$

as the derivatives are determined from (2.11). Consequently $\mathbf{u}_c \sim \nabla p_t$, i.e., the isalobaric gradient determines the direction of the pressure-center movement.

The next differentiation of (2.12) with respect to t would give us an acceleration of pressure centers $d\mathbf{u}_c/dt$. Having computed $\mathbf{u}_c$ and $d\mathbf{u}_c/dt$, an extrapolation of the future position of the pressure centers can be made in principle. However, this is a kinematic, rather formal extrapolation. Besides, the acceleration formulae involve higher-order derivatives of p, which ordinarily cannot be approximated with sufficient accuracy from observations. That is why this method has a limited use. Let us also point out that $\mathbf{u}_c \neq \mathbf{u}$ (the wind velocity).

c. Evolution of Pressure Systems

Again we examine pressure centers only. Consider the relation

$$\frac{dp}{dt} = \frac{\partial p}{\partial t} + \mathbf{u}_c \cdot \nabla p. \tag{2.13}$$

The quantity $\dot{p} = dp/dt$ expresses the *intensification rate* of the center. For a cyclonic (anticyclonic) center it will be said to intensify when $\dot{p} < 0$ (> 0). Otherwise, the center will be said to weaken. In a particular case, $\dot{p} = 0$ – the center being translated without change of shape. In another particular case when $(\mathbf{u}_c \cdot \nabla p) = 0$, i.e., $u_c = 0$ or $\nabla p = 0$ or $\mathbf{u}_c \perp \nabla p$, $\dot{p}$ may be noted directly from the observed pressure tendency alone – $\dot{p} \equiv \partial p/\partial t$.

Let us now form the quantity

$$Y = \nabla \cdot \nabla p = \nabla^2 p = p_{xx} + p_{yy}. \tag{2.14}$$

Then, similar to (2.13), the intensification of the pressure center (cyclone if $\nabla^2 p > 0$ and anticyclone if $\nabla^2 p < 0$) may be characterized by

$$\dot{Y} = \left(\frac{\partial}{\partial t} + \mathbf{u}_c \cdot \nabla\right) \nabla^2 p. \tag{2.15}$$

Again, a special choice of the coordinate axes, as shown in Figure 1.2, may annul all third derivatives in the second term of (2.15), so that finally

$$\dot{Y} = \frac{\partial \nabla^2 p}{\partial t} = \nabla^2 \frac{\partial p}{\partial t}. \tag{2.16}$$

Formulae (2.11) – (2.16) may be applied to the weather map by expressing the derivatives in terms of finite differences.

3. GEOMETRICAL CHARACTERISTICS OF THE WIND FIELD

a. Streamlines

The streamline is a line drawn in the flow in such a way that the velocity vectors of all the fluid particles lying on it in a given instant t should be tangential to it at the corresponding points (Figure 1.3). Consequently, the streamline is a purely geometrical characteristic referring to a proper instant t. The streamlines constructed at $t = t_1$ and $t = t_2$, generally do not coincide.

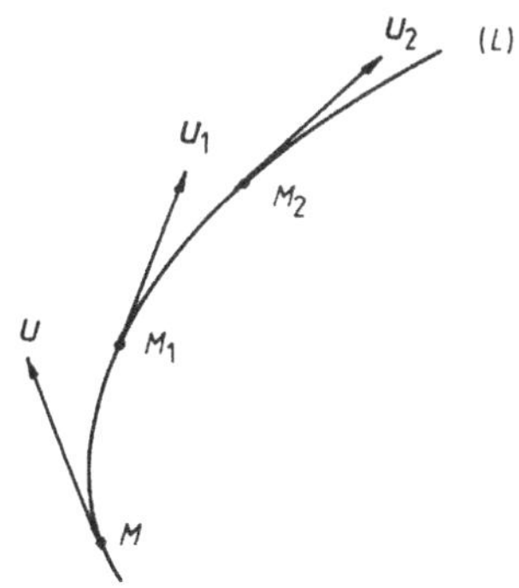

Fig. 1.3. A streamline (L).

It is easy to derive the differential equation of the streamlines. Let

$$\mathbf{dr} = \mathbf{i}\, dx + \mathbf{j}\, dy, \qquad \mathbf{u} = \mathbf{i}u + \mathbf{j}v. \tag{3.1}$$

According to the definition and Figure 1.4 $\mathbf{u} \times \mathbf{dr} = 0$ or

$$\frac{dx}{u(x, y, t)} = \frac{dy}{v(x, y, t)}, \tag{3.2}$$

which is, in fact, the required equation in which t is a parameter. If u, v are known functions of x, y, it could be integrated to yield $y = y(x, t)$.

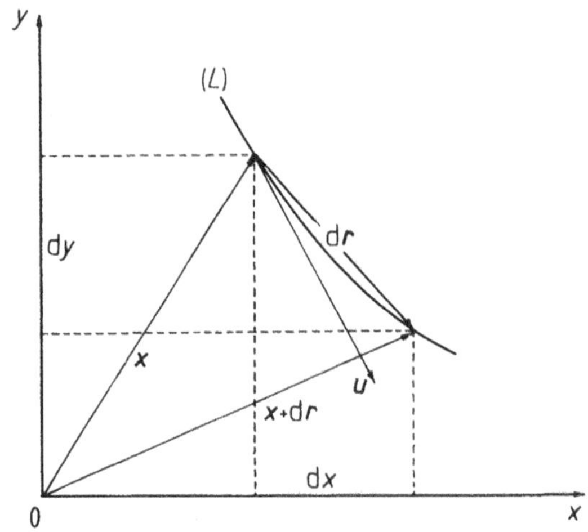

Fig. 1.4. The derivation of formula (3.2).

In one special case, this equation can be derived in another way too. This is the case of a nondivergent wind field when $u_x + v_y = 0$. Actually, from here and from (3.2)

$$\frac{\partial(-u)}{\partial x} = \frac{\partial v}{\partial y}, \qquad v\,dx + (-u)\,dy = 0. \tag{3.3}$$

Hence, it follows that Equation (3.2) may be represented as a perfect differential of some function $\psi(x, y, t)$:

$$u = -\psi_y, \qquad v = \psi_x, \tag{3.4}$$

$$d\psi(x, y, t) = 0, \qquad \psi(x, y, t) = \text{const.} \tag{3.5}$$

Obviously $\psi(x, y, t)$ should be called a *stream function*. Equated to 'const', it becomes the equation of the corresponding streamline. The stream function can be introduced for two-dimensional motion only.

b. Trajectory

The trajectory is the line described by the consecutive positions of a given fluid-particle, i.e., this is the path of the particle. Generally, streamlines and trajectories do not coincide. However, in the case of steady motion, they do.

Actually with a steady wind field, the differential equation (3.2) of the streamlines reads

$$\frac{dx}{u(x, y)} = \frac{dy}{v(x, y)}, \tag{3.6}$$

The differential equations of the trajectories in Eulerian variables are

$$\frac{dx}{dt} = u(x, y), \qquad \frac{dy}{dt} = v(x, y) \tag{3.7}$$

or

$$dt = \frac{dx}{u(x, y)}, \qquad dt = \frac{dy}{v(x, y)},$$

which is identical to (3.6). Thus, the geometrical analogues of Equations (3.6) and (3.7), the streamlines and trajectories, will also coincide. Moreover, a crowding together of the streamlines denotes an increase in the velocity of the air flow.

Generally, only one streamline passes through an arbitrary point in the space occupied by the flow. However, there may exist *special points* through which more than one streamline can pass. At the point of intersection of two or more streamlines the fluid cannot possess a finite velocity, since this would mean that it was moving in two or more directions at the same time. At such a critical point, the velocity is at zero or infinity. Such examples found in the problems of dynamic meteorology are discussed in Section 5 (see Figure 1.7).

The streamlines and trajectories are, by nature, vector lines and a positive direction can be attached to them, namely the one determined by **u**. Besides, they can be characterized by their curvature K. If, in its positive direction, the lines turn to the left, then $K > 0$ (cyclonic curvature). Otherwise, $K < 0$ (anticyclonic curvature). By definition, $K = R^{-1}$, where R is the radius of the curvature.

4. DIFFERENTIAL AND INTEGRAL CHARACTERISTICS OF THE WIND FIELD

a. Divergence

By definition, the divergence

$$D = \begin{cases} \nabla \cdot \mathbf{v} = \operatorname{div} \mathbf{v} = u_x + v_y + w_z & \text{in the space} \\ \nabla \cdot \mathbf{u} = \operatorname{div} \mathbf{u} = u_x + v_y & \text{on the plane.} \end{cases} \tag{4.1}$$

Let $\delta S = \delta x \, \delta y$ be an infinitely small rectangular surface element where δx, δy are the material linear elements. Due to the differences in the motion of the particles belonging to δx and δy, the latter may stretch or shrink following the motion. Consequently, one can write

$$\frac{d\delta x}{dt} = \delta \frac{dx}{dt} = \delta u = u_x \, dx$$

$$\frac{d\delta x}{dt} = \delta \frac{dy}{dt} = \delta v = v_y \, dy$$

$$\frac{d(\delta S)}{dt} = \delta y \frac{d\delta x}{dt} + \delta x \frac{d\delta y}{dt} = (u_x + v_y) \, dS$$

or

$$u_x + v_y = \nabla \cdot \mathbf{u} = \frac{1}{\delta S} \frac{d\delta S}{dt}. \tag{4.2}$$

Similarly, it can be proved that

$$u_x + v_y + w_z = \nabla \cdot \mathbf{v} = \frac{1}{\delta V} \frac{d\delta V}{dt} \tag{4.3}$$

where $\delta V = \delta x \; \delta y \; \delta z$ is an elementary volume element of the fluid. Therefore, the divergence D characterizes the relative change of the surface (volume) element following the motion. If δS or δV are constant, the fluid is said to be *incompressible*. Otherwise the fluid would be compressible. In this case, changes of δV will cause changes of the density ρ of the fluid. We can write

$$\delta m = \rho \delta V \qquad \text{at } t = t_0,$$

$$\delta m = (\rho + d\rho)(\delta V + d\delta V) \qquad \text{at } t = t_0 + dt.$$

Hence,

$$\rho \, d\delta V + d\rho \delta V + \underline{d\rho \cdot d\delta V} = 0. \tag{4.4}$$

The underlined term can be neglected as infinitely small of a higher order. Then we obtain the equation

$$\frac{1}{\delta V} \frac{d\delta V}{dt} = -\frac{1}{\rho} \frac{d\rho}{dt}, \tag{4.5}$$

which has a clear physical meaning.

The divergence D can be positive or negative. The negative divergence is called *convergence*. As is usual, the case $D \neq 0$ is connected with divergent or convergent streamlines on maps of the wind field. It is possible, however, that $D \neq 0$ at parallel streamlines. The configuration of the streamlines shown in Figure 1.5 is frequently found in the atmosphere. That is why, as a rule, $u_x + v_y = D \sim 10^{-5}$–$10^{-6}\,s^{-1}$ which is much smaller than the order of u_x or v_y separately. Regardless of its very small value, the divergence plays an important role in many problems of atmospheric dynamics.

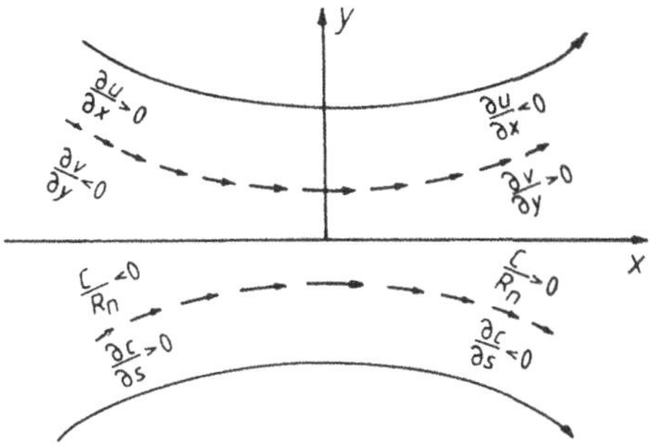

Fig. 1.5. Divergence and convergence of the wind velocity.

b. Vorticity

By definition, the vorticity

$$\Omega = \text{curl } \mathbf{v} = \nabla \times \mathbf{v} \tag{4.6}$$

is a vector with components

$$\xi = w_y - v_z, \qquad \eta = u_z - w_x, \qquad \zeta = v_x - u_y. \tag{4.7}$$

With the horizontal motion ($w = 0$) and (u, v) depending on x, y, z,

$$\xi = -v_z, \qquad \eta = u_z, \qquad \zeta = v_x - u_y. \tag{4.8}$$

Henceforth, we will be concerned mainly with the third component ζ, which may also be written as

$$\zeta = \mathbf{k} \cdot (\nabla \times \mathbf{v}) = -\nabla \cdot \mathbf{s}, \quad \mathbf{s} = \mathbf{k} \times \mathbf{v} = (-v, u, 0), \tag{4.9}$$

where $\mathbf{k} = (0, 0, 1)$ is a vertical unit vector.

The physical meaning of ζ becomes more clear if one uses the natural (curvilinear) coordinates (s, n), where s-coordinate lines coincide with the streamlines while n-coordinate lines are normal to them at a fixed instant (Figure 1.6). In such a system formula (4.9) reads

$$\zeta = \frac{c}{R_s} - \frac{\partial c}{\partial n} \tag{4.10}$$

where $c = |\mathbf{u}|$, R_s is the radius of curvature of the streamline. The term c/R_s gives the angular velocity of the rotation of the fluid (air) particle as a solid body around a vertical axis passing through the instantaneous center of the curvature. With the cyclonic rotation in the Northern Hemisphere, its contribution is positive and, in the opposite case, negative. As to the second term (the lateral gradient of the scalar wind velocity), it may also be positive or negative. It is seen from (4.10) that there can exist a straightforward ($R_s = \infty$) with a vorticity ($\zeta \neq 0$) motion, as well as a curvilinear ($R_s < \infty$) irrotational ($\zeta = 0$) motion when both terms in (4.10) compensate, although the latter rarely happens in the atmosphere.

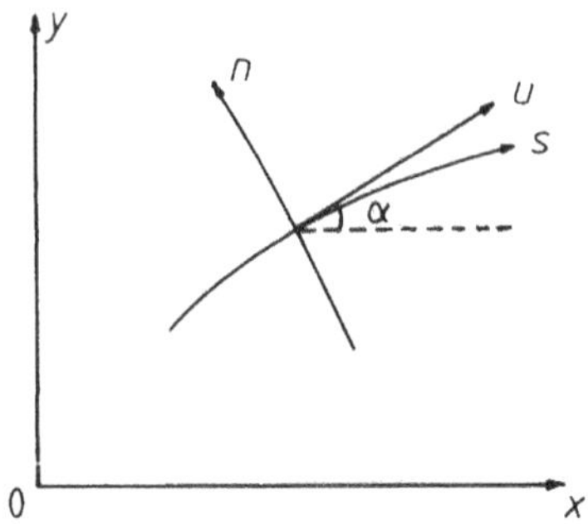

Fig. 1.6. Natural coordinates.

Similar to (4.10),

$$D = \frac{c}{R_n} + \frac{\partial c}{\partial s}. \tag{4.11}$$

can be obtained for the plane divergence. The interpretation of both terms is obvious. As seen from Figure 1.5, very frequently they have opposite signs and approximately compensate each other, i.e., $D \approx 0$. However, if $\nabla \cdot \mathbf{u} = 0$, as follows from (3.4)

$$\mathbf{u} = \mathbf{k} \times \nabla \psi, \tag{4.12}$$

so that

$$\zeta = \mathbf{k} \cdot (\nabla \times \mathbf{u}) = \nabla^2 \psi. \tag{4.13}$$

On the other hand, if $\nabla \times \mathbf{u} = 0$ we can write

$$\mathbf{u} = \nabla \phi \tag{4.14}$$

where ϕ is a scalar velocity potential. Then $\nabla \cdot \mathbf{u} = \nabla^2 \phi$. In the general case, the wind velocity vector $\mathbf{u}$ may be decomposed to potential ($\mathbf{u}_\phi$) and solenoidal ($\mathbf{u}_\psi$) components

$$\mathbf{u} = \mathbf{u}_\phi + \mathbf{u}_\psi = \nabla \phi + \mathbf{k} \times \nabla \psi. \tag{4.15}$$

Obviously $\mathbf{u}_\phi$ is an irrotational component while $\mathbf{u}_\psi$ is nondivergent.

c. Vortex Lines and Tubes

A vortex line is defined as a line drawn in the fluid so that the tangent to it at any one point has the direction of vorticity vector $\boldsymbol{\Omega}$. Then, similarly to (3.2), the differential equations of the vortex line will be

$$\frac{dx}{\xi} = \frac{dy}{d\eta} = \frac{dz}{\zeta} \tag{4.16}$$

The vortex lines through every point of a small closed curve form a tube called a *vortex tube*. For such a vortex tube of a small cross-section, it may be shown that the product of the magnitude of the vorticity and the area of a normal cross-section at any point has the same value all along the tube. This constant product is called the *strength* of the tube. As a consequence of the constancy of the strength, vortex tubes and, therefore, also vortex lines cannot begin or end in the fluid. The vortex lines form closed curves or begin or end on the boundaries of the fluid or, if we suppose the fluid to be unbounded, go to infinity. The stretching of the vortex tubes will tend to produce more vorticity. This mechanism is responsible for the vorticity generation in a three-dimensional turbulent flow and we will return to it later in Chapter 8.

The quantity

$$\Gamma = \oint_L (\mathbf{v} \cdot \mathrm{d}\mathbf{r}) = \int_S (\boldsymbol{\Omega} \cdot \mathbf{n})\, \mathrm{d}S \tag{4.17}$$

where S is the surface bounded by the contour L and is called the *circulation* of the velocity. The relationship between Γ and the vorticity Ω is evident.

d. Deformation

Let us consider again the rectangular element having sides δx and δy. The ratio $\delta x/\delta y$ characterizes its form; it may or may not be equal to one. It is easy to prove that

$$\frac{1}{\delta x/\delta y}\,\frac{\mathrm{d}}{\mathrm{d}t}\left(\frac{\delta x}{\delta y}\right) = \frac{\partial u}{\partial x} - \frac{\partial v}{\partial y} = F. \tag{4.18}$$

The quantity F is called the *stretching* deformation while $H = u_y + v_x$ is the *shearing* deformation. If a stream function is introduced, then $F = -2\psi_{xy}$, and $H = \psi_{xx} - \psi_{yy}$.

In order to characterize the deformation of a fluid volume in the space, the so-called tensor of deformation is introduced

$$\Phi_{ij} = \frac{1}{2}\left(\frac{\partial u_i}{\partial x_j} + \frac{\partial u_j}{\partial x_i}\right). \tag{4.19}$$

Obviously this is a symmetric tensor with six different components

$$\begin{matrix} u_x & \frac{1}{2}(u_y + v_x) & \frac{1}{2}(u_z + w_x) \\ & v_y & \frac{1}{2}(v_z + w_y) \\ & & w_z \end{matrix} \tag{4.20}$$

5. A LINEAR WIND FIELD AROUND AN ARBITRARY POINT

a. Decomposition of the Velocity

Let $\mathbf{x}_0$ be an arbitrary point on the plane xOy in which the origin of the coordinate system is located. Around this point the wind velocity $\mathbf{u}(x, y)$ can be represented as

$$\mathbf{u}(x, y) = \mathbf{u}_0 + \mathbf{u}_x^0\, x + \mathbf{u}_y^0\, y + \mathbf{O}(r^2) \tag{5.1}$$

where $\mathbf{u}_0$ and its derivatives refer to the point $\mathbf{x}_0$ and the term $\mathbf{O}(r^2)$ vanishes at the rate $r^2 = x^2 + y^2$ as $r \to 0$ (linear approximation). Written in components, (5.1) reads

$$\begin{aligned} u &= u_0 + \tfrac{1}{2}(D + F)\,x + \tfrac{1}{2}(H - \zeta)\,y, \\ v &= v_0 + \tfrac{1}{2}(H + \zeta)\,x + \tfrac{1}{2}(D - F)\,y, \end{aligned} \tag{5.2}$$

where D, F, ζ, H refer to the point $\mathbf{x}_0$, as well as u_0, v_0. Following [34], we note that $F^2 + H^2$ is invariant with respect to the rotation of the coordinate system. By rotation of the axes, this allows $F > 0$ and $H = 0$, i.e., $v_x = -u_y$. Thus, (5.2) becomes

$$\begin{aligned} u &= u_0 + \tfrac{1}{2}(Fx + Dx - \zeta y) = u_0 + u_1 + u_2 + u_3, \\ v &= v_0 + \tfrac{1}{2}(-Fy + Dy + \zeta x) = v_0 + v_1 + v_2 + v_3. \end{aligned} \tag{5.3}$$

These expressions describe a linear velocity field in the small vicinity of the point $\mathbf{x}_0$. The various terms represent the pure types of motion as follows: (u_0, v_0) – translation, (u_1, v_1) – deformation, (u_2, v_2) – divergence, and (u_3, v_3) – rotation. All of them are illustrated in Figure 1.7.

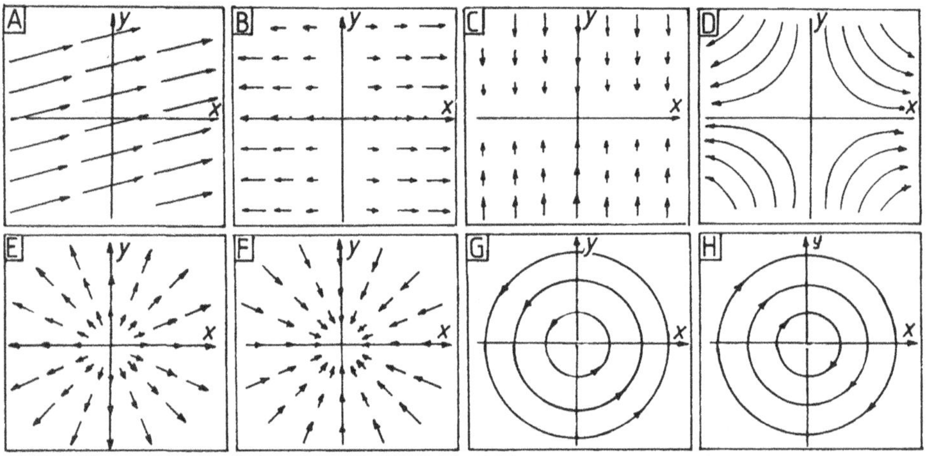

Fig. 1.7. Elementary types of motions corresponding to linear representation (5.3) (after [56]).

It follows from (5.3) that

$$(u_3^2 + v_3^2)^{1/2} = c_3 = \tfrac{1}{2}\zeta r, \quad r = (x^2 + y^2)^{1/2}.$$

Thus, $\zeta = 2c_3/r = 2\omega$, where $\omega = c_3/r$ is the angular velocity of the rotation of the particles around the vertical axis. Various combinations among the above-mentioned pure types of motion can produce the streamline configurations most frequently found on wind velocity maps. We shall deal with those in which the rotation and deformation is dominant, in comparison with divergence and which are maintained by pressure distribution.

In (5.3) we let $u = v = 0$ and solve the system of algebraic equations for $x = x_c$, $y = y_c$:

$$x_c = -2\,\frac{u_0(D - F) + v_0\zeta}{D^2 - F^2 + \zeta^2}, \qquad y_c = 2\,\frac{-v_0(D + F) + u_0\zeta}{D^2 - F^2 + \zeta^2}. \tag{5.4}$$

If $(x_c, y_c) < \infty$, this will mean that the wind field (5.3) has a kinematic center and the condition necessary for the existence of such a center will be

$$D^2 - F^2 + \zeta^2 \neq 0 \quad (<, > 0). \tag{5.5}$$

Translating the origin of the coordinate system into the point (x_c, y_c), we write expressions (5.3) in the form

$$2u = (D + F)\,x - \zeta y, \qquad 2v = \zeta x + (D - F)\,y. \tag{5.6}$$

Then the equations

$$\frac{dy}{dx} = \frac{v}{u}\ (= \tan\alpha) \tag{5.7}$$

will express the fact that straight streamlines exist which pass through the center. Substituting u, v from (5.6) we obtain

$$\tan\alpha = \frac{\zeta + (D - F)\tan\alpha}{D + F - \zeta\tan\alpha}.$$

Hence,

$$\tan\alpha = \zeta^{-1}\,[F \pm (F^2 - \zeta^2)^{1/2}] \tag{5.8}$$

i.e., the angle α between $0x$ and the streamlines does not depend on D.

b. Analysis of the Results

The following conclusions can be made by analyzing formula (5.8):

(i) At $F = \zeta = 0$ (pure divergence, Figures 1.7e, f), innumerable streamlines exist which pass through the center. This case is not observed in real conditions.

(ii) $F^2 > \zeta^2$. There exist two straight streamlines (two real roots in (5.8)) which pass through the center. They divide the plane into four sectors with hyperbolic streamlines (Figure 1.8b, c). Obviously, they are the result of the superposition of rotation upon pure deformation (Figure 1.8a).

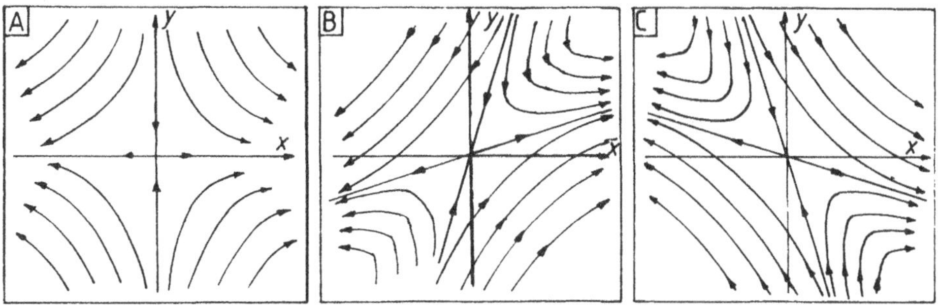

Fig. 1.8. Idealized hyperbolic streamlines (after [56]).

(iii) $\zeta^2 > F^2$. There is no real solution for α. Consequently, there will be no straight streamlines passing through the center. Typical configurations corresponding to this case are shown in Figure 1.9: pure rotation – a, d; rotation plus deformation – b, e; divergence superimposed on the previous case – c, f. These configurations of the streamlines resemble cyclones and anticyclones in the free atmosphere or at ground level.

If $F^2 - \zeta^2 = D^2$, according to (5.4), the kinematic center would not exist. This case can be analyzed in a similar way (see [56]).

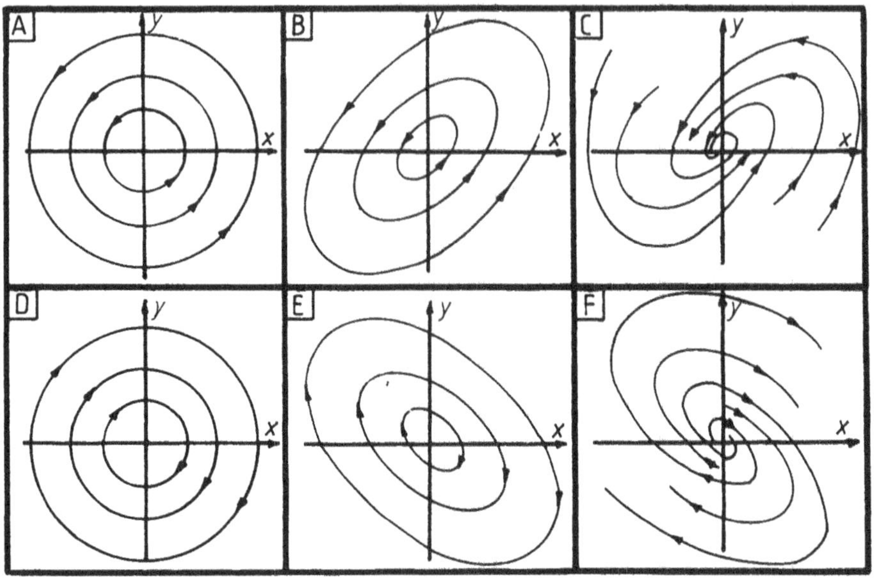

Fig. 1.9. Idealized streamlines corresponding to cyclones or anticyclones (after [56]).

6. THE CONTINUITY EQUATION

a. Derivation of the Equation

The main feature of fluids determining their different character of motion, as compared to solid bodies, is their great mobility. During the motion of the fluid volumes, a redistribution of the mass inside them is observed. This fact by itself exerts influence on the distribution of the speed of the particles and this dependence is analytically expressed by the so-called equation of continuity.

As mentioned in Section 1 of this chapter, we can ignore the discrete, molecular structure of the air in the problems of dynamic meteorology and consider the atmosphere as a continuous medium. In fact, we have already come to the continuity equation in Section 4. Thus, comparing (4.3) and (4.5) we obtain

$$\frac{d\rho}{dt} + \rho \nabla \cdot \mathbf{v} = 0, \tag{6.1}$$

which is the required equation. Here, we shall present another derivation based on the same mass conservation law.

Let V be a closed volume and S its surface. The quantity

$$Q_m = \int_S \rho v_n \, dS = \int_V \nabla \cdot \rho \mathbf{v} \, dV, \tag{6.2}$$

where v_n is the normal to the surface velocity component, will give the mass outflux (influx) from (in) the volume depending on the orientation of the normal n. For definiteness, let Q_m be the outflux. This will cause a decrease in the fluid mass in the volume

$$Q'_m = -\int_V \left(\frac{\partial \rho}{\partial t}\right) \mathrm{d}V. \tag{6.3}$$

Since $Q_m = Q'_m$, we obtain

$$\int_V \left(\frac{\partial \rho}{\partial t} + \nabla \cdot \rho \mathbf{v}\right) \mathrm{d}V = 0. \tag{6.4}$$

Hence,

$$\frac{\partial \rho}{\partial t} + \nabla \cdot \rho \mathbf{v} = 0 \tag{6.5}$$

because volume V is arbitrary and the integrand is a continuous function. But Equations (6.1) and (6.5) are equivalent.

b. Analysis and Particular Cases

The alternative forms (6.1) and (6.5) of the continuity equation hold for the general case of a *compressible* fluid. The case of a stationary ($\partial\rho/\partial t = 0$) but nonhomogeneous ($\nabla\rho \neq 0$) density field is of particular interest. Then, according to (6.5)

$$\nabla \cdot \rho \mathbf{v} = 0. \tag{6.6}$$

Another particular case is also possible – a homogeneous ($\nabla\rho = 0$) but nonstationary density field. Then $\dot{\rho} = \rho_t$.

If $\mathrm{d}\rho/\mathrm{d}t = \dot{\rho} = 0$, i.e., the density of the fluid parcels remains constant following the motion and the fluid is said to be *incompressible*. Since $\rho \neq 0$, it follows from (6.1) that

$$\nabla \cdot \mathbf{v} = 0, \tag{6.7}$$

i.e., the velocity vector is nondivergent (solenoidal). Both Equations (6.6) and (6.7) will be used very often in the future analysis.

By the definition of an incompressible fluid

$$\dot{\rho} = \frac{\mathrm{d}\rho}{\mathrm{d}t} = \frac{\partial \rho}{\partial t} + \mathbf{v} \cdot \nabla\rho = 0. \tag{6.8}$$

Two particular cases are possible:

(i) Stationary and homogeneous fluid ($\partial\rho/\partial t = \nabla\rho = 0$). Then the density is a physical constant (ρ = const), given the characteristic of the fluid. Evidently, $\dot{\rho} = 0$. This is the simplest (trivial) case.

(ii) Nonstationary ($\partial\rho/\partial t \neq 0$) and nonhomogeneous ($\nabla\rho \neq 0$) fluid. In this general case, Equation (6.1) splits into the two independent Equations (6.7) and (6.8) which could be used together. Examples of such use could be found in the theory of wave motions of the stratified geophysical fluids (the atmosphere and the ocean).

If $\nabla\rho = 0$ (6.8) yields $\partial\rho/\partial t = 0$. However, if $\partial\rho/\partial t = 0$ it follows only that $\mathbf{v} \cdot \nabla\rho = 0$ and not necessarily $\nabla\rho = 0$.

With few exceptions in the problems of dynamic meteorology, the feature 'incompressibility' is understood as the absolute constancy of density

$$\rho_t = \nabla\rho = 0 \quad \text{or} \quad \rho(x, y, z, t) = \rho = \text{const.} \tag{6.9}$$

Strictly speaking, the atmosphere is a compressible fluid. The air parcels may expand or contract considerably following the motion due to thermal and other reasons (humidity changes, for instance). This yields the changes in the vertical and horizontal pressure distribution and, hence, leads to changes in the air currents. Consequently, the compressibility is related to the conversion of the internal energy to kinetic energy and vice-versa and should be taken into consideration when these phenomena are of interest. The incompressibility is a good approximation for fluids such as sea water. However, it is a reasonable approximation for certain atmospheric conditions too. Moreover, it is very convenient from a mathematical point of view.

7. BAROTROPY AND BAROCLINICITY OF THE ATMOSPHERE

a. Barotropy

In a compressible medium, the density is a variable quantity and, except for (6.2), may satisfy other equations. When ρ depends only on the pressure p

$$\rho = \rho(p), \tag{7.1}$$

the fluid is said to be *barotropic*. For instance

$$\rho = ap^m, \tag{7.2}$$

where a = const. The power $m = 1$ during isothermal processes and $m = c_p/c_v = \kappa$ during adiabatic processes.

Geometrically, for barotropy, condition (7.1) implies that the surfaces of constant ρ and p coincide. Actually, (7.1) yields

$$\nabla\rho = \frac{\partial\rho}{\partial p}\nabla p, \qquad \nabla\rho \times \nabla p \sim \nabla p \times \nabla p \equiv 0 \tag{7.3}$$

which is a necessary and sufficient condition for the coincidence of both families of surfaces. Two independent scalar conditions follow from (7.3)

$$\rho_x p_y - \rho_y p_x = J_{xy}(\rho, p) = 0, \qquad \rho_z p_x - \rho_x p_z = J_{zx}(\rho, p) = 0. \tag{7.4}$$

The atmosphere, on the whole, is not a barotropic medium. Some parts of it for some interval of time may, however, be regarded as barotropic. Then all consequences and simplifications resulting from this can be applied to them. Such problems will be discussed in Chapters 5–7.

On the other hand, for the atmospheric air the thermodynamic equation of the state holds

$$p = \rho RT, \tag{7.5}$$

where $R = c_p - c_v$ is the specific gas constant of the air. Comparing (7.5) with (7.1), we conclude that both equations are compatible (they can be used simultaneously) if $p = p(T)$, i.e., $T = T(p)$. Then condition (7.3) reads

$$\nabla p \times \nabla T = 0 \tag{7.6}$$

and means the coincidence of isobaric and isothermic surfaces. On the plane $x0y$

$$J_{xy}(p, T) = p_x T_y - p_y T_x = 0, \tag{7.7}$$

the isobars and isotherms coincide.

b. Baroclinicity

When a more general relationship than (7.1) holds, e.g.,

$$F(\rho, p, T, \ldots) = 0, \tag{7.8}$$

the fluid is said to be *baroclinic*. The single Equation (7.5) without (7.1) is an indication for baroclinicity of a dry air. If q is the specific humidity of the moist air, Equation (7.5) should be replaced by

$$p = \rho RT(1 + 0.6q). \tag{7.9}$$

For the sea water, the salinity s plays the role of q.

In a baroclinic medium, the surfaces of constant pressure and temperature do not coincide any more. Generally, the atmosphere is a baroclinic medium with a variable (in space and time) degree of baroclinicity. There are phenomena and corresponding problems for which this property is important and must be taken into consideration. Examples will be met in the next chapters.

8. ON THE USE OF SCALAR, VECTOR, AND TENSOR NOTATIONS

So far, scalar and vector notations have been used. However, a tensor notation may also prove to be useful. In any particular case, the choice should be based upon considerations of compactness of the formulae (economy of space) and clearness in their interpretation. To facilitate their use, we give a brief review below.

A point in space will be specified by its radius vector

$$\mathbf{x} = (x, y, z) \quad \text{or} \quad (x_1, x_2, x_3) = x_i, \quad i = 1, 2, 3.$$

The velocity vector $\mathbf{v}$ will have components (u, v, w) or $(v_1, v_2, v_3) = v_i$, $i = 1, 2, 3$. Similarly,

$$\nabla = \left(\frac{\partial}{\partial x}, \frac{\partial}{\partial y}, \frac{\partial}{\partial z}\right) \quad \text{or} \quad \frac{\partial}{\partial x_i}, \quad i = 1, 2, 3.$$

We can now rewrite some of the already-obtained equations using the tensor notation. Let us start with (1.7):

$$\frac{dv_i}{dt} = \frac{\partial v_i}{\partial t} + \sum_{k=1}^{n} v_k \frac{\partial v_i}{\partial x_k}, \quad i = 1, 2, (3) \tag{8.1}$$

where $n = 3$ in space, while $n = 2$ on the plane. Henceforth, Σ will be omitted but should be taken as implicit. We come to the so-called *summation convention* which requires us to sum on repeated indices when they appear in two quantities that are multiplied by each other. For instance, instead of (8.1) we can write

$$\frac{dv_i}{dt} = \frac{\partial v_i}{\partial t} + v_k \frac{\partial v_i}{\partial x_k}. \tag{8.1$'$}$$

Generally

$$\frac{d}{dt} = \frac{\partial}{\partial t} + v_k \frac{\partial}{\partial x_k}, \qquad \nabla^2 = \frac{\partial^2}{\partial x_i \partial x_i} \text{ etc.,} \tag{8.2}$$

and yet

$$\text{div } \mathbf{v} = \nabla \cdot \mathbf{v} = \frac{\partial v_k}{\partial x_k} \tag{8.3}$$

$$\Omega = \text{curl } \mathbf{v} = \nabla \times \mathbf{v} = \epsilon_{ijk} \frac{\partial v_k}{\partial x_j}, \tag{8.4}$$

where ϵ_{ijk} is called an alternating unit tensor or permutation symbol:

$$\epsilon_{ijk} = \begin{cases} 0 & \text{if } i = j \text{ or } j = k \text{ or } i = k \\ 1 & \text{if } i, j, k \text{ are an even permutation of } 1, 2, 3 \\ -1 & \text{if } i, j, k \text{ are an odd permutation of } 1, 2, 3. \end{cases}$$

For two-dimensional motion in the plane

$$\zeta = \frac{\partial u_2}{\partial x_1} - \frac{\partial u_1}{\partial x_2} = \epsilon_{ik3} \frac{\partial u_k}{\partial x_i}. \tag{8.5}$$

This formula is a tensor analogue of (4.9).

In connection with the nonlinear term in (1.7), the following vector identity will be used in the future transformations

$$(\mathbf{v} \cdot \nabla)\, \mathbf{v} = \nabla(\tfrac{1}{2}\, \mathbf{v} \cdot \mathbf{v}) + \Omega \times \mathbf{v}. \tag{8.6}$$

In tensor notation this reads

$$v_k \frac{\partial v_i}{\partial x_k} = \frac{\partial(\frac{1}{2} v_k{}^2)}{\partial x_i} + \epsilon_{ijk} \Omega_j v_k. \tag{8.6$'$}$$

Let us also mention that $\nabla \cdot \mathbf{v}$ and $\nabla \mathbf{v}$ (without a point) denote different things. Namely

$$\nabla \cdot \mathbf{v} = \frac{\partial v_k}{\partial x_k}, \qquad \nabla \mathbf{v} = \frac{\partial v_k}{\partial x_i}, \tag{8.7}$$

i.e., $\nabla \cdot \mathbf{v}$ is a scalar while $\nabla\mathbf{v}$ is a tensor. Then the left-hand side of (8.6) can also be written as

$$(\mathbf{v} \cdot \nabla)\,\mathbf{v} = \mathbf{v} \cdot \nabla\mathbf{v}, \tag{8.7'}$$

or

$$\left(v_k \frac{\partial}{\partial x_k}\right) v_i = v_k \frac{\partial v_i}{\partial x_k}. \tag{8.7''}$$

We will also need the following identities

$$(\mathbf{A} \times \mathbf{B}) \times \mathbf{C} = (\mathbf{C} \cdot \mathbf{A})\,\mathbf{B} - (\mathbf{C} \cdot \mathbf{B})\,\mathbf{A} \tag{8.8}$$

$$\mathbf{A} \cdot (\mathbf{B} \times \mathbf{C}) = \mathbf{B} \cdot (\mathbf{C} \times \mathbf{A}) - \mathbf{C} \cdot (\mathbf{A} \times \mathbf{B}) \tag{8.9}$$

$$\mathbf{A} \times (\mathbf{B} \times \mathbf{C}) = (\mathbf{A} \cdot \mathbf{C})\,\mathbf{B} - (\mathbf{A} \cdot \mathbf{B})\,\mathbf{C} \tag{8.10}$$

$$\nabla \cdot (\mathbf{B} \times \mathbf{C}) = -\,\mathbf{B}\,(\nabla \times \mathbf{C}) + \mathbf{C}\,(\nabla \times \mathbf{B}) \tag{8.11}$$

$$\nabla \times (\mathbf{B} \times \mathbf{C}) = \mathbf{B}\,(\nabla \cdot \mathbf{C}) - \mathbf{C}\,(\nabla \cdot \mathbf{B}) - \mathbf{B} \cdot \nabla\mathbf{C} + \mathbf{C} \cdot \nabla\mathbf{B}\,. \tag{8.12}$$

According to (8.7), the last two terms in (8.12) can also be written as $(\mathbf{B} \cdot \nabla)\,\mathbf{C}$ and $(\mathbf{C} \cdot \nabla)\,\mathbf{B}$.

Evidently

$$\nabla \times \nabla\alpha = 0, \qquad \nabla \cdot (\nabla \times \mathbf{B}) = 0, \tag{8.13}$$

but

$$\begin{aligned} \nabla \cdot (\alpha\nabla\beta) &= \nabla\alpha \cdot \nabla\beta + \alpha\nabla^2\beta, \\ \nabla \times (\alpha\nabla\beta) &= \nabla\alpha \times \nabla\beta \end{aligned} \tag{8.14}$$

where α, β are scalars.

In this book preference is given to vector and tensor notations.

PROBLEMS

1. Analyze the solution (5.4) when the denominator vanishes. Note: see [56].
2. Show that

$$\nabla \cdot (\nabla \times \mathbf{A}) = \epsilon_{ijk} \frac{\partial^2 A_k}{\partial x_i \partial x_j} = 0\,.$$

3. Show that $\epsilon_{ijk}\epsilon_{imn} = \delta_{jm}\delta_{kn} - \delta_{km}\delta_{jn}$ where $\delta_{ij} = 1$ (0) when $i = j$ ($i \neq j$) is the Kronecker delta.
4. Prove the identities (8.6) and (8.7).

PART I

THE DYNAMICS OF AN IDEAL (WITHOUT FRICTION) ATMOSPHERE

In this part, the dynamic and thermodynamic processes in the free atmosphere are studied, i.e., above the so-called planetary boundary layer which, on average, extends up to a 1–1.5 km height. This is referred to as an *ideal* atmosphere because the influence of the turbulent friction force – the main dissipative force for the atmospheric motions – is ignored. As will be shown further on, although the air is a viscous fluid, its molecular viscosity only slightly influences its dynamics with the exception of the rather special cases to be considered later. That is why in almost all problems of dynamic meteorology (and, more generally, of geophysical fluid dynamics because a similar conclusion also holds for the ocean), the molecular friction force can be omitted and the air can be regarded as a nonviscous (ideal) fluid. But as far as this can be done everywhere, the turbulent (eddy) friction force is nonessential in the free atmosphere but should be taken into account in the planetary boundary layer.

The second part of our course in dynamic meteorology is, on the whole, dedicated to this problem. The inclusion of the planetary boundary layer and the turbulent exchange makes the model of the atmosphere, which is developed in Part II, closer to reality. For this reason, we may distinguish between the dynamics of *ideal* and *real* atmospheres.

CHAPTER 2

Equations of Thermo-Hydrodynamics of the Atmosphere (Weather Equations)

1. THE THERMODYNAMIC ENERGY EQUATION

a. General Form

The thermodynamic energy equation, also known as the heat influx equation, is a mathematical expression of the first principle. It is well known from general courses in physics and meteorology and in dynamic meteorology it is written for a moving air parcel.

Let $\mathcal{E}$ be the amount of heat received by a unit volume of air per unit time. Then a finite volume V for a finite interval of time $\tau = t_2 - t_1$ would receive (give) the amount of heat

$$\mathcal{E}' = \int_{t_1}^{t_2} \mathrm{d}t \int_V \mathcal{E}\, \mathrm{d}V. \tag{1.1}$$

Part of it, namely

$$\mathcal{E}_1' = \int_{t_1}^{t_2} \mathrm{d}t \int_V \rho c_v \frac{\mathrm{d}T}{\mathrm{d}t} \mathrm{d}V, \tag{1.2}$$

is spent on the temperature change of the moving air mass, and another part,

$$\mathcal{E}_2' = \int_{t_1}^{t_2} \mathrm{d}t \int_V p \frac{\mathrm{d}}{\mathrm{d}t}(\mathrm{d}V), \tag{1.3}$$

for changing of the volume. But according to formulae (4.3) and (6.1) of Chapter 1,

$$\frac{\mathrm{d}}{\mathrm{d}t} \mathrm{d}V = \operatorname{div} \mathbf{v} \cdot \mathrm{d}V = -\frac{1}{\rho}\frac{\mathrm{d}\rho}{\mathrm{d}t} \cdot \mathrm{d}V. \tag{1.4}$$

Since the energy conservation law requires $\mathcal{E}' = \mathcal{E}_1' + \mathcal{E}_2'$ then

$$\int_{t_1}^{t_2} \mathrm{d}t \int_V \left[\mathcal{E} - \left(\rho c_v \frac{\mathrm{d}T}{\mathrm{d}t} - \frac{p}{\rho}\frac{\mathrm{d}\rho}{\mathrm{d}t}\right)\right] \mathrm{d}V = 0.$$

Hence,

$$\rho c_v \frac{\mathrm{d}T}{\mathrm{d}t} - \frac{p}{\rho}\frac{\mathrm{d}\rho}{\mathrm{d}t} = \mathcal{E}, \tag{1.5}$$

which is the general form of the required equation.

b. Alternative Forms and Particular Cases

The density ρ in (1.5) can be replaced with the help of the equation of state

$$p = \rho RT. \tag{1.6}$$

We obtain

$$\frac{dT}{dt} - \frac{\kappa - 1}{\kappa}\frac{T}{p}\frac{dp}{dt} = \frac{Q}{c_p}, \quad Q = \frac{\mathcal{E}}{\rho}, \tag{1.7}$$

where $\kappa = c_p/c_v$, $R = c_p - c_v$, Q is the rate of heating/cooling per unit mass per unit time. Introducing the potential temperature

$$\theta = T\left(\frac{p_{00}}{p}\right)^{\lambda}, \quad \lambda = \frac{R}{c_p}, \quad p_{00} = 1000 \text{ mb}, \tag{1.8}$$

we write (1.7) in the more compact form

$$\frac{1}{\theta}\frac{d\theta}{dt} = \frac{d}{dt}\ln\theta = \frac{Q}{Tc_p}. \tag{1.9}$$

Finally, since the entropy

$$s = c_p \ln\theta + \text{const}, \tag{1.10}$$

we write our basic equation (1.5) in the most compact form

$$\frac{ds}{dt} = \frac{Q}{T}. \tag{1.11}$$

The particular case of adiabatic processes ($Q = 0$) is of considerable theoretical interest. Then instead of (1.7)–(1.11), we obtain

$$\frac{dT}{dt} = \frac{\lambda T}{p}\frac{dp}{dt} \quad \text{or} \quad \frac{dp}{dt} = \frac{\kappa p}{\rho}\frac{d\rho}{dt} = a_s^2\frac{d\rho}{dt}, \tag{1.12}$$

where $a_s = (\kappa RT)^{1/2}$ is the speed of sound. Also

$$\frac{d\theta}{dt} = 0, \quad \frac{ds}{dt} = 0, \tag{1.13}$$

i.e., the potential temperature and the specific entropy (as well as the total one $S = \int s\rho \, dV$) are adiabatic invariants following the motion.

Thus, so far we have constructed three equations – of continuity (1.4), of the heat influx (1.7) and of the state (1.6). In the case $Q = 0$ they contain six unknown functions: T, p, ρ, u, v, w. The missing three equations for completing the system are the equations of motion from the next section.

c. Heat Sources

The assumption that, in Equation (1.7), the heat influx function $Q \approx 0$ is known as a quasi-adiabatic hypothesis. It introduces a considerable simplification in the theory of meteorological forecasting but, unfortunately, has a rather restricted application. Such a simplification is only possible for short-range forecasts (up to 1 day) and even then, not always.

Strictly speaking, in the real atmosphere $\mathcal{E} \neq 0$ and represents an extra unknown quantity in the so-far deduced equations. General physical considerations lead us to think that it is an extremely complex function of $\mathbf{x}$ and t. Its physical sense as an integral heat influx (referred to a given air mass) allows us, however, to decompose $\mathcal{E}$ into components which may be studied one by one.

Thus, we can write that in Equation (1.7)

$$\mathcal{E} = \mathcal{E}_d + \mathcal{E}_m + \mathcal{E}_L + \mathcal{E}_R \tag{1.14}$$

where $\mathcal{E}_d$ is the heat influx due to dissipation of the kinetic energy of the motion under the influence of the molecular viscosity, $\mathcal{E}_m$ – due to molecular diffusion (heat conductivity), $\mathcal{E}_L$ – due to the water phase transition and latent heat release (absorbtion), $\mathcal{E}_R$ – due to radiation processes.

Exact expressions exist for the first two terms in (1.14). For instance,

$$\epsilon = \frac{\mathcal{E}_d}{\rho} = \frac{\nu}{2}\left(\frac{\partial v_i}{\partial x_k} + \frac{\partial v_k}{\partial x_i}\right)^2, \quad \nu = \frac{\eta}{\rho}, \tag{1.15}$$

is called the rate of dissipation, η is the molecular viscosity coefficient. Thus, new unknown functions are not introduced. It turns out, however, that the heat-effect (the rate of heating $\mathcal{E}_d/\rho c_p$) related to the dissipation of kinetic energy in the atmosphere in the majority of dynamic meteorology problems is very small and can be neglected.

As to the contribution of the molecular heat conductivity mechanism, we can write

$$\mathcal{E}_m = \lambda_m \nabla^2 T, \quad \lambda_m = \rho c_p \kappa_m, \tag{1.16}$$

where κ_m is an analogue of ν from (1.15) and has the same dimension: $[\kappa_m] = [\nu] = L^2 T^{-1}$, ∇^2 is the three-dimensional Laplacian.

The problem of finding explicit expressions for the remaining two terms ($\mathcal{E}_L$ and $\mathcal{E}_R$) in (1.14) has not yet been solved satisfactorily. The theoretical approach leads to the construction of a great number of new equations connecting $\mathcal{E}_L$ and $\mathcal{E}_R$ with various atmospheric parameters. This naturally complicates the problem in a mathematical respect. Of course, $\mathcal{E}_L$ and $\mathcal{E}_R$ can also be introduced into (1.14) and (1.7) by means of empirical formulae or tables.

Substituting (1.14) into (1.7), we find the general form of the thermodynamic energy equation

$$\frac{\mathrm{d}T}{\mathrm{d}t} - \frac{\kappa - 1}{\kappa}\,\frac{T}{p}\,\frac{\mathrm{d}p}{\mathrm{d}t} = \frac{\epsilon}{c_p} + \kappa_m \nabla^2 T + \frac{1}{\rho c_p}\,(\mathcal{E}_L + \mathcal{E}_R). \tag{1.17}$$

One of the most popular simplified versions of this equation, having a great application in theory, reads

$$\frac{\partial T}{\partial t} + v_k \frac{\partial T}{\partial x_k} = \kappa_m \nabla^2 T, \tag{1.18}$$

which coincides exactly with the equation

$$\frac{\partial S}{\partial t} + v_k \frac{\partial S}{\partial x_k} = \kappa_s \nabla^2 S \tag{1.19}$$

describing the field of concentration $S(\mathbf{x}, t)$ of a *passive* substance with a coefficient of molecular diffusion $\kappa_s (\neq \kappa_m)$. But the heat (temperature) is not a passive admixture in the flow. It could be considered as passive only in the case of small temperature variations in space. At the same time, they should be large enought to consider the dissipation effect as being negligible. Then the term with dp/dt in (1.17) can also be neglected to obtain (1.18).

For the time being, the lack of a satisfactory solution to the problem concerning the diabatic heating function Q (or $\mathcal{E}$) in Equation (1.7) stimulates the use of simple empirical formulae, which despite not having a good physical basis nevertheless describe well some important phenomena of theoretical and practical interest. The most suitable expression of this kind, known as Newtonian heating (cooling), reads

$$Q = -\delta c_p (T - T_E), \tag{1.20}$$

where δ is a constant and T_E is some reference temperature. One of the most frequent applications of (1.20) is in the theory of climate and climate modelling. Then T is the temperature of the air close to the ocean surface, while T_E is the mean temperature of the water surface itself. Due to the thermal inertia of the ocean, T_E is a representable characteristic.

2. THE EQUATIONS OF MOTION

a. General Form

Two of the conservation laws known in physics, namely the law of mass conservation and the law of energy conservation, have already been used in the derivation of the equations of continuity and heat influx. The equations of motion are a consequence of the law of momentum conservation.

We start with a rather general form of these equations, well-known from courses in fluid dynamics

$$\frac{\partial \rho v_i}{\partial t} = -\frac{\partial}{\partial x_k} (\rho v_i v_k + P_{ik}) + \rho F_i, \quad i = 1, 2, 3, \tag{2.1}$$

where

$$P_{ik} = p\delta_{ij} - \sigma_{ik}. \tag{2.2}$$

Here p is the hydrostatic pressure, δ_{ik} is the Kronecker delta, and σ_{ik} is the molecular stress tensor

$$\sigma_{ik} = \rho\nu\left(\frac{\partial v_i}{\partial x_k} + \frac{\partial v_k}{\partial x_i}\right) - \frac{2}{3}\rho\nu\delta_{ik}\frac{\partial v_\alpha}{\partial x_\alpha}, \tag{2.3}$$

where F_i is the external force. With the help of the continuity equation

$$\frac{\partial\rho}{\partial t} + \frac{\partial\rho v_k}{\partial x_k} = 0 \tag{2.4}$$

the following identity can be proved

$$\rho\frac{dv_i}{dt} = \frac{\partial\rho v_i}{\partial t} + \frac{\partial\rho v_i v_k}{\partial x_k}. \tag{2.5}$$

Then instead of (2.1) we can write

$$\rho\frac{dv_i}{dt} = -\frac{\partial p}{\partial x_i} + \rho\nu\nabla^2 v_i + \rho F_i + \frac{1}{3}\rho\nu\frac{\partial^2 v_\alpha}{\partial x_\alpha \partial x_i}. \tag{2.6}$$

For an incompressible fluid $\partial v_\alpha/\partial x_\alpha = 0$ and (2.6) simplifies to

$$\frac{dv_i}{dt} = -\alpha\frac{\partial p}{\partial x_i} + \nu\nabla^2 v_i + F_i, \quad \alpha = \frac{1}{\rho}. \tag{2.7}$$

When the external force is absent ($F_i = 0$)

$$\frac{\partial v_i}{\partial t} + \frac{\partial v_i v_k}{\partial x_k} = -\alpha\frac{\partial p}{\partial x_i} + \nu\nabla^2 v_i, \quad i = 1, 2, 3. \tag{2.8}$$

These are known as the Navier–Stokes equations.

b. Application to the Atmosphere

Atmospheric motions are realized with the presence of external forces ($F_i \neq 0$). These are the gravitational force $\mathbf{g}_*$, the centrifugal force $\mathbf{g}_c = \omega^2\mathbf{R}$, and the Coriolis force $\mathbf{K} = -2\boldsymbol{\omega} \times \mathbf{v}$. The latter two are caused by the rotation of the Earth around its axis with an angular velocity $\boldsymbol{\omega}$. It is convenient to define an effective gravity acceleration $\mathbf{g}$ such that $\mathbf{g} = \mathbf{g}_* + \mathbf{g}_c$. Since $g_c \ll g_*$, we have $g \approx g_*$. Let us also point out that in meteorology the atmospheric motions are observed and studied in a noninertial reference frame rotating with the Earth. Thus, additional apparent forces are required so that Newton's second law remains valid. Consequently, Equation (2.7) applied to the atmosphere will read

$$\frac{dv_i}{dt} = -\alpha\frac{\partial p}{\partial x_i} + \nu\nabla^2 v_i - 2\epsilon_{ijk}\omega_j v_k + g_i, \tag{2.9}$$

or in vector form

$$\frac{\mathrm{d}\mathbf{v}}{\mathrm{d}t} = -\alpha\nabla p + \nu\nabla^2\mathbf{v} - 2\boldsymbol{\omega} \times \mathbf{v} + \mathbf{g}, \tag{2.10}$$

where

$$\frac{\mathrm{d}}{\mathrm{d}t} = \frac{\partial}{\partial t} + (\mathbf{v}\cdot\nabla) = \frac{\partial}{\partial t} + v_k\frac{\partial}{\partial x_k}. \tag{2.11}$$

However, as will be shown later, in the majority of problems encountered in dynamic meteorology, the molecular friction force can be neglected. Thus, we obtain the fundamental equations of atmospheric dynamics

$$\frac{\mathrm{d}\mathbf{v}}{\mathrm{d}t} = -\alpha\nabla p - 2\boldsymbol{\omega} \times \mathbf{v} + \mathbf{g} \tag{2.12}$$

named after Euler. Identities (8.6) or (8.7) in Chapter 1, allow us to rewrite (2.12) in the following alternative forms

$$\frac{\partial v_i}{\partial t} + \frac{\partial}{\partial x_i}\left(\frac{1}{2}v_k^2 + \frac{p}{\rho} + \Phi\right) + \epsilon_{ijk}(\Omega_j + 2\omega_j)v_k = \nu\nabla^2 v_i \tag{2.13}$$

or

$$\frac{\partial\mathbf{v}}{\partial t} + \nabla\left(\frac{1}{2}|\mathbf{v}|^2 + \frac{p}{\rho} + \Phi\right) + (\boldsymbol{\Omega} + 2\boldsymbol{\omega}) \times \mathbf{v} = \nu\nabla^2\mathbf{v}, \tag{2.14}$$

where $\nabla\Phi = -\mathbf{g}$, i.e., Φ is a geopotential. The sum

$$\boldsymbol{\Omega} + 2\boldsymbol{\omega} = \boldsymbol{\Omega}_a \tag{2.15}$$

is called the absolute vorticity.

c. Boussinesq Approximation

Let us return to Equation (2.12) and write it in the form

$$\rho\frac{\mathrm{d}\mathbf{v}}{\mathrm{d}t} = -\nabla p - 2\rho\boldsymbol{\omega} \times \mathbf{v} + \rho\mathbf{g}. \tag{2.16}$$

With strictly incompressible fluid $\dot{\rho} = 0$ and $\nabla \cdot \mathbf{v} = 0$. But when $\dot{\rho} \neq 0$ even small variations of density ρ, due to thermal (or other) reasons, give rise to Archimedean (buoyancy) force in the direction determined by the gravity acceleration vector **g** and this force may considerably influence the motion. To account for this effect we replace ρ in (2.16) by some standard density ρ_s, except in the last term in which ρ multiplies **g**. The result is

$$\frac{\mathrm{d}\mathbf{v}}{\mathrm{d}t} = -\frac{\nabla p}{\rho_s} - 2\boldsymbol{\omega} \times \mathbf{v} + \frac{\rho}{\rho_s}\mathbf{g}. \tag{2.17}$$

Evidently, an approximation is made here. It was first proposed by Boussinesq and so bears his name, and has a wide application in geophysical fluid dynamics. We shall return to it later in Section 4.

There is also, however, another way of considering the changes of ρ. Preserving $\nabla \cdot \mathbf{v} = 0$, we let $d\rho/dt = \mathcal{E}_\rho$ be the rate of density change due to thermal or other factors. This new equation may replace the equation of the first principle. Such an approach can be met in some specific problems in the theory of nonlinear waves in the atmosphere, where the Boussinesq approximation is not precise enough.

3. WEATHER EQUATIONS IN SPHERICAL COORDINATES

a. Preliminary Preparation

The so-far derived Equations (6.1) of Chapter 1, (1.6), (1.11) and (2.12), constitute the system of equations of thermo-fluid dynamics of the atmosphere or simply the *weather equations*. Let us write them together in scalar form

$$\frac{du}{dt} = -\alpha \frac{\partial p}{\partial x} - 2(\omega_y w - \omega_z v) + g_x, \tag{3.1}$$

$$\frac{dv}{dt} = -\alpha \frac{\partial p}{\partial y} - 2(\omega_z u - \omega_x w) + g_y, \tag{3.2}$$

$$\frac{dw}{dt} = -\alpha \frac{\partial p}{\partial z} - 2(\omega_x v - \omega_y u) + g_z, \tag{3.3}$$

$$\rho T \frac{ds}{dt} = \mathcal{E}, \qquad p = \rho R T, \quad \left(\alpha = \frac{1}{\rho}\right), \tag{3.4}$$

$$\frac{d\rho}{dt} + \rho\left(\frac{\partial u}{\partial x} + \frac{\partial v}{\partial y} + \frac{\partial w}{\partial z}\right) = 0, \tag{3.5}$$

where

$$\begin{gathered} \frac{d}{dt} = \frac{\partial}{\partial t} + u\frac{\partial}{\partial x} + v\frac{\partial}{\partial y} + w\frac{\partial}{\partial z}, \\ (u, v, w) = \mathbf{v}, \quad (\omega_x, \omega_y, \omega_z) = \boldsymbol{\omega}. \end{gathered} \tag{3.6}$$

The equations are written in a Cartesian coordinate system $0xyz$ rotating with the Earth, as shown in Figure 2.1. Such a reference frame, whose origin is located in the Earth's center ($0x$ and $0y$ axes lie in the equatorial plane while $0z$ is an axis of rotation), is not suitable, however, for practical applications. It does not reflect any of the characteristic features of the Earth's atmosphere. Thus, for instance, in Equations (3.1)–(3.3) the gravity force has components on all axes. It is known, however, that it considerably exceeds the remainder of the acting forces in the atmosphere with the exception of the pressure-gradient force. That is why the features of the atmospheric motions in the

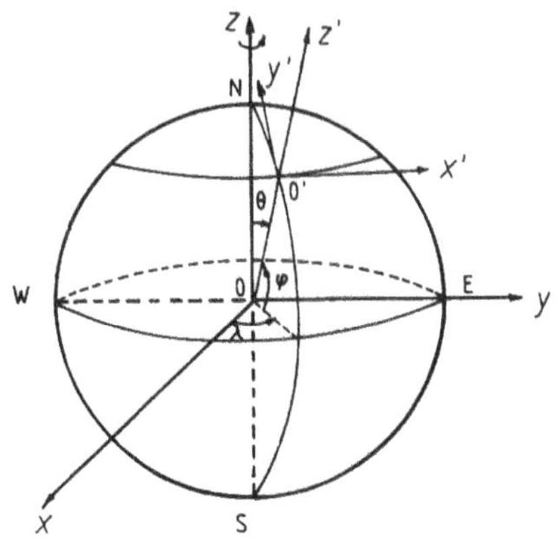

Fig. 2.1. Spherical and local (standard) coordinates.

vertical direction (determined by the vector **g**) differ considerably from their features in the horizontal direction, i.e., in a plane normal to **g**. Moreover, the atmosphere is a spherical gaseous cover of a spherical (in the first and sufficient for meteorological applications approximation) Earth, and this fact must be taken into consideration, especially when studying the atmospheric motions of planetary scales – for instance, the general circulation of the atmosphere.

The necessity of writing down weather equations in spherical coordinates is naturally imposed.

b. Introduction of Spherical Coordinates

We have to replace the independent variables (x, y, z) by (λ, θ, r) as defined in Figure 2.1. Obviously

$$x = r \sin\theta \cos\lambda, \qquad y = r \sin\theta \sin\lambda, \qquad z = r\cos\theta, \tag{3.7}$$

where λ is the geographical longitude and $\varphi = (\pi/2) - \theta$ is the geographical latitude. The Earth's surface may be approximated to a coordinate surface $r = r_0 =$ const. (Here r_0 is the radius of the Earth.)

In order to define the velocities $(v_\lambda, v_\theta, v_r)$ we consider the linear elements on the sphere $r =$ const and along r:

$$\mathrm{d}s_\lambda = r\sin\theta\,\mathrm{d}\lambda, \qquad \mathrm{d}s_\theta = r\,\mathrm{d}\theta, \quad \mathrm{d}s_r = \mathrm{d}r. \tag{3.8}$$

Then

$$v_\lambda = \frac{\mathrm{d}s_\lambda}{\mathrm{d}t} = r\sin\theta\frac{\mathrm{d}\lambda}{\mathrm{d}t}, \qquad v_\theta = \frac{\mathrm{d}s_\theta}{\mathrm{d}t} = r\frac{\mathrm{d}\theta}{\mathrm{d}t}, \qquad v_r = \frac{\mathrm{d}s_r}{\mathrm{d}t} = \frac{\mathrm{d}r}{\mathrm{d}t}. \tag{3.9}$$

Let F be whichever dependent scalar variable $(p, T, \rho$, etc.)

$$F = F(t, r, \theta, \lambda).$$

Evidently

$$\frac{\mathrm{d}F}{\mathrm{d}t} = \frac{\partial F}{\partial t} + \frac{\partial F}{\partial r}\frac{\mathrm{d}r}{\mathrm{d}t} + \frac{\partial F}{\partial \theta}\frac{\mathrm{d}\theta}{\mathrm{d}t} + \frac{\partial F}{\partial \lambda}\frac{\mathrm{d}\lambda}{\mathrm{d}t}.$$

But according to (3.9)

$$\frac{\mathrm{d}r}{\mathrm{d}t} = v_r, \qquad \frac{\mathrm{d}\theta}{\mathrm{d}t} = \frac{v_\theta}{r}, \qquad \frac{\mathrm{d}\lambda}{\mathrm{d}t} = \frac{v_\lambda}{r \sin\theta}. \tag{3.10}$$

Therefore, in spherical coordinates

$$\frac{\mathrm{d}}{\mathrm{d}t} = \frac{\partial}{\partial t} + v_r\frac{\partial}{\partial r} + \frac{v_\theta}{r}\frac{\partial}{\partial \theta} + \frac{v_\lambda}{r\sin\theta}\frac{\partial}{\partial \lambda}. \tag{3.11}$$

On the other hand, the velocity vector v can be written in the form $v = v_r i_r + v_\theta i_\theta + v_\lambda i_\lambda$, where $(i_r, i_\theta, i_\lambda)$ are unit vectors whose directions are not constant but contrary to $(\mathbf{i}, \mathbf{j}, \mathbf{k})$ in the Cartesian system $0xyz$ and are functions of position on the sphere. Taking this fact into consideration, after tedious calculations, the following final result can be reached

$$\frac{\mathrm{d}v_\lambda}{\mathrm{d}t} + \frac{1}{r} v_\lambda v_r + \frac{1}{r}\operatorname{cotan}\theta \cdot v_\theta v_\lambda = -\frac{\partial p/\partial\lambda}{\rho r\sin\theta} - 2\omega(v_r \sin\theta + v_\theta\cos\theta), \tag{3.12}$$

$$\frac{\mathrm{d}v_\theta}{\mathrm{d}t} + \frac{1}{r} v_\theta v_r - \frac{1}{r}\operatorname{cotan}\theta \cdot v_\lambda^2 = -\frac{1}{\rho r}\frac{\partial p}{\partial \theta} + 2\omega v_\lambda \cos\theta, \tag{3.13}$$

$$\frac{\mathrm{d}v_r}{\mathrm{d}t} - \frac{1}{r}(v_\theta^2 + v_\lambda^2) = -g - \frac{1}{\rho}\frac{\partial p}{\partial r} + 2\omega v_\lambda \sin\theta. \tag{3.14}$$

Here, as well as in Equation (3.4) which formally remains unchanged, $\mathrm{d}/\mathrm{d}t$ has to be substituted from (3.11). The continuity equation transforms into one of the following alternative forms

$$\frac{\mathrm{d}\rho}{\mathrm{d}t} + \rho\left[\frac{1}{r^2}\frac{\partial}{\partial r}(r^2 v_r) + \frac{1}{r}\frac{\partial v_\theta}{\partial\theta} + \frac{1}{r\sin\theta}\frac{\partial v_\lambda}{\partial\lambda} + \frac{1}{r}v_\theta\operatorname{cotan}\theta\right] = 0, \tag{3.15}$$

or

$$\frac{\partial\rho}{\partial t} + \frac{1}{r^2}\frac{\partial}{\partial r}(\rho r^2 v_r) + \frac{1}{r\sin\theta}\left(\frac{\partial}{\partial\theta}\rho v_\theta\sin\theta + \frac{\partial}{\partial\lambda}\rho v_\lambda\right) = 0. \tag{3.16}$$

Without loss of accuracy, a considerable simplification of the equations can be made taking into account the quasi-two-dimensional character of the atmosphere, i.e., its characteristic thickness Z is much smaller than the radius of the Earth ($Z \ll r_0$). Introducing a new vertical variable $z = r - r_0$, we can write

$$\frac{1}{r} \approx \frac{1}{r_0}, \qquad \frac{\partial}{\partial r} = \frac{\partial}{\partial z}, \qquad v_r = v_z = w. \tag{3.17}$$

Then instead of (3.11)

$$\frac{\mathrm{d}}{\mathrm{d}t} = \frac{\partial}{\partial t} + w\frac{\partial}{\partial z} + \frac{v_\theta}{r_0}\frac{\partial}{\partial \theta} + \frac{v_\lambda}{r_0 \sin\theta}\frac{\partial}{\partial \lambda}. \tag{3.18}$$

In the same manner

$$\begin{aligned} &\frac{1}{r}\frac{\partial}{\partial r}(rF) \approx \frac{\partial F}{\partial r}, \qquad F = v_\theta, v_\lambda, \\ &\frac{1}{r^2}\frac{\partial}{\partial r}(r^2 v_r) \approx \frac{\partial v_r}{\partial r} = \frac{\partial w}{\partial z}, \\ &v_r\frac{\partial v_\theta}{\partial_r} + \frac{1}{r}v_r v_\theta = v_r\frac{1}{r}\frac{\partial}{\partial r}(r v_\theta) \approx w\frac{\partial v_\theta}{\partial z}. \end{aligned} \tag{3.19}$$

Bearing this in mind, we conclude that the terms in (3.12) and (3.13) marked with overbars could be omitted. Thus we obtain simpler equations

$$\frac{\mathrm{d}}{\mathrm{d}t}v_\lambda + \frac{1}{r_0}\cot\text{an}\,\theta\cdot v_\theta v_\lambda = -\frac{\partial p/\partial\lambda}{\rho r_0 \sin\theta} - 2\omega(w\sin\theta + v_\theta\cos\theta), \tag{3.20}$$

$$\frac{\mathrm{d}}{\mathrm{d}t}v_\theta - \frac{1}{r_0}\cot\text{an}\,\theta\cdot v_\lambda^2 = -\frac{1}{\rho r_0}\frac{\partial p}{\partial\theta} + 2\omega v_\lambda\cos\theta, \tag{3.21}$$

$$\frac{\mathrm{d}}{\mathrm{d}t}w - \frac{1}{r_0}(v_\lambda^2 + v_\theta^2) = -\frac{1}{\rho}\frac{\partial p}{\partial r} - g + 2\omega v_\lambda\sin\theta, \tag{3.22}$$

$$\frac{\mathrm{d}}{\mathrm{d}t}\rho + \rho\left(\frac{\partial w}{\partial z} + \frac{1}{r_0}\frac{\partial}{\partial\theta}v_\theta + \frac{1}{r_0\sin\theta}\frac{\partial}{\partial\lambda}v_\lambda + \frac{1}{r_0}v_\theta\cot\text{an}\,\theta\right) = 0. \tag{3.23}$$

One can go further with the simplification and neglect the terms with the vertical velocity w in (3.20) and (3.21) because they are small compared to the others. A justification for this is given in Chapter 3. Then, after some rearrangements, these equations can be written in the form

$$\frac{\partial}{\partial t}v_\lambda + \frac{1}{r_0\sin\theta}\frac{\partial E}{\partial\lambda} = -\frac{\partial p/\partial\lambda}{\rho r_0\sin\theta} - (2\omega\cos\theta + \zeta)\,v_\theta, \tag{3.24}$$

$$\frac{\partial}{\partial t}v_\theta + \frac{1}{r_0}\frac{\partial E}{\partial\theta} = -\frac{1}{\rho r_0}\frac{\partial p}{\partial\theta} + (2\omega\cos\theta + \zeta)\,v_\lambda, \tag{3.25}$$

where

$$E = \tfrac{1}{2}(v_\lambda^2 + v_\theta^2), \qquad \zeta = \frac{1}{r_0\sin\theta}\left(\frac{\partial}{\partial\theta}v_\lambda\sin\theta - \frac{\partial}{\partial\lambda}v_\theta\right), \tag{3.26}$$

are, correspondingly, the kinetic energy of the two-dimensional motion and the z component of the vorticity. These equations are very often used in the theory of general atmospheric circulation.

Further simplifications are also possible in continuity Equation (3.23) and are written for that purpose in the form

$$\left(\frac{\partial \rho}{\partial t} + \frac{v_\theta}{r_0}\frac{\partial \rho}{\partial \theta} + \frac{v_\lambda}{r_0 \sin\theta}\frac{\partial \rho}{\partial \lambda}\right) + \frac{\partial \rho w}{\partial z} + \frac{\rho}{r_0 \sin\theta}\left(\frac{\partial}{\partial \lambda} v_\lambda + \frac{\partial}{\partial \theta} v_\theta \sin\theta\right) = 0.$$

The first three terms in the parentheses are small and can be neglected. Thus,

$$\frac{\partial}{\partial z}\rho w + \frac{\rho}{r_0 \sin\theta}\left(\frac{\partial}{\partial \lambda} v_\lambda + \frac{\partial}{\partial \theta} v_\theta \sin\theta\right) = 0. \tag{3.27}$$

4. WEATHER EQUATIONS IN LOCAL (STANDARD) COORDINATES: BOUNDARY CONDITIONS

a. Introduction of Local Coordinates

Let us now return to the vector form of momentum Equations (2.12)

$$\frac{d\mathbf{v}}{dt} = -\alpha\nabla p - 2\omega \times \mathbf{v} + \mathbf{g}, \quad \alpha = \frac{1}{\rho}. \tag{4.1}$$

To a considerable degree, this record is universal, but it should be applied to a particular coordinate system – say spherical. The use of this system is compulsory in global-scale problems requiring consideration of the Earth's sphericity. However, when the processes are to be investigated in a limited area, its use is not expedient because of the complex form of the equations. Thus, we come to realize the necessity of a 'plane approximation'. This is achieved by the introduction of the so-called *local* or *standard* coordinate system (Figure 2.1), the origin $0'$ of which is located in a point on the surface of the Earth, the axis $0'x'$ is directed eastward tangential to the parallel, $0'y'$ – northward tangential to the meridian and $0'z'$ – vertically upward. Henceforth, we shall omit the prime, writing (x, y, z) instead of (x', y', z'). Obviously, in the new system

$$\omega = (0, \omega\sin\theta, \omega\cos\theta), \qquad \mathbf{g} = (0, 0, -g). \tag{4.2}$$

In order to transform Equations (3.12)–(3.14), we write them in a slightly different form

$$\begin{aligned}
\frac{d}{dt} v_\lambda &= -\left(2\omega + \frac{v_\lambda}{r\sin\theta}\right)(v_r \sin\theta + v_\theta\cos\theta) - \frac{\partial p/\partial\lambda}{\rho r \sin\theta},\\
\frac{d}{dt} v_\theta &= \left(2\omega + \frac{v_\lambda}{r\sin\theta}\right) v_\lambda\cos\theta - \frac{1}{\rho r}\frac{\partial p}{\partial \theta} - \frac{1}{r}v_\theta v_r,\\
\frac{d}{dt} v_r &= \left(2\omega + \frac{v_\lambda}{r\sin\theta}\right) v_\lambda \sin\theta - \frac{1}{\rho}\frac{\partial p}{\partial r} + \frac{1}{r}v_\theta^2 - g.
\end{aligned} \tag{4.3}$$

The quantity

$$\alpha_a = \frac{v_\lambda}{r\sin\theta} \tag{4.4}$$

represents the angular velocity of the atmosphere's rotation with respect to the Earth's surface (v_λ is the zonal wind velocity and $r \sin \theta$ is the distance from the axis of rotation) and is called *index of circulation*. It is an important characteristic of westerlies in the middle latitude atmosphere (see Chapter 11, Section 1).

In agreement with the choice of a local coordinate system, we can write

$$\begin{aligned} dx &= r_0 \sin\theta\, d\lambda, \quad & dy &= -r_0\, d\theta, \quad & dr &= dz, \\ v_\lambda &= u, & v_\theta &= -v, & v_r &= w. \end{aligned} \tag{4.5}$$

The introduction of local coordinates into Equations (4.3) consists of substituting (4.5) into (4.3) and a subsequent transition $r_0 \to \infty$ (plane-Earth approximation). This is equivalent to neglecting the nonlinear terms at the right-hand sides of Equation (4.3), respectively α_a compared to 2ω. The result is

$$\left.\begin{aligned} \frac{du}{dt} &= \dot{u} = -\alpha \frac{\partial p}{\partial x} - 2\omega(w \sin\theta - v\cos\theta) \\ \frac{dv}{dt} &= \dot{v} = -\alpha \frac{\partial p}{\partial y} - 2\omega u \cos\theta \end{aligned}\right\} \tag{4.6}$$

$$\frac{dw}{dt} = \dot{w} = -g - \alpha \frac{\partial p}{\partial z} + 2\omega u \cos\theta. \tag{4.7}$$

Henceforth, (**i**, **j**, **k**) will denote the unit vectors directed eastward, northward, and upward, respectively. Then

$$\mathbf{u} = u\mathbf{i} + v\mathbf{j}, \qquad \mathbf{v} = \mathbf{u} + w\mathbf{k}$$

and Equation (4.6) for the horizontal motion can be written in vector form

$$\frac{d\mathbf{u}}{dt} = -\alpha \nabla p - f\mathbf{k} \times \mathbf{u} - f_1 w\mathbf{i} \tag{4.8}$$

where

$$f = 2\omega_z = 2\omega\cos\theta = 2\omega\sin\varphi, \tag{4.9}$$

is what is called the 'parameter of Coriolis', $f_1 = 2\omega\cos\varphi$ is an analogous parameter which is rarely used and, according to (4.5) and (3.11), $d/dt = \partial/\partial t + (\mathbf{v} \cdot \nabla)$.

As to the equations of first principle (3.4) and continuity (3.5), formally they remain unchanged in the form

$$\frac{ds}{dt} = \frac{Q}{T}, \qquad \frac{d\rho}{dt} = -\rho(\nabla \cdot \mathbf{v}). \tag{4.10}$$

The name 'local system' reflects the fact that to each point on the Earth's surface a coordinate system 'of its own' corresponds, with a new orientation of the axes in the space. The latter is a consequence of the Earth's sphericity, but if we consider a limited area with horizontal dimensions $L \lesssim 10^3$ km, then the Earth's surface can be approximated with a plane and the above-mentioned effect can be neglected. Actually, since the Coriolis parameter $f(y)$ can be written as

$$f(y) = 2\omega\left(\sin\varphi_0 + \frac{\cos\varphi_0}{1!}\frac{y}{r_0} - \frac{\sin\varphi_0}{2!}\frac{y^2}{r_0^2} + \cdots\right), \tag{4.11}$$

where φ_0 is some reference latitude, we see that in a first (f-plane) approximation

$$f(y) \approx 2\omega\sin\varphi_0 = f_0 \quad \text{at } y \ll r_0(= 6.37 \times 10^3\,\text{km}). \tag{4.12}$$

In a second (β-plane) approximation

$$f(y) \approx 2\omega\sin\varphi_0 + \frac{2\omega}{r_0}\cos\varphi_0 \cdot y = f_0 + \beta y, \tag{4.13}$$

where β is the Rossby parameter. In the middle latitudes

$$f \sim 10^{-4}\,\text{s}^{-1}, \qquad \beta \sim 10^{-11}\,\text{m}^{-1}\,\text{s}^{-1}. \tag{4.14}$$

Strictly observed, 'the independent' variables (x, y) of the local system are not completely independent. In reality, according to (3.8) and (4.5), they are interrelated by means of θ and a direct consequence of this is the inequality of the mixed derivatives $\mathbf{u}_{xy} \neq \mathbf{u}_{yx}$ though the function $\mathbf{u}(x, y)$ is continuous. It can be proved that the error (the deviation from the equality $\mathbf{u}_{xy} = \mathbf{u}_{yx}$) is small at $\varphi \to 0°$ and increases considerably at $\varphi \to 90°$. Thus, the local coordinate system cannot be used in the polar regions.

b. The Boussinesq Approximation

When using the local coordinate system, as it follows from (4.2), the gravitational force does not enter in Equations (4.8) for the horizontal motion. Consequently, the spatial inhomogeneities in the density field and interconnected Archimedean (buoyancy) force would be manifested directly in the vertical (convective) motion and indirectly in the horizontal one as far as (4.6) and (4.7) are interrelated equations. That is why it is relevant to apply the Boussinesq approximation from Section 2b to the third equation

$$\rho_s\dot{w} = -\frac{\partial p}{\partial z} - \rho g + \rho_s f_1 u. \tag{4.15}$$

In the general case $\rho = \rho(p, T)$ which is the baroclinic atmosphere and we let $F = F_s(z) + F'$, where $F = T, p, \rho$ and

$$dp_s = -g\rho_s\,dz, \qquad \rho_s = \rho(p_s, T_s).$$

Further we assume $p', T' \ll p_s, T_s$ so that $\rho' \ll \rho_s$. Then for an arbitrary z, one can write

$$\rho = \rho_s + \left(\frac{\partial\rho}{\partial T_s}\right)_p \cdot T' + \left(\frac{\partial\rho}{\partial p_s}\right)_T \cdot p' = \rho_s(1 - \beta_0 T' + \alpha_0 p').$$

However, as a rule $|\alpha_0 p'| \ll |\beta_0 T'|$. Thus,

$$\rho \approx \rho_s(1 - \beta_0 T') \quad \text{or} \quad \rho' = -\rho_s \beta_0 T'.$$

Then (4.15) yields

$$\frac{\mathrm{d}w}{\mathrm{d}t} = -\alpha_s \frac{\partial p'}{\partial z} + f_1 u + \beta_0 g T', \quad \alpha_s = \frac{1}{\rho_s}. \tag{4.16}$$

Some particular problems in the theory require a joint-recording of (4.16) and (4.6) with ρ being replaced by ρ_s:

$$\frac{\mathrm{d}\mathbf{v}}{\mathrm{d}t} = -\alpha_s \nabla p' - 2\omega \times \mathbf{v} + \beta_0 g T' \mathbf{k}, \tag{4.17}$$

or in tensor form

$$\frac{\mathrm{d}v_i}{\mathrm{d}t} = -\alpha_s \frac{\partial p'}{\partial x_i} - 2\epsilon_{ijk}\omega_j v_k + \beta_0 g T' \delta_{3i}, \tag{4.18}$$

where the components of ω are given by (4.2) and $\mathbf{k} = (0, 0, 1)$. The requirement for small density deviations $\rho' \ll \rho_s$ allows us to write the continuity equation as in the case of an incompressible fluid

$$\nabla \cdot \mathbf{v} = \frac{\partial v_\alpha}{\partial x_\alpha} = 0 \tag{4.19}$$

which is an element of the Boussinesq approximation. A complete analysis shows that the latter is possible in case of a shallow convection, developing in an atmospheric layer with a thickness $\Delta \sim 1$ km. Then, as a rule, $|\dot{w}| \ll g$ and in Equation (4.15) one can write $\rho_s \dot{w}$ instead of $\rho \dot{w}$.

A typical example of a meteorological phenomenon well described by this approximation is the sea breeze. However, in the case of powerful, deeply penetrating cumulus convection $\Delta \sim 10$ km and $\dot{w} \sim g$ so that the Boussinesq approximation cannot be applied. Field observations made under appropriate conditions and laboratory experiments support the theoretical results obtained using this approximation.

c. Boundary Conditions

The system of weather equations (4.1) and (4.10) consists of nonlinear partial differential equations. In the adiabatic approximation ($Q = 0$) it is closed – the number of equations equals those of the unknown functions (u, v, w, p, ρ, T) and, in principal, could be solved. In order to provide a description of a particular situation in the atmosphere, however, both boundary and initial conditions associated with the equations are needed, regardless of the methods of solution (analytical or numerical). The formulation of such conditions depends, to a considerable degree, on the peculiarities of each particular problem. That is why we shall be concerned here only with some most-general questions referring to flow velocity.

First of all, the atmosphere has a *rigid* lower boundary – the continents on the Earth's surface. Such a kind of boundary prevents the air from flowing past it. Thus, the velocity normal to the surface must vanish, that is

$$\mathbf{v} \cdot \mathbf{n} = 0, \tag{4.20}$$

where $\mathbf{n}$ is a vector, normal to the boundary, directed into the flow. The remaining conditions as well as the further transformations of (4.20) depend on whether molecular (frictional and conductivity) effects are important at the boundary:

(i) Let us suppose that the latter effects are unimportant. This case is known as *free-slip* boundary flow [11]. In such a frictionless (ideal) flow, the air particles lying on the surface follow its form during their movement. If $z = H(x, y)$ is the equation of the Earth's topography (Figure 2.2) and $x(t)$, $y(t)$, $z(t)$ are the coordinates of a moving air particle on the boundary, then at any instant t one can write

$$z(t) = H[x(t), y(t)]. \tag{4.21}$$

Differentiating, we obtain $\dot{z} = \dot{x}H_x + \dot{y}H_y$ or

$$w = u\frac{\partial H}{\partial x} + v\frac{\partial H}{\partial y} = (\mathbf{u} \cdot \nabla H) \quad \text{at } z = H. \tag{4.22}$$

It is easily seen that (4.22) is equivalent to (4.20). If $\nabla H = 0$ (horizontal boundary), then at $z = 0$, $w = 0$ but the horizontal components u, v may not be zero.

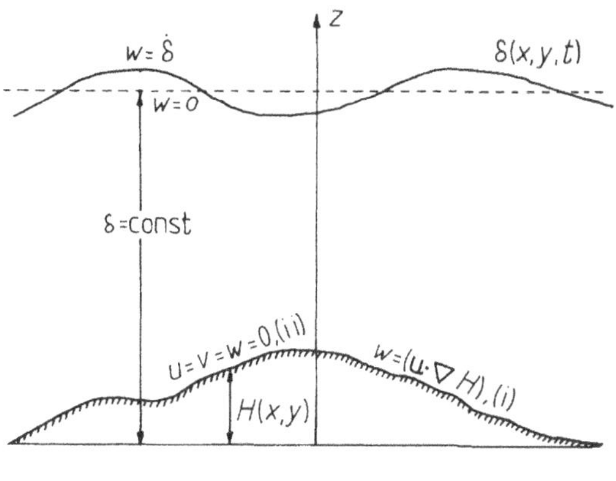

Fig. 2.2. Choice of boundary conditions.

In addition to (4.20) we can also require that

$$\frac{\partial \mathbf{v}}{\partial n} \cdot \mathbf{t} = 0 \tag{4.23}$$

where $\mathbf{t}$ is a vector tangential to the boundary (i.e., normal to $\mathbf{n}$) and n is a distance along $\mathbf{n}$. Equation (4.23) means that there is no shear at the boundary itself. The condition for the temperature corresponding to (4.20) reads [11]

$$\nabla T \cdot \mathbf{n} = 0. \tag{4.24}$$

(ii) Frictional effects are important at the boundary. This case is known as the *nonslip* boundary flow. Then the so-called *adherence condition* holds, in addition to (4.20)

$$\mathbf{v} \cdot \mathbf{t} = 0. \tag{4.25}$$

However, if (4.20) and (4.25) are fulfilled simultaneously, this means that at the boundary the wind velocity vanishes

$$u = v = w = 0 \quad \text{at } z = H. \tag{4.26}$$

As to the upper boundary condition, two possibilities exist. The simpler one consists of assuming an imaginary rigid horizontal boundary (ceiling) so that $w = 0$ at $z = \delta = \text{const}$, but more realistic is the assumption for a free-surface upper boundary which can be considered as a material surface of the flow $z = \delta(x, y, t)$ (Figure 2.2). Then similarly to (4.22), we obtain

$$w = \frac{\partial \delta}{\partial t} + (\mathbf{u} \cdot \nabla\delta) \quad \text{at } z = \delta. \tag{4.27}$$

By nature, the actual atmosphere has no lateral boundaries. That is why, when written in spherical coordinates, the weather equations need only lower and upper boundary conditions. However, when local (Cartesian) coordinates are used, the suitable lateral boundary conditions should be established depending on the nature of the problem under study. For instance, free-slip rigid boundary zonally oriented simulating walls at constant values of y (zonal channel) can be used in many theoretical problems. Then the condition of *cyclic continuity* is imposed. The latter means that the solutions are required to be periodic over some basic length L.

The initial conditions are taken from the observations.

5. EQUATIONS OF MOTION IN CYLINDRICAL AND NATURAL COORDINATES: THE SHALLOW-WATER APPROXIMATION

a. Introduction of Cylindrical Coordinates

The nature of some atmospheric phenomena and corresponding laboratory models requires a cylindrical coordinate system to be used for their easier description. Typical examples for such phenomena are the convective jets, the tornado, etc.

The transition from local (x, y, z) to cylindrical coordinates (r, θ, z) (Figure 2.3), is a pure mathematical problem. We define the radial, tangential and vertical velocity components

$$u = \frac{dr}{dt}, \qquad v = r\frac{d\theta}{dt}, \qquad w = \frac{dz}{dt}. \tag{5.1}$$

Then similarly to (3.11)

$$\frac{d}{dt} = \frac{\partial}{\partial t} + u\frac{\partial}{\partial r} + w\frac{\partial}{\partial z} + \frac{v}{r}\frac{\partial}{\partial \theta}. \tag{5.2}$$

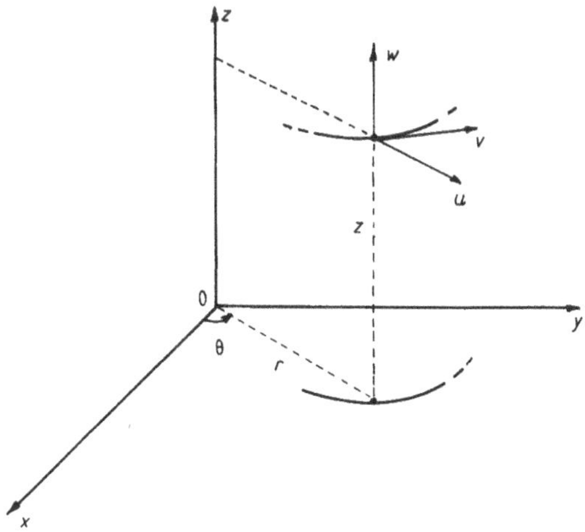

Fig. 2.3. The transition from local (x, y, z) to cylindrical coordinates (r, θ, z).

The transformed equations of motion in the Boussinesq approximation (4.17) take the form

$$\frac{du}{dt} - \frac{v^2}{r} - fv = -\alpha_s \frac{\partial p'}{\partial r},$$

$$\frac{dv}{dt} + \frac{uv}{r} + fu = -\alpha_s \frac{\partial p'}{r\,\partial \theta}, \tag{5.3}$$

$$\frac{dw}{dt} = -\alpha_s \frac{\partial p'}{\partial z} + \beta_0 g T',$$

where $\alpha_s = 1/\rho_s$. The equation of continuity $d\rho/dt + \rho \nabla \cdot \mathbf{v} = 0$ now reads

$$\frac{d\rho}{dt} + \rho\left(\frac{\partial u}{\partial r} + \frac{u}{r} + \frac{\partial v}{r\,\partial \theta} + \frac{\partial w}{\partial z}\right) = 0. \tag{5.4}$$

Two particular cases are of special interest for the applications:

(i) Let $z = 0$ and $w = 0$ in Equation (5.3) and (5.4). We obtain the corresponding equations for a horizontal motion in the polar coordinates (r, θ).

(ii) Stationary motion $(\partial/\partial t = 0)$ with cylindrical symmetry $(\partial/\partial\theta = 0)$. The terms in (5.2), marked by overbars, vanish. Then the second equation of motion (5.3) may be written in the form

$$\frac{d}{dt}\left(rv + \tfrac{1}{2} fr^2\right) = 0, \tag{5.5}$$

provided that f = const. This means that the angular momentum of the particles is conserved following the motion.

b. The Natural (*s*, *n*) Coordinates

For simplicity, let us consider the horizontal motion ($w = 0$) described by the equation

$$\frac{\mathrm{d}_2\mathbf{u}}{\mathrm{d}t} = -\alpha\nabla p + f\mathbf{u} \times \mathbf{k}, \quad \alpha = \frac{1}{\rho}. \tag{5.6}$$

Let **t** be a unit vector oriented in the direction of the flow at each point and **n** be a unit vector normal to **t** which is positive to the left of the flow direction (Figure 1.4). Consequently, the stream *s*-lines and the normal to them *n*-lines stand for coordinate lines (in the general case curvilinear) (see also Figure 1.6). The so-chosen coordinates (*s*, *n*) are called *natural* or *intrinsic* because they are closely related to the motion itself.

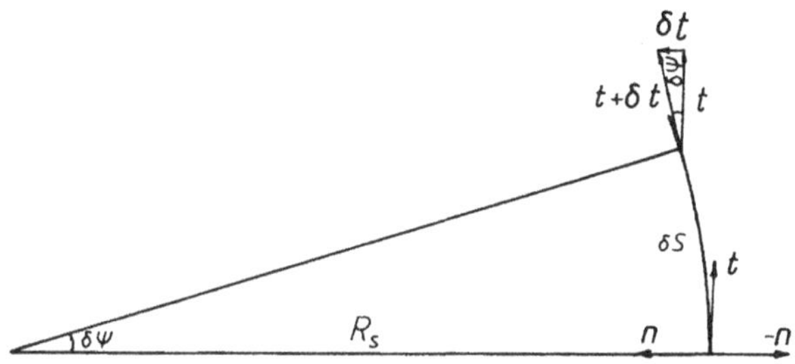

Fig. 2.4. The transformation of equations in natural coordinates (after [27]).

Obviously,

$$\mathbf{u} = c\mathbf{t}, \qquad c = |\mathbf{u}| = \frac{\mathrm{d}s}{\mathrm{d}t}, \qquad |\mathbf{t}| = 1.$$

Hence,

$$\frac{\mathrm{d}\mathbf{u}}{\mathrm{d}t} = \mathbf{t}\,\frac{\mathrm{d}c}{\mathrm{d}t} + c\,\frac{\mathrm{d}\mathbf{t}}{\mathrm{d}t}. \tag{5.7}$$

Moreover,

$$\delta\psi = \frac{\delta s}{R_s} = |\delta\mathbf{t}|, \qquad \delta\mathbf{t} \times \mathbf{n} = 0.$$

Letting $\delta s \to 0$ we obtain

$$\frac{\mathrm{d}\mathbf{t}}{\mathrm{d}s} = \frac{\mathbf{n}}{R_s}, \qquad \frac{\mathrm{d}\mathbf{t}}{\mathrm{d}t} = \frac{\mathrm{d}\mathbf{t}}{\mathrm{d}s}\,\frac{\mathrm{d}s}{\mathrm{d}t} = \frac{c}{R_s}\,\mathbf{n}.$$

Therefore

$$\frac{\mathrm{d}\mathbf{u}}{\mathrm{d}t} = \mathbf{t}\,\frac{\mathrm{d}c}{\mathrm{d}t} + \mathbf{n}\,\frac{c^2}{R_s}. \tag{5.8}$$

Similar to expressions (4.10) and (4.11) from Chapter 1,

$$\zeta = \frac{c}{R_s} - \frac{\partial c}{\partial n}, \qquad D = \frac{c}{R_n} + \frac{\partial c}{\partial s}. \tag{5.9}$$

The equality (5.8) shows that the acceleration of the air particles consists of two parts, the second one being related to the curvilinear character of the motion ($R_s < \infty$). At the cyclonic curvature of the streamlines, as shown in Figure 2.4, $R_s > 0$.

The terms at the right-hand side of Equation (5.6) transform as follows

$$\nabla p = \mathbf{t}\frac{\partial p}{\partial s} + \mathbf{n}\frac{\partial p}{\partial n}, \qquad f\mathbf{u} \times \mathbf{k} = -fc\mathbf{n}.$$

Then

$$\mathbf{t}\frac{\mathrm{d}c}{\mathrm{d}t} + \mathbf{n}\frac{c^2}{R_s} = -\alpha\left(\mathbf{t}\frac{\partial p}{\partial s} + \mathbf{n}\frac{\partial p}{\partial n}\right) - fc\mathbf{n}.$$

Hence, we obtain two scalar equations

$$\frac{\mathrm{d}c}{\mathrm{d}t} = -\alpha\frac{\partial p}{\partial s}, \tag{5.10}$$

$$\frac{c^2}{R_s} + fc = -\alpha\frac{\partial p}{\partial n}, \tag{5.11}$$

where

$$\frac{\mathrm{d}c}{\mathrm{d}t} = \frac{\partial c}{\partial t} + c\frac{\partial c}{\partial s}. \tag{5.12}$$

They express the force balances parallel to and normal to the direction of the flow respectively, and could also be simply constructed from physical considerations.

Equations (5.10) and (5.11) describe an arbitrary curvilinear motion. If the streamlines coincide with the isobars, then

$$\frac{\partial p}{\partial s} = 0 \quad \text{and} \quad \frac{\mathrm{d}c}{\mathrm{d}t} = 0. \tag{5.13}$$

This is the case of *gradient* wind (motion). Its velocity can easily be determined as a solution of the quadratic equation (5.11) (see Section 6 of Chapter 3). Obviously $\mathrm{d}c/\mathrm{d}t = 0$ is a necessary and sufficient condition for the existence of a gradient flow. If, in addition, the motion is stationary, i.e., $\partial c/\partial t = 0$, then from (5.13) and (5.12) it follows that

$$\frac{\partial c}{\partial s} = 0. \tag{5.14}$$

In other words, alongside the trajectories (which in this particular case coincide with the streamlines and isobars), the scalar wind velocity remains constant. According to (5.9), the divergence $D = c/R_n$. Consequently, a nondivergent ($D = 0$) gradient flow is possible only at $R_n = \infty$, which means that the n lines are straight lines, i.e., the s lines are circuits ($R_s = \text{const}$).

In summary, the stationary gradient wind without vertical motion ($w = 0$ and $D = 0$) is the wind blowing along the isobars. This is the simplest dynamical model of a cyclone or anticyclone in the free atmosphere. When $R_s = 0$ (straight-line isobars) the gradient wind is called the *geostrophic* wind.

c. The 'Shallow-Water' Approximation

The shallow-water approximation in dynamic meteorology problems is borrowed from the dynamics of the ocean. The actual atmosphere is generally a compressible medium and does not have a well-defined upper boundary, while sea water is practically incompressible and the world ocean has a free-surface upper boundary. Within the framework of the hypothetical model for the 'homogeneous atmosphere', however, this essential difference disappears. It is well known that in this mode $\rho(z) = \text{const}$ and the atmosphere extends up to a height $H_s = p_0/g\rho$ (~8 km) where p_0 is the surface pressure, i.e., it has a sharply-defined upper boundary (free-surface) at which $p_h = 0$ (Figure 2.5). Besides, the hydrostatic equation

$$\mathrm{d}p = -\rho g\,\mathrm{d}z \tag{5.15}$$

holds. Therefore, the assumptions in the model are:

(i) Constant density $\rho(x, y, z, t) = \rho_0 = \text{const}$.
(ii) Hydrostatic equilibrium (5.15).
(iii) Independent of z horizontal velocity, i.e., $\mathbf{u} = \mathbf{u}(x, y, t)$.

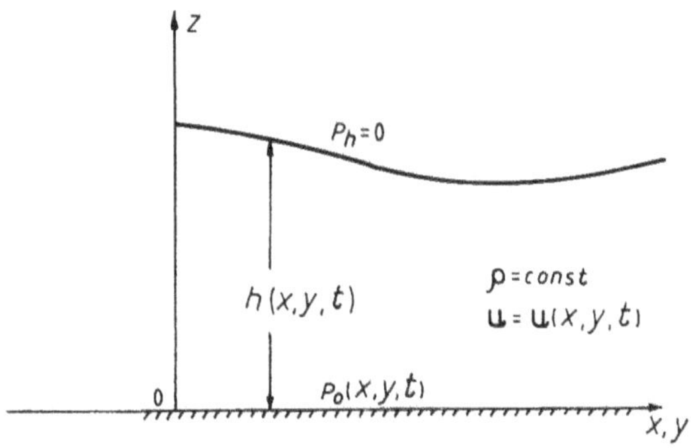

Fig. 2.5. Homogeneous atmosphere.

As will be shown later, this implies a barotropic atmosphere (see Section 5 of Chapter 3).

From (5.15) and assumption (i), we derive

$$p(x, y, z, t) = g\rho_0 [h(x, y, t) - z],$$

so that

$$\frac{-\nabla p}{\rho_0} = -g\nabla h. \tag{5.16}$$

Integrating the continuity equation $\nabla \cdot \mathbf{u} + \partial w/\partial z = 0$ under conditions (ii) and $w = 0$ at $z = 0$, we obtain

$$w_h = -\int_0^h (\nabla \cdot \mathbf{u})\,\mathrm{d}z = -h(\nabla \cdot \mathbf{u}). \tag{5.17}$$

On the other hand, since $h(x, y, t)$ is a material surface,

$$w_h = \frac{\mathrm{d}h}{\mathrm{d}t} = \frac{\partial h}{\partial t} + (\mathbf{u} \cdot \nabla h). \tag{5.18}$$

Substituting (5.18) into (5.17) and (5.16) into (5.6), we obtain the required system

$$\frac{\partial \mathbf{u}}{\partial t} + (\mathbf{u} \cdot \nabla)\,\mathbf{u} = -g\nabla h - f\mathbf{k} \times \mathbf{u}, \tag{5.19}$$

$$\frac{\partial h}{\partial t} = -(\nabla \cdot h\mathbf{u}). \tag{5.20}$$

This consists of three equations with the same number of unknown functions – (u, v, h), i.e., it is a closed system and, in principle, can be solved. The one-dimensional analogue, for instance,

$$u_t + vu_y = fv, \qquad v_t + vv_y = -gh_y - fu, \qquad h_t + vh_y = -hv_y, \tag{5.21}$$

which preserves the nonlinearity, is of interest because of its simplicity (see Section 1 of Chapter 6).

6. WEATHER EQUATIONS IN GENERALIZED VERTICAL COORDINATES

a. Preliminary Formulae

The usual independent vertical variable in the previously defined coordinate systems is the geometrical height z above the Earth's surface. This role, however, could be played by whichever other quantity $q = q(z)$ or, more generally,

$$q = q(x, y, z, t), \tag{6.1}$$

is synonymously related to z. The observational and theoretical experience in studying the Earth's atmosphere suggests that it might be useful in many problems to replace the independent variable z with one of the thermodynamic variables (pressure p, potential temperature θ, etc.) and then to use the constant value surface heights of that variable as a dependent variable. Following Kasahara (1974), we shall transform our equations for general case (6.1) Later in Section 3 of Chapter 3, identifying q with p, θ, etc., it will be easy to derive the corresponding equations for any particular case.

The equations to be transformed are

$$\frac{d\mathbf{u}}{dt} + f\mathbf{k} \times \mathbf{u} = -\frac{\nabla p}{\rho}, \tag{6.2}$$

$$\frac{d \ln \rho}{dt} + \nabla \cdot \mathbf{u} + \frac{\partial w}{\partial z} = 0, \tag{6.3}$$

$$\frac{d \ln \theta}{dt} = \frac{Q}{c_p T}, \tag{6.4}$$

where

$$\frac{d}{dt} = \frac{\partial}{\partial t} + (\mathbf{u} \cdot \nabla) + w \frac{\partial}{\partial z} \tag{6.5}$$

The 'old' independent variables (x, y, z, t) are to be replaced by the new independent variables (x, y, q, t) on the basis of (6.1). Thus, the new coordinates to be introduced have the same variables x and y as the original ones and they have the same unit vector $\mathbf{k}$ in the vertical. The time variable t, of course, is the same in both systems.

Let us assume that Equation (6.1) gives a single-valued monotonic relationship between q and z, when z, y, t are fixed. Then we can write

$$z = z(x, y, q, t) \tag{6.1'}$$

and also

$$A(x, y, z, t) = A[x, y, z(x, y, q, t), t], \tag{6.6}$$

where A is any scalar function. According to the chain rule for differentiation,

$$\left(\frac{\partial A}{\partial t}\right)_q = \left(\frac{\partial A}{\partial t}\right)_z + \frac{\partial A}{\partial z}\left(\frac{\partial z}{\partial t}\right)_q, \tag{6.7}$$

where the subscripts q and z indicate which particular vertical coordinate has been held constant for partial differentiation. However,

$$\frac{\partial A}{\partial z} = \frac{\partial A}{\partial q}\left(\frac{\partial q}{\partial z}\right), \tag{6.8}$$

so that (6.7) takes the form

$$\left(\frac{\partial A}{\partial t}\right)_q = \left(\frac{\partial A}{\partial t}\right)_z + \frac{\partial q}{\partial z}\left(\frac{\partial z}{\partial t}\right)_q \frac{\partial A}{\partial q}. \tag{6.9}$$

Two similar formulae can be derived if we differentiate with respect to x and y instead of t. In vector form, they read

$$\nabla_q A = \nabla_z A + \frac{\partial q}{\partial z} (\nabla_q z) \frac{\partial A}{\partial q}. \tag{6.10}$$

These relationships can be derived from pure geometrical considerations (Figure 2.6). We denote by A_1, A_2 and A_3 the values of A in the points M_1, M_2 and M_3. Evidently

$$\frac{A_3 - A_1}{\Delta x} = \frac{A_2 - A_1}{\Delta x} + \frac{A_3 - A_2}{\Delta z}\,\frac{\Delta z}{\Delta x}.$$

Letting Δx, $\Delta z \to 0$, we obtain

$$\left(\frac{\partial A}{\partial x}\right)_z = \left(\frac{\partial A}{\partial x}\right)_q - \frac{\partial A}{\partial z}\left(\frac{\partial z}{\partial x}\right)_q$$

which coincides with (6.7) at $t \equiv x$.

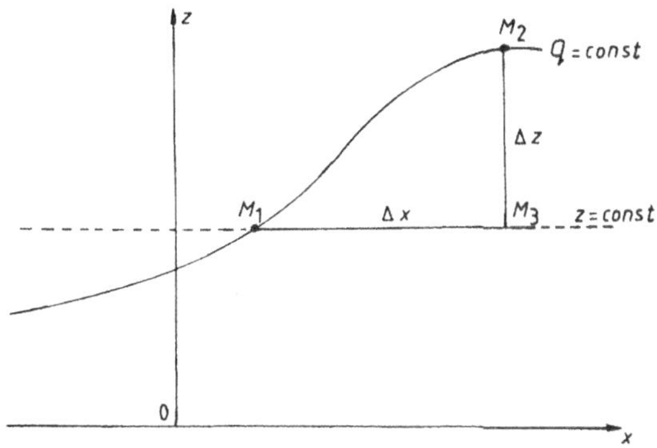

Fig. 2.6. The generalized vertical coordinate.

We now transform the total derivative $\mathrm{d}/\mathrm{d}t$ in the z system (6.5) into a q system, making use of (6.9) and (6.10). The result is

$$\frac{\mathrm{d}}{\mathrm{d}t} = \left(\frac{\partial}{\partial t}\right)_q + \mathbf{u}\cdot\nabla_q + \left[w - \left(\frac{\partial z}{\partial t}\right)_q - \mathbf{u}\cdot\nabla_q z\right]\frac{\partial q}{\partial z}\,\frac{\partial}{\partial q}. \tag{6.11}$$

On the other hand, by the definition of the total derivative in the q system we have

$$\frac{\mathrm{d}}{\mathrm{d}t} = \left(\frac{\partial}{\partial t}\right)_q + \mathbf{u}\cdot\nabla_q + \dot{q}\,\frac{\partial}{\partial q} \tag{6.12}$$

where

$$w_q = \dot{q} = \frac{\mathrm{d}q}{\mathrm{d}t} \tag{6.13}$$

is interpreted as a generalized vertical velocity, an analogue of

$$w = \dot{z} = \frac{\mathrm{d}z}{\mathrm{d}t}. \tag{6.14}$$

Comparing (6.11) and (6.12), we obtain the relationship between w and w_q:

$$w_q = \left[w - \left(\frac{\partial z}{\partial t}\right)_q - \mathbf{u}\cdot\nabla_q z\right]\frac{\partial q}{\partial z} \tag{6.15}$$

b. Transformation of Equations

Let us start with the equation of motion (6.2). If we let $A \equiv p$, then expression (6.10) yields

$$\nabla_z p = \nabla_q p - \frac{\partial q}{\partial z} (\nabla_q z) \frac{\partial p}{\partial q} \tag{6.16}$$

Substituting into the right-hand side of (6.2), we obtain the transformed equation in a most general form

$$\frac{\mathrm{d}\mathbf{u}}{\mathrm{d}t} + f\mathbf{k} \times \mathbf{u} = -\alpha \nabla_q p + \alpha \frac{\partial q}{\partial z} (\nabla_q z) \frac{\partial p}{\partial q}, \tag{6.17}$$

where $\mathrm{d}/\mathrm{d}t$ is given by (6.11).

If, in addition to (6.2), we assume that the hydrostatic equation holds ($\partial p/\partial z = -\rho g$), hence it follows that

$$g\rho \frac{\partial z}{\partial q} = -\frac{\partial p}{\partial q}. \tag{6.18}$$

Then (6.16) simplifies

$$\nabla_z p = \nabla_q p + \rho g \nabla_q z, \tag{6.19}$$

and (6.17) becomes

$$\frac{\mathrm{d}\mathbf{u}}{\mathrm{d}t} + f\mathbf{k} \times \mathbf{u} = -\alpha \nabla_q p - g \nabla_q z, \quad \alpha = \frac{1}{\rho}. \tag{6.20}$$

This is the horizontal momentum equation in a hydrostatic q system which is of great interest for the theory.

The next step is to transform the equation of continuity (6.3). From (6.15) we determine

$$w = \left(\frac{\partial z}{\partial t}\right)_q + \mathbf{u} \cdot \nabla_q z + \dot{q} \frac{\partial z}{\partial q}. \tag{6.21}$$

Subsequent differentiation yields

$$\begin{aligned} \frac{\partial w}{\partial q} &= \frac{\partial}{\partial q}\left(\frac{\partial z}{\partial t}\right)_q + \frac{\partial \mathbf{u}}{\partial q} \cdot \nabla_q z + \mathbf{u} \cdot \frac{\partial}{\partial q} \nabla_q z + \frac{\partial \dot{q}}{\partial q}\frac{\partial z}{\partial q} + \dot{q}\frac{\partial^2 z}{\partial q^2}, \\ \frac{\partial w}{\partial z} &= \frac{\partial w}{\partial q}\frac{\partial q}{\partial z} = \frac{\partial \dot{q}}{\partial q} + \frac{\partial q}{\partial z}\left[\frac{\mathrm{d}}{\mathrm{d}t}\left(\frac{\partial z}{\partial q}\right) + \frac{\partial \mathbf{u}}{\partial q} \cdot \nabla_q z\right], \end{aligned} \tag{6.22}$$

where the identity

$$\frac{\mathrm{d}}{\mathrm{d}t}\left(\frac{\partial z}{\partial q}\right) = \left(\frac{\partial^2 z}{\partial t\, \partial q}\right)_q + \mathbf{u} \cdot \nabla_q \left(\frac{\partial z}{\partial q}\right) + \dot{q}\frac{\partial^2 z}{\partial q^2}$$

following from (6.12) has been used. With the aid of (6.10), it can be proved that

$$\nabla_z \cdot \mathbf{u} = \nabla_q \cdot \mathbf{u} - \frac{\partial q}{\partial z} (\nabla_q z) \cdot \frac{\partial \mathbf{u}}{\partial q}. \tag{6.23}$$

Substituting (6.22) and (6.23) into (6.3), we obtain the continuity equation in the q system

$$\left(\frac{\mathrm{d}}{\mathrm{d}t}\right) \ln \left(\rho \frac{\partial z}{\partial q}\right) + \nabla_q \cdot \mathbf{u} + \frac{\partial \dot{q}}{\partial q} = 0 \tag{6.24}$$

or, in an alternative form,

$$\left[\frac{\partial}{\partial t}\left(\rho \frac{\partial z}{\partial q}\right)\right]_q + \nabla_q \cdot \left(\rho \mathbf{u} \frac{\partial z}{\partial q}\right) + \frac{\partial}{\partial q}\left(\rho \dot{q} \frac{\partial z}{\partial q}\right) = 0. \tag{6.25}$$

In an hydrostatic approximation, $\rho(\partial z/\partial q)$ may be taken from (6.18) to obtain the final form of the continuity equation

$$\frac{\partial}{\partial q}\left(\frac{\partial p}{\partial t}\right)_q + \nabla_q \cdot \left(\mathbf{u} \frac{\partial p}{\partial q}\right) + \frac{\partial}{\partial q}\left(\dot{q} \frac{\partial p}{\partial q}\right) = 0. \tag{6.26}$$

In the end, the thermodynamic equation (6.4) remains formally unchanged except that the total derivative is defined by (6.12), i.e.,

$$\left[\left(\frac{\partial}{\partial t}\right)_q + \mathbf{u} \cdot \nabla_q + \dot{q} \frac{\partial}{\partial q}\right] \ln \theta = \frac{Q}{c_p T}. \tag{6.27}$$

c. Boundary Conditions

The boundary conditions attached to Equation (6.2)–(6.3) must also be transformed. As before (see Section 4), let us consider only the lower and upper boundary conditions. If $z = 0$ corresponds to the sea level, while $z = H(x, y)$ defines the Earth's surface, then in the q system, according to (6.1), the lower boundary will be defined by

$$q = q_H = q(x, y, H, t) \tag{6.28}$$

and, in the general case, q_H will be variable in both space and time. Since the air cannot penetrate through the Earth's surface but may move only along it, similarly to (4.27) we get

$$\dot{q} = \frac{\partial q_H}{\partial t} + \mathbf{u}_H \cdot \nabla q_H \quad \text{at } q = q_H \tag{6.29}$$

where u_H denotes the horizontal wind velocity at $z = H$. If the Earth's surface coincides with a constant q surface, then (6.29) reduces to

$$\dot{q} = 0 \quad \text{at } q = q_H. \tag{6.30}$$

As to the upper boundary condition, it is convenient to select some vertical coordinate surface $q = q_c = \text{const}$ and to assume that

$$\dot{q} = 0 \quad \text{at } q = q_c. \tag{6.31}$$

Examples of a concrete choice of the vertical coordinate q will be discussed in Section 3 of Chapter 3. As we shall show, considerable simplifications result from this procedure.

PROBLEMS

1. Derive expression (4.11).
2. Derive Equation (5.5).
3. Solve Equation (5.11) and analyze the solution if $\partial p/\partial s = 0$, $R_s = \text{const}$ and $f < 0$ = the Southern Hemisphere (see Section 6 of Chapter 3).
4. Show that in the case of mountains with a variable height $H(x, y) > 0$ (Figure 2.2), Equation (5.20) should be replaced by

$$\frac{\partial h}{\partial t} = -\nabla \cdot [h - H(x, y)]\, \mathbf{u} \tag{P.1}$$

where $h - H$ is the real thickness of the homogeneous atmosphere.

CHAPTER 3

Simplification of Weather Equations

1. METHODS FOR SIMPLIFICATION OF WEATHER EQUATIONS

a. General Characteristics

One of the main difficulties encountered in dynamic meteorology is the construction of differential equations which describe the atmospheric processes under investigation. But even when this obstacle has been overcome, we have not progressed very far because these equations are so complex that they cannot be solved without a preliminary simplification. There is no universal prescription for this purpose. In each particular case, a special analysis is needed for choosing the most appropriate method of simplification. That is why it seems necessary to have some actual information about the atmospheric processes under investigation at the very beginning of a theoretical study.

If the simplification has been carried out correctly, it not only facilitates the mathematical solution but also a physical interpretation of this solution that quantitatively reflects the interaction of the main factors. We should also point out that in modern mathematics there are no methods for giving an exact (analytical) solution to the complete system of weather equations. However, each simplification of the original system of equations means the introduction of a mathematical submodel whose value depends on the degree of simplification.

Out of the various approaches to the problem of simplifying weather equations, four will be discussed here and applied hereafter:

(i) *Special choice of the coordinate system*. When appropriately chosen, the coordinate system may yield considerable simplification and, what is more, without any loss of the information contained in the initial equations. For examples, see below, Sections 3 and 4.

(ii) *Statistical evaluations*. With few exceptions, weather equations are of the type $a_1 = a_2 + a_3$ where a_1, a_2, a_3 are differential or even integro-differential terms. Suppose we have enough observational data in order to statistically evaluate the order of each term, say the mean absolute values $|\overline{a_i}|$ or the mean square values $\sigma_i = (\overline{a_i^2})^{1/2}$. Then if, for instance, $\sigma_i \ll \sigma_k$, one could neglect second-order terms in the above equation compared to the main terms. Very often this procedure yields a significant simplification of the equations. To facilitate this technique, special tables have been prepared and widely used in the past – for instance, the tables of Hesselberg and Friedman drawn up in 1914, or those of Yudin (1950) for the fundamental meteorological elements and their spatial and time derivatives. These tables are now outdated but they have played a considerable role in the past development of theoretical investigations.

(iii) *Spectral representation with subsequent truncation of the equations.* An idea of the application of this powerful method with some examples will be given in Section 9 of this chapter and in Section 2 of Chapter 7.

(iv) *Scale analysis.* As in the case of the statistical method, the purpose is to evaluate the order of the various terms in the equations with the subsequent neglecting of the second-order terms. This can be realized with much less initial information concerning the various meteorological elements and their space-time distribution than in the previous case and with much more physical insight into the required procedure, which is a great advantage of this method. To a large extent, it has universal character.

Unfortunately, the fact that a given term in an equation is evaluated as a second-order term does not necessarily mean that it can be neglected. In some cases, such a term may prove to exert considerable influence upon the character of the solution of the differential equation (see, for example, Equation (2.18), Chapter 7). In spite of this danger, in practice it is done in just the same way. The more precise method, namely the filtering of the unnecessary partial solutions from the total one rather than from the equations, is connected with great difficulties of analytical and computational character.

We now turn to the last method for a detailed description and the application of a simple example.

b. Scale Analysis and Similarity

The main aspect of this method is the choice of appropriate scales for the dependent and independent variables. Let $A_s = \partial A/\partial s$, $s = x, y, z, t$ and $O(A_s)$ denotes the characteristic (typical) value of A_s (say, the mean absolute value). Then M_A and S_A are said to be the corresponding scales of A and s if

$$O(A_s) = \frac{M_A}{S_A}, \qquad O(A_{ss}) = \frac{M_A}{S_A{}^2} \quad \text{etc.} \tag{1.1}$$

Therefore, M_A and S_A represent some typical values of A and s, such that their ratios, according to (1.1), should determine the typical values of A_s, A_{ss}, etc.

On the other hand, we shall write $A = O(1)$ if $0 \leqslant |A| \leqslant 1$. More generally, if $A_0 \leqslant |A| \leqslant A_0 + \delta A$, then obviously $(|A - A_0|)/\delta A = O(1)$ or, which is the same, $O(\Delta A) = \delta A = M_A$ where $\Delta A = A - A_0$. Thus, for example, the difference between the values of p in the center and the periphery, rather than the value of p in some particular point, should be taken as a scale M_p for the pressure p in cyclones (anticyclones) when estimating the horizontal pressure gradient. The sea–land temperature difference determines the intensity of the sea-breeze circulation and must be taken as a scale rather than a mean temperature, etc.

According to the definitions given above, if we let

$$A = M_A A', \qquad s = S_A s', \tag{1.2}$$

where A' and s' are dimensionless quantities, then

$$A_s = \left(\frac{M_A}{S_A}\right) A'_s, \qquad A_{ss} = \left(\frac{M_A}{S_A{}^2}\right) A'_{ss}, \quad \text{etc.} \tag{1.3}$$

where

$$A_s' = O(1), \qquad A_{ss}' = O(1), \text{etc.} \tag{1.4}$$

When such 'nondimensionalization' is carried out in all the terms of an equation, then the order of each term would be determined by the scale-coefficient in front of it, because the remaining part, according to (1.4), would be of the order of one. Then it becomes clear why the equation in which the scales for the dependent and independent variables have been chosen, in accordance with the nature of the particular phenomenon under study, begins to reflect the regularities of this phenomenon more clearly. By this procedure, we introduce additional information into the equation which, figuratively speaking, 'concentrates its attention' upon a particular phenomenon and makes it 'insensitive' to other phenomena. The initial (dimensional) equations do not show such preferences. The features of dimensionless equations will be demonstrated in Section 2.

Now let us turn to the Navier–Stokes Equations (2.8) of Chapter 2

$$\frac{\partial \mathbf{v}}{\partial t} + (\mathbf{v} \cdot \boldsymbol{\nabla})\mathbf{v} = -\boldsymbol{\nabla}\pi + \nu\nabla^2\mathbf{v}, \qquad \pi = \alpha p \tag{1.5}$$

and the equation of continuity $\boldsymbol{\nabla} \cdot \mathbf{v} = 0$. In accordance with (1.2), we let

$$\mathbf{v} = M_v\mathbf{v}', \quad x = S_x x', \qquad \pi = M_\pi \pi', \quad t = S_t t'. \tag{1.6}$$

Popular symbols for the scales in this case are $L = S_x$, $T = S_t$, $V = M_v$. Substituting (1.6) into (1.5), we obtain

$$\frac{V}{T}\frac{\partial \mathbf{v}'}{\partial t'} + \frac{V^2}{L}(\mathbf{v}' \cdot \boldsymbol{\nabla}')\mathbf{v}' = \frac{M_\pi}{L}\boldsymbol{\nabla}'\pi' + \frac{\nu V}{L^2}\boldsymbol{\nabla}'^2 v' \tag{1.7}$$

where $\boldsymbol{\nabla}' = \partial/\partial x_\alpha'$ All scalecoefficients in (1.7) have dimensions of acceleration. Dividing, say to V^2L^{-1}, we obtain the dimensionless equation

$$\mathrm{S}\frac{\partial \mathbf{v}'}{\partial t'} + (\mathbf{v}' \cdot \boldsymbol{\nabla}')\mathbf{v}' = -\mathrm{Eu}\,\boldsymbol{\nabla}'\pi' + \mathrm{Re}^{-1}\boldsymbol{\nabla}'^2\mathbf{v}' \tag{1.8}$$

where

$$\begin{aligned} \mathrm{S} &= \frac{L}{VT} \text{ is the Strouhal number,} \\ \mathrm{Eu} &= \frac{M_\pi}{V^2} \text{ is the Euler number, and} \\ \mathrm{Re} &= \frac{LV}{\nu} \text{ is the Reynolds number.} \end{aligned} \tag{1.9}$$

The continuity equation has the same form: $\boldsymbol{\nabla}' \cdot \mathbf{v}' = 0$. The boundary conditions should also be written in a dimensionless form.

Two fluid flows (1) and (2) are called *hydrodynamically similar* if they are:

(i) *geometrically* similar – the space volumes in which they are realized are similar (e.g., cylindrical tubes, spherical containers, etc.);

(ii) *kinematically* similar – in corresponding space–time points $\mathbf{v}_2 = k\mathbf{v}_1$, k = const;

(iii) *dynamically* similar – the same proportionality exists between the vectors of the mass forces and the stress tensors.

Further on, when discussing similar motions, we shall consider hydrodynamical (full) similarity.

With the aid of Equation (1.8), the necessary conditions for similarity are easily established. Actually, if two motions are similar, then the corresponding nondimensional equations of the type (1.8) should coincide. Hence, it follows that $S_1 = S_2$, $Eu_1 = Eu_2$, $Re_1 = Re_2$ since the prime terms, by definition, coincide. Symbolically, this result can be written as

$$S = \text{idem}, \qquad Eu = \text{idem}, \qquad Re = \text{idem}, \tag{1.10}$$

where the Latin 'idem' means 'one and the same'. In other words, we have found the necessary criteria for similarity. In some special cases they prove to be enough but in the general case they are not.

Interesting conclusions can be made from an analysis of the derived criteria (1.9) and Equation (1.5). Let us consider, for example the Reynolds number Re. According to (1.5) the molecular friction force can be neglected when $\nu \to 0$ (ideal, nonviscous, inviscid fluid). Then in (1.8) Re $\to \infty$. However, according to (1.9), Re may tend to infinity when $\nu \neq 0$, $V < \infty$ (natural limitation) but $L \to \infty$. From a dynamical point of view, the final result will be the same. Consequently, the influence of the molecular friction force is unessential for the motion of real, viscous ($\nu \neq 0$) fluids at very large Reynold numbers (Re $\to \infty$). This fundamental conclusion is of great importance for atmospheric dynamics.

2. SCALE ANALYSIS OF WEATHER EQUATIONS

a. Choice of Scales

As has been mentioned several times before, the vertical is a fundamental direction in the atmosphere and the features of its motions in horizontal and vertical directions differ considerably. Thus, different length and velocity scales should be introduced in both directions

$$O(x) = O(y) = L, \qquad O(z) = Z, \tag{2.1}$$

$$O(u) = O(v) = V, \qquad O(w) = W. \tag{2.2}$$

Due to the fact that the effective thickness of the Earth's atmosphere is small compared to its horizontal dimensions, the inequality $Z < L$ will hold for some motions. Really $Z_{\max} \lesssim H_s \sim 10$ km (height of the homogeneous atmosphere), while $L_{\max} \lesssim 2\pi r_0 \approx 4.10^4$ km (r_0 is the Earth's radius). Therefore, for some classes of motions, it is possible that $Z \ll L$. One might expect, observations confirm that in this case $W \ll V$.

Synoptic motions are those motions whose scale $L \sim 10^3$ km. Since the whole troposphere and the lower stratosphere are involved in this kind of motion, then $Z \sim 10$ km. As to the velocity, the observations show that $V \sim 10$ m s^{-1}, while $W \sim 10^{-2}$ m s^{-1}. The inequality $W \ll V$ at $Z \ll L$ also follows from the continuity equation $w_z = -\nabla \cdot \mathbf{u}$. Actually, introducing the scales, we obtain

$$\left(\frac{W}{Z}\right) w'_{z'} = -\left(\frac{V}{L}\right) \nabla' \cdot u'.$$

Hence,

$$\frac{W}{Z} \sim \frac{V}{L}, \quad \text{or} \quad W \sim \left(\frac{Z}{L}\right) V. \tag{2.3}$$

As far as $Z \ll L$, then $W \ll V$ too. In the real atmosphere, however, a strong inequality

$$W \ll \left(\frac{Z}{L}\right) V \tag{2.3'}$$

holds, rather than (2.3), due to the fact that according to observations, u_x and v_y have opposite signs and their sum $D = u_x + v_y$ is a small difference of large numbers. The correct relationship, consistent with the observations, would be

$$D \sim 10^{-1} \left(\frac{V}{L}\right) \sim 10^{-6}\ \mathrm{s}^{-1}. \tag{2.3''}$$

The atmospheric motions are nonstationary and the corresponding time scale $T = S_t$ should be chosen. Meteorological practice shows that at synoptic scale motions $T \sim 24$ h. At $L \sim 10^3$ km and $V \sim 10\ \mathrm{m\ s^{-1}}$ this implies that

$$T \sim \frac{L}{V}. \tag{2.4}$$

Therefore, for the synoptic scale motions, the Strouhal number (1.9) S = 1.

On the other hand, the equations of motion possess a natural timescale related to the Coriolis parameter $T_f = f^{-1}$. In the middle latitudes $f \sim 10^{-4}\ \mathrm{s}^{-1}$ so that $T_f \sim 3$ h. The combination

$$\mathrm{Ki} = \frac{1}{fT} \tag{2.5}$$

is called Kibel's number. At $T \sim 24$ h*, $\mathrm{Ki} \sim 0.1$ and decreases with T increasing. When (2.4) is substituted into (2.5), the so-called Rossby number is obtained

$$\mathrm{Ro} = \frac{V}{fL}. \tag{2.6}$$

Obviously, at synoptic scale motions $\mathrm{Ro} \sim 10^{-1}$, but when studying motions with $\mathrm{Ro} \ll 1$, we restrict the possible values of L from below and in this way we separate the synoptic motions from the so-called subsynoptic or mesoscale motions.

---

* At $T = 2\pi/\omega$ and $f = 2\omega \sin \varphi$ we obtain $\mathrm{Ki} = 1/4\pi \sin \varphi$. If $\varphi > 5$–$10°$ then $\mathrm{Ki} \ll 1$.

In a local coordinate system, according to (4.11) in Chapter 2,

$$f(y) = f(y_0) + \beta(y - y_0) + \cdots, \tag{2.7}$$

where $\beta = df/dy \sim 10^{-11}\ \mathrm{m^{-1}\ s^{-1}}$ (in the middle latitudes) is the Rossby parameter. Letting $y - y_0 = L$, we form the ratio of the second term to the first in (2.7)

$$B = \frac{\beta L}{f(y_0)} = \left(\frac{L}{r_0}\right) \text{cotan}\,\varphi_0. \tag{2.8}$$

The condition for retaining two terms in (2.7) would be $B \lesssim 0.1$. At $L \sim 10^3$ km, $B \sim 0.1$ everywhere except in the tropical atmosphere ($|\varphi_0| < 5$–$10°$), although assuming $B \lesssim 0.1$, we restrict L from above, thus separating the synoptic motions from the planetary motions having horizontal scales $L \sim 10^4$ km. As can be seen, for the synoptic motions Ro $\sim B$.

For the air $\nu = 1.3 \times 10^{-5}\ \mathrm{m^2\ s^{-1}}$. Then at $L \sim 10^3$ km and $V \sim 10\ \mathrm{m\ s^{-1}}$, Re $\sim 10^{12}$. Moreover, at $L \sim 10^3$ km and $V \sim 1\ \mathrm{m\ s^{-1}}$ Re $\sim 10^8$. Therefore, the Reynolds number Re is very large in the atmospheric motions responsible for the weather, thus permitting us to neglect the influence of the molecular friction force on atmospheric dynamics.

b. Simplification of the Equations

Let us begin with the equations of the horizontal motion

$$\frac{\partial \mathbf{u}}{\partial t} + (\mathbf{u} \cdot \nabla)\mathbf{u} + w \frac{\partial \mathbf{u}}{\partial z} + f\mathbf{k} \times \mathbf{u} = \frac{-\nabla p}{\rho}. \tag{2.9}$$

The magnitude of the terms from left to right is $V/T = V^2/L$, V^2/L, WV/Z, fV and M_p/LM_ρ. But, according to (2.3′), $WV/Z \ll V^2/L$ so that the convective derivative in (2.9) can be neglected. Then, in a dimensionless form, Equation (2.9) reads

$$\mathrm{Ro}\left[\frac{\partial \mathbf{u}'}{\partial t'} + (\mathbf{u}' \cdot \nabla')\mathbf{u}'\right] + \mathbf{k} \times \mathbf{u}' = -\,\mathrm{Ro}\,\mathrm{Eu}\,\frac{\nabla' p'}{\rho'}. \tag{2.10}$$

Comparing this with (1.8), we conclude that in rotating fluids (atmosphere, ocean, laboratory model) the Rossby number Ro is an additional criterion for similarity.

At synoptic scale motions Ro ~ 0.1 so that the main term on the left-hand side of (2.10) is the second one. Thus, it follows that the Coriolis and pressure gradient terms are of the same order of magnitude, i.e., we can write

$$f\mathbf{k} \times \mathbf{u} \approx \frac{-\nabla p}{\rho}. \tag{2.11}$$

instead of (2.10). Hence, we determine

$$\mathbf{u} \approx \frac{1}{\rho f}\,\mathbf{k} \times \nabla p \equiv \mathbf{u}_g \tag{2.12}$$

where $\mathbf{u}_g$ is the geostrophic wind velocity. Thus, as a result of the simplifications, we have reached the so-called *quasi-geostrophic* relationships (2.12). According to them, in synoptic scale motions the real wind is very close to the hypothetical geostrophic one – a fact well known from the observations.

With a decreasing of the horizontal scale L, the value of Ro increases and Equation (2.11) becomes more and more inaccurate. At $L \sim V/f \sim 10^5$ m according to (2.6), Ro ~ 1 and so a simplification of (2.9) cannot be made (except by neglecting the w-term). At still smaller scales ($L \lesssim 10^4$ m) the Coriolis term can be neglected in (2.9) and contrary to (2.11), however, the equations of motion remain essentially nonlinear in the last two cases.

To summarize, a classification of the horizontal motions in the atmosphere can be proposed as follows

(i) Synoptic motions: $L \sim 10^3$ km, Ro $\ll 1$.
(ii) Subsynoptic (mesoscale) motions: $L \sim 10^2$ km, Ro ~ 1.
(iii) Small-scale motions: $L \lesssim 10$ km, Ro $\gg 1$.

To investigate them, f- or β-plane approximations are used, while for larger, global-scale motions the use of a spherical coordinate system is compulsory.

The next step is to simplify the third equation of motion

$$\frac{dw}{dt} + f_1 u + g = -\alpha \frac{\partial p}{\partial z}, \quad \alpha = \frac{1}{\rho}, \tag{2.13}$$

where $f_1 = 2\omega \cos \varphi = f \operatorname{cotan} \varphi$. Since the pressure and density changes along the vertical distance $Z \sim 10$ km are comparable with their surface (standard) values p_0, ρ_0, the latter can be considered as their scales. Thus, Equation (2.13) takes the form

$$\frac{VW}{L}\frac{dw'}{dt'} + \operatorname{cotan} \varphi \, Vfu' + g = -\frac{p_0}{\rho_0 Z}\alpha' \frac{\partial p'}{\partial z'}.$$

It is easily seen that outside the equatorial atmosphere the magnitudes of the terms on the left are, correspondingly, 10^{-7}, 10^{-3} and 10 m s^{-2}. Consequently, the main term on the left is the third one. As to the pressure gradient term on the right, at $Z \sim H_s = p_0/\rho_0 g$ we see that it balances the acceleration. Therefore, for synoptic-scale motions the *quasi-static* relationship holds with great accuracy

$$g \approx -\alpha \frac{\partial p}{\partial z}. \tag{2.14}$$

With less but sufficient accuracy, it can be used for subsynoptic scale motions ($L \gtrsim 10^2$ km). However, at $L \lesssim 10$ km, Equation (2.14), which is also known as the hydrostatic approximation, is inapplicable. A typical example of the latter case is a powerful convective cloud in which $dw/dt \sim g$.

Let us now consider the continuity equation

$$\frac{1}{\rho}\left(\frac{\partial \rho}{\partial t} + \mathbf{u} \cdot \nabla \rho\right) + \nabla \cdot \mathbf{u} = -\frac{1}{\rho}\frac{\partial \rho w}{\partial z}. \tag{2.15}$$

The main term on the left is the third one. Actually, it follows from (2.15) that

$$\frac{M_\rho}{\rho}\frac{V}{L}\frac{d_2 \rho'}{dt'} + \frac{V}{L}(\nabla' \cdot \mathbf{u}') = -\frac{1}{\rho}\frac{\partial \rho w}{\partial z}, \tag{2.16}$$

where M_ρ is a scale, determined from the horizontal variations of density at an arbitrary level in the atmosphere. Intuitively, one can expect that $M_\rho/\rho \ll 1$. The observations, indirectly, support this assumption. They suggest that $M_p/p \ll 1$. But in the case of adiabatic processes $p\rho^{-\kappa}$ = const and, hence, $M_\rho/\rho = (1/\kappa)M_p/p$, $\kappa = c_p/c_v$. Thus, instead of (2.16) we can write

$$\nabla \cdot \mathbf{u} + \frac{\partial w}{\partial z} + \frac{w}{\rho}\frac{\partial \rho}{\partial z} = 0, \tag{2.17}$$

or alternatively

$$\rho\nabla \cdot \mathbf{u} + \frac{\partial \rho w}{\partial z} = 0. \tag{2.17'}$$

At 'deep' disturbances, penetrating into the entire layer of vertical scale $Z \sim H_s \sim 10$ km, $M_\rho \sim \rho$ and the second and third terms on the left-hand side of Equation (2.17) are of one and the same order of magnitude. Contrary to this, at $Z \ll H_s$ $M_\rho/\rho \ll 1$ and the approximation for quasi-incompressibility holds

$$\nabla \cdot \mathbf{u} + \frac{\partial w}{\partial z} = 0 \tag{2.18}$$

The last simplification of the continuity equation consists in writing $\nabla \cdot \mathbf{u} \approx 0$, i.e., $u_x \approx -v_y$ (quasi-nondivergent horizontal flow) which does not necessarily mean that $\partial w/\partial z = 0$, i.e., $w = 0$ (see the comments after formula (2.3) and also later in Section 4).

Finally, let us turn to the equation of the first principle of thermodynamics (1.7) in Chapter 2

$$\frac{\mathrm{d}_2 T}{\mathrm{d}t} + w\frac{\partial T}{\partial z} - \frac{R}{c_p}\frac{T}{p}\left(\frac{\mathrm{d}_2 p}{\mathrm{d}t} + w\frac{\partial p}{\partial z}\right) = \frac{Q}{c_p}, \tag{2.19}$$

where

$$\frac{\mathrm{d}_2 p}{\mathrm{d}t} = \frac{\partial p}{\partial t} + (\mathbf{u} \cdot \nabla p) \approx \frac{\partial p}{\partial t}$$

because, due to (2.12), $\mathbf{u} \cdot \nabla p \approx 0$. On the other hand, according to (2.14), $\partial p/\partial z \approx -\rho g$. Also introducing the dry adiabatic lapse rate $\gamma_a = g/c_p$, we write Equation (2.19) in the form

$$\frac{\mathrm{d}_2 T}{\mathrm{d}t} - \underline{\frac{\kappa - 1}{\kappa}\frac{T}{p}\frac{\partial p}{\partial t}} + \left(\frac{\partial T}{\partial z} + \gamma_a\right) w = \frac{Q}{c_p}. \tag{2.20}$$

The underlined term is small and can be neglected in the majority of cases.

c. Some Remarks

(1) The quasi-geostrophic relationships (2.11) are of a diagnostic type. Although small, the acceleration terms in (2.10) are to be retained if we want to forecast the wind field. Therefore, the simplified system of weather equations written together reads

$$\frac{d_2\mathbf{u}}{dt} = \frac{-\nabla p}{\rho} - f\mathbf{k} \times \mathbf{u}, \tag{2.21}$$

$$\rho g = -\frac{\partial p}{\partial z}, \quad p = \rho R T, \tag{2.22}$$

$$\rho \nabla \cdot \mathbf{u} + \frac{\partial \rho w}{\partial z} = 0, \tag{2.23}$$

$$\frac{d_2 T}{dt} + w\left(\frac{\partial T}{\partial z} + \gamma_a\right) = \frac{Q}{c_p}. \tag{2.24}$$

Three of them are prognostic and three are diagnostic.

(2) If the exact continuity Equation (2.15) is integrated into the hydrostatic approximation (2.14), the so-called tendency equation is easily derived

$$\frac{\partial p_0}{\partial t} = -g \int_0^\infty (\nabla \cdot \rho \mathbf{u})\, dz, \tag{2.25}$$

where p_0 is the surface pressure. Due to the proximity of the real wind to the geostrophic one (2.12), $\nabla \cdot \rho\mathbf{u} \approx 0$ and Equation (2.25) cannot be used as the prognostic one for p_0, because $\partial p_0/\partial t$ is determined as a small difference of large numbers (right-hand terms), known from the observations with random and/or systematic errors.

(3) A similar situation arises when (2.18) is integrated so as to obtain the vertical velocity

$$w(z) = -\int_0^z (\nabla \cdot \mathbf{u})\, dz. \tag{2.26}$$

If observational data for $\mathbf{u}$ are used, w cannot be calculated with the required accuracy. Two possibilities exist – either derive new equations for $\partial p_0/\partial t$ and w or substitute $\mathbf{u}$ in (2.25) and (2.26) from other equations rather than from the observations.

(4) Unfortunately, the same difficulty is faced when the momentum equations in the form (2.21) (also called *primitive* equations), are to be used for prognostic purposes with respect to the wind field. Due to (2.11), the left-hand side of (2.21) has at least one order of magnitude smaller than each term on the right. Because of this, direct use of (2.21) for the above-mentioned purpose is associated with serious difficulties.

(5) It becomes clear that the proximity of the real wind to the geostrophic one in synoptic scale motions causes more complications than simplifications to the problems of dynamic meteorology. Nevertheless, one of the most exciting periods in the contemporary history of this discipline, coinciding with the development and consolidation of numerical methods for weather forecasting, is related to the appropriate utilization of the quasi-geostrophic relationships (2.11). Chapter 7 is devoted to this problem.

3. WEATHER EQUATIONS IN p, σ, θ, AND OTHER VERTICAL COORDINATES

Considerable simplification of weather equations and/or their corresponding boundary conditions can be achieved by a suitable choice of coordinate system and, in particular, of the vertical coordinate. Let us now refer back to Section 6 of Chapter 2, in which the system of weather equations was transformed from (x, y, z, t) to (x, y, q, t) coordinates, where $q = q(x, y, z, t)$ was an unspecified but single-valued monotonic function of z.

In a quasi-static approximation, the fundamental equations of motion, continuity and heat influx read

$$\frac{\mathrm{d}\mathbf{u}}{\mathrm{d}t} + f\mathbf{k} \times \mathbf{u} = -\alpha \nabla_q p - g \nabla_q z, \quad \left(\alpha = \frac{1}{\rho}\right). \tag{3.1}$$

$$\frac{\mathrm{d}}{\mathrm{d}t}\left(\rho \frac{\partial z}{\partial q}\right) + \nabla_q \cdot \mathbf{u} + \frac{\partial \dot{q}}{\partial q} = 0, \tag{3.2}$$

$$\left[\left(\frac{\partial}{\partial t}\right)_q + \mathbf{u} \cdot \nabla_q + \dot{q}\frac{\partial}{\partial q}\right] \ln \theta = \frac{Q}{c_p T}. \tag{3.3}$$

They will serve as the initial equations for a particular choice of q.

a. Isobaric p System

As is well known, atmospheric pressure p is always observed to decrease monotonically with height. This fact makes it a suitable candidate for a vertical coordinate instead of z. The single-valued relationship $p = p(x, y, z, t)$ between z and p, required for deduction of Equations (3.1)–(3.3), is assured by the quasi-static equation $\partial p/\partial z = -\rho g$, valid for synoptic and, eventually, subsynoptic scale motions.

Thus, we let $q \equiv p$. Then

$$\frac{\mathrm{d}}{\mathrm{d}t} = \left(\frac{\partial}{\partial t}\right)_p + \mathbf{u} \cdot \nabla_p + w_p \frac{\partial}{\partial p}, \quad w_p = \dot{p} = \frac{\mathrm{d}p}{\mathrm{d}t} \tag{3.4}$$

where w_p is the 'isobaric' vertical velocity (another notation used in the literature is ω), and the subscript p denotes that the pressure is held constant when differentiating. The horizontal wind $\mathbf{u}$ is the same in both systems as are all three unit vectors $\mathbf{i}, \mathbf{j}, \mathbf{k}$,

When pressure p becomes an independent variable, then the height $z(x, y, p, t)$ of the constant pressure surface p = const or the corresponding absolute geopotential $\Phi = gz$ stands for a dependent variable. Due to the fundamental feature of the isobaric surfaces $\nabla_p p = 0$ and, consequently, the first term on the right-hand side of Equation (3.1) vanishes and takes the form

$$\frac{\mathrm{d}\mathbf{u}}{\mathrm{d}t} + f\mathbf{k} \times \mathbf{u} = -\nabla_p \Phi. \tag{3.5}$$

The first term in the equation of continuity (3.2) also vanishes because $\rho\ \partial z/\partial p = -1/g$ = const. It takes the form

$$\nabla_p \cdot \mathbf{u} + \frac{\partial w_p}{\partial p} = 0. \tag{3.6}$$

The hydrostatic equation and the equation of state can be written together as follows

$$\frac{\partial \Phi}{\partial p} = -\frac{RT}{p} = -\frac{1}{\rho} = -\alpha. \tag{3.7}$$

As to Equation (3.3), it could be rewritten in exactly the same form with p instead of q. It is more convenient, however, to eliminate the potential temperature

$$\theta = T\left(\frac{p_{00}}{p}\right)^{\lambda}, \quad p_{00} = 1000 \text{ mb}, \qquad \lambda = \frac{R}{c_p}, \quad T = \frac{\alpha p}{R}. \tag{3.8}$$

Obviously,

$$\ln \theta = \begin{cases} \ln T - \lambda \ln p + \text{const}, \\ \ln \alpha + (1 - \lambda) \ln p + \text{const}. \end{cases} \tag{3.9}$$

Hence,

$$\nabla_p \ln \theta = \nabla_p \ln T = \nabla_p \ln \alpha, \quad \nabla_p \ln p \equiv 0. \tag{3.10}$$

Substituting into (3.3) we obtain

$$\left(\frac{\partial T}{\partial t}\right)_p + (\mathbf{u} \cdot \nabla_p T) - m_T w_p = \frac{Q}{c_p}. \tag{3.11}$$

Alternatively,

$$\left(\frac{\partial \alpha}{\partial t}\right)_p + (\mathbf{u} \cdot \nabla_p \alpha) - m_\alpha w_p = \alpha \frac{Q}{T c_p}, \tag{3.12}$$

where

$$\begin{aligned} m_T &= -\frac{T}{\theta}\frac{\partial \theta}{\partial p} = \frac{RT}{gp}(\gamma_a - \gamma), \\ m_\alpha &= \frac{\alpha}{T} m_T = \frac{R^2 T}{g p^2}(\gamma_a - \gamma), \end{aligned} \tag{3.13}$$

are parameters of static stability, since their sign depends on the difference $\gamma_a - \gamma$. Here, $\gamma = -\partial T/\partial z$ is the actual lapse rate in the atmosphere and γ_a is the dry adiabatic one. The parameters m_T and m_α are variable quantities which make Equations (3.11) and (3.12) quite complicated. Usually they are considered as constants, equal to some averaged values $\bar{m}_T$ and $\bar{m}_\alpha$. Then the equations simplify considerably.

Thus, the transformed equations are six in number: (3.5)–(3.7) and (3.11) for seven unknown functions: u, v, w_p, Φ, T, ρ and Q. At adiabatic processes $Q = 0$ and the system becomes closed. Knowing w_p, the 'geometrical' vertical velocity w can be determined from the obvious equation

$$w_p = \frac{\mathrm{d}p}{\mathrm{d}t} = \left(\frac{\mathrm{d}_2 p}{\mathrm{d}t}\right)_z + w\frac{\partial p}{\partial z} = \left(\frac{\mathrm{d}_2 p}{\mathrm{d}t}\right)_z - wg\rho, \tag{3.14}$$

where the subscript z reminds us that the total derivative is taken on the plane z = const. A precise evaluation shows that, with great accuracy,

$$w_p \approx -wg\rho. \tag{3.15}$$

Therefore, $w_p > 0$ corresponds to downward motion ($w < 0$) and vice-versa.

An alternative to (3.14) follows if we start from the general formula (6.12) in Chapter 2

$$w = \frac{1}{g}\left[\left(\frac{d_2\Phi}{dt}\right)_p - \frac{RT}{p} w_p\right] \tag{3.16}$$

which is consistent with (3.15) since $d_2\Phi/dt$ is negligibly small.

The isobaric p system has a number of advantages:

(i) The weather equations simplify considerably. Actually, the coefficient in front of $\nabla_p\Phi$ in Equation (3.5) is 1 instead of $1/\rho$ as in (2.9). Hence, the quasi-geostrophic relationship $\mathbf{u} = f^{-1}\mathbf{k} \times \nabla\Phi$ has an universal form for all isobaric levels which facilitates the calculation. The continuity Equation (3.6) becomes diagnostic and looks like $\nabla \cdot \mathbf{u} + \partial w/\partial z = 0$ which is valid for an incompressible fluid. This is the most significant simplification. The heat influx Equation (3.11) remains diagnostic but is simpler than the starting one when assuming m_T = const and $Q = 0$.

(ii) As a result of the simplifications, the number of the equations in the system can be reduced to three. Actually, substituting α from (3.7) into (3.12) at $Q = 0$ and m_α = const we obtain

$$\left(\frac{d_2}{dt}\right)_p \frac{\partial\Phi}{\partial p} + m_\alpha w_p = 0 \tag{3.17}$$

where $(d_2/dt)_p = (\partial/\partial t)_p + \mathbf{u} \cdot \nabla_p$. Having determined w_p from here and substituted into (3.6) and (3.5) (w_p enters in d/dt), we obtain three equations for three dependent variables – u, v, Φ. Such a reduction of the system in standard coordinates is impossible.

(iii) For many years, aerological (radiosonde) observations have actually been taken in the (x, y, p) system rather than in (x, y, z). The height $z(x, y, p, t)$ of a given isobaric surface p = const is derived from the temperature and is a dependent variable. Thus, the use of the (x, y, p) system permits a direct comparison of theoretical predictions with observations, which is a very important advantage.

But the isobaric p system has an important disadvantage too, which is that the Earth's surface $z = H(x, y)$ is not a coordinate surface in this system. This yields serious difficulties and introduces errors when formulating and using the lower boundary conditions over strongly mountainous terrain.

Actually, if $p = p_c$ = const at the upper boundary, we assume $w_p^s = 0$, then according to (6.29) in Chapter 2

$$w_p = w_p^H = \frac{\partial p_H}{\partial t} + \mathbf{u}_H \cdot \nabla p_H \quad \text{at } p = p_H(x, y, t), \tag{3.18}$$

where the surface pressure p_H is an unknown function for which a prognostic equation can be derived. For this purpose, we integrate (3.6) and obtain

$$w_p = -\int_{p_s}^{p} (\nabla_p \cdot \mathbf{u})\, dp, \qquad w_p^H = -\int_{p_s}^{p_H} (\nabla_p \cdot \mathbf{u})\, dp. \tag{3.19}$$

Comparing the second expression with (3.18), we obtain the required prognostic equation

$$\frac{\partial p_H}{\partial t} = -\mathbf{u}_H \cdot \nabla p_H - \int_{p_s}^{p_H} (\nabla_p \cdot \mathbf{u})\, dp. \tag{3.20}$$

Finally, from (3.7) it follows that

$$\Phi = \Phi_H + R \int_p^{p_H} T \, \mathrm{d} \ln p \tag{3.21}$$

where $\phi_H = gH$.

At given initial conditions for the meteorological element $\mathbf{u}$, T and p_H, prognostic Equations (3.5), (3.11) (at $Q = 0$) and (3.20) in principle allow the forecasting of these elements at $t > t_0$. The diagnostic Equations (3.19) and (3.21) determine w_p and Φ.

b. Isobaric σ System

To avoid the above-mentioned shortcoming of the p system, the so-called σ system has been introduced in the following way

$$\sigma = \frac{p - p_c}{p_H - p_c} \quad \text{or} \quad \sigma = \frac{p}{p_H} \quad \text{at } p_c = 0, \tag{3.22}$$

where

$$p_H = p[x, y, H(x, y), t] = p_H(x, y, t)$$

is the surface pressure.

Evidently $0 \leqslant \sigma \leqslant 1$ and the Earth's surface coincides with the coordinate surface $\sigma = 1$, thus allowing an exact formulation of the lower-boundary conditions. Unfortunately, the equations of motion become more complex than they were in the p system.

Let us refer back to the general Equation (3.1). Since $\nabla_\sigma \sigma \equiv 0$ then (3.22) yields

$$\nabla_\sigma \sigma = \frac{\nabla_\sigma p}{p_H} - \frac{p \nabla p_H}{p_H^2} = \frac{\nabla_\sigma p - \sigma \nabla p_H}{p_H} = 0$$

so that $\nabla_\sigma p/\rho = (RT/p_H) \nabla p_H$ and Equation (3.1) takes the form

$$\frac{\mathrm{d}\mathbf{u}}{\mathrm{d}t} + f\mathbf{k} \times \mathbf{u} = -\nabla_\sigma \Phi - \frac{RT \nabla p_H}{p_H}. \tag{3.23}$$

On the other hand, $\partial/\partial p = \partial\sigma/\partial p \cdot \partial/\partial\sigma = (1/p_H)\, \partial/\partial\sigma$. Then the hydrostatic equation (3.7) becomes

$$\frac{\partial \Phi}{\partial \sigma} = -\frac{RT}{\sigma}. \tag{3.24}$$

We let $q = \sigma$ in Equation (6.21) of Chapter 2 and substitute $\partial p/\partial\sigma = p_H$. Then the equation of continuity becomes

$$\frac{\partial p_H}{\partial t} + \nabla_\sigma \cdot (\mathbf{u} p_H) + p_H \frac{\partial w_\sigma}{\partial \sigma} = 0 \tag{3.25}$$

where $w_\sigma = \dot{\sigma} = \mathrm{d}\sigma/\mathrm{d}t$.

Further we consider the heat influx equation

$$\frac{dT}{dt} = \frac{RT}{c_p \sigma p_H} \frac{dp}{dt} + \frac{Q}{c_p}. \tag{3.26}$$

To determine $w_p = dp/dt$ we proceed in the following way:
Using the boundary conditions

$$w_\sigma = 0 \quad \text{at } \sigma = 1 \text{ and } \sigma = 0 \tag{3.27}$$

and integrating (3.25) we obtain

$$\sigma \frac{\partial p_H}{\partial t} + \int_0^\sigma \nabla_\sigma \cdot (\mathbf{u} p_H)\, d\sigma + p_H w_\sigma = 0, \tag{3.28}$$

and also

$$\frac{\partial p_H}{\partial t} = -\int_0^1 \nabla_\sigma \cdot (\mathbf{u} p_H)\, d\sigma. \tag{3.29}$$

Eliminating $\partial p_H/\partial t$ from the last two equations we find

$$w_\sigma = -\frac{1}{p_H} \int_0^\sigma \nabla_\sigma \cdot (\mathbf{u} p_H)\, d\sigma + \frac{\sigma}{p_H} \int_0^1 \nabla_\sigma \cdot (\mathbf{u} p_H)\, d\sigma. \tag{3.30}$$

Moreover, since $\sigma = p/p_H$ then

$$w_\sigma = \frac{d\sigma}{dt} = p_H^{-1} \left(\frac{dp}{dt} - \sigma \frac{dp_H}{dt} \right), \tag{3.31}$$

where, according to (3.18),

$$\frac{dp_H}{dt} = \frac{\partial p_H}{\partial t} + \mathbf{u}_H \cdot \nabla p_H. \tag{3.32}$$

A suitable combination of Equations (3.29)–(3.32) yields

$$\frac{dp}{dt} = \sigma(\mathbf{u}_H \cdot \nabla p_H) - \int_0^\sigma \nabla_\sigma \cdot (\mathbf{u} p_H)\, d\sigma. \tag{3.33}$$

This expression should be substituted into Equation (3.26). With this, the transformation of the weather equations in the σ-isobaric system is completed. We shall point out in addition that in the left-hand sides of Equations (3.23) and (3.26)

$$\frac{d}{dt} = \left(\frac{\partial}{\partial t} \right)_\sigma + \mathbf{u} \cdot \nabla_\sigma + w_\sigma \frac{\partial}{\partial \sigma}. \tag{3.34}$$

Besides, formula (3.31) connects w_σ with w_p

$$p_H w_\sigma = w_p - \sigma \frac{dp_H}{dt} \tag{3.31'}$$

and by means of (3.15) with w, too.

The prognostic equations are (3.23), (3.29) and (3.26) in combination with (3.33). They allow the determination of **u**, T and p_H. Then from the diagnostic equation (3.30) and

$$\Phi = \Phi_H + R \int_\sigma^1 T \, \mathrm{d} \ln \sigma \tag{3.35}$$

which follows from (3.24) and corresponds to (3.21), w_σ and Φ can be determined.

The σ isobaric system, first introduced by N. A. Phillips, overcomes the main defect of the p system, but at the expense of the appearance of an additional term in momentum equation (3.23). Besides, the isobaric surfaces on which the meteorological elements are known from the observations are no longer coordinate surfaces in the σ system. This fact requires the additional interpolation of the initial data as well as of the theoretically-calculated values for the purpose of comparing them with the observations. These are real disadvantages of the σ system which restrict its application.

When applying the p and σ systems, it is necessary to keep in mind that p_H and T_H depend on (x, y), explicitly and implicitly, by means of $H(x, y)$, i.e.,

$$F_H = F_H[x, y, H(x, y), t], \quad F = p, T.$$

Hence, for example,

$$\nabla p_H = \tilde{\nabla} p_H + \frac{\partial p_H}{\partial H} \nabla H = \tilde{\nabla} p_H - \frac{g p_H}{R T_H} \nabla H, \tag{3.36}$$

where $\tilde{\nabla} p_H$ is the gradient taken with respect to the explicit variables $(\underset{\sim}{x, y})$. Besides, the barometric formula $\partial p_H/\partial H = -g p_H/R T_H$ has been used. Obviously, $\tilde{\nabla} p_H$ reflects the horizontal inhomogeneities due to some processes taking place in the atmosphere above the mountains, while the second term in (3.36) reflects the topography. In practice, $H(x, y)$ is a rather smoothed function. But, nevertheless, even at $\partial H/\partial x \sim 50 \text{ m } 100 \text{ km}^{-1}$ the second term in (3.36) dominates. Therefore, the additional terms in (3.23) and in the other equations are mainly caused by orographic effects.

c. Insentropic θ System

Now the potential temperature θ (see (1.8) and (1.9) in Chapter 2) is chosen as a vertical coordinate

$$\theta = T\left(\frac{p_{00}}{p}\right)^\lambda, \qquad \frac{\mathrm{d} \ln \theta}{\mathrm{d}t} = \frac{Q}{c_p T}, \qquad \lambda = \frac{R}{c_p}. \tag{3.37}$$

Obviously, $w_\theta = \mathrm{d}\theta/\mathrm{d}t = \theta Q/c_p T$ and in the case of adiabatic processes ($Q = 0$), we find $w_\theta = 0$, so that in the expression for the total derivative

$$\frac{\mathrm{d}}{\mathrm{d}t} = \left(\frac{\partial}{\partial t}\right)_\theta + \mathbf{u} \cdot \nabla_\theta + w_\theta \frac{\partial}{\partial \theta}, \tag{3.38}$$

the last term vanishes. This simplification makes the system convenient for studying adiabatic processes.

Since $\nabla_\theta \theta \equiv 0$, then according to (3.37)

$$\nabla_\theta \ln \theta = \nabla_\theta \ln T - \frac{R}{c_p} \nabla_\theta \ln p,$$

so that $\nabla_\theta p/\rho = c_p \nabla_\theta T$. Substituting in (3.1), we obtain the equation of motion in the θ system

$$\frac{d\mathbf{u}}{dt} + f\mathbf{k} \times \mathbf{u} = -\nabla_\theta M, \tag{3.39}$$

where

$$M = c_p T + gz = c_p T + \Phi. \tag{3.40}$$

is called the isentropic stream function or the Montgomery acceleration potential. The hydrostatic equation $\partial p/\partial z = -\rho g$ takes the form

$$\frac{\partial M}{\partial \theta} = \frac{c_p T}{\theta}. \tag{3.41}$$

The continuity equation is obtained from (6.26) in Chapter 2, by replacing q with θ

$$\frac{\partial}{\partial \theta}\left(\frac{\partial p}{\partial t}\right)_\theta + \nabla_\theta \cdot \left(\mathbf{u}\,\frac{\partial p}{\partial \theta}\right) + \frac{\partial}{\partial \theta}\left(w_\theta \frac{\partial p}{\partial \theta}\right) = 0 \tag{3.42}$$

If $w_\theta = 0$ (adiabatic process), the last term vanishes.

The simplification of the equations is a serious advantage of the θ system. However, once again the Earth's surface $H(x, y)$ is not a coordinate surface and one faces the same difficulties as in the p system.

d. Other Vertical Coordinates

In the pursuit of finding 'better' coordinate systems, a great number of other systems have been proposed. Two of them will be mentioned here.

(i) The *Geometrical* $\tilde{z}$ *system*, proposed by N. A. Phillips

$$\tilde{z} = \frac{z_c - z}{z_c - H} \quad \text{or} \quad \tilde{z}' = \frac{z - H}{z_c - H}. \tag{3.43}$$

Evidently, the Earth's surface $H(x, y)$ and the 'ceiling' z_c = const of the atmospheric layer under consideration are coordinate surfaces. By these features $\tilde{z}$ and $\tilde{z}'$ resemble the σ system. However, contrary to the latter, $\tilde{z}$ and $\tilde{z}'$ do not depend on time (t) and do not need the hydrostatic approximation for their introduction. Unfortunately, as a result, the weather equations become complex.

(ii) The ζ *isobaric system*:

$$\zeta = -H_s \ln \frac{p}{p_{00}}, \quad H_s = \frac{RT_0}{g}, \tag{3.44}$$

where $p_{00} = 1000$ mb. In an isothermic atmosphere $\zeta \equiv z$, while in the general case $\zeta \approx z$. In this system

$$\frac{d\zeta}{dt} = w_\zeta = -\frac{H_s}{p} w_p$$

so that

$$\frac{\partial w_p}{\partial p} = \frac{\partial w_\zeta}{\partial \zeta} - \frac{w_\zeta}{H_s}. \tag{3.45}$$

Substituting into (3.6), we obtain the continuity equation

$$\frac{\partial u}{\partial x} + \frac{\partial v}{\partial y} + \frac{\partial w_\zeta}{\partial \zeta} - \frac{w_\zeta}{H_s} = 0. \tag{3.46}$$

The remaining equations are transformed in a similar way

$$\begin{aligned} &\frac{d\mathbf{u}}{dt} = -\nabla\phi - f\mathbf{k} \times \mathbf{u}, \\ &\frac{\partial \phi}{\partial \zeta} = \frac{RT}{H_s}, \\ &\frac{d_2 T}{dt} + w_\zeta \Gamma = \frac{Q}{c_p}, \end{aligned} \tag{3.47}$$

where $d/dt = d_2/dt + w_\zeta \partial/\partial\zeta$ and

$$\Gamma = \frac{\partial T}{\partial \zeta} + \frac{RT}{c_p H_s} = \frac{T}{\theta}\frac{\partial \theta}{\partial \zeta} \tag{3.48}$$

is a parameter of static stability similar to m_α or m_T from (3.13). With a larger accuracy than in the latter case, Γ can be considered as a constant in the entire atmosphere – $\Gamma \approx$ const $\approx 3°\text{C km}^{-1}$.

The ζ isobaric system finds application in the theory of atmospheric gravity waves (see Section 4 of Chapter 5).

4. AGEOSTROPHIC AND THERMAL WINDS

a. Ageostrophic Wind

It was shown in Section 2 that at synoptic scale motions the real wind $\mathbf{u}$ is very close to the geostrophic one – $\mathbf{u} \approx \mathbf{u}_g$. Therefore, we can write

$$\mathbf{u} = \mathbf{u}_g + \mathbf{u}', \tag{4.1}$$

where $\mathbf{u}'$ is called the *ageostrophic* wind. Having introduced (4.1) and the geostrophic relationship

$$\nabla\Phi = -f\mathbf{k} \times \mathbf{u}_g \quad \text{or} \quad \mathbf{u}_g = f^{-1}\mathbf{k} \times \nabla\Phi \tag{4.2}$$

into the equation of motion

$$\frac{d\mathbf{u}}{dt} = -\nabla\Phi - f\mathbf{k} \times \mathbf{u}, \tag{4.3}$$

we obtain

$$\frac{d\mathbf{u}}{dt} = -f\mathbf{k} \times \mathbf{u}' \quad \text{or} \quad \mathbf{u}' = f^{-1}\mathbf{k} \times \frac{d\mathbf{u}}{dt} \tag{4.4}$$

Therefore, in our model of an ideal atmosphere, the deviation of the real wind $\mathbf{u}$ from the geostrophic one $\mathbf{u}_g$ is directly related to the acceleration $d\mathbf{u}/dt$ of the air particles. Evidently $(\mathbf{u}' \cdot d\mathbf{u}/dt) = 0$ which reveals that both vectors are perpendicular. In the Northern Hemisphere ($f > 0$) $d\mathbf{u}/dt$ is on the right of $\mathbf{u}'$ (Figure 3.1).

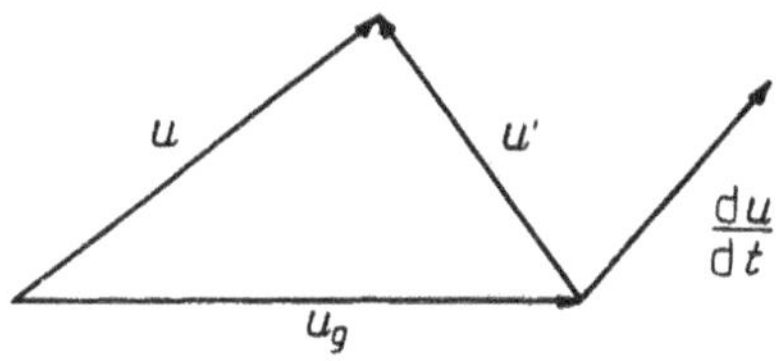

Fig. 3.1. Ageostrophic wind.

Let us estimate the magnitude of $\mathbf{u}'$, introducing the corresponding scale $V' = O(u') = O(v')$. Then Equations (4.4) and (2.4) yield

$$V' = \frac{V^2}{fL} \quad \text{or} \quad \frac{V'}{V} = \text{Ro} \sim 10^{-1}. \tag{4.5}$$

Since $V \approx V_g$, then $V'/V_g \approx \text{Ro}$ too. Moreover, from (4.1) and (4.2)

$$D = D_g + D' = D' - \frac{\beta}{f^2}\Phi_x = D' - \frac{\beta}{f}v_g \tag{4.6}$$

and

$$\zeta = \zeta_g + \zeta' = f^{-1}\nabla^2\Phi + \frac{\beta}{f}u_g + \zeta'. \tag{4.7}$$

Scale analysis of divergence and vorticity expressions (4.6) and (4.7), with the use of (4.5), yields some interesting conclusions. Actually

$$\frac{O(u'_x, v'_y)}{O(u_x, v_y)} \sim \text{Ro} \sim \frac{O(D')}{O(u_x, v_y)}$$

where $D' = u'_x + v'_y$. Also

$$\frac{O(\beta v_g/f)}{O(u_x, v_y)} \sim B \sim \text{Ro} \sim \frac{O(D)}{O(u_x, v_y)}$$

where $D = u_x + v_y$. Consequently, in (4.6), $O(D')/O(\beta v_g/f) \sim 1$.

Further, from (4.7) we find $O(\beta u_g/f)/O(\nabla^2\Phi/f) \sim B$ so that

$$\zeta_g \approx \frac{\nabla^2\Phi}{f}. \tag{4.8}$$

Besides $O(v_x, u_y)/O(\nabla^2\Phi/f) \sim 1$ and ζ' in (4.7) is much smaller than ζ or ζ_g, i.e.,

$$\frac{O(\zeta')}{O(\zeta, \zeta_g)} \sim 10^{-1}, \qquad \frac{O(\zeta)}{O(v_x, u_y)} \sim 1, \tag{4.9}$$

while $O(D)/O(u_x, v_y) \sim 10^{-1}$. This is an essential distinguishing feature of vorticity ζ, as compared to divergence D. At the same time

$$\frac{O(\zeta)}{f} = \frac{V}{Lf} = \mathrm{Ro} \sim 10^{-1}, \tag{4.10}$$

but

$$\frac{O(\mathrm{d}\zeta/\mathrm{d}y)}{O(\mathrm{d}f/\mathrm{d}y)} = \frac{V}{L^2\beta} = \frac{\mathrm{Ro}}{B} \sim 1. \tag{4.11}$$

Therefore, in the Northern Hemisphere ($f > 0$), with rare exceptions, the absolute vorticity $\zeta_a = \zeta + f$ is positive (cyclonic).

Now we are in a position to give a more precise justification of inequality (2.3′), namely $W/Z \ll V/L$. From the continuity equation $\partial w/\partial z = -D$ and the relationships (4.5) and (4.6) it follows that

$$\frac{W}{Z} = \frac{V'}{L} = \mathrm{Ro} \cdot \frac{V}{L}. \tag{4.12}$$

This result agrees with (2.3″). For the vertical velocity itself, we can write

$$w(z) = -\int_0^z D'(z)\,\mathrm{d}z + \frac{\beta}{f}\int_0^z v_g(z)\,\mathrm{d}z. \tag{4.13}$$

The main difficulty consists in evaluating the ageostrophic divergence.

According to definition (4.1), the ageostrophic wind can be determined as a vector difference $\mathbf{u}' = \mathbf{u} - \mathbf{u}_g$, where $\mathbf{u}$ is measured while $\mathbf{u}_g$ is calculated on pressure (geopotential) data. But because of (4.5), the acceptable percentage errors, say 10%, in measuring $\mathbf{u}$ or calculating $\mathbf{u}_g$ on aerological data can cause the estimated value of $\mathbf{u}'$ and $\nabla \cdot \mathbf{u}'$ entering into (4.13) to be in error by 100% or more. That is why, for the sake of accuracy, a question arises whether it should not be possible to calculate $\mathbf{u}'$ directly, having observational data on the pressure (geopotential) field only, without using the measured values of the real wind. The answer is positive and was first given by Kibel in 1940.

Let us start with the equations of motion written in the form

$$\left(\frac{\mathrm{d}_2}{\mathrm{d}t} + w\,\frac{\partial}{\partial z}\right)\mathbf{u} = -f\mathbf{k} \times (\mathbf{u} - \mathbf{u}_g). \tag{4.14}$$

After the usual procedure of nondimensionalization and using the same notations for dimensionless quantities as in the previous equation, as well as the important relationship (4.12), we obtain

$$\left(\mathrm{Ro}\,\frac{\mathrm{d}_2}{\mathrm{d}t} + \mathrm{Ro}^2\, w\,\frac{\partial}{\partial z}\right)\mathbf{u} = -\mathbf{k} \times (\mathbf{u} - \mathbf{u}_g). \tag{4.14'}$$

We assume a solution in the form of a power series of the small parameter Ro ($\sim 10^{-1}$)

$$\mathbf{u} = \mathbf{u}_0 + \mathrm{Ro}\,\mathbf{u}_1 + \mathrm{Ro}^2\,\mathbf{u}_2 + \cdots \tag{4.15}$$

Upon substituting (4.15) into (4.14') and equating the corresponding terms, we obtain an infinite set of equations

$$\mathbf{u}_0 = \mathbf{u}_g, \qquad -\mathbf{k} \times \mathbf{u}_1 = \frac{\mathrm{d}_2 \mathbf{u}_0}{\mathrm{d}t}, \ldots$$

which are immediately solved. Returning to the dimensional form we get

$$\mathbf{u}_0 = \mathbf{u}_g, \qquad \mathbf{u}_1 = f^{-1}\mathbf{k} \times \frac{\mathrm{d}_2 \mathbf{u}_g}{\mathrm{d}t}, \ldots \tag{4.16}$$

Assuming that at Ro $\sim 10^{-1}$, the series (4.15) converges fast enough and we can write $\mathbf{u} \approx \mathbf{u}_g + \mathbf{u}_1$ so that the ageostrophic wind $\mathbf{u}' \approx \mathbf{u}_1$, i.e.,

$$\mathbf{u}' \approx f^{-1}\mathbf{k} \times \left[\frac{\partial \mathbf{u}_g}{\partial t} + (\mathbf{u}_g \cdot \nabla)\mathbf{u}_g\right]. \tag{4.17}$$

The same result can be obtained starting from (4.3) instead of from (4.14).

Inserting in (4.17) $\mathbf{u}_g$ from (4.2) one obtains

$$\mathbf{u}' = -f^{-2}\nabla\Phi_t - f^{-3}\mathbf{k} \times \{[(\mathbf{k} \times \nabla\Phi) \cdot \nabla]\,(\mathbf{k} \times \nabla\Phi)\}, \tag{4.18}$$

or in scalar form

$$u' = -f^{-2}\Phi_{xt} - f^{-3}J(\Phi, \Phi_x), \qquad v' = -f^{-2}\Phi_{yt} - f^{-3}J(\Phi, \Phi_y), \tag{4.18'}$$

where f has been considered independent of y, $J = J_{xy}$ is the Jacobian, the subscripts denote derivatives. The linear part of (4.15) is called the *isalobaric* wind. The formulae (4.18'), first derived by Kibel, allow the calculation of $\mathbf{u}'$, although only having data for Φ. They have also been used by many other authors, including Charney and Eliassen. Moreover, Eliassen (1959) proposed a slightly different approach to the problem which could be considered as a modification of Kibel's method. It is now known as the *geostrophic momentum approximation* and consists of (i) replacing the advected momentum (u, v) in the equation for the horizontal motion

$$\frac{\partial \mathbf{u}}{\partial t} + (\mathbf{u} \cdot \nabla)\mathbf{u} = -f\mathbf{k} \times (\mathbf{u} - \mathbf{u}_g)$$

by the geostrophic momentum (u_g, v_g) and (ii) keeping $\mathbf{u}$ in the remaining terms. Thus, one obtains

$$\frac{\partial \mathbf{u}_g}{\partial t} + (\mathbf{u} \cdot \nabla)\mathbf{u}_g = -f\mathbf{k} \times (\mathbf{u} - \mathbf{u}_g) \tag{4.19}$$

or, upon introducing $\mathbf{u} = \mathbf{u}_g + \mathbf{u}'$,

$$(\mathbf{u}' \cdot \nabla \mathbf{u}_g) + f\mathbf{k} \times \mathbf{u}' = -\frac{\mathrm{d}_2 \mathbf{u}_g}{\mathrm{d}t}. \tag{4.20}$$

In scalar form, this equation is equivalent to a system of two algebraic equations with respect to (u', v') which can be solved in a straightforward way.

b. Thermal Wind

Let us summarize the formulae for the geostrophic wind $\mathbf{u}_g$ in geometrical (z), isobaric (p), and isentropic (θ) coordinates.

$$\begin{aligned} &\text{(a)} \qquad \mathbf{u}_g = \frac{1}{\rho f}\mathbf{k} \times \nabla_z p = \frac{RT}{fp}\mathbf{k} \times \nabla_z p, \\ &\text{(b)} \qquad \mathbf{u}_g = f^{-1}\mathbf{k} \times \nabla_p \Phi, \\ &\text{(c)} \qquad \mathbf{u}_g = f^{-1}\mathbf{k} \times \nabla_\theta (c_p T + \Phi). \end{aligned} \tag{4.21}$$

Hereafter we shall write ∇ instead of ∇_z but shall keep ∇_p and ∇_θ.

Obviously $\mathbf{u}_g$ is a function of the horizontal coordinates x, y and, parametrically, it can also depend on time t. The dependence of $\mathbf{u}_g$ on the vertical coordinate requires special attention.

Let us first consider expression (4.21a) in combination with the hydrostatic equation

$$\frac{\mathbf{u}_g}{T} = \frac{R}{f}\mathbf{k} \times \nabla \ln p, \qquad \frac{g}{T} = -R\frac{\partial \ln p}{\partial z}.$$

By correspondingly applying the operators $\partial/\partial z$ and ∇ and eliminating $\ln p$, we obtain

$$\begin{aligned} \frac{\partial}{\partial z}\mathbf{u}_g &= \frac{1}{T}\frac{\partial T}{\partial z}\mathbf{u}_g + \frac{g}{fT}\mathbf{k} \times \nabla T \\ &= \frac{R}{fp}\left(\frac{\partial T}{\partial z}\mathbf{k} \times \nabla p - \frac{\partial p}{\partial z}\mathbf{k} \times \nabla T\right). \end{aligned} \tag{4.22}$$

It is now easily seen that the term in brackets vanishes when $\nabla p \times \nabla T = 0$, i.e., the atmosphere is barotropic and $\partial \mathbf{u}_g/\partial z = 0$, $\mathbf{u}_g(z) = \text{const}$. Conversely, in a baroclinic atmosphere $\partial \mathbf{u}_g/\partial z \neq 0$, $\mathbf{u}_g = u_g(z)$. If z_0 and z are two levels in the atmosphere, then the vector difference

$$\mathbf{u}_T = \mathbf{u}_g(z) - \mathbf{u}_g(z_0) = \mathbf{u}_g - \mathbf{u}_g^0 \tag{4.23}$$

is called the *thermal* wind.

The explicit dependence of $\mathbf{u}_g$ on z is completely determined by the vertical temperature profile. From the hydrostatic equation in integral form

$$p(x, y, z) = p_0(x, y) \exp\left[-\frac{g}{R}\int_0^z \frac{dz}{T(x, y, z)}\right],$$

we obtain

$$\nabla p = \frac{p}{p_0}\nabla p_0 + \frac{gp}{R}\int_0^z \frac{1}{T^2}\nabla T\, dz,$$

where $p_0 = p(x, y, 0)$. Re-evaluating ∇p and ∇p_0 in geostrophic velocities, we find

$$\mathbf{u}_g = \frac{T}{T_0}\mathbf{u}_g^0 + \frac{gT}{f}\int_0^z \frac{1}{T^2}\mathbf{k} \times \nabla T\, dz, \tag{4.24}$$

where

$$\mathbf{u}_g = \frac{RT}{fp}\mathbf{k} \times \nabla p, \qquad \mathbf{u}_g^0 = \frac{RT_0}{fp_0}\mathbf{k} \times \nabla p_0.$$

Let

$$T(x, y, z) = T_0(x, y) - \gamma z, \qquad \gamma = \text{const.} \tag{4.25}$$

Inserting in (4.24), and integrating, we obtain

$$\mathbf{u}_g(z) = \mathbf{u}_g^0 + \mathbf{u}_T' z, \qquad \mathbf{u}_T' = -\frac{\gamma R}{fp_0}\mathbf{k} \times \nabla p_0 + \frac{g}{fT_0}\mathbf{k} \times \nabla T_0. \tag{4.26}$$

The linear approximation (4.26) for $\mathbf{u}_g(z)$ is often used in studies on tropospheric dynamics including the planetary boundary layer (see Section 2 and Chapter 10).

A more general approximation than (4.25) would be

$$\begin{aligned} T(x, y, z) &= T_\infty - A(x, y)\, e^{-mz}, \\ A(x, y) &= T_\infty - T_0(x, y), \qquad m, T_\infty = \text{const.} \end{aligned} \tag{4.27}$$

For the result of integration see Problem 3 at the end of this chapter.

Let us now reconsider the thermal wind equation (4.22). A scale analysis shows that the dominant term is the second one, i.e.,

$$\frac{\partial \mathbf{u}_g}{\partial z} \approx \frac{g}{fT}\mathbf{k} \times \nabla T. \tag{4.28}$$

On the other hand, upon differentiating (4.21b) with respect to p and eliminating Φ through the hydrostatic equation $\partial\Phi/\partial p = -RT/p$, we obtain

$$p\frac{\partial \mathbf{u}_g}{\partial p} = \frac{\partial \mathbf{u}_g}{\partial \ln p} = -\frac{R}{f}\mathbf{k} \times \nabla_p T. \tag{4.29}$$

In comparison with (4.28) it can be written also as

$$\frac{\partial \mathbf{u}_g}{\partial z} = \frac{g}{fT} \mathbf{k} \times \nabla_p T. \tag{4.30}$$

It is clearly seen that in the barotropic atmosphere, $\nabla_p T = 0$ and $\mathbf{u}_g$ does not change with height.

The thermal wind equations (4.28)–(4.30), as well as the corresponding one in the isentropic coordinate

$$\frac{\partial \mathbf{u}_g}{\partial \theta} = \frac{c_p}{f\theta} \mathbf{k} \times \nabla_\theta T \tag{4.31}$$

are extremely useful diagnostic relationships which are often used to check analyses of observed wind and temperature fields for consistency. For instance, formula (4.28) implies that, at $f > 0$, the cold areas on the $x0y$ plane are to the left of the vector $\partial \mathbf{u}_g/\partial z$. For synoptic practice it is more convenient to use nonlocal relationships instead ((4.28)–(4.31)).

We write (4.20) for two isobaric surfaces p_0 = const and p_1 = const and subtract, according to (4.23)

$$\mathbf{u}_T = \mathbf{u}_g^1 - \mathbf{u}_g^0 = f^{-1}\mathbf{k} \times \nabla\, \delta\Phi, \quad \delta\Phi = \Phi_1 - \Phi_0. \tag{4.32}$$

The integration of the hydrostatic equation $\partial\Phi/\partial p = -RT/p$ between p_0 and p_1 yields an expression for the relative geopotential

$$\delta\Phi = R\bar{T} \ln (p_0/p_1), \tag{4.33}$$

where $\bar{T}(x, y)$ is the averaged temperature for the layer (p_0, p_1). Then

$$\mathbf{u}_T = \frac{R}{f} \ln \frac{p_0}{p_1} \mathbf{k} \times \nabla\bar{T}. \tag{4.34}$$

Evidently, $(\mathbf{u}_T \cdot \nabla\bar{T}) = 0$, i.e., the thermal wind 'blows' parallel to the isotherms of the averaged temperature $\bar{T}$. One must keep in mind, however, that $\mathbf{u}_T$ is not a real wind – according to (4.34) it is a measure of the horizontal temperature gradients (inhomogeneities), re-evaluated in velocities. On the other hand $\mathbf{u}_T \neq 0$ means a baroclinic atmosphere. Thus $\nabla\bar{T}$ is a measure of baroclinicity. The areas of strong horizontal gradients ∇T (fronts, jet streams, etc.) are also areas of baroclinicity concentration.

The rotation of the vector $\mathbf{u}_g(z)$ with the height is related to the temperature advection type in the layer under consideration (Figure 3.2). It can be determined by means of aerological data in a single station since $\mathbf{u}(z) \approx \mathbf{u}_g(z)$ in the free atmosphere. The advective changes of temperature can be calculated in the following way

$$\frac{\partial T}{\partial t} = -(\mathbf{u} \cdot \nabla T) \approx -(\mathbf{u}_g \cdot \nabla T). \tag{4.35}$$

Eliminating ∇T through (4.22) or (4.28) we find

$$\frac{\partial T}{\partial t} = -\frac{fT}{g}\left(\mathbf{u}_g \times \frac{\partial}{\partial z}\mathbf{u}_g\right) \cdot \mathbf{k} = -\frac{fT}{g}\left(u_g \frac{\partial}{\partial z} v_g - v_g \frac{\partial}{\partial z} u_g\right). \tag{4.36}$$

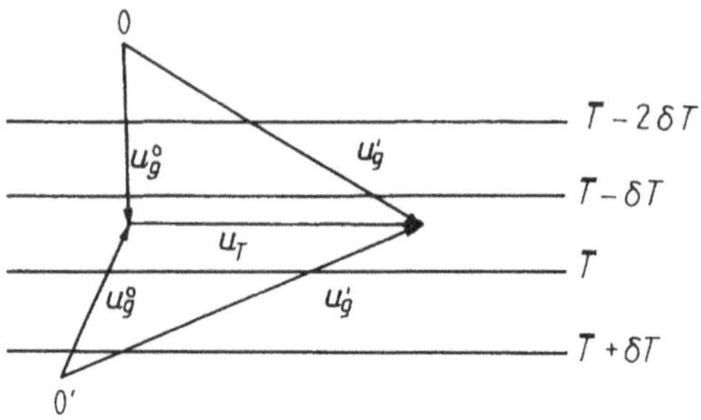

Fig. 3.2. Thermal wind and geostrophic heat advection.

Let

$$u_g = c_g \cos \delta, \qquad v_g = c_g \sin \delta, \quad c_g = |\mathbf{u}_g|,$$

where $\delta(z)$ is the angle between $\mathbf{u}_g$ and the $0x$ axis, oriented along $\mathbf{u}_g^0$. Then

$$\frac{\partial T}{\partial t} = -\frac{fT}{g} c_g^2 \frac{\partial \delta}{\partial z}. \tag{4.37}$$

It follows from here that at $\partial\delta/\partial z > 0$ (the left rotation of $\mathbf{u}_g$ with height) cold advection will be observed and vice-versa (Figure 3.2).

5. VORTICITY, DIVERGENCE AND BALANCE EQUATIONS

Having written the equations of motion in one or another coordinate system, one can derive from them new equations about differential or some other motion characteristics. The new equations, in principle, may have advantages and shortcomings different from those of the initial equations. Examples will be discussed below.

a. The Vorticity Equation

In order to derive this equation let us again write Equation (2.9) for the horizontal motion with $\alpha = 1/\rho$

$$\frac{\partial \mathbf{u}}{\partial t} + (\mathbf{u} \cdot \nabla)\mathbf{u} + w \frac{\partial \mathbf{u}}{\partial z} = -\alpha \nabla p - f\mathbf{k} \times \mathbf{u}. \tag{5.1}$$

But according to Equation (8.6) in Chapter 1,

$$(\mathbf{u} \cdot \nabla)\mathbf{u} = \nabla(\tfrac{1}{2}\, \mathbf{u} \cdot \mathbf{u}) + \zeta\mathbf{k} \times \mathbf{u}. \tag{5.2}$$

Then Equation (4.1) takes the form

$$\frac{\partial \mathbf{u}}{\partial t} = -\nabla(\tfrac{1}{2}\, \mathbf{u} \cdot \mathbf{u}) - w \frac{\partial \mathbf{u}}{\partial z} - \alpha \nabla p - (\zeta + f)\mathbf{k} \times \mathbf{u}. \tag{5.3}$$

Since the vertical component of the vorticity $\zeta = \mathbf{k} \cdot (\nabla \times \mathbf{u})$ and $\nabla \times \nabla a = 0$ (a is a scalar quantity), then it follows from (4.3) that

$$\frac{\partial \zeta}{\partial t} + \mathbf{u} \cdot \nabla(\zeta + f) + w \frac{\partial \zeta}{\partial z}$$

$$= -(\zeta + f)\nabla \cdot \mathbf{u} + \mathbf{k} \cdot \left(\frac{\partial \mathbf{u}}{\partial z} \times \nabla w\right) - \mathbf{k} \cdot (\nabla \alpha \times \nabla p). \tag{5.4}$$

But $f = f(y)$ and $\mathrm{d}f/\mathrm{d}t = v\, \mathrm{d}f/\mathrm{d}y = \beta v$. Then the left-hand side of (5.4) may be written as $\mathrm{d}\zeta_a/\mathrm{d}t = \dot{\zeta}_a$ where $\zeta_a = \zeta + f$ is the absolute vorticity. In scalar notations, Equation (5.4) reads

$$\frac{\mathrm{d}\zeta}{\mathrm{d}t} + \beta v = -\zeta_a(u_x + v_y) - (w_x v_z - w_y u_z) - (\alpha_x p_y - \alpha_y p_x), \tag{5.5}$$

where the subscripts x, y, z, as usual, denote partial derivatives.

The vorticity equation has a simpler mathematical form in isobaric p -variables. Actually, according to (3.5)

$$\frac{\mathrm{d}\mathbf{u}}{\mathrm{d}t} = -\nabla_p \Phi - f\mathbf{k} \times \mathbf{u}. \tag{5.6}$$

Hence, instead of (5.4), one obtains

$$\frac{\mathrm{d}\zeta_a}{\mathrm{d}t} = -\zeta_a \nabla_p \cdot \mathbf{u} + \mathbf{k} \cdot \left(\frac{\partial \mathbf{u}}{\partial p} \times \nabla_p w_p\right), \tag{5.7}$$

where $\zeta_a = \zeta_p + f$, $\zeta_p = \mathbf{k} \cdot (\nabla_p \times \mathbf{u})$. Obviously, $\zeta_p \neq \zeta$ because all horizontal derivatives are evaluated at constant pressure. Following the rules for differentiation ((6.10) in Chapter 2), at $q \equiv p$, $A = u$, v, consecutively, and $\partial p/\partial z = -\rho g$ it can be shown that

$$\zeta_p = \zeta - \mathbf{k} \cdot \left(\nabla_p \Phi \times \frac{\partial \mathbf{u}}{\partial p}\right) \rho. \tag{5.8}$$

At synoptic scale motion, the second term is always much smaller so that $\zeta_p \approx \zeta$. Equations (5.4) and (5.7) coincide in form only in the case of a horizontally-barotropic atmosphere $-\mathbf{k} \cdot (\ \alpha \times\ \ p) = 0$. The former, however, is easier for interpretation. It states that the rate of change of the absolute vorticity following the motion is given by the sum of the three terms on the right:

(i) *The divergence term*. Suppose that the other two terms vanish. Then $\dot{\zeta}_a = -\zeta_a \nabla \cdot \mathbf{u}$. At horizontal divergence ($\nabla \cdot \mathbf{u} > 0$) the signs of ζ_a and $\dot{\zeta}_a$ are opposite, while at horizontal convergence ($\nabla \cdot \mathbf{u} < 0$) they coincide (note that in the middle latitudes $\zeta \sim \pm 10^{-5}\ \mathrm{s}^{-1}$ so that $\zeta_a > 0$). But according to formula (4.2) in Chapter 1, $\nabla \cdot \mathbf{u} = (1/\delta S)\, \mathrm{d}\delta S/\mathrm{d}t$ where δS may be interpreted as a cross-section of an elementary vortex tube with a horizontal plane. Then the equality

$$\frac{1}{\zeta_a} \frac{\mathrm{d}\zeta_a}{\mathrm{d}t} + \frac{1}{\delta S} \frac{\mathrm{d}\delta S}{\mathrm{d}t} = 0$$

means that

$$\frac{d}{dt}(\zeta_a \cdot \delta S) = 0, \tag{5.9}$$

i.e., the vortex intensity is conserved following the motion. This mechanism for changing vorticity is very important in synoptic scale disturbances.

(ii) *The tilting or twisting term* $(w_x v_z - w_y u_z)$. This describes the generation of vertical vorticity $\zeta = v_x - u_y$ due to the interaction of the horizontal vorticities $\xi = w_y - v_z$ and $\eta = u_z - w_x$ (components of $\nabla \times \mathbf{v}$) with the horizontally nonhomogeneous field of vertical velocity ($\nabla w \neq 0$). Actually, as can be seen, $w_x v_z - w_y u_z = \xi w_x + \eta w_y$. Geometrically, this means that a horizontally-varying vertical motion tends to tilt the vorticity about the horizontal axes into the vorticity about the vertical axes. Conversely, vertical shear of the horizontal velocity $\mathbf{u}$ creates vorticity ξ or η by tilting vertical vortices. The contribution of this factor may be positive or negative.

(iii) *The baroclinic (solenoidal) term*. With the help of the equation of state $\alpha = RT/p$ and the quasi-geostrophic relationships (2.12), it can be written in the form

$$J(p, \alpha) = \mathbf{k} \cdot (\nabla p \times \nabla \alpha) = \frac{R}{p}(\nabla p \times \nabla T) = \frac{f}{T}(\mathbf{u}_g \cdot \nabla T). \tag{5.10}$$

It is seen that in the case of a barotropic atmosphere (zero geostrophic temperature advection), this term vanishes. Otherwise, the sign of $J(p, \alpha)$ may be plus or minus.

The baroclinic term is missing explicitly in the corresponding Equation (5.7). It can be shown that it is almost assimilated by the divergence term $\zeta_a \nabla_p \cdot \mathbf{u}$ (see Problem 1). This fact must be kept in mind when interpreting Equation (4.7). However, the simplest form the vorticity equation takes is in the θ-isentropic coordinate.

Actually, starting from Equations (3.38)–(3.40) with $w_\theta = d\theta/dt = 0$ for the isentropic (adiabatic) processes one obtains

$$\frac{d}{dt}(\zeta_\theta + f) = -(\zeta_\theta + f)\,\nabla_\theta \cdot \mathbf{u}. \tag{5.11}$$

without difficulty. Thus, in this case, the vorticity over the isentropic surfaces will change only through horizontal convergence (compare with (5.4) and (5.7)).

b. Vorticity Conservation Laws

Let us return to Equation (5.9). This holds provided that $J(p, \alpha) = 0$ (barotropic atmosphere) and $\mathbf{u}_z \times \nabla w = 0$ which implies the existence of layers without shear ($\partial \mathbf{u}/\partial z = 0$). If, in addition, the incompressibility condition holds $\delta S = \text{const}$, then we obtain

$$\frac{d\zeta_a}{dt} = 0 \quad \text{or} \quad \zeta + f = \text{const}, \tag{5.12}$$

i.e., in a barotropic, incompressible atmosphere the absolute vorticity $\zeta_a = \zeta + f$ conserves following the motion. Similarly, if $\partial \mathbf{u}/\partial p = 0$, then Equation (5.7) yields

$$\frac{d}{dt}(\zeta_a \cdot \delta S_p) = 0, \tag{5.13}$$

where $\zeta_a = \zeta_p + f$ and δS_p is the cross-section of an elementary vortex tube with the isobaric surface p = const. Again if δS_p = const, then $\zeta_p + f$ = const following the motion of an air parcel over the surface p = const. Finally, if $\nabla_\theta \cdot \mathbf{u} = 0$, Equation (5.11) yields $\zeta_\theta + f$ = const directly over the isentropic surfaces.

It is remarkable that under some conditions the vorticity equation admits a new quantity ζ_* which is more conservative than vorticity itself. It is called the *potential vorticity*. Let us now consider a few modifications of the potential vorticity conservation theorem.

(i) *Barotropic incompressible atmosphere*

$$\frac{\mathrm{d}_2\zeta_a}{\mathrm{d}t} = \zeta_a \frac{\partial w}{\partial z}, \quad \zeta_a = \zeta + f. \tag{5.14}$$

Due to barotropy, $\mathbf{u}$ and ζ do not depend on z. Integrating (5.14) between two material surfaces $z_1(x, y, t)$ and $z_2(x, y, t)$ and having in mind that $w_i = \mathrm{d}z_i/\mathrm{d}t$, $i = 1, 2$, we obtain

$$\frac{1}{\zeta_a}\frac{\mathrm{d}_2}{\mathrm{d}t}\zeta_a = \frac{1}{\Delta}\frac{\mathrm{d}_2}{\mathrm{d}t}\Delta \quad \text{or} \quad \frac{\mathrm{d}_2}{\mathrm{d}t}\left(\frac{\zeta_a}{\Delta}\right) = 0,$$

where $\Delta = z_2 - z_1$. Hence, it follows that

$$\zeta_* = \frac{\zeta(x, y, t) + f(y)}{\Delta(x, y, t)} = \text{const}, \tag{5.15}$$

i.e., the potential vorticity ζ_* remains constant, following the motion of an air parcel. In a particular case, z_1 may coincide with the Earth's surface – $z_1 = H(x, y)$.

(ii) *Barotropic, compressible and hydrostatic atmosphere*

$$\frac{\mathrm{d}_2\rho}{\mathrm{d}t} = -\rho\nabla\cdot\mathbf{u} - \frac{\partial\rho w}{\partial z}, \tag{5.16}$$

where $\mathbf{u} = \mathbf{u}(x, y, t)$ does not depend on z (or p). Upon integrating (5.16) and noting that

$$\int_0^\infty \rho\,\mathrm{d}z = -\frac{1}{g}\int_{p_0}^0 \mathrm{d}p = \frac{p_0}{g} \quad \text{and} \quad (\rho w) = 0 \text{ at } z = 0, \infty,$$

we obtain

$$\nabla\cdot\mathbf{u} = -\frac{1}{p_0}\frac{\mathrm{d}_2 p_0}{\mathrm{d}t}, \tag{5.17}$$

where $p_0 = p(x, y, 0, t)$ is the surface pressure. On the other hand, (5.14) can be written as

$$\nabla\cdot\mathbf{u} = -\frac{\partial w}{\partial z} = -\frac{1}{\zeta_a}\frac{\mathrm{d}_2\zeta_a}{\mathrm{d}t}. \tag{5.18}$$

Equating the right-hand sides of (5.17) and (5.18) we obtain

$$\frac{\mathrm{d}_2}{\mathrm{d}t}\left(\frac{\zeta_a}{p_0}\right) = 0 \quad \text{or} \quad \zeta_* = \frac{\zeta(x, y, t) + f(y)}{p(x, y, 0, t)} = \text{const}. \tag{5.19}$$

(iii) *Baroclinic, adiabatic hydrostatic atmosphere*. It is convenient to start from the vorticity equation (5.11) in the θ-coordinate

$$\nabla_\theta \cdot \mathbf{u} = -(\zeta_\theta + f)^{-1} \frac{\mathrm{d}_2(\zeta_\theta + f)}{\mathrm{d}t}. \tag{5.20}$$

Due to the adiabaticity $w_\theta = \mathrm{d}\theta/\mathrm{d}t = 0$ and the continuity equation (3.42) becomes

$$\left(\frac{\partial}{\partial t} + \mathbf{u} \cdot \nabla_\theta\right) \frac{\partial p}{\partial \theta} + \frac{\partial p}{\partial \theta} \nabla_\theta \cdot \mathbf{u} = 0.$$

Hence

$$\nabla_\theta \cdot \mathbf{u} = \left(\frac{\partial p}{\partial \theta}\right)^{-1} \frac{\mathrm{d}_2(\partial p/\partial \theta)}{\mathrm{d}t}. \tag{5.21}$$

Equating the right-hand sides of (5.20) and (5.21), we obtain

$$\zeta_* = (\zeta_\theta + f) \frac{\partial \theta}{\partial p} = \text{const.} \tag{5.22}$$

Let us remember that in this case θ = const following the motion.

It is seen that the three slightly different mathematical expressions for potential vorticity (5.15), (5.19) and (5.22) are, in some sense, a measure of the ratio of the absolute vorticity to the effective depth of the vortex – Δ in (5.15), $p_0 = p_0 - p_\infty$ where $p_\infty = 0$ in (5.19) and $\partial\theta/\partial p = (\partial p/\partial\theta)^{-1}$. At a fixed finite value of $\delta\theta$, (5.22) and (5.21) will read

$$\frac{\zeta_\theta + f}{\delta p} = \text{const}, \qquad \frac{1}{\delta p} \frac{\mathrm{d}_2}{\mathrm{d}t} (\delta p) = -\nabla_\theta \cdot \mathbf{u} \tag{5.23}$$

(compare with (5.19) and (5.17)). As a conservative quantity following the motion, the potential vorticity can be used as a tracer to tag air parcels. The corresponding conservation laws, derived above, represent powerful constraints on the large-scale atmospheric motions.

In the most general case of motion described by the set of equations

$$\begin{aligned} &\frac{\mathrm{d}\mathbf{v}}{\mathrm{d}t} = -\alpha\nabla p - 2\omega \times \mathbf{v} + \mathbf{g} + \mathbf{F}, \\ &\frac{\mathrm{d}\alpha}{\mathrm{d}t} = \alpha\nabla \cdot \mathbf{v}, \quad \left(\alpha = \frac{1}{\rho}\right), \\ &\frac{\mathrm{d}s}{\mathrm{d}t} = \frac{Q}{T}, \quad (s = c_p \ln \theta), \end{aligned} \tag{5.24}$$

Obukhov (1962, 1964) derived the so-called potential vorticity evolution equation

$$\frac{\mathrm{d}}{\mathrm{d}t} (\alpha\Omega_a \cdot \nabla s) = \nabla \cdot \mathbf{A}, \quad \mathbf{A} = \frac{Q}{T} \Omega_a + s\nabla \times \mathbf{F}, \tag{5.25}$$

where $\mathbf{F}$ is an additional (say, frictional) force, $\Omega_a = \nabla \times \mathbf{v} + 2\boldsymbol{\omega}$ is the three-dimensional vector of the absolute vorticity, and $\mathrm{d}/\mathrm{d}t$ and ∇ are also three-dimensional operators. Thus, in this case,

$$\Omega_* = \alpha \Omega_a \cdot \nabla s \tag{5.26}$$

stands for a potential vorticity. Evidently

$$\frac{\mathrm{d}\Omega_*}{\mathrm{d}t} = 0 \quad \text{at } \nabla \cdot \mathbf{A} = 0, \tag{5.27}$$

which is the general condition for conservation of Ω_*. Two decades earlier, Ertel (1942) derived the conservation law $\Omega_* = \text{const}$ for the case of adiabatic ($Q = 0$) and nonviscous flow ($\mathbf{F} = 0$), known as Ertel's theorem.

c. Simplification of the Vorticity Equation

We are now in a position to considerably simplify the vorticity equation. With the aid of (5.10) and (4.12), we can write

$$\begin{array}{cccccc}
\dfrac{\partial \zeta}{\partial t} & -\ \mathbf{u}\cdot\nabla\zeta\ - & \beta v & -\ w\dfrac{\partial \zeta}{\partial z}\ - & (\zeta + f)\,\nabla\cdot\mathbf{u} & + \\
V^2L^{-2} & V^2L^{-2} & \beta V & \dfrac{WV}{ZL} & (VL^{-1} + f)\,\mathrm{Ro}\,VL^{-1} & \\
1 & 1 & \dfrac{B}{\mathrm{Ro}} \sim 1 & \mathrm{Ro} & (1 + \mathrm{Ro}^{-1})\,\mathrm{Ro} &
\end{array}$$

$$\begin{array}{ccc}
+\ \mathbf{k}\cdot & \left(\dfrac{\partial \mathbf{u}}{\partial z} \times \nabla w\right) & +\left(\dfrac{f}{T}\right)(\mathbf{u}_g \cdot \nabla T) \\
 & \dfrac{WV}{ZL} & \left(\dfrac{M_T}{T}\right)\left(\dfrac{fV}{L}\right) \\
 & \mathrm{Ro} & n_T\,\mathrm{Ro}
\end{array} \tag{5.28}$$

where $B = \beta L/f \sim 10^{-1} \sim \mathrm{Ro}$. The second and the third lines indicate the order of each term before and after multiplication by L^2V^{-2}. Assuming $M_T \sim 10°\mathrm{C}$ and $T \sim 300\mathrm{K}$ we evaluate $n_T = M_T/T \sim 3 \times 10^{-2}$. The analysis of (5.28) yields:

(i) It is justified to let $(\zeta + f)\,\nabla \cdot \mathbf{u} \approx f\,\nabla \cdot \mathbf{u}$.

(ii) The smallest term is the last (barotropic) one. It can always be neglected. The remaining terms, having an order of $\mathrm{Ro} \sim 10^{-1}$ can either be retained or neglected. In the maximum simplification approach we will have

$$\frac{\partial \zeta}{\partial t} = -\,\mathbf{u}\cdot\nabla\zeta - f\,\nabla\cdot\mathbf{u} - \beta v \tag{5.29}$$

or

$$\frac{\mathrm{d}_2}{\mathrm{d}t}(\zeta + f) = -\,f\,\nabla\cdot\mathbf{u} = f\,\frac{\partial w}{\partial z} \tag{5.30}$$

where the continuity equation $w_z = -\nabla \cdot \mathbf{u}$ has been used.

(iii) A remarkable feature of Equation (5.29), as well as of (5.28), is that the left-hand side $\partial\zeta/\partial t$ and the main terms on the right (all retained in (5.29)) have one and the same order of magnitude. This was not the case with the equations of motion (5.1). One of the most important periods in the theoretical development of meteorological forecasting is based on this feature of the vorticity equation and will be discussed in Chapter 7.

d. Divergence and Balance Equations

Similar to (5.8),

$$\nabla_p \cdot \mathbf{u} = \nabla_z \cdot \mathbf{u} - \left(\nabla_z p \cdot \frac{\partial \mathbf{u}}{\partial p}\right)$$

and the second term is negligibly small, so that we can simply write $\nabla \cdot \mathbf{u}$ and denote it by D. Let us derive an equation for D. For that purpose, we write (5.6) with the aid of (5.2) as

$$\frac{\partial \mathbf{u}}{\partial t} = -\nabla(\Phi + \tfrac{1}{2}c^2) - \zeta_a \mathbf{k} \times \mathbf{u} - w_p \frac{\partial \mathbf{u}}{\partial p} \tag{5.31}$$

where $c = |\mathbf{u}|$, $\zeta_a = \zeta + f$. Upon applying the operator ∇ we obtain

$$\frac{\partial D}{\partial t} = -\nabla^2(\Phi + \tfrac{1}{2}c^2) - \nabla \cdot (\mathbf{k} \times \mathbf{u}\zeta_a) - w_p \frac{\partial D}{\partial p} - \nabla w_p \cdot \frac{\partial \mathbf{u}}{\partial p} \tag{5.32}$$

which is the required equation.

Identical transformations in (5.23) or direct application of ∇ to Equation (5.6) yield the following alternative forms of the divergence equation

$$\frac{\mathrm{d}D}{\mathrm{d}t} + \nabla w_p \cdot \frac{\partial \mathbf{u}}{\partial p} + D^2 - 2\mathbf{k} \cdot (\nabla u \times \nabla v) = -\nabla^2\Phi + f\zeta - \beta u \tag{5.33}$$

or

$$\frac{\mathrm{d}D}{\mathrm{d}t} + \nabla w_p \cdot \frac{\partial \mathbf{u}}{\partial p} + \nabla u \cdot \frac{\partial \mathbf{u}}{\partial x} + \nabla v \cdot \frac{\partial \mathbf{u}}{\partial y} = -\nabla^2\Phi + f\zeta - \beta u, \tag{5.34}$$

where

$$\begin{aligned} \frac{\mathrm{d}D}{\mathrm{d}t} &= \frac{\partial D}{\partial t} + \mathbf{u} \cdot \nabla D + w_p \frac{\partial D}{\partial p}, \\ D^2 - 2\mathbf{k} \cdot (\nabla u \times \nabla v) &= u_x^2 + v_y^2 + 2u_y v_x, \end{aligned} \tag{5.35}$$

$u_x = \partial u/\partial x$, etc., but $w_p = \dot{p} = \mathrm{d}p/\mathrm{d}t$. The scale analysis of (5.33) or (5.34) shows that, contrary to vorticity equation (5.28), the main terms are the first two on the right-hand side, so that to a first approximation we can write

$$0 \approx -\nabla^2\Phi + f\zeta \quad \text{or} \quad \zeta \approx \frac{\nabla^2\Phi}{f}. \tag{5.36}$$

Evidently, this is equivalent to the quasi-geostrophic relationship $\mathbf{u} \approx \mathbf{k} \times \nabla\Phi/f$ at $f =$ const.

In further approximation the smallest terms, namely dD/dt and $\nabla w_p \cdot \partial \mathbf{u}/\partial p$, should be neglected. Thus, the so-called *balance equation* is obtained

$$u_x^2 + v_y^2 + 2u_y v_x = -\nabla^2 \Phi + f\zeta - \beta u. \tag{5.37}$$

If, in addition, the motion is nondivergent ($D = 0$), one can introduce streamfunction $\psi_y = -u$, $\psi_x = v$, so that Equation (5.37) takes the form

$$f\nabla^2 \psi + 2(\psi_{xx}\psi_{yy} - \psi_{xy}^2) + \beta\psi_y = \nabla^2 \Phi. \tag{5.38}$$

As seen, this equation, as well as (5.36), is a diagnostic one (the time derivative is missing). It can be used to check the independently-observed velocity and geopotential fields for consistency or to determine ψ if Φ is known from the observations. In the latter case, (5.38) is a highly nonlinear partial differential equation of a known type (Monge–Ampère equation). Nevertheless, its solution is difficult to handle.

6. GRADIENT WIND AT CURVILINEAR ISOBARS

a. Natural Coordinates

Let us consider Equation (5.11) in Chapter 2,

$$\frac{c^2}{R_s} + fc = -\alpha \frac{\partial p}{\partial n} = fc_g, \tag{6.1}$$

where $c_g = -(\alpha/f)\partial p/\partial n$ is the geostrophic wind velocity in natural coordinates (c is the real wind velocity). Let the radius of curvature R_s = const, i.e., the streamlines (coinciding with the isobars and the trajectories) are circuits. Then Equation (6.1) can be written as

$$\frac{c_g}{c} = 1 + \mathrm{Ro}, \quad \mathrm{Ro} = \frac{c}{fR_s}. \tag{6.2}$$

The Rossby number Ro may be positive or negative. Since $f > 0$ (Northern Hemisphere), for cyclones ($R_s > 0$) c_g is larger than c, while for $R_s < 0$, $c_g < c$. Because for the cyclones and anticyclones of the mid-latitudes Ro $\sim 10^{-1}$, the difference between c and c_g rarely exceeds to–15%. Two particular cases of (6.1) are of preliminary interest.

(i) Unform pressure field ($\partial p/\partial n = 0$). The equation $c^2/R_s + fc = 0$ has a nonzero root $R_s = -c/f < 0$, i.e., at $dc/dt = 0$ and $f(y)$ = const, the air particles follow circular paths with anticyclonic (negative, $R_s < 0$) curvature. Obviously, the motion is of an oscillatory, pure inertial type. The period of this oscillation is

$$T_i = -\frac{2\pi R_s}{c} = \frac{2\pi}{f} \quad \text{or} \quad \nu_i = T_i^{-1} = \frac{f}{2\pi}, \tag{6.3}$$

i.e., the Coriolis parameter f plays the role of inertial frequency. At $f \sim 10^{-4}$ s^{-1} and $c \sim 10$ m s^{-1} we find $T_i \approx 17$ h, $R_s \sim 10^2$ km. Though observed, the pure inertial oscillations are not of importance in the atmosphere. This is not the case in the ocean, where the current velocities are small and the radius of the inertial circles $R_s = -c/f$ may be much smaller than in the atmosphere, even in the equatorial areas. Observations confirm this conclusion.

(ii) Negligible Coriolis force (Ro ≫ 1). Since f and c vary little, this requires that the radius of curvature R_s should be very small. Then (6.1) yields an expression for the velocity of the *cyclostrophic wind.*

$$c = \left(-\alpha R_s \frac{\partial p}{\partial n}\right)^{1/2}. \tag{6.4}$$

Figure 3.3 illustrates a case when the flow, governed by Equation (6.4), may be either cyclonic or anticyclonic with low pressure in the center. The cyclostrophic flow may serve as a simple model for some meso- and microscale atmospheric vortices (tornadoes, dust devils, etc).

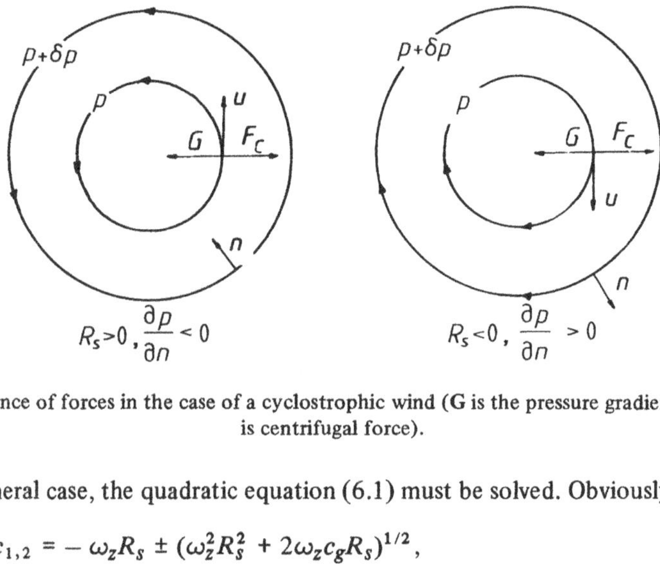

Fig. 3.3. Balance of forces in the case of a cyclostrophic wind (G is the pressure gradient force and F_c is centrifugal force).

In the general case, the quadratic equation (6.1) must be solved. Obviously

$$c_{1,2} = -\omega_z R_s \pm (\omega_z^2 R_s^2 + 2\omega_z c_g R_s)^{1/2}, \tag{6.5}$$

where $\omega_z = f/2 = \omega \sin \varphi$. The condition $c \to c_g$ at $R_s \to \infty$ requires only the plus sign to be retained. As an example of analysis with the aid of (6.5), we consider an anticyclone in the Northern Hemisphere ($R_s = -|R_s| < 0$, $\omega_z > 0$). Then

$$\frac{c}{\omega_z |R_s|} = 1 - \left(1 - \frac{2c_g}{\omega_z |R_s|}\right)^{1/2}. \tag{6.6}$$

Since $c > 0$, it then follows that

$$\omega_z |R_s| \geqslant 2c_g, \quad c \leqslant |R_s| \, \omega_z = c_{\max}, \tag{6.7}$$

i.e., at a given ω_z and c_g, the dimensions of the anticyclones must be limited from below ($|R_s| \geqslant R_{\min} = 2c_g/\omega_z$), while the velocity of the wind must be limited from above ($c \leqslant c_{\max}$). Qualitatively, these conclusions agree with observations. For cyclones, there are no such limitations.

b. Cartesian Coordinates

Let us consider the horizontal stationary motion ($\partial \mathbf{u}/\partial t = 0$) governed by the equations

$$uu_x + vu_y = -\Phi_x + fv, \qquad uv_x + vv_y = -\Phi_y - fu, \qquad u_x + v_y = 0. \tag{6.8}$$

Let $\Phi(x, y) = \Phi_\infty + \phi(x, y)$, where $\Phi_\infty = \Phi(\infty, \infty)$ while $\phi(x, y)$ is a localized disturbance in the geopotential field of the cyclonic ($\phi < 0$) or anticyclonic ($\phi > 0$) type. System (6.8) is closed.

The third equation ($u_x + v_y = 0$) implies a solution of the type (Chakalov and Panchev (1978))

$$u(x, y) = M(x)\, N'(y), \qquad v(x, y) = -M'(x)\, N(y), \tag{6.9}$$

where ($'$) = d/dx or d/dy. Inserting (6.9) into (6.8), we obtain

$$MM'(N'^2 - NN'') = -\phi_x - fM'N,$$

$$NN'(M'^2 - MM'') = -\phi_y - fMN'.$$

A cross-differentiation with respect to y and x at f = const and the subsequent elimination of ϕ yields

$$\frac{1}{MM'}(M'M'' - MM''') = \frac{1}{NN'}(N'N'' - NN''') = 4a^2 = \text{const}.$$

Consequently, the functions $M(x)$ and $N(y)$ satisfy one and the same kind of nonlinear, ordinary, differential equations

$$PP''' - P'P'' + 4a^2PP' = 0, \qquad P = M, N. \tag{6.10}$$

Under physically-reasonable conditions $M(\infty) = N(\infty) = 0$, the solutions of (4.10) are

$$M(x) = A \exp[-a^2(x - x_0)^2], \qquad N(y) = B \exp[-a^2(y - y_0)^2] \tag{6.11}$$

where A, B, x_0, y_0 are constants of integration. Hence, as well as from (6.9), for the velocity components we find

$$u(x, y) = -2a^2G(y - y_0) \exp[-a^2r^2], \qquad v(x, y) = -2a^2G(x - x_0) \exp[-a^2r^2] \tag{6.12}$$

where $G = A \cdot B$, $r^2 = (x - x_0)^2 + (y - y_0)^2$.

Having $u(x, y)$ and $v(x, y)$ determined by (6.12), each of Equations (6.8) can be integrated once to yield

$$\phi(x, y) = \Phi - \Phi_\infty = -a^2G^2 \exp[-2a^2r^2] - fG \exp[-a^2r^2] \tag{6.13}$$

Obviously, (6.13) describes a circle disturbance in the geopotential field. Its center is at the point (x_0, y_0).

Let us assume that $x_0 = y_0 = 0$, i.e., $r^2 = x^2 + y^2$. Then

$$\phi(0,0) = \phi_m = \max \begin{cases} < 0 & \text{for cyclone} \\ > 0 & \text{for anticyclone} \end{cases} \tag{6.14}$$

and ϕ_m could be considered as a given parameter. Then, the solutions (6.12) and (6.13) include only one integration constant – G. Letting $r = 0$ in (6.13), we obtain a quadratic equation for G:

$$G^2 + 2\omega_z L^2 G + \phi_m L^2 = 0, \quad L = \frac{1}{a},$$

with a solution

$$\frac{G}{\omega_z L^2} = -1 + \left(1 - \frac{\phi_m}{\omega_z^2 L^2}\right)^{1/2}, \tag{6.15}$$

ensuring $G \to 0$ at $L \to \infty$. In such a way ω_z and ϕ_m enter into (6.12) through G. In cyclones $\phi_m < 0$ and $G > 0$, while in anticyclones $\phi_m > 0$ and $G < 0$ (Figure 3.4). Since G is a real quantity, in the second case the condition

$$\phi_m \leqslant (\omega_z L)^2 = \phi_{\max}, \left(L^2 \geqslant \frac{\phi_m}{\omega_z} = L^2_{\min}\right) \tag{6.16}$$

is observed. If $\omega_z \to 0$, then $\phi_{\max} \to 0$, $L_{\min} \to \infty$, i.e., stationary anticyclones with a circular symmetry cannot exist in the equatorial atmosphere.

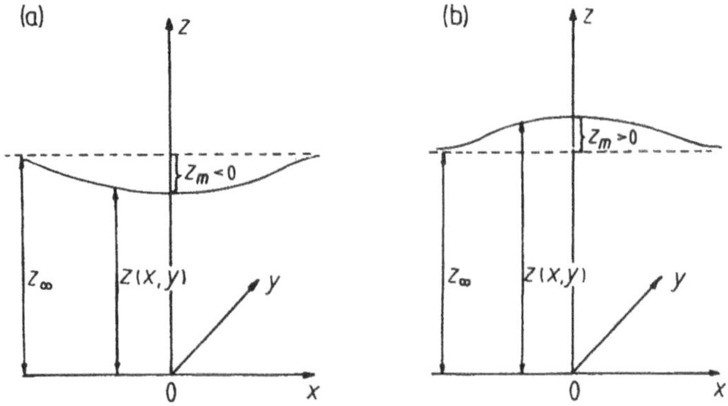

Fig. 3.4. Deformation of the isobaric surfaces in the case of a cyclone (a) and an anticyclone (b) ($\phi = gz$, $\phi_m = gz_m$).

The velocity field (6.12) is with nonzero vorticity

$$v_x - u_y = \zeta(r) = 4a^2 G(1 - a^2 r^2) \exp[-a^2 r^2]. \tag{6.17}$$

Hence, $\zeta_m = \zeta(0) = 4a^2 G$, so that $\zeta_m > 0$ in cyclones and vice-versa. Moreover, $\zeta(r_0) = 0$ at $r_0 = L = 1/a$, which clarifies the sense of the constant a. Besides,

$$c^2(r) = u^2 + v^2 = 4a^4 G^2 r^2 \exp[-a^2 r^2] \tag{6.18}$$

Obviously, $c(0) = c(\infty) = 0$ and $c = c_{\max}$ at $r = r_m = L/\sqrt{2}$. The geometry of the velocity field is presented in Figure 3.5 by isolines of u, v in their relative units.

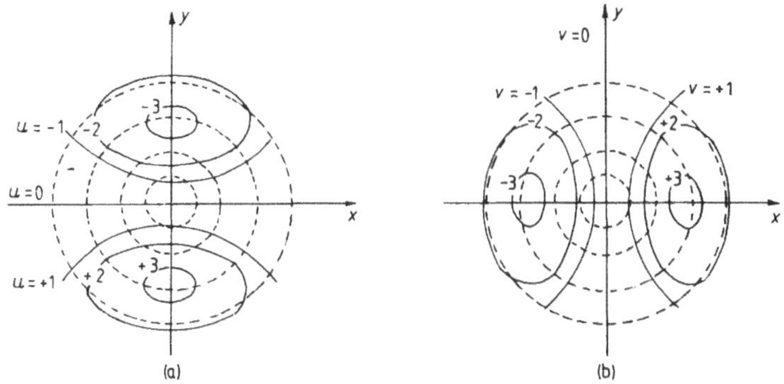

Fig. 3.5. Geometrical structure of the wind field in a cyclone ($G > 0$) corresponding to (6.12) (u and v are in relative units).

However, (6.12)–(6.13) is not the only solution of this type inherent to Equations (6.8). Actually, assuming

$$u(x, y) = -yF(r), \qquad v(x, y) = xF(r), \tag{6.19}$$

where $r^2 = x^2 + y^2$, by similar mathematical operations (see Chakalov and Panchev, 1977) one may obtain

$$\begin{aligned} \phi(x, y) &= \phi_m \exp[-a^2 r^2], \\ F(r) &= -\omega_z [1 - (1 - m \exp\{-a^2 r^2\})^{1/2}], \end{aligned} \tag{6.20}$$

where $m = 2a^2 \phi_m / \omega_z^2$ and for ϕ_m (6.14) holds.

This result is not surprising since Equations (6.8) admit a first integral of the $\nabla^2 \psi = A(\psi)$ type – an arbitrary function, ψ is the streamfunction such that at circular symmetry $\psi = B(r)$, where B is an arbitrary function of $r = (x^2 + y^2)^{1/2}$. Therefore, any particular form of the solution should be compared with observations in order to judge whether, or not, it provides a good mathematical approximation to the velocity and geopotential field.

7. PRESSURE–VELOCITY RELATIONSHIPS IN THE LOW-LATITUDES ATMOSPHERE

a. Linear Approach

The quasi-geostrophic formulae

$$u = -\frac{1}{\rho f} \frac{\partial p}{\partial y}, \qquad v = \frac{1}{\rho f} \frac{\partial p}{\partial x}, \tag{7.1}$$

theoretically motivated and practically confirmed, are very good approximations to the synoptic scale motions in the mid-latitude atmosphere. On approaching the tropical and equatorial areas, they become less and less precise and at the equator ($\varphi = 0$) they are not applicable, since $f = 0$. At the same time, although not so well manifested as in the mid-latitudes, the atmospheric pressure field is horizontally nonuniform – $\nabla p \neq 0$. As observations show, this nonuniformity is primarily meridional rather than along the parallels. In other words, the averaged pressure exhibits almost a zonal distribution – $\partial p/\partial x \approx 0$, $\partial p/\partial y \neq 0$. If Equations (7.1) are valid, one would expect the existence of an almost-zonal flow in the free atmosphere. Calculations show that for $f \sim 10^{-5}$ s^{-1} ($\varphi \sim 5$–$7°$) and the generally small observed values of $\partial p/\partial y$, the geostrophic relationship

$$u = -\frac{1}{\rho f}\frac{\partial p}{\partial y} \tag{7.1'}$$

gives reasonable values for $u \sim 5$–10 m s^{-1}.

Close to the equator, $y \to 0$ and $f = \beta_0 y$, where $\beta_0 = 2\omega/r_0$, r_0 is the Earth's radius. Assume that at the equator the pressure $p = p(y)$ has an extremum ($\partial p/\partial y = 0$) and that (7.1′) can be extended to $y = 0$. Then we can write

$$u = -\frac{1}{\rho\beta_0}\frac{\partial^2 p}{\partial y^2}\bigg|_{y=0}. \tag{7.2}$$

In case of maximum at the equator ($p_{yy} < 0$), this flow will be westerly ($u > 0$) and vice-versa. As a rough approximation to reality, this conclusion is true.

The geostrophic relationships (7.1) and (7.2) are very simple and useful for rough estimates but they do not reflect the real balance of the forces in the low-latitude atmosphere, where the nonlinear terms in the momentum equations become more and more important as the Coriolis force decreases. Thus, the dynamics of the equatorial atmosphere is essentially nonlinear and this fact must be taken into consideration in mathematical models.

b. Two-Dimensional Nonlinear Approach

Out of the narrow equatorial belt ($\varphi = \pm 5$–$7°$), as a next approximation to the problem, equations and methods of investigation still can be used which have been developed for the mid-latitude atmosphere. For example, if we ignore the vertical motions, we may use Equation (5.3) written in isobaric coordinates

$$\frac{\partial \mathbf{u}}{\partial t} = -\nabla(\Phi + E) - \zeta_a \mathbf{k} \times \mathbf{u} \tag{7.3}$$

where $E = c^2/2$ is the kinetic energy of the motion, $c = |u|$, $\zeta_a = \zeta + f$ and $\Phi + E = B$ is the total energy. Let $\partial \mathbf{u}/\partial t = 0$, then (7.3) can be solved with respect to $\mathbf{u}$ to yield

$$\mathbf{u} = \frac{1}{\zeta_a}\mathbf{k} \times \nabla B. \tag{7.4}$$

Compare (7.4) with the geostrophic relationship

$$\mathbf{u}_g = \frac{1}{f}\mathbf{k} \times \nabla\Phi. \tag{7.5}$$

Both expressions have identical mathematical structures. However, in principle, (7.4) may be applied at $f \to 0$, where $\zeta_a \to \zeta$, while (7.5) is inapplicable there. In the middle latitudes $O(\zeta/f) \sim \mathrm{Ro} \sim 10^{-1}$ and $\nabla B \approx \nabla\Phi$, so that (7.4) coincides with (7.5).

Incidentally, (7.4) is not a new relationship between the velocity and the geopotential field specific for the low-latitude atmosphere only. It can easily be shown that it is equivalent to the equation of balance (5.38)

$$f\nabla^2\psi + 2(\psi_{xx}\psi_{yy} - \psi_{xy}^2) + \beta_0\psi_y = \nabla^2\Phi. \tag{7.6}$$

The two first integrals can be written together as

$$(f + \nabla^2\psi)\,\nabla\psi = \nabla(\Phi + \tfrac{1}{2}\,|\nabla\psi|^2). \tag{7.7}$$

Since $\mathbf{u} = \mathbf{k} \times \nabla\psi$ and $\zeta = \nabla^2\psi$, it then immediately follows that (7.7) is equivalent to (7.4).

c. Three-Dimensional Nonlinear Approach

The vertical motions in the equatorial atmosphere play an extremely important role in its dynamics, making the motion essentially three-dimensional. In horizontal momentum equations, this role is reflected not only by the presence of the convective derivative $w\partial\mathbf{u}/\partial z$ but also by the second Coriolis term:

$$\frac{\mathrm{d}_2\mathbf{u}}{\mathrm{d}t} + w\frac{\partial\mathbf{u}}{\partial z} = -\alpha\nabla p - f\mathbf{k} \times \mathbf{u} - f_1 w\mathbf{i}, \tag{7.8}$$

where $f_1 = 2\omega\cos\varphi$, $\alpha = 1/\rho$.

Let us consider, following Dobrishman (1964), the simplest case of a stationary ($\partial\mathbf{u}/\partial t = 0$) and zonal ($\partial/\partial x = 0$) circulation of an incompressible ($\rho = \mathrm{const}$) and hydrostatic ($\mathrm{d}p = -\rho g\,\mathrm{d}z$) equatorial atmosphere in a standard coordinate system, located at the equator ($\varphi_0 = 0$). Thus

$$f_1 = 2\omega, \quad f = \beta_0 y,\ \beta_0 = \frac{2\omega}{r_0},\ \pi = \alpha p, \tag{7.9}$$

and in scalar form, the governing equations are

$$vu_y + wu_z = \beta_0 yv - 2\omega w, \tag{7.10}$$

$$vv_y + wv_z = -\pi_y - \beta_0 yu, \tag{7.11}$$

$$v_y + w_z = 0. \tag{7.12}$$

The first one implies a conservation of the angular momentum, since it can be written in the form

$$\frac{\mathrm{d}}{\mathrm{d}t}\left(u - \frac{\omega}{r_0}y^2 + 2z\omega\right) = 0, \tag{7.13}$$

where $\mathrm{d}/\mathrm{d}t = v\,\partial/\partial y + w\,\partial/\partial z$, $\partial/\partial t = 0$.

Let Z be the typical depth of the equatorial atmosphere (~ 15–20 km), L – the typical width of the equatorial belt (~ ± 500 km), and $U \equiv V$ and W – the corresponding velocity scales. In order that all three terms in (7.13) have one and the same order of magnitude, it is necessary that

$$L = (2r_0 Z)^{1/2}, \qquad U = 2\omega Z \tag{7.14}$$

Hence, and from the left-hand sides of (7.10)–(7.12), we obtain

$$W = U\frac{Z}{L} = \frac{2\omega}{\sqrt{2r_0}} Z^{3/2}. \tag{7.15}$$

Moreover, $O(\pi_y) = \beta_0 LU$. With this scaling, the governing equations take the form (we use the same notations)

$$vu_y + wu_z = yv - w, \qquad vv_y + wv_z = -yu - \pi_y,$$
$$v_y + w_z = 0. \tag{7.16}$$

The pressure field $\pi(y, z)$ given this system allows the determination of the velocity field (u, v, w). An idea for its solution will only be given here.

Obviously, a streamfunction can be introduced: $v = -\psi_z$, $w = \psi_y$. Then the first equation, according to (7.13), becomes

$$\left(-\frac{\partial\psi}{\partial z}\frac{\partial}{\partial y} + \frac{\partial\psi}{\partial y}\frac{\partial}{\partial z}\right)(u - \tfrac{1}{2}y^2 + z) = 0,$$

or

$$J(u - \tfrac{1}{2}y^2 + z, \psi) = 0. \tag{7.17}$$

Hence,

$$u - \tfrac{1}{2}y^2 + z = \delta(\psi), \tag{7.18}$$

where $\delta(\psi)$ is an arbitrary function. The simplest assumption concerning $\delta(\psi)$ would be $\delta(\psi) = u_0 = \text{const}$. Then

$$u(y, z) = u_0 + \tfrac{1}{2}y^2 - z, \tag{7.19}$$

where $u_0 = u(0, 0)$ is the surface wind velocity at the equator. Inserting (7.19) into (7.16), we obtain the fundamental equation of the problem

$$\psi_z\psi_{zy} - \psi_y\psi_{zz} = -y(\tfrac{1}{2}y^2 - z + u_0) - \pi_y, \tag{7.20}$$

which is a nonlinear, nonhomogeneous, partial differential equation. The choice of method for its solution (analytical or numerical) depends on the form of the function $\pi(y, z)$. In a number of cases, analytical solutions are possible [10].

8. LAGRANGIAN APPROACH TO THE PROBLEM OF SIMPLIFICATION

a. Middle Latitudes

The main difficulty in atmospheric dynamics is related to the nonlinearity of the momentum equations which, in the Eulerian formulation of the problems, are partial differential equations. Having this fact in mind, the Lagrangian treatment of the motion, governed by the same mathematical equations but with the particles (parcels) coordinates $x(t), y(t)$ as unknown functions, appears to be much more attractive. Then, the equations of motion

$$\frac{\mathrm{d}\mathbf{u}}{\mathrm{d}t} + f\mathbf{k} \times \mathbf{u} = -\alpha\nabla p = f\mathbf{k} \times \mathbf{u}_g = \mathbf{F}(x, y, t) \tag{8.1}$$

become *ordinary* differential equations ($\cdot = \mathrm{d}/\mathrm{d}t$)

$$\ddot{x} - f\dot{y} = F_1(x, y, t), \qquad \ddot{y} + f\dot{x} = F_2(x, y, t), \tag{8.2}$$

which is the principal simplification, although generally the equations remain nonlinear. The nonlinearity comes, first of all, from the arbitrary form of the functions $F_i(x, y, t)$ and secondly, because $f = f(y)$. Moreover, the system is of fourth order with respect to time t and needs four initial conditions specifying (x_0, y_0) and $u_0 = \dot{x}(0)$, $v_0 = \dot{y}(0)$. Then the solution of (8.2) would give the trajectories of the particles $x = x(t)$, $y = y(t)$ and they could then be compared with the observations.

The system of Equations (8.2) cannot be solved analytically with arbitrary functions F_1 and F_2. We shall consider here the case when

$$F_1 = b_1 + a_{11}x + a_{12}y, \qquad F_2 = b_2 + a_{21}x + a_{22}y, \tag{8.3}$$

where b_i and a_{ij} may depend on t and $f(y) = \text{const}$. Only two examples will be discussed in detail.

(i) Let $a_{ij} = 0$, $b_i(t) = b_i = \text{const}$. Then we have to solve the equations

$$\dot{u} - fv = b_1, \qquad \dot{v} + fu = b_2, \tag{8.4}$$

where $u = \dot{x}(t)$ and $v = \dot{y}(t)$ are the particle velocities. Integrating (8.4) we obtain

$$\begin{aligned} u(t) &= \left(u_0 - \frac{b_2}{f}\right)\cos ft + \left(v_0 + \frac{b_1}{f}\right)\sin ft + \frac{b_2}{f}, \\ v(t) &= \left(v_0 + \frac{b_1}{f}\right)\cos ft - \left(u_0 - \frac{b_2}{f}\right)\sin ft - \frac{b_1}{f}. \end{aligned} \tag{8.5}$$

One more integration would yield expressions for $x(t)$ and $y(t)$. It is seen from (8.5) that the particles move oscillatorily under the action of the constant-pressure gradient and inertial (Coriolis) forces. If $b_1 = b_2 = 0$, the oscillations will be purely inertial

$$\begin{aligned} x(t) &= x_0 + [u_0 \sin ft + v_0(1 - \cos ft)]f^{-1}, \\ y(t) &= y_0 + [v_0 \sin ft - u_0(1 - \cos ft)]f^{-1}. \end{aligned} \tag{8.6}$$

The trajectories are circuits (inertial circuits). Actually, eliminating t from (8.6) we obtain

$$(x - x_c)^2 + (y - y_c)^2 = R_i^2 \tag{8.7}$$

where

$$x_c = x_0 + \frac{v_0}{f}, \qquad y_c = y_0 - \frac{u_0}{f}, \qquad R_i = \frac{c_0}{f}, \; c_0^2 = u_0^2 + v_0^2. \tag{8.8}$$

This result was obtained earlier, in Section 6.

(ii) Let now choose the coefficients in (8.3) such that $u_g = \lambda y$, $v_g = -\lambda x$, i.e., $\zeta_g = \partial v_g/\partial x - \partial u_g/\partial y = -2\lambda = \text{const}$. Then we have to solve the system

$$\ddot{x} - f\dot{y} = f\lambda x, \qquad \ddot{y} + f\dot{x} = f\lambda y. \tag{8.9}$$

We introduce $z = x + iy$, $i = \sqrt{-1}$. Hence,

$$\ddot{z} + if\dot{z} - f\lambda z = 0.$$

This equation simplifies if we let $z(t) = \omega(t) \exp[\frac{1}{2}(ift)]$:

$$\ddot{\omega} + \mu^2 \omega = 0, \qquad \mu^2 = \frac{1}{4} f^2 - f\lambda. \tag{8.10}$$

Thus, if $\mu^2 > 0$, i.e.,

$$4\lambda < f \quad \text{or} \quad \zeta_g > -f/2 \tag{8.11}$$

the solution of (8.10) is periodic. Therefore, (8.11) is a condition for stability of the solution. Otherwise ($4\lambda > f$), the solution will be of an exponential type. But in the Northern Hemisphere, as a rule, $|\zeta|/f \approx |\zeta_g|/f \sim \text{Ro} \sim 0.1$, so that not only $f + \zeta_g$ but also $f + 2\zeta_g$ will be positive and the motion will be oscillatory.

b. Low Latitudes

Close to the equator $f \approx \beta_0 y$. Based on the discussions in Section 7, we can also assume that in (8.2) $F_1 = 0$ and $F_2 = F_2(y)$. Then the equations to be solved are

$$\ddot{x} - \beta_0 y\dot{y} = 0, \qquad \ddot{y} + \beta_0 y\dot{x} = F_2(y). \tag{8.12}$$

Integrating the first one, we obtain

$$\dot{x} = \frac{1}{2}\beta_0 y^2 - A, \tag{8.13}$$

where A is the constant of integration. Then the second equation takes the form

$$\ddot{y} - m^2 y + ny^3 = F_2(y), \tag{8.14}$$

where $m^2 = A\beta_0$, $n = \beta_0^2/2$.

The free right-hand side of (8.14) allows the analytical modelling of the pressure gradient force and the subsequent solution of this equation. Two cases are of special interest:

(i) $F_2 = 0$ – pure inertial motion. Then

$$\ddot{y} - m^2 y + n y^3 = 0, \tag{8.15}$$

known as Dufing's equation, has an exact solution through the elliptic functions of Jacobi, describing a purely periodic motion. Following Zui and Yudin (1972), we shall try to estimate the period of the oscillations only. For that purpose, let $\dot{x}(t) \approx \dot{x}(0) = u_0$ in the second equation of (8.12), which is a good approximation for the narrow equatorial zone. Then at $F_2 = 0$, we obtain the linear equation $\ddot{y} + u_0\beta_0 y = 0$. At $u_0 > 0$ (eastward motion), the solution of (8.16) is periodic with a period $T_i' = 2\pi/(u_0\beta_0)^{1/2}$.

Let us remember that in the middle latitudes $T_i = 2\pi/f \sim 1$ day. At $u_0 \sim 1$ m s^{-1} and $\beta_0 = 2.3 \times 10^{-11}$ m^{-1} s^{-1} we obtain $T_i' \sim 15$ days, while at $u_0 \sim 10$ m s^{-1}, $T_i' \sim 5$ days. Therefore, the pure inertial oscillations in the equatorial atmosphere (and ocean, too) are long-period (low-frequency) oscillations. Qualitatively, this conclusion is confirmed by observations.

(ii) $F_2(y) = a_1 y + a_2 y^3$. The motion is noninertial but governed by the same Equation (8.15), but with different parameters:

$$\ddot{y} - (m^2 + a_1)y + (n - a_2)y^3 = 0.$$

Thus, again, it is a periodic motion.

c. Global-Scale Oscillations

Let us now consider the horizontal inertial motions on the rotating spherical Earth. The governing equations can be obtained from (3.20)–(3.21) of Chapter 2 by letting

$$\begin{gathered} u = v_\lambda, \qquad v = -v_\theta, \qquad \varphi = \frac{\pi}{2} - \theta, \qquad w = 0, \\ u = r_0 \cos\varphi \frac{d\lambda}{dt}, \qquad v = r_0 \frac{d\varphi}{dt}. \end{gathered} \tag{8.16}$$

The result is

$$\frac{du}{dt} - \frac{uv}{r_0} \tan\varphi - 2\omega v \sin\varphi = 0, \tag{8.17}$$

$$\frac{dv}{dt} + \frac{u^2}{r_0} \tan\varphi + 2\omega u \sin\varphi = 0. \tag{8.18}$$

These equations are ordinary but highly nonlinear differential equations – the algebraic and trigonometric nonlinearities are present. Nevertheless, two first integrals can be found without difficulty:

$$u^2 + v^2 = u_0^2 + v_0^2 = c_0^2, \tag{8.19}$$

$$u \cos\varphi + \omega r_0 \cos^2\varphi = u_0 \cos\varphi_0 + \omega r_0 \cos^2\varphi_0, \tag{8.20}$$

expressing the energy (E) and angular momentum ($AM = (u + \omega r)r$, $r = r_0 \cos\varphi$) conservation. Actually, differentiating (8.20) we obtain (8.17).

Let us consider trajectories passing through the pole and accept $\cos\varphi_0 = 0$ as the initial condition. Then (8.19) and (8.20) yield

$$u = -\omega r_0 \cos\varphi, \qquad v^2 = c_0^2 - \omega^2 r_0^2 \cos^2\varphi, \tag{8.21}$$

under the condition that $\cos\varphi \leqslant c_0/\omega r_0 \leqslant 1$. Since $c_0 \sim 10\ \mathrm{m\,s^{-1}}$, the latter inequality is fulfilled. Thus, trajectories passing through the pole do not cross the equator ($\cos\varphi < 1$). Inserting (8.21) into (8.16) and integrating, we find $\lambda(t) = \lambda_0 - \omega t$ and $t = t(\varphi)$, which cannot be converted into $\varphi = \varphi(t)$. However, Equation (8.17) can be integrated exactly to produce the dependence $u = u(\varphi)$ (Ingel and Lepikash, 1981).

By replacing in (8.17) v by $r_0\,\mathrm{d}\varphi/\mathrm{d}t$, we obtain a linear equation with respect to u:

$$\frac{\mathrm{d}u}{\mathrm{d}t} - u\tan\varphi\cdot\frac{\mathrm{d}\varphi}{\mathrm{d}t} = 2\omega r_0 \sin\varphi\cdot\frac{\mathrm{d}\varphi}{\mathrm{d}t}. \tag{8.22}$$

Hence

$$u = \exp\left(\int_0^t \tan\varphi\frac{\mathrm{d}\varphi}{\mathrm{d}t'}\,\mathrm{d}t'\right)$$

$$\left[C_1 + 2\omega r_0 \int_0^t \sin\varphi\frac{\mathrm{d}\varphi}{\mathrm{d}t'}\exp\left(-\int_0^{t'}\tan\varphi\frac{\mathrm{d}\varphi}{\mathrm{d}t''}\,\mathrm{d}t''\right)\mathrm{d}t'\right], \tag{8.23}$$

where the constant C_1 can be determined from the initial condition $u = u_0$, $\varphi = \varphi_0$ at $t = 0$. But

$$\exp\left(\int_0^t \tan\varphi\frac{\mathrm{d}\varphi}{\mathrm{d}t'}\,\mathrm{d}t'\right) = \exp\left(\int_0^t \tan\varphi\,\mathrm{d}\varphi\right)$$

$$= \exp\left(-\ln\cos\varphi\,\Big|_0^t\right) = \frac{\cos\varphi_0}{\cos\varphi}$$

After some rearrangements, (8.23) takes the final form

$$u(\varphi) = \frac{\cos\varphi_0}{\cos\varphi}\left[u_0 - \frac{\omega r_0}{\cos\varphi_0}(\cos^2\varphi - \cos^2\varphi_0)\right] \tag{8.24}$$

which is consistent with (8.21) at $\cos\varphi_0 = 0$.

Now let us consider the equations

$$\begin{aligned} &\frac{\mathrm{d}u}{\mathrm{d}t} - 2\omega v\sin\varphi = 0,\\ &\frac{\mathrm{d}v}{\mathrm{d}t} + 2\omega u\sin\varphi = -\frac{1}{\rho r_0}\frac{\partial p}{\partial\varphi}, \end{aligned} \tag{8.25}$$

governing the noninertial motion on the rotating sphere in the linear approximation. Let us assume, following [37], that

$$p(\varphi) = \bar{p} + b \cos 2\varphi. \tag{8.26}$$

This expression approximates rather well the observed pressure distribution in the middle troposphere, which is responsible for changing the northward air flow at the level (~5 km) into eastward (zonal) in the middle latitudes. Obviously, $p(10^\circ) > p(45^\circ) > p(90^\circ)$.

Since $\mathrm{d}y = r_0 \,\mathrm{d}\varphi$, Equations (8.25) can also be written as

$$\begin{aligned} \ddot{x} &= 2\omega r_0 \dot{\varphi} \sin \varphi, \\ r_0 \ddot{\varphi} &= -2\omega \dot{x} \sin \varphi + \frac{2b}{\rho r_0} \sin 2\varphi. \end{aligned} \tag{8.27}$$

Under the condition $\dot{x} = 0$ at $\varphi = 0^\circ$, we obtain

$$\dot{x} = u(\varphi) = 2\omega r_0 (1 - \cos \varphi), \tag{8.28}$$

and substitute it into the second equation of (8.27). Bearing in mind that $v = y = r_0 \varphi$ and assuming $v = 0$ at $\varphi = 0^\circ$, we obtain the solution

$$\dot{y} = v(\varphi) = 2 \left[\sin^2 \varphi \left(\frac{b}{\rho} - \omega^2 r_0^2 \frac{1 - \cos \varphi}{1 + \cos \varphi} \right) \right]^{1/2}. \tag{8.29}$$

At a given amplitude b of the pressure meridional variations (8.26), one can find the critical value φ_c at which $v(\varphi_c) = 0$

$$\cos \varphi_c = \frac{\omega^2 \rho r_0^2 - b}{\omega^2 \rho r_0^2 + b}. \tag{8.30}$$

Solved for b, Equation (8.30) will give the necessary amplitude in (8.26) so that a given particle, starting from the equator, should reach the geographical parallel φ_c

$$b = \omega^2 \rho r_0^2 \frac{1 - \cos \varphi_c}{1 + \cos \varphi_c}. \tag{8.31}$$

Averaged data from January pressure maps at the 5 km level suggest values $b \approx 35$ mb, $\rho \approx 0.65$ kg m^{-3}. Then Equation (8.30) yields $\varphi_c \approx 18^\circ$N – the latitude at which the subtropical belt of high pressure begins, reaching its maximum at about 30°N. It is of dynamical origin – an accumulation of air masses draining away from the equatorial region.

9. SPECTRAL APPROACH TO THE PROBLEM OF SIMPLIFICATION

a. General Discussion

At the end of this chapter we shall return to our initial consideration – the need for simplifying weather equations and the methods of achieving this. The choice of simplification is dictated by the particular problem to which the equations are to be applied. All the methods fall into two groups:

(i) *Physical simplifications*. For example, omission of mountains and replacement of the ocean and land areas with a homogeneous underlying surface; introduction of an infinite or bounded ground surface instead of a spherical one; treating the atmosphere as a barotropic, incompressible ideal gas and utilization of the quasi-adiabatic approximation (sometimes replaced with Newtonian cooling); specifying in advance the large-scale motions on which the smaller-scale ones are superposed when we are interested in the latter; replacement of the vertical equation of motion with the hydrostatic equation and the horizontal momentum equation with the quasi-geostrophic relationships, etc. An extremely important physical simplification is the disregarding of the water vapour and liquid water as atmospheric constituents as well as the evaporation from the underlying surface as a fundamental process. In such a way we lost the possibility of studying the water-phase transition and the related phenomena: clouds and precipitations. Moreover, we cannot accurately specify the incoming and outgoing solar and terrestrial radiation or take into account the latent heat release. Thus, in this case the energetics of the atmosphere is highly influenced.

(ii) *Mathematical simplifications*. Inevitably, they accompany the physical simplifications. However, pure mathematical methods of simplification can also be mentioned. For example, the special choice of coordinate system (p, σ, θ, . . .); derivation of equations with desirable features (vorticity equation (5.30), for instance), allowing subsequent slight mathematical modifications of some terms in order to make the equations easily solvable (see Sections 2 and 3 of Chapter 7); linearization of originally nonlinear equations (see Chapter 5), etc. When computers became available, the well-established method of finite difference approximation of partial and ordinary differential equations received a powerful stimulus for further development, since the great number of algebraic equations, to which the original ones were reduced, could now be easily solved by computers. As a most spectacular result, a new discipline was born – numerical weather prediction, a child of dynamic meteorology and electronics (computer technics).

Although the most common method in meteorological theory and practice for solving partial differential equations governing the atmospheric processes during the past three decades has been the use of finite differences, other mathematical methods are possible too. An alternative one is the spectral expanding of the field of a dependent variable into a series of orthogonal functions, usually chosen to be eigenfunctions of some equation. The most familiar example of such a method of analysis in one dimension is the Fourier series in which the orthogonal functions are sines and cosines. The coefficients in the Fourier series become the new dependent variables.

Weather equations, as we know, are essentially nonlinear and the most prominent nonlinear terms are the advective ones which are quadratic, containing products of the advected quantities (velocity $\mathbf{v}$, temperature T, vorticity ζ, density ρ, etc.) with the advecting wind $\mathbf{v}$. They cannot be removed by any mathematical transformation of the independent or dependent variables. Let us point out that when radiation and condensation processes are included into the model, nonlinearities which are not quadratic appear.

The idea of the method, originally known as Galerkin's method, consists of the following: Let F be a function, say, of two independent variables x and t. We expand it into a series of eigenfunctions of x (say, sines) and a coefficient, depending on t, and substitute it into the equation. Due to its nonlinearity, an infinite set of ordinary nonlinear differential equations is obtained. Then we truncate this system by retaining a limited number of coefficients, thus closing the system. If a minimum number of equations for the coefficients are retained, we obtain the so-called *low order system*. Obviously, it corresponds to a maximum simplification of the original partial-differential equations. As we shall see, in some cases the low-order system is simple enough to be solved analytically. Anyway, it is much less time-consuming than the original system when numerical methods are applied.

b. Examples of Low-Order Systems

Much knowledge about the behaviour of the atmosphere has been gained by studies of greatly simplified models. Two of them will be discussed here in connection with their approximation by low order systems.

(i) One-dimensional advection equation

$$\frac{\partial u}{\partial t} + u\,\frac{\partial u}{\partial x} = 0. \tag{9.1}$$

Obviously, this is the simplest equation preserving some features of the complete momentum equations – quadratic nonlinearity and first-order time derivatives. According to Platzman (1964), its general solution in implicit form is

$$u = F(x - ut), \tag{9.2}$$

where $F(y)$ is an arbitrary function, determined by the initial x-profile $u_0 = F(x)$ at $t = 0$. For instance, if $u(x, 0) = -U \sin kx$, then $u(x, t) = -U \sin k(x - ut)$. Using U and k^{-1} as scaling parameters for velocity and length, we can write simply

$$u(x, t) = -\sin(x - ut). \tag{9.3}$$

At $t < 1$ (9.3) has a well-defined Fourier sine-series representation

$$-u(x, t) = \sum_{n=1}^{\infty} u_n(t) \sin nx, \tag{9.4}$$

where

$$u_n(t) = -\frac{1}{\pi}\int_{-\pi}^{\pi} u(x, t) \sin nx \, dx. \tag{9.5}$$

As shown by Platzman, (9.5) and (9.4) yield the unexpectedly simple result

$$u_n(t) = \frac{2J_n(nt)}{nt}, \tag{9.6}$$

where $J_n(y)$ is Bessel function. Therefore

$$u(x, t) = -2 \sum_{n=1}^{\infty} (nt)^{-1} J_n(nt) \sin nx \tag{9.7}$$

represents the explicit solution of (9.3).

If we now take the Fourier transformation of the original Equation (9.1) after substituting expression (9.4), the result will be an infinite system of nonlinear ordinary differential equations.

$$\frac{2}{n}\frac{du_n}{dt} = \frac{1}{2}\sum_{m=1}^{n-1} u_m u_{n-m} - \sum_{m=1}^{\infty} u_m u_{n+m}, \tag{9.8}$$

where $n = 1, 2, 3, \ldots$ Obviously, the simplest (low-order) system following from (9.8) is that which consists of two components u_1, u_2 only. Neglecting all u_s with $s > 2$, we obtain the truncated system

$$\frac{du_1}{dt} = -\tfrac{1}{2} u_1 u_2, \qquad \frac{du_2}{dt} = \tfrac{1}{2} u_1^2. \tag{9.9}$$

Handbooks on differential equations or direct computations convince us that the exact solution of (9.9) under initial conditions $u_1(0) = 1$, $u_2(0) = 0$ is

$$u_1(t) = \operatorname{sech}\frac{t}{2}, \qquad u_2(t) = \tanh\frac{t}{2}. \tag{9.10}$$

According to (9.6) the exact solutions for $u_n(t)$ are

$$u_1(t) = 2t^{-1} J_1(t), \qquad u_2(t) = t^{-1} J_2(2t). \tag{9.11}$$

For $t < 1$ the agreement between (9.10) and (9.11) is rather satisfactory.

(ii) Two-dimensional Rayleigh convection, governed by the equations

$$\frac{\partial \nabla^2 \psi}{\partial t} + J(\psi, \nabla^2 \psi) - \beta_0 g \frac{\partial \theta}{\partial x} = \nu \nabla^4 \psi, \tag{9.12}$$

$$\frac{\partial \theta}{\partial t} + J(\psi, \theta) - \left(\frac{\Delta T}{T}\right)\frac{\partial \psi}{\partial x} = \kappa_m \nabla^2 \theta, \tag{9.13}$$

and the boundary conditions

$$\psi = 0, \qquad \nabla^2 \psi = 0, \qquad \theta = 0 \quad \text{at } z = 0, H, \tag{9.14}$$

where $J = J_{xz}$ means the Jacobian, $\mathbf{\nabla} = (\partial/\partial x, 0, \partial/\partial z)$. They were first obtained by Saltzman (1962), starting from Equations (4.17), (4.19) and (1.18) of Chapter 2

$$\begin{gathered} \frac{\partial \mathbf{v}}{\partial t} + (\mathbf{v} \cdot \mathbf{\nabla})\mathbf{v} = -\alpha_s \mathbf{\nabla} p' + g\beta_0 T'\mathbf{k} + \nu \nabla^2 \mathbf{v}, \\ \mathbf{\nabla} \cdot \mathbf{v} = 0, \qquad \frac{\partial T'}{\partial t} + (\mathbf{v} \cdot \mathbf{\nabla})T' = \kappa_m \nabla^2 T'. \end{gathered} \tag{9.15}$$

Let us remember that T' denotes a temperature deviation from the standard (averaged over the volume) value T_s. The physical phenomenon to be described by Equations (9.15) is taken to be the following: A fluid layer of uniform depth H, extending in an horizontal direction to infinity, is confined between the lower and upper boundary (Figure 3.6). The layer is uniformly heated from below and a constant positive temperature difference

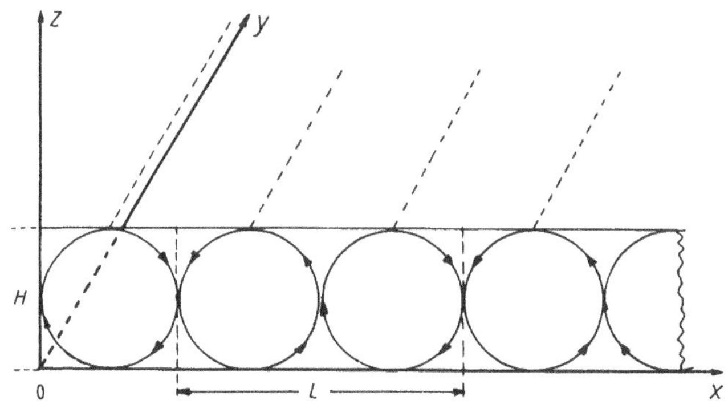

Fig. 3.6. Fluid motion in cylindrical cells at a slightly supercritical Rayleigh number.

$\Delta T = \bar{T}(H) - \bar{T}(0) > 0$ between the horizontally-averaged temperature is maintained. Due to this, an equilibrium vertical temperature profile $\bar{T}(z) = \bar{T}(0) - (\Delta T/H)z$ is created. Then

$$T'(x, z, t) = \bar{T}(z) + \theta(x, z, t). \tag{9.16}$$

Besides, $\psi_x = w$, $\psi_z = -u$, i.e., $\psi(x, z, t)$ denotes the streamfunction. To simplify the problem, the convective motion arising in the fluid layer is constrained to develop only in the form of two-dimensional 'rolls' in the x, z plane (i.e., $v = 0$, $\partial/\partial y = 0$) (Figure 3.6). The boundaries from below and above may be free or rigid (no-slip), or combined. The conditions (9.14) correspond to the free boundaries (vertical velocity and tangential stress vanishing – $w = \partial u/\partial z = 0$). Finally, the condition of cyclic continuity along the x-axis is imposed, with L being the period (Figure 3.6).

It is convenient to write system (9.12) and (9.13) in dimensionless variables

$$x' = \frac{x}{H}, \qquad z' = \frac{z}{H}, \qquad t' = \frac{\kappa_m}{H^2} t, \qquad \psi' = \frac{\psi}{\kappa_m},$$
$$\theta' = \frac{g\beta_0 H^3}{\nu\kappa_m}\theta, \qquad \nabla'^2 = H^2\nabla^2. \tag{9.17}$$

The result is (primes omitted)

$$\frac{\partial\nabla^2\psi}{\partial t} + J(\psi, \nabla^2\psi) - \sigma\frac{\partial\theta}{\partial x} - \sigma\nabla^4\psi = 0, \tag{9.18}$$

$$\frac{\partial\theta}{\partial t} + J(\psi, \theta) - R\frac{\partial\psi}{\partial x} - \nabla^2\theta = 0, \tag{9.19}$$

where

$$\sigma = \frac{\nu}{\kappa_m}, \qquad R = \frac{g\beta_0 H^3 \Delta T}{\nu\kappa_m}, \tag{9.20}$$

are the corresponding Prandtl and Rayleigh numbers.

Based on Rayleigh's classical studies on the convection phenomenon under consideration here and on his concept for maximum simplification of the original equations (in our case (9.18) and (9.19), Lorenz (1963) proposed to approximate $\psi(x, z, t)$ and $\theta(x, z, t)$ in the very beginning by the following truncated double Fourier series

$$a(1 + a^2)^{-1}\psi = \sqrt{2}X(\tau) \sin(\pi a x) \sin \pi z,$$
$$\pi R_c^{-1}\theta = \sqrt{2}\,Y(\tau) \cos(\pi a x) \sin \pi z - Z(\tau) \sin 2\pi z, \tag{9.21}$$

where

$$a = \frac{2H}{L}, \qquad R_c = \pi^4 a^{-2}(1 + a^2)^3, \qquad \tau = \pi^2(1 + a^2)t', \tag{9.22}$$

R_c is, in some sense, the critical value of R from (9.20). According to Lorenz's physical interpretation, X is proportional to the intensity of the convective motion, Y is proportional to the temperature difference between the ascending and descending currents (Figure 3.6), and Z is proportional to the distortion of the vertical temperature profile from linearity – $Z > 0$ means that the strongest gradients occur near the boundaries.

When (9.21) is substituted into (9.18) and (9.19) and the trigonometric terms which appear, other than those occurring in (9.21), are omitted, one obtains a set of nonlinear ordinary differential equations

$$\dot{X} = -\sigma X + \sigma Y, \qquad \dot{Y} = -XZ + rX - Y, \qquad \dot{Z} = XY - bZ, \tag{9.23}$$

now known as the Lorenz system. In Equation (9.22), $r = R/R_c$, $b = 4(1 + a^2)^{-1}$, $(\dot{\cdot}) = \mathrm{d}/\mathrm{d}\tau$. Equation (9.22) cannot be solved analytically. Nevertheless, with some delay, they eventually became the subject of great attention on the part of physicists and mathematicians as well, due to their ability to describe the stochastic behaviour of dynamical systems, observed by Lorenz himself, and other remarkable features (for some results see Section 8 or Chapter 5). Note that with (9.21) $J(\psi, \nabla^2\psi) \equiv 0$, i.e., the nonlinear self-interactions in the velocity field are 'lost' in the Lorenz model. Six coefficients in (9.21) and, correspondingly, six equations instead of (9.22) are needed in order to include them in the model (Shirer and Dutton, 1979).

PROBLEMS

1. Show that the infinite series

$$w = w_g - \frac{1}{if}\frac{\mathrm{d}}{\mathrm{d}t} w_g + \frac{1}{(if)^2}\frac{\mathrm{d}^2}{\mathrm{d}t^2} w_g - \frac{1}{(if)^3}\frac{\mathrm{d}^3}{\mathrm{d}t^3} w_g + \cdots$$

is a solution of the equation $\mathrm{d}w/\mathrm{d}t + if(w - w_g) = 0$, which is an analogue of Equation (4.4) in complex variables $w = u + iv$, $w_g = u_g + iv_g$, $i = \sqrt{-1}$ (Herbert, 1978).

2. Show that the term $\zeta_a \nabla_p \cdot \mathbf{u}$ in Equation (5.7) assimilates in itself the barotropic term $J(p, \alpha)$ in (5.5).
Solution: Equations (5.10), (4.35) and (4.36) yield

$$J(p, \alpha) = \frac{f^2}{g}\left(\mathbf{u}_g \times \frac{\partial \mathbf{u}_g}{\partial z}\right) \cdot \mathbf{k}.$$

On the other hand, the rules for differentiation give

$$f\nabla_p \cdot \mathbf{u} = f\nabla_z \cdot \mathbf{u} - \frac{f^2}{g}\left(\mathbf{u} \times \frac{\partial \mathbf{u}}{\partial z}\right) \cdot \mathbf{k}.$$

As far as $\mathbf{u} \approx \mathbf{u}_g$ and $\zeta = \zeta_p$, the above statement is true.

3. Show that the substitution of (4.27) into (4.24) results in

$$\mathbf{u}_g(z) = \mathbf{u}_g^\infty - \mathbf{u}_T\, e^{-mz}$$

where

$$\mathbf{u}_T = \frac{AR}{fp_0}\,\mathbf{k} \times \nabla p_0 + \frac{g}{mfT_0}\,\mathbf{k} \times \nabla T_0,$$

$$\mathbf{u}_g^\infty = \mathbf{u}_g^0 + \mathbf{u}_T = \frac{RT_\infty}{fp_0}\,\mathbf{k} \times \nabla p_0 + \frac{g}{mfT_0}\,\mathbf{k} \times \nabla T_0.$$

4. Calculate the work done by the gradient force $A = (\mathbf{u} \cdot -\nabla\Phi)$ in the case of a non-zero ageostrophic (cross-isobaric) motion.
Answer: $A = -(\mathbf{u}' \cdot \nabla\Phi) = fc'c_g \sin(\mathbf{u}_g, \mathbf{u}')$, $\quad c' = |\mathbf{u}'|$

5. Starting from (4.3) and (4.4) show that

$$\frac{\mathrm{d}c}{\mathrm{d}t} = \left|\frac{\mathrm{d}\mathbf{u}}{\mathrm{d}t}\right| \sin(\mathbf{u}, \mathbf{u}'), \quad c = |\mathbf{u}|.$$

6. Solve for (u', v') and analyze the scalar analogue of Equation (4.20)

$$u'\,\frac{\partial u_g}{\partial x} + v'\left(\frac{\partial u_g}{\partial y} - f\right) = -\,\frac{\mathrm{d}_2 u_g}{\mathrm{d}t},$$

$$u'\left(\frac{\partial v_g}{\partial x} + f\right) + v'\,\frac{\partial v_g}{\partial y} = -\,\frac{\mathrm{d}_2 v_g}{\mathrm{d}t}.$$

Note: See Wu and Blumen (1982).

7. Perform a scale analysis of the divergence Equation (5.34).

8. Show that after two subsequent substitutions $P'^2 = F(P)$ and $P = e^Q$, the variables in (6.10) split and the solution (6.11) follows.
Note: After the first substitution two integrations can be performed with a free constant $C_1 = 0$ in order to satisfy $P(\infty) = 0$.

9. Prove that (7.7) represents two first integrals of (7.6).

10. Study the effect of the linear friction added to the system (8.4):

$$\dot{u} - fv = b_1 - ku, \qquad \dot{v} + fu = b_2 - kv.$$

Note: Introduce complex variable $w = u + iv$ or $x + iy$. Find also $u(\infty)$ and $v(\infty)$.

11. Solve the system (8.1)

$$\ddot{x} - f\dot{y} = -fv_g, \qquad \ddot{y} + f\dot{x} = fu_g$$

for $v_g = 0$, $u_g = U - \frac{1}{2}\gamma y^2$, U and γ-arbitrary constants.

12. Solve the system (8.9) under the conditions

$$x(0) = x_0, \qquad y(0) = y_0, \qquad \dot{x}(0) = u_0, \qquad \dot{y}(0) = v_0.$$

13. Show that the generalization of (9.1)

$$\frac{\partial u}{\partial t} + u\frac{\partial u}{\partial x} = \nu\frac{\partial^2 u}{\partial x^2}$$

known as Burger's equation, with the aid of the substitution

$$u = -\frac{2\nu}{w}\frac{\partial w}{\partial x}$$

can be reduced to the linear equation

$$\frac{\partial w}{\partial t} = \nu\frac{\partial^2 w}{\partial x^2}.$$

14. Derive the spectral analogue of (9.8) for Burger's equation. Show the difference if the viscous term is replaced by linear friction $-ku$.
Note: See the paper by Wiin-Nielsen (1975).

15. As a generalization of Problem 13 and simplification of Equations (9.15), consider the one-dimensional set of equations:

$$\frac{\partial w}{\partial t} + w\frac{\partial w}{\partial z} = \sigma\theta + \sigma\frac{\partial^2 w}{\partial z^2}$$

$$\frac{\partial \theta}{\partial t} + w\frac{\partial \theta}{\partial z} = rw + \frac{\partial^2 \theta}{\partial z^2}$$

with parameters r, $\sigma > 0$ and boundary conditions $w = \theta = 0$ at $z = 0, \pi$. Assuming

$$F(z, t) = \sum_{n=1}^{\infty} F_n(t)\sin nz, \qquad F = w, \theta,$$

derive the low order system (Panchev, 1983)

$$2\dot{w}_1 = w_1w_2 - 2\sigma w_1 + 2\sigma\theta_1 \quad ,$$

$$2\dot{w}_2 = -w_1^2 - 8\sigma w_2 + 2\sigma\theta_2 \quad ,$$

$$2\dot{\theta}_1 = 2w_1\theta_2 - w_2\theta_1 - 2\theta_1 + 2rw_1,$$

$$2\dot{\theta}_2 = -w_1\theta_1 - 8\theta_2 + 2rw_2 \quad .$$

CHAPTER 4

Energetics of the Atmosphere

1. TYPES OF ENERGY AND ENERGY CONVERSIONS

The atmosphere is in ceaseless motion and practically the only source maintaining this motion is the Sun. However, the mechanism of converting the Sun's radiation into various other kinds of energies, including the kinetic energy of the motions, has not yet been completely studied.

It is well known that the atmosphere is a very poor 'heat machine' – only 2% of the Sun's radiation which is received on the Earth's surface converts into kinetic energy of the atmospheric motions. The rest of the Sun's radiation undergoes various other conversions and their investigation is of interest, not only for its own sake. It is also an important link in general dynamic considerations, based on the equations of motion and thermodynamic equations. Moreover, because of the great complexity of these equations, quite often the only way to obtain rather qualitative conclusions about the character of the processes in the system under investigation, proves to be that based on energetic considerations and evaluations. First of all, this requires some definitions of the basic energy types encountered in the meteorological problems to be given and some relationships between them to be established.

a. Definitions

(1) *Internal energy* E_i. In the case of a perfect gas, the internal energy is determined by the kinetic energy of the molecules and depends on the temperature T only. Since the atmosphere is a good approximation to a perfect gas, we can write

$$dE_i = c_v T \, dm, \tag{1.1}$$

where $dm = \rho \, dz$ is the mass of an elementary volume of unit horizontal cross-section and height dz. Assuming the hydrostatic atmosphere ($dp = -\rho g \, dz$) and integrating (1.1) between z_1 and z_2 (p_1 and p_2), we obtain the internal energy of the air column

$$E_i = c_v \int_{z_1}^{z_2} \rho T \, dz = \frac{c_v}{g} \int_{p_2}^{p_1} T(p) \, dp. \tag{1.2}$$

(2) *Gravitational potential energy* E_p. This is the energy relating to the position of air particles in the gravity force field

$$dE_p = gz \, dm. \tag{1.3}$$

For a column with a finite thickness

$$E_p = \int_{z_1}^{z_2} g\rho z \, dz = \int_{p_2}^{p_1} z(p) \, dp = \frac{1}{g}\int_{p_2}^{p_1} \Phi \, dp. \tag{1.4}$$

Integrating by parts

$$E_p = p_1 z_1 - p_2 z_2 + \int_{z_1}^{z_2} p(z) \, dz$$

and using the equation of state $p = \rho RT$, we get

$$E_p = p_1 z_1 - p_2 z_2 + \frac{R}{g}\int_{p_2}^{p_1} T(p) \, dp. \tag{1.5}$$

The formulae simplify for an infinitely extending air column – $z_1 = 0, z_2 = \infty$ $(p_1 = p_0,$ $p_2 = 0)$. Since $zp(z) \to 0$ at $z \to \infty$, then

$$E_p = \frac{R}{g}\int_0^{p_0} T(p) \, dp, \qquad E_i = \frac{c_v}{g}\int_0^{p_0} T(p) \, dp \tag{1.6}$$

or

$$\frac{E_i}{E_p} = \frac{c_v}{R} = \frac{c_v}{c_p - c_v} = (\kappa - 1)^{-1} \approx 2.5, \tag{1.7}$$

where $\kappa = c_p/c_v \approx 1.41 \approx \sqrt{2}$. Therefore, in an hydrostatic atmosphere, the internal energy (E_i) and the gravitational potential energy (E_p) are proportional to each other and, therefore, it is not necessary to consider them separately. They may be combined into a single term $E_i + E_p = E_{ip}$.

(3) *Total potential energy* E_{ip}. By definition,

$$E_{ip} = E_i + E_p = \frac{c_p}{g}\int_0^{p_0} T(p) \, dp = \frac{1}{g}\int_0^{p_0} e(p) \, dp, \tag{1.8}$$

where $e = c_p T$ is the enthalpy. Obviously

$$E_{ip} = \frac{c_p}{c_v} E_i = \frac{c_p}{R} E_p. \tag{1.9}$$

After some rearrangements we can write (1.8) in the form

$$E_{ip} = \frac{1}{g(\kappa - 1)}\int_0^{p_0} a_s^2(p) \, dp, \tag{1.10}$$

where $a_s = (\kappa RT)^{1/2}$ is the speed of sound.

(4) *Kinetic energy* E_k. By definition,

$$dE_k = \tfrac{1}{2}\,|\mathbf{v}|^2\,dm \tag{1.11}$$

so that

$$E_k = \frac{1}{2}\int_0^\infty \rho|\mathbf{v}|^2\,dz = \frac{1}{2g}\int_0^{p_0} |\mathbf{v}|^2\,dp. \tag{1.12}$$

An interesting conclusion can be made comparing (1.12) with (1.10). With the help of the mean values theorem we can write

$$E_{ip} = \frac{2.5p_0}{g}\,\langle a_s^2\rangle, \qquad E_k = \frac{p_0}{2g}\,\langle|\mathbf{v}|^2\rangle$$

where $\langle\ \rangle$ denotes a mean value of the corresponding quantity in the interval $(0, p_0)$ in agreement with the cited theorem. In the atmosphere, however, $|\mathbf{v}|/a_s \sim 1/20$ so that

$$\frac{E_k}{E_{ip}} = \frac{1}{5}\,\frac{\langle|\mathbf{v}|^2\rangle}{\langle a_s^2\rangle} \sim 0.5\;10^{-3} \tag{1.13}$$

Consequently, only 0.05% of the total potential energy converts into kinetic energy. This part of E_{ip} is sometimes called 'useful' potential energy. The reason for the ratio (1.13) becomes clear if one bears in mind the concept for a new type of energy, introduced in meteorology by Lorenz in 1955 (see also [40]).

(5) *Available potential energy* E_{ap}. This is the difference between the total potential energy E_{ip} and the minimum E_{ip}^* which could result from an adiabatic rearrangement of the temperature field yielding a stable horizontal stratification of the potential temperature field. In other words, E_{ap} is that portion of E_{ip} that may be released by adiabatically relaxing all potential temperature surfaces to a horizontal configuration. In this case, θ, T and p surfaces coincide and this is a potential minimum state of the atmosphere. That is why E_{ip}^* could be called *unavailable* (for conversion) potential energy. To determine E_{ap} and E_{ip}^*, it is convenient to use θ as a vertical coordinate – $\theta = T(p_{00}/p)^\lambda$, $p_{00} =$ 1000 mb, $\lambda = R/c_p$.

In the case of adiabatic processes, the total mass of the air above the surface, $\theta =$ const, will not change in time so that the atmospheric pressure $p^*(\theta)$ averaged over the whole surface will also be constant in time. An integration of (1.8) by parts yields

$$E_{ip} = \frac{c_p}{g}\,p_{00}^{-\lambda}\int_0^{p_0} \theta p^\lambda\,dp = \frac{c_p p_{00}^{-\lambda}}{g(1+\lambda)}\left[\theta p^{\lambda+1}\Big|_0^{p_0} - \int_\infty^0 p^{\lambda+1}\,d\theta\right].$$

Conditionally assuming that $\theta = 0$ at $p = p_0$ we can write

$$E_{ip} = \frac{c_p p_{00}^{-\lambda}}{g(1+\lambda)}\int_0^\infty p^{\lambda+1}\,d\theta. \tag{1.14}$$

Then E_{ip}^* will be given by the same formula but with p^* instead of p, so that

$$E_{ap} = E_{ip} - E_{ip}^* = \frac{c_p p_{00}^{-\lambda}}{g(1+\lambda)} \int_0^\infty (p^{\lambda+1} - p^{*\lambda+1})\, d\theta. \tag{1.15}$$

On the basis of this formula, more convenient expressions are derived for a calculation of E_{ap}, using observational data. It follows from the definition that the available potential energy is a *global* characteristic, since the reference state E_{ip}^* is an averaged one. It is also a quasi-geostrophic characteristic, since it neglects the changes in potential energy that could, in principle, take place via a purely vertical mass exchange in nongeostrophic motion.

Theoretical and empirical evaluations result in the conclusion that $E_{ap}/E_{ip} \sim 1/200$, so that (1.13) yields

$$\frac{E_k}{E_{ap}} \sim \frac{1}{10}. \tag{1.16}$$

Thus, only about 0.5% of E_{ip} is available and only about 10% of E_{ap} is actually converted into the kinetic energy of the atmospheric motions. In reality, the atmosphere is a rather inefficient heat engine. If the kinetic energy E_k is not realized, this does not necessarily mean a lack of E_{ap} in the atmosphere. Other reasons may also exist.

(6) *Latent energy* E_L. This is related to the water phase transitions in the atmosphere: evaporation/condensation (freezing/melting). If q is the specific humidity, then

$$dE_L = Lq\, dm, \tag{1.17}$$

where L is the latent heat of vaporization – the amount of heat necessary to change unit mass of liquid into vapour at the same temperature. Hence, as before,

$$E_L = \int_{z_1}^{z_2} Lq\rho\, dz = \frac{1}{g} \int_{p_2}^{p_1} Lq\, dp. \tag{1.18}$$

Some other types of energies could also be mentioned, for instance, the electromagnetic energy that might have a special meaning for some particular atmospheric phenomena. For synoptic processes, however, they are of no importance. That is why we may introduce the concept of *total energy* as a sum of the following

$$E_\Sigma = E_i + E_p + E_k + (E_L). \tag{1.19}$$

In a dry atmosphere $E_L = 0$.

Since the works of Lorenz (1955), (1960), the investigation of energy exchange has been a central tool in the interpretation of atmospheric dynamical processes. Much about the achievements in this direction can be found in [66] and [40]. Recently, a valuable contribution to the problem was made by Plumb (1982) who reviewed the traditional derivation of the energy cycle and showed some paradoxical properties of energy conversion and flux terms under steady conditions. He derived an alternative scheme free of these shortcomings. Here we will only give an idea of the possible energy transitions in the atmosphere. This problem will be discussed later in more detail in connection with particular questions.

b. Energy Conversions

Continuous conversions of one type of energy into another take place in the atmosphere. As far as the prevailing portion of the heat coming from the Sun, in the form of radiation energy, does not directly heat the atmosphere but only the underlying surface, the system 'atmosphere – underlying surface' (A-US) should be considered as a whole when studying these conversions. The heat accumulated by the surface is transferred through heat conductivity, convective heat exchange and evaporation to the higher layers of the atmosphere, thus changing the internal energy (enthalpy). The role of evaporation should be emphasized, not only as a source of water vapour, but of latent energy too, which is transported in the atmosphere by advection, convection and diffusion and can be realized at considerable distances from its source of entry. The conversion of latent energy into internal energy takes place at condensation. A continuous conversion of total potential energy into kinetic energy and vice-versa is carried out. Thus, the conversion $E_p \rightleftarrows E_k$ is mainly realized through vertical motions, while with horizontal motions, the conversion $E_i \rightleftarrows E_k$ prevails. Both conversions could be considered as adiabatic and thermodynamically reversible. Moreover, in the second case, E_k changes only at a motion with a component perpendicular to the isobars, i.e., at ageostrophic motion (see Section 4 of Chapter 3). In this connection, we shall recall that the Coriolis force does not do any work, thus it does not contribute to the changes in E_k. Friction forces, however, have dissipative natures and lead to the conversion $E_k \rightarrow E_i$ – the process being nonreversible.

The problem of energy conversion acquires some interesting aspects when the synoptic and global-scale motions are expressed as a sum of an averaged (e.g., zonal) motion with velocity $\mathbf{U}$ and a fluctuational motion (often called an eddy motion) with velocity $\mathbf{u}'$, i.e., $\mathbf{u} = \mathbf{U} + \mathbf{u}'$. The corresponding two types of kinetic energy – *zonal* and *eddy*, are introduced by

$$E_{kz} = \frac{1}{2g}\int_0^{p_0} |\mathbf{U}|^2 \, dp, \qquad E_{ke} = \frac{1}{2g}\int_0^{p_0} |\mathbf{u}'| \, dp, \tag{1.20}$$

as well as zonal-available potential energy E_{apz} and eddy-available potential energy E_{ape}, so that $E_{apz} + E_{ape} = E_{ap}$. Such partitioning of E_k and E_{ap} allows a more detailed investigation of the conversion processes concerning one or the other component. It is clear, for example, that the global temperature difference between pole and equator is decisive enough for the generation of E_{apz}, while its relevance to E_{ape} is physically much more screened and indefinite.

In this connection, we shall note that the conversions between E_{kz} and E_{ke} are of great interest and, in particular, the conversion of E_{ke} to E_{kz}, discovered and extensively investigated during the last decade.

Lorenz (1955) summarized these results in four equations as follows:

$$\frac{\partial E_{apz}}{\partial t} = -C_a - C_z + G_z, \qquad \frac{\partial E_{ape}}{\partial t} = C_a - C_e + G_e,$$

$$\frac{\partial E_{kz}}{\partial t} = C_z - C_k - D_z, \qquad \frac{\partial E_{ke}}{\partial t} = C_e + C_k - D_e. \tag{1.21}$$

where C, G and D denote conversion, generation and dissipation, respectively. Thus, C_a is the conversion from zonal- to eddy-available potential energy, C_e is the conversion from eddy-available potential to eddy kinetic energy, C_k is the conversion from zonal to eddy kinetic energy, and C_z is the conversion from zonal-available potential energy to zonal kinetic energy, as illustrated in Figure 4.1. Note that each quantity C_α, $\alpha = a, z, e$ enters in two equations with opposite signs. G_z and G_e represent the generation of zonal- and eddy-available potential energy, while D_z and D_e denote the dissipation of zonal and eddy kinetic energy.

We shall face these problems of atmospheric energetics again in relation to the general circulation of the atmosphere, which is the subject of Chapter 11.

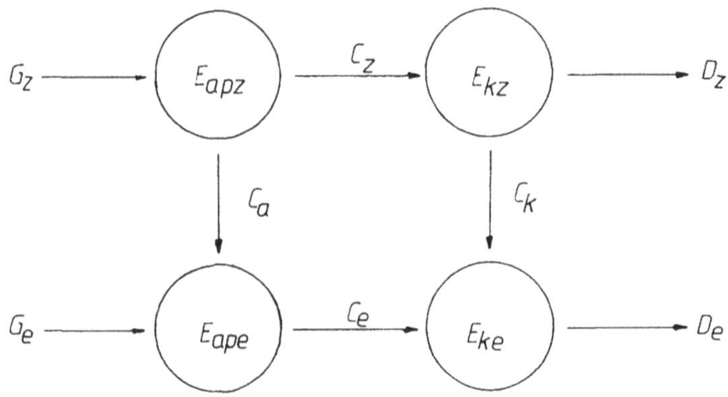

Fig. 4.1. Energy conversion as described by Equation (1.21).

2. THE ENERGY BALANCE EQUATION PER UNIT AIR MASS

The various energy conversions discussed in Section 1 are subject to rules, the mathematical expression of which are the so-called energy balance equations for separate kinds of energies or their sum. These equations are particular cases of the general physical law of energy conservation. We shall derive the equations for an air parcel with unit mass ($dm = 1$), so that instead of (1.1), (1.3) and (1.11) we shall have

$$e_i = c_v T, \qquad e_p = gz = \Phi, \qquad e_k = \frac{1}{2}\,|\mathbf{v}|^2. \tag{2.1}$$

Since the geometrical (z) and pressure (p) coordinates are equally made use of, we shall derive these equations in both coordinates.

a. The Energy Balance Equation in z Coordinates

The starting equations are:

$$\frac{d\mathbf{v}}{dt} = -\alpha\nabla p - f\mathbf{k} \times \mathbf{v} - g\mathbf{k} + \mathbf{F}, \tag{2.2}$$

$$c_v \frac{dT}{dt} + p \frac{d\alpha}{dt} = Q, \tag{2.3}$$

$$\nabla \cdot \mathbf{v} = -\frac{1}{\rho}\frac{d\rho}{dt} = \frac{1}{\alpha}\frac{d\alpha}{dt}, \tag{2.4}$$

where **F** is the frictional force. Scalar multiplication of (2.2) by **v** yields the kinetic energy balance equation

$$\frac{de_k}{dt} = -\alpha(\nabla p \cdot \mathbf{v}) - gw + (\mathbf{F} \cdot \mathbf{v}). \tag{2.5}$$

But $gw = g\, dz/dt = de_p/dt$, so that

$$\frac{d}{dt}(e_k + e_p) = -\alpha(\nabla p \cdot \mathbf{v}) + (\mathbf{F} \cdot \mathbf{v}). \tag{2.6}$$

In Equation (2.5), the quantity $\alpha(\nabla p \cdot \mathbf{v})$ appears as work done by the pressure gradient force gw – as a potential-to-kinetic energy conversion and $(\mathbf{F} \cdot \mathbf{v})$ – as a dissipative sink (this term is always negative). Their balance determines the individual changes of the kinetic energy (de_k/dt). The interpretation of Equation (2.6) is analogous.

We rearrange the right-hand side of (2.6) as follows: From the identity

$$(\nabla p \cdot \mathbf{v}) = (\nabla \cdot p\mathbf{v}) - p(\nabla \cdot \mathbf{v}) \tag{2.7}$$

and (2.4) it follows

$$\alpha(\nabla p \cdot \mathbf{v}) = \alpha(\nabla \cdot p\mathbf{v}) - p\frac{d\alpha}{dt}. \tag{2.8}$$

On the other side

$$\alpha(\nabla \cdot p\mathbf{v}) = \frac{d\alpha p}{dt} - \alpha\frac{\partial p}{\partial t} \tag{2.9}$$

Then (2.3), (2.8) and (2.9) yield

$$\alpha(\mathbf{v} \cdot \nabla p) = \frac{d}{dt}(\alpha p + c_v T) - \alpha\frac{\partial p}{\partial t} + Q. \tag{2.10}$$

But $\alpha p = RT$ and $R = c_p - c_v$ so that

$$\alpha p + c_v T = c_p T. \tag{2.11}$$

Combining Equation (2.10) with (2.6) yields the balance equation for the total energy in a dry atmosphere

$$\frac{d}{dt}\left(\frac{1}{2}|\mathbf{v}|^2 + gz + c_pT\right) - \alpha\frac{\partial p}{\partial t} = Q + (\mathbf{F}\cdot\mathbf{v}), \tag{2.12}$$

where Q is a thermal source/sink of energy. In the case of a stationary pressure field ($\partial p/\partial t$), adiabatic process ($Q = 0$) and a lack of dissipative force ($F = 0$), it follows from (2.12) that during the motion of the air particles the total energy remains constant

$$\frac{1}{2}|\mathbf{v}|^2 + gz + c_pT = E_k + E_p + E = \text{const}, \tag{2.13}$$

where $E = c_pT$ is the enthalpy. The invariant (2.13) is also referred to as Bernoulli's equation. In the case of pure horizontal motion z = const, $\mathbf{v} = (\mathbf{u}, 0) = (u, v, 0)$ and Equation (2.13) simplifies

$$\frac{1}{2}|\mathbf{u}|^2 + c_pT = \text{const}. \tag{2.14}$$

Bernoulli's equation has been successfully applied in meteorological theory and practice (including instrumentation). The simplest example may be that related to the vertical ascent of air parcels. Actually, from (2.13)

$$\frac{dT}{dt} = -\frac{1}{c_p}\frac{d}{dt}\left(\frac{1}{2}|\mathbf{v}|^2 + gz\right).$$

But in the case of vertical motion $de_p/dt \gg de_k/dt$ so that with great accuracy

$$\frac{dT}{dt} = -\frac{g}{c_p}\frac{dz}{dt} \quad \text{or} \quad -\frac{dT}{dz} = \gamma_a = \frac{g}{c_p},$$

where γ_a is the dry adiabatic lapse rate.

If T_0, z_0, $\mathbf{v}_0$ and T, z, $\mathbf{v}$ are the values of these quantities in two successive points on the trajectory of the air particle, an alternative and useful form of Bernoulli's equation follows from (2.13)

$$\frac{1}{2}|\mathbf{v}|^2 + gz + c_pT = \frac{1}{2}|\mathbf{v}_0|^2 + gz_0 + c_pT_0, \tag{2.15}$$

or

$$c_p(T - T_0) = -g\left[(z - z_0) + \frac{1}{2g}(|\mathbf{v}|^2 - |\mathbf{v}_0|^2)\right]. \tag{2.16}$$

b. The Energy Balance Equation in p Coordinates

Since the third equation of motion in this system is replaced by the quasi-static equation, the energy balance equation will refer to horizontal motion. The starting equations are

$$\frac{\partial \mathbf{u}}{\partial t} + (\mathbf{u}\cdot\nabla)\mathbf{u} + w_p\frac{\partial \mathbf{u}}{\partial p} = -\nabla\Phi - f\mathbf{k}\times\mathbf{u} + \mathbf{F}, \tag{2.17}$$

$$\nabla\cdot\mathbf{u} + \frac{\partial w_p}{\partial p} = 0, \tag{2.18}$$

$$\frac{\partial T}{\partial t} + (\mathbf{u}\cdot\nabla T) + w_p\left(\frac{\partial T}{\partial p} - \frac{RT}{pc_p}\right) = \frac{Q}{c_p}. \tag{2.19}$$

The last equation is equivalent to (3.11) in Chapter 3, which can easily be seen with the help of (3.8) and (3.13) of Chapter 3.

A scalar multiplication of (2.17) by **u** yields

$$\frac{\partial e_k}{\partial t} + \mathbf{u} \cdot \nabla e_k + w_p \frac{\partial e_k}{\partial p} = -\mathbf{u} \cdot \nabla \Phi + \mathbf{u} \cdot \mathbf{F} \tag{2.20}$$

where $e_k = \frac{1}{2} |\mathbf{u}|^2$. Further, we multiply (2.18) by e_k and add to (2.20):

$$\frac{\partial e_k}{\partial t} + \nabla \cdot e_k \mathbf{u} + \frac{\partial w_p e_k}{\partial p} = -\mathbf{u} \cdot \nabla \Phi + \mathbf{u} \cdot \mathbf{F}. \tag{2.21}$$

A new multiplication of (2.18), this time by Φ, and summing with (2.21), yields alternative forms of the energy balance equation

$$\frac{\partial e_k}{\partial t} + \nabla \cdot e_k \mathbf{u} + \Phi \nabla \cdot \mathbf{u} + \frac{\partial w_p e_k}{\partial p} + \Phi \frac{\partial w_p}{\partial p} = -\mathbf{u} \cdot \nabla \Phi + \mathbf{u} \cdot \mathbf{F}, \tag{2.22}$$

or (divergent form)

$$\frac{\partial e_k}{\partial t} + \nabla \cdot (e_k \mathbf{u} + \Phi \mathbf{u}) + \frac{\partial}{\partial p} (w_p e_k + \Phi w_p)$$
$$= w_p \frac{\partial \Phi}{\partial p} + \mathbf{u} \cdot \mathbf{F} = -\alpha w_p + \mathbf{u} \cdot \mathbf{F}, \tag{2.23}$$

where the hydrostatic equation $\partial \Phi / \partial p = -RT/p = -\alpha$ has been used.

The divergent form of the energy balance equation (2.23) is convenient for its physical interpretation. Bearing in mind that $\mathbf{u}(e_k + e_p)$ and $w_p(e_k + e_p)$ are horizontal and vertical kinetic plus potential energy fluxes, we then see that the local changes of kinetic energy $\partial e_k / \partial t$ are caused by four factors – the horizontal and vertical divergence/convergence of these fluxes and convection and dissipation (the two terms of the right-hand side of (2.23)).

Further, we multiply Equation (2.19) by c_p and (2.18) by $e = c_p T$ and sum the result. We get

$$\frac{\partial e}{\partial t} + \nabla \cdot e\mathbf{u} + \frac{\partial e w_p}{\partial p} = \alpha w_p + Q. \tag{2.24}$$

The term αw_p, which enters in the right-hand sides of (2.24) and (2.23) with opposite signs, appears as potential-to-kinetic energy conversion. When summing (2.23) and (2.24), it cancels out to obtain the balance equation

$$\frac{\partial}{\partial t} (e_k + e) + \nabla \cdot (e_k + e_p + e)\mathbf{u} + \frac{\partial}{\partial p} (e_k + e_p + e) w_p = Q + \mathbf{u} \cdot \mathbf{F}$$

with a similar interpretation as Equation (2.23).

Analogous energy balance equations can be derived in other coordinate systems (σ, θ, etc).

3. INTEGRAL FORMS OF THE ENERGY BALANCE EQUATIONS

The equations derived in Section 2 are in differential form. By integration over a closed volume V or over the vertical only, the corresponding integral equations can be found. For this purpose we need, first of all, some formulae.

a. Subsidiary Formulae

Let $Y(x, y, z, t)$ be a meteorological element. Then from the expression

$$\frac{dY}{dt} = \frac{\partial Y}{\partial t} + \mathbf{v} \cdot \nabla Y \tag{3.1}$$

and the continuity equation

$$\frac{d\rho}{dt} = -\rho \nabla \cdot \mathbf{v} \tag{3.2}$$

the identity

$$\rho \frac{dY}{dt} = \partial \rho Y + \nabla \cdot \rho \mathbf{v} Y \tag{3.3}$$

follows. If V is a closed volume with a vanishing normal to the surface S velocity component ($v_n = 0$), then

$$\int_V \nabla \cdot \rho \mathbf{v} Y \, dV = \int_S \rho v_n Y \, dS = 0, \tag{3.4}$$

so that

$$\int_V \rho \frac{dY}{dt} dV = \int_V \frac{\partial \rho Y}{\partial t} dV = \frac{\partial}{\partial t} \int_V \rho Y \, dV. \tag{3.4'}$$

We shall use this formula below.

In this kind of investigation, however, a vertical air column with unit cross-section and free boundaries is to be frequently considered. In particular, the lower boundary may coincide with the Earth's surface $H = H(x, y)$. As to the upper one, it may be free – $\delta = \delta(x, y, t)$ or $\delta = \infty$. This problem is more complicated and requires other subsidiary formulae. To obtain them we write (3.3) in the form

$$\rho \frac{dY}{dt} = \frac{\partial \rho Y}{\partial t} + \nabla \cdot \rho \mathbf{u} Y + \frac{\partial \rho w}{\partial z}, \tag{3.5}$$

and then integrate between $H(x, y)$ and $\delta(x, y, t)$ (Figure 2.3). Making use of the rule for the differentiation of integrals whose boundaries are functions, we obtain

$$\frac{\partial}{\partial t} \int_H^\delta \rho Y \, dz = \int_H^\delta \rho \frac{dY}{dt} dz - \nabla \cdot \int_H^\delta \rho \mathbf{u} Y \, dz + \rho_\delta Y_\delta \left(\frac{\partial \delta}{\partial t} + \mathbf{u}_\delta \cdot \nabla \delta - w_\delta \right) -$$

$$-\rho_H Y_H (\mathbf{u}_H \cdot \nabla H - w_H) = a_1 + a_2 + a_3 + a_4. \tag{3.6}$$

In other words, the local change of the 'stores' of the specific quantity Y in the column under consideration (Y may be some kind of energy) is caused by four factors: a_1 – non-conservativity of Y during the motion, a_2 – advection by the horizontal wind, a_3 – flux through the upper boundary, a_4 – flux through the lower boundary. If the boundaries are nonpenetrating, then according to formulae (4.25) and (4.27) of Chapter 2, $w_\delta = \mathrm{d}\delta/\mathrm{d}t$ and $W_H = \mathrm{d}H/\mathrm{d}t = \mathbf{u}_H \cdot \nabla H$, so that $a_3 = a_4 = 0$ and formula (3.6) simplifies considerably. This is the case also when $\delta = \infty$, $H(x, y) = \text{const} = 0$ (no mountains) and $w_0 = 0$.

If p-coordinates are used, then instead of (3.3) and (3.6)

$$\frac{\mathrm{d}Y}{\mathrm{d}t} = \frac{\partial Y}{\partial t} + \nabla \cdot \mathbf{u}Y + \frac{\partial w_p Y}{\partial p}, \tag{3.7}$$

$$\frac{\partial}{\partial t}\int_0^{p_H} Y\,\mathrm{d}p = \int_0^{p_H} \frac{\mathrm{d}Y}{\mathrm{d}t}\,\mathrm{d}p - \nabla \cdot \int_0^{p_H} \mathbf{u}Y\,\mathrm{d}p + Y_H\left(\frac{\mathrm{d}p_H}{\mathrm{d}t} - w_p^H\right). \tag{3.8}$$

But $w_p^H = \mathrm{d}p_H/\mathrm{d}t$ (see Equation (3.18) of Chapter 3), so that (3.8) also simplifies.

b. Closed Air Mass

Let us write again Equation (2.6), (2.8) and (2.3)

$$\frac{\mathrm{d}}{\mathrm{d}t}(e_k + e_p) = -\alpha(\nabla p \cdot \mathbf{v}) + \mathbf{v} \cdot \mathbf{F},$$

$$\alpha(\nabla p \cdot \mathbf{v}) = \alpha(\nabla \cdot p\mathbf{v}) - p\,\frac{\mathrm{d}\alpha}{\mathrm{d}t}, \tag{3.9}$$

$$p\,\frac{\mathrm{d}\alpha}{\mathrm{d}t} = Q - c_v\,\frac{\mathrm{d}T}{\mathrm{d}t},$$

where $\alpha = 1/\rho$. Hence,

$$\frac{\mathrm{d}}{\mathrm{d}t}(e_k + e_p + e_i) = \alpha(\nabla \cdot p\mathbf{v}) + \mathbf{v} \cdot \mathbf{F}. \tag{3.10}$$

We multiply (3.10) by ρ, integrate over a closed volume V, make use of (3.4) and (3.4′), and obtain

$$\frac{\partial}{\partial t}(E_k + E_p + E_i) = \int_V \rho(Q + \mathbf{v} \cdot \mathbf{F})\,\mathrm{d}V, \tag{3.10′}$$

where

$$E_m = \int_V \rho e_m\,\mathrm{d}V.$$

Here $m = k, p, i$ are the integrated kinetic, potential and internal energies for the closed air mass under consideration. It is seen that the total energy $E_\Sigma = E_k + E_p + E_i$ varies with time under the influence of nonadiabatic and dissipative factors only. If they are absent, then

$$\frac{\partial E_\Sigma}{\partial t} = 0, \tag{3.11}$$

and the energy E_Σ remains constant with time. This is the so-called integral of energy or Margules's equation.

The energy balance Equations (3.10) and (3.11), as well as those derived in Section 2 are exact consequences of the fundamental equations of atmospheric fluid dynamics and thermodynamics. Therefore, whatsoever simplification of these equations, made for some specific purpose including their transformation into a finite difference form for numerical weather prediction and experimentation, must be consistent with the corresponding form of the energy balance. Otherwise, the simplification of the starting equations (of momentum, continuity and of the first law of thermodynamics) will introduce fictitious energy sources which will make a physical interpretation of the results difficult, if not impossible.

Integral energy balance equations can also be derived for separate kinds of energies. Thus, for the kinetic energy of horizontal motion, instead of (2.5) we will have

$$\rho \frac{\mathrm{d}_2 e_k}{\mathrm{d}t} = -\nabla p \cdot \mathbf{u} + \rho(\mathbf{u} \cdot F).$$

After integration

$$\frac{\partial}{\partial t} \int_V \rho e_k \,\mathrm{d}V = -\int_V \mathbf{u} \cdot \nabla p \,\mathrm{d}V + \int_V \rho \mathbf{u} \cdot \mathbf{F} \,\mathrm{d}V, \tag{3.12}$$

where (3.4) has been used. From (3.9)

$$\rho \frac{\mathrm{d}_2 e_i}{\mathrm{d}t} = Q + \mathbf{u} \cdot \nabla p - \nabla \cdot p\mathbf{u},$$

so that

$$\frac{\partial}{\partial t} \int_V \rho e_i \,\mathrm{d}V = \int_V \mathbf{u} \cdot \nabla p \,\mathrm{d}V + \int_V \rho Q \,\mathrm{d}V. \tag{3.13}$$

As it should be expected, the term $\int \mathbf{u} \cdot \nabla p \,\mathrm{d}V$, which describes the work done by the pressure gradient force, enters the right-hand side of Equations (3.12) and (3.13) with an opposite sign. As known, the sign and magnitude of this term depend on the deviation of the wind from the geostrophic one (formula (4.1), Chapter 3). In a nondivergent flow, $\mathbf{u} \cdot \nabla p = \nabla \cdot p\mathbf{u}$ and vanishes. Its interpretation is facilitated in the isobaric p-system.

If for simplicity we assume that $Q = 0$, $\mathbf{F} = 0$ and integrate Equations (2.23) and (2.24) over the whole atmosphere under the condition of no-energy flux through its boundaries, instead of (3.12) and (3.13), we shall obtain

$$\frac{\partial}{\partial t}\int_V e_k \, \mathrm{d}V = -R\int_V \frac{w_p T}{p}\, \mathrm{d}V = -\frac{\partial}{\partial t}\int e \, \mathrm{d}V. \tag{3.14}$$

The sign of the 'middle' integral depends on that of w_p. In the case of upward vertical motion $w_p < 0$, the kinetic energy will increase and the potential will decrease in time. If $w_p > 0$, the opposite process will take place.

c. Vertical Air Column

Aside from the differential and integral forms of the energy balance equations considered so far in Sections 2 and 3, a third one exists which can be obtained by integration on the vertical coordinate (z, p, σ, . . .) only. In this case, the balance equations refer to an air column of unit cross-section and free (nonpenetrating) vertical (side) boundaries.

Let us first derive the kinetic energy balance equation. The starting equation will be (2.21) written in the form

$$\frac{\partial}{\partial t} e_k + \nabla\cdot e_k\mathbf{u} + \frac{\partial}{\partial p} e_k w_p = \frac{\mathrm{d}}{\mathrm{d}t} e_k = -\mathbf{u}\cdot\nabla\Phi + \mathbf{u}\cdot\mathbf{F}, \tag{3.15}$$

where (3.7) has been used. Integrating between 0 and p_H and using (3.8), we obtain

$$\frac{\partial}{\partial t}\int_0^{p_H} e_k \, \mathrm{d}p = -\nabla\cdot\int_0^{p_H} e_k\mathbf{u}\, \mathrm{d}p + \int_0^{p_H} (-\mathbf{u}\cdot\nabla\Phi + \mathbf{u}\cdot\mathbf{F})\, \mathrm{d}p. \tag{3.16}$$

With a coefficient of proportionality $1/g$ the left-hand side of (3.16) is equal to $\partial E_k/\partial t$ for the whole column. The divergent term on the right gives the influx of kinetic energy into the column, while the integral to which ∇ acts represents the energy flux through the vertical walls of the column. The interpretation of the second term is clear.

Now let us start from Equation (2.24) for the enthalpy $e = c_p T$

$$\frac{\partial e}{\partial t} + \nabla\cdot e\mathbf{u} + \frac{\partial}{\partial p} e w_p = \frac{\mathrm{d}e}{\mathrm{d}t} = \alpha w_p + Q. \tag{3.17}$$

Procedures analogous to (3.16) yield

$$\frac{\partial}{\partial t}\int_0^{p_H} e \, \mathrm{d}p = -\nabla\cdot\int_0^{p_H} e\mathbf{u}\, \mathrm{d}p + R\int_0^{p_H} w_p T\frac{\mathrm{d}p}{p} + \int_0^{p_H} Q\, \mathrm{d}p. \tag{3.18}$$

Hence, the role of the upward ($w_p < 0$) and downward ($w_p > 0$) vertical motions in the column for changes of its total enthalpy is clearly seen.

PROBLEMS

1. Starting from the Poisson's equation $T = T_0(p/p_0)^\lambda$, $\lambda = R/c_p$ and Bernoulli's Equation (2.16) find a connection between p, z and $|\mathbf{v}|^2$ and give it a physical interpretation.
2. Prove that a single level z exists in the atmosphere at which the ratio between the internal and potential energies is the same as for the whole column (1.7), i.e., $e_i/e_p = c_v/R$.
   Note: Using the hydrostatic equation investigate the function $\psi(z) = zp(z)$ for extremum.
3. Using the equations of continuity and the first law in the σ-system (Section 3 of Chapter 3), namely

$$\frac{\partial p_H}{\partial t} + \nabla \cdot p_H \mathbf{u} + \frac{\partial}{\partial \sigma} p_H w_\sigma = 0,$$

$$\frac{\partial \theta}{\partial t} + \mathbf{u} \cdot \nabla \theta + w_\sigma \frac{\partial \theta}{\partial \sigma} = 0$$

   find the conditions under which

$$\frac{\partial}{\partial t} \int_V (p_H \theta)\, \mathrm{d}V = 0$$

   where the integration is over the whole atmosphere.
   Note: See Kasahara (1974).

CHAPTER 5

Waves and Instabilities in the Atmosphere

1. GENERAL INFORMATION ON WAVE MOTIONS: THE PERTURBATION METHOD

a. Mathematical Description

The theory of wave motions in a continuous media is one of the best-developed theories in classical fluid dynamics and special mathematical methods have been created for the purpose. A short review is given below

Let $\delta(x, t)$ be a function of the distance x and time t and $\delta = \psi(x)$ at $t = 0$. Then, if at $x = \theta \pm ct$ the equality

$$\delta = \psi(\theta) = \psi(x \pm ct) \tag{1.1}$$

holds, this will mean that the initial profile $\psi(x)$ moves in the positive or negative direction with velocity c preserving its shape. Evidently, a function of the type (1.1) is a solution of the one-dimensional wave equation $\psi_{tt} = c^2 \psi_{xx}$.

The simplest example of a constant shape wave is the harmonic one (Figure 5.1)

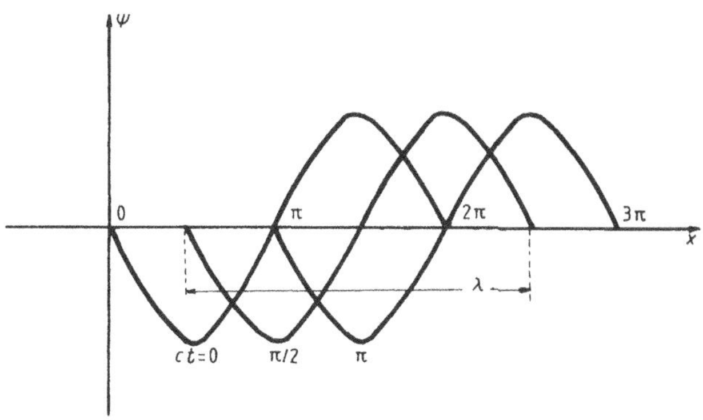

Fig. 5.1. Harmonic wave propagating in the positive direction of the x-axis.

$$\psi(x, t) = \hat{\psi}(\cos k\theta \quad \text{or} \quad \sin k\theta), \tag{1.2}$$

where $\theta = x - ct$ is called the phase of the wave, $\hat{\psi}$ – amplitude, k – wavenumber, $\lambda = 2\pi/k$ – wavelength, c – phase velocity and $T = \lambda/c$ – wave period. Formula (1.2) can also be written in the form

$$\psi(x, t) = \hat{\psi}[\cos(kx - \omega t) \quad \text{or} \quad \sin(kx - \omega t)], \tag{1.3}$$

where

$$\omega = \frac{2\pi}{T} = \frac{2\pi c}{\lambda} = kc \tag{1.4}$$

is the wave frequency.

In theory, the complex notation for wave functions is preferred:

$$\psi(x, t) = \hat{\psi} \exp[ik(x - ct)] = \hat{\psi} \exp[i(kx - \omega t)] \tag{1.5}$$

Since ψ is a real physical quantity, the real part of (1.5) is taken in the final result. But when (1.5) is used, both $\hat{\psi}$ and c could be complex quantities – $\hat{\psi} = \hat{\psi}_R + i\hat{\psi}_I$, $c = c_R + ic_I$, $i = \sqrt{-1}$. Then

$$\psi(x, t) = \hat{\psi} \exp[kc_I t] \cdot \exp[ik(x - c_R t)]. \tag{1.6}$$

This expression describes a travelling wave at constant velocity c_R and variable amplitude

$$\hat{\psi}(t) = \hat{\psi} \exp[kc_I t]. \tag{1.7}$$

Depending on the sign of c_I, the following classification applies:

$$\begin{aligned} &c_I > 0 \text{ amplified waves,} \\ &c_I = 0 \text{ neutral (stable) waves,} \\ &c_I < 0 \text{ damped waves.} \end{aligned} \tag{1.8}$$

As we shall see later, the proof that a given equation has a wave-type solution is reduced to obtaining the phase velocity from the so-called characteristic (dispersion) equation. If the so-defined velocity c turns out to be complex, then the sign of its imaginary part will determine the type of the motion according to (1.8).

For two- and three-dimensional cases, Equation (1.6) may be generalized by writing

$$\psi(\mathbf{x}, t) = \hat{\psi}\, e^{i\theta}, \quad \theta = \mathbf{k} \cdot \mathbf{x} - \omega t, \tag{1.9}$$

where $\mathbf{k} = [k_x, k_y, (k_z)]$ or $[k_1, k_2, (k_3)]$ is a wave vector in the plane (space) and $\mathbf{k} \cdot \mathbf{x}$ means a scalar product. Evidently, the equation

$$\mathbf{k} \cdot \mathbf{x} - \omega t = -\theta_1 = \text{const} \tag{1.10}$$

represents a straight line (plane), which is also called a phase line (plane). Let $k = |\mathbf{k}|$. Then (1.10) can be rewritten as

$$k^{-1}(k_i x_i) - k^{-1}(\omega t - \theta_1) = 0,$$

showing that $h = (\omega t - \theta_1)k^{-1}$ gives the distance of the phase line (plane) θ_1 = const from the origin of the coordinate system. Hence, $c = \mathrm{d}h/\mathrm{d}t = \omega/k$ and for the vector propagation velocity we can write

$$\mathbf{c} = \frac{c\mathbf{k}}{k} = \frac{\omega}{k^2}\mathbf{k} \quad \text{or} \quad c_i = \frac{ck_i}{k} = \frac{\omega}{k^2} k_i, \tag{1.11}$$

where $i = 1, 2, (3)$. On the other hand, the velocities at which the one-dimensional waves will propagate along the axes (Figure 5.2) will be

$$c_x = \frac{2\pi/k_x}{2\pi/\omega} = \frac{\omega}{k_x}, \qquad c_y = \frac{2\pi/k_y}{2\pi/\omega} = \frac{\omega}{k_y} \tag{1.12}$$

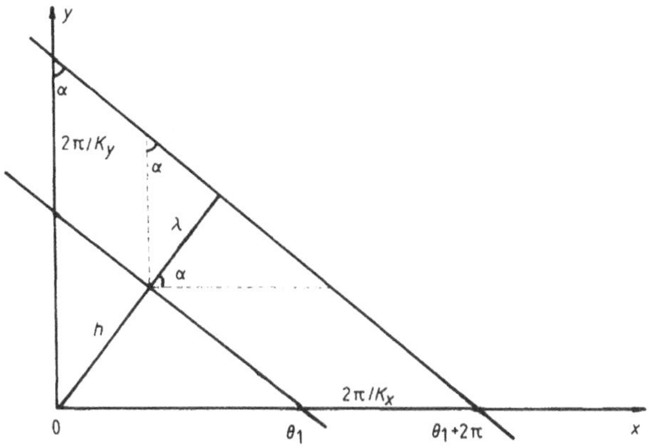

Fig. 5.2. Phase lines and other wave characteristics in the xOy-plane (after [45]).

Along with the wavenumbers k_i and $k = (k_i k_i)^{1/2}$, their reciprocal values $L_i = k_i^{-1}$ and $L = k^{-1}$ (called the *scales of motion*) are often used. Obviously, $L^{-2} = L_x^{-2} + L_y^{-2} + L_z^{-2}$. Besides, using (1.9), the following expressions can be derived

$$L_i = \frac{O(\psi)}{O(\partial\psi/\partial x_i)}, \qquad L = \frac{O(\psi)}{O(|\nabla\psi|)},$$

which are in agreement with the concept for 'scale' given in (1.1) of Chapter 3.

Usually, a functional dependence

$$c = c(\lambda), \tag{1.13}$$

known as a dispersion relation, exists. Since $k = 2\pi/\lambda$ and $\omega = kc$, (1.13) also implies that

$$\omega = \omega(k). \tag{1.13'}$$

Consequently, a dispersion of waves with different lengths is observed. This phenomenon can arise from two distinct reasons: (i) physical – structural properties of the medium (atmosphere, ocean) itself and (ii) geometrical – interferential effect due to reflection at the boundaries of the medium. As a rule, the first reason is responsible for the dispersion of the atmospheric waves. More precisely, it is connected with the existence of atmospheric internal reasonance frequencies such as the Coriolis frequency (Section 8, Chapter 3), the Brunt–Vaisala frequency (Section 4 of this chapter), and others. One consequence of dispersion is that the wave energy propagates at a different speed and in a different direction to the phase propagation. It should also be remembered that the phase velocity has nothing in common with the motion of the fluid particles. In fact, the wave motions in the atmosphere are not of simple periodicity. Instead of (1.9) the general expression

$$\psi(x, t) = \sum \hat{\psi} \exp[i(\mathbf{k} \cdot \mathbf{x} - \omega t)] \tag{1.9'}$$

holds, where a summing is made over all possible values of k and ω on which $\hat{\psi}$ may depend. At two or more Fourier components in (1.9′) with nearly equal phase velocities, we have a group of harmonic waves. If the phase velocity of all the waves in a group does not depend on the wavelength, then the group will propagate with the phase velocity without changing shape. Otherwise, the propagation speed of the wave group (packet) is generally different from that of the individual Fourier components. Accordingly, the concept for *group velocity* c_{gr} is introduced (Figure 5.3), representing the velocity at which the observable disturbance and, hence, the energy propagates. The following relationship can be derived

$$c_{gr} = \frac{\partial \omega}{\partial k} = c + k\,\frac{\partial c}{\partial k} = c - \lambda\,\frac{\partial c}{\partial \lambda}\,. \tag{1.14}$$

Obviously, c_{gr} may be less, larger or equal to c. If, for example, $c > c_{gr}$ then an individual wave moving behind the group can be seen catching up with the group and passing through it.

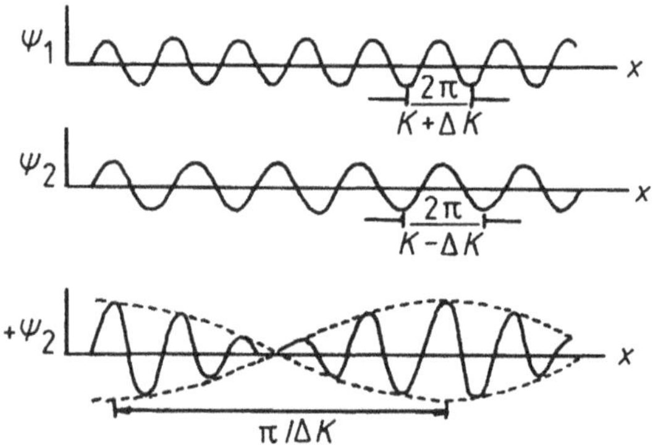

Fig. 5.3. Superposition of two harmonic waves ψ_1 and ψ_2 with slightly different wavenumbers and frequencies and the formation of a wave packet travelling with a group velocity c_{gr} (after [45]).

In a highly dispersive medium, the phase velocities of the individual waves differ significantly from each other. This prevents the formation of wave packets, thus allowing each component to travel individually at its own phase velocity.

b. The Perturbation Method

Observations in the atmosphere and the ocean convincingly show that a great part of the processes and motions that occur in these two geophysical fluids are of a wave nature. Theoretical confirmation of these facts would be proof that the corresponding set of differential equations has wave-type solutions (1.9′). Thus, other types of waves in the atmosphere or in the ocean, still not observed, could be predicted.

A principal difficulty in the theoretical approach is the nonlinearity of the equations. That is why only the linear wave theory, based on linearization with the aid of the so-called perturbation method, has received considerable attention. The essence of the method consists of the following:

(i) A basic state of the atmosphere is assumed. The physical quantities that characterize this state are denoted by ψ_*. Then we put

$$\psi(\mathbf{x}, t) = \psi_*(\mathbf{x}, t) + \psi'(\mathbf{x}, t) \tag{1.15}$$

where ψ' is a disturbance superimposed on the basic state.

(ii) The assumption is made that both ψ and ψ_* satisfy the same equations (e.g., the equations of motion, continuity, etc.); ψ is successively identified with $\mathbf{v}$, p, T, ρ, etc.

(iii) The perturbations are small enough ($|\psi'| / \psi_* \ll 1$) so that all terms in the governing equations which involve products and squares of ψ' can be neglected.

We shall illustrate the linearization technique on the equations of motion

$$\frac{\partial \mathbf{v}}{\partial t} + (\mathbf{v} \cdot \nabla)\mathbf{v} = -\alpha \nabla p - 2\boldsymbol{\omega} \times \mathbf{v} + g\mathbf{k}.$$

Following the above procedure, we write

$$\frac{\partial}{\partial t}(\mathbf{v}_* + \mathbf{v}') + [(\mathbf{v}_* + \mathbf{v}') \cdot \nabla]\,(\mathbf{v}_* + \mathbf{v}')$$

$$= -(\alpha_* + \alpha')\nabla(p_* + p') - 2\boldsymbol{\omega} \times (\mathbf{v}_* + \mathbf{v}') + g\mathbf{k},$$

and also

$$\frac{\partial \mathbf{v}_*}{\partial t} + (\mathbf{v}_* \cdot \nabla)\mathbf{v}_* = -\alpha_* \nabla p_* - 2\boldsymbol{\omega} \times \mathbf{v}_* + g\mathbf{k}.$$

Subtracting the last two equations and omitting $(\mathbf{v}' \cdot \nabla)\mathbf{v}'$ and $\alpha' \nabla p'$, one obtains the linearized equations

$$\frac{\partial \mathbf{v}'}{\partial t} + (\mathbf{v}_* \cdot \nabla)\mathbf{v}' + (\mathbf{v}' \cdot \nabla)\mathbf{v}_* = -\alpha_* \nabla p' - \alpha' \nabla p_* - 2\boldsymbol{\omega} \times \mathbf{v}'. \tag{1.16}$$

At $\alpha' = 0$ (incompressible atmosphere), they form a complete set together with the continuity equation $\nabla \cdot \mathbf{v}' = 0$. Otherwise, the rest of the weather equations

$$c_v \frac{\mathrm{d}T}{\mathrm{d}t} + p \frac{\mathrm{d}\alpha}{\mathrm{d}t} = Q, \qquad \alpha \nabla \cdot \mathbf{v} = \frac{\mathrm{d}\alpha}{\mathrm{d}t}, \quad p\alpha = RT, \tag{1.17}$$

are submitted to linearization too.

An important stage in building linear wave theory is the specification of the basic state in (1.16) and in other linearized equations. Many possibilities exist and in the simplest case, it is assumed that

$$\mathbf{v}_* = 0, \qquad \psi_* = \psi_*(z), \qquad \frac{\mathrm{d}p_*}{\mathrm{d}z} = -g\rho_* \tag{1.18}$$

where $\psi = p, T, \rho$, i.e., the basic state of the atmosphere is that at rest and the hydrostatic distribution of pressure, density and temperature with height. In this case $\mathbf{v} \equiv \mathbf{v}'$ and the model excludes the existence of a nonoscillating velocity.

Another possibility is to assume that

$$w_* = 0, \qquad -\alpha_* \nabla p_* - f\mathbf{k} \times \mathbf{u}_* = 0, \tag{1.19}$$

i.e., a geostrophic relationship which is closer to real situations in the atmosphere. Then $\mathbf{u}_*$ = const is assumed by considering the analytical solution of Equations (1.16).

c. Types of Waves in the Atmosphere

In a comprehensive classification of wave motions in the atmosphere, more than two dozen waves can be numbered. However, from a meteorological point of view, only some of these are of interest and they are usually the object of study in dynamic meteorology courses. Some others are interpreted as 'noises' in solving weather equations, especially when applying numerical methods. In order to remove these noises, it is necessary to know their properties. This extends the list of wave motions in which we are interested.

Independent of their physical nature, atmospheric waves can be divided into three groups (Figure 5.4):

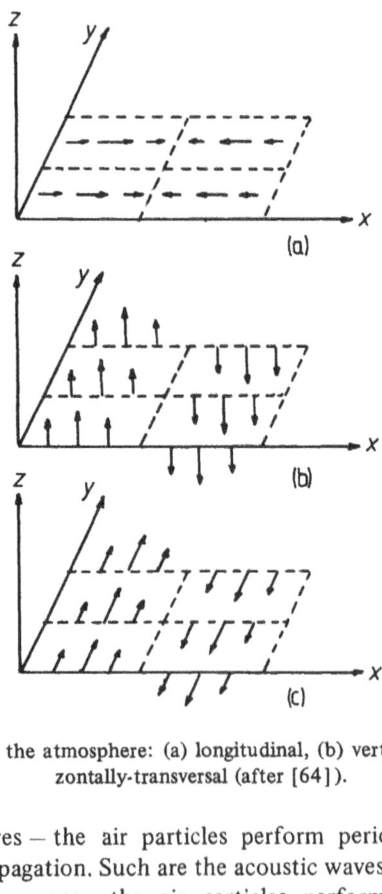

Fig. 5.4. Types of waves in the atmosphere: (a) longitudinal, (b) vertically-transversal, and (c) horizontally-transversal (after [64]).

(a) Longitudinal waves – the air particles perform periodic oscillations along the direction of the wave propagation. Such are the acoustic waves.

(b) Vertical-transverse waves – the air particles perform oscillations in a vertical plane as the waves propagate in a horizontal direction. Such are gravity waves in their numerous varieties.

(c) Horizontal-transverse waves – the air particles perform oscillations in a horizontal plane, transverse to the direction of wave propagation. Inertial waves and Rossby waves are included here.

Particular waves of each group or their combinations can exist as very small disturbances of the basis state of the atmosphere and can be described by linearized equations. That means that separate perturbations of this kind, possessing different amplitudes, frequencies and wavelengths, can be summed without interacting with each other (the superposition principle). On the contrary – a complex wave motion can be decomposed into elementary wave components (1.9′), called Fourier components (modes), and they can be investigated. For neutral or damped (because of dissipative factors) wave motions this exhausts the solution of the problem. In the case of amplifying amplitudes, however, it is impossible to come to a conclusion about the behaviour of the wave motion, based on the linear theory, because it no longer holds. Nonlinear effects of the interaction between waves arise which principally require a new approach. A concept of nonlinear waves will be given at the end of this chapter.

Though the separate problems to be described later have their specific peculiarities, a principal scheme for their solution can be given, consisting of the following: Suppose we study a rather general case of motion, governed by five equations (the three equations of motion and those of continuity and the first law of thermodynamics) for five unknown quantities – u, v, w, p, ρ. When no external forces and heat fluxes are present (the adiabatic case), it is a problem of *free* oscillations of the physical system and the equations, as well as the boundary conditions being homogeneous. The unknown quantities are represented in the form (1.9′) and then substituted into the equations. If the latter allow wave-type solutions, they become a homogeneous system of algebraic equations for the amplitude functions $\hat{\psi}(\psi = u, v, w, p, \rho)$. For a nonzero solution of the amplitudes $\hat{\psi}$, the determinant of the coefficients in the system must vanish. This leads to a fifth-order dispersion equation for the permissible values of the frequency ω in terms of other parameters describing the basic state and wavenumber (wavelength) of the perturbation. Each of the five roots $\omega_1, \omega_2, \ldots, \omega_5$ will correspond to a particular type of wave motion.

Another approach is also possible – through the successive elimination of the unknown functions in the linearized initial system to reduce the system to a differential equation of a higher order in time (fifth, in this case) and for this equation a wave-type solution of the form (1.9′) can be looked for. The results should be the same.

Both approaches are illustrated in the next sections.

2. SOUND WAVES IN THE ATMOSPHERE

It is an unquestionable fact that sound oscillations are generated and propagated in the atmosphere. This takes place under exceptionally complex conditions for which a complete theoretical treatment is not possible. Only some idealized cases will be considered here.

a. Constant Basic State

In order to derive the acoustic waves in the purest possible form as solutions of the weather equations, we neglect the Coriolis force, the gravity force, the molecular friction, and the conductivity effects. Then our starting equations read

$$\frac{d\mathbf{v}}{dt} = -\alpha \nabla p, \qquad \frac{dp}{dt} = -\kappa p \nabla \cdot \mathbf{v} \tag{2.1}$$

where $\pi = c_p/c_v$, $\alpha = 1/\rho$ and the second equation has been obtained as a combination between the continuity and thermodynamic equations

$$\frac{d\alpha}{dt} = \alpha \nabla \cdot \mathbf{v}, \qquad \frac{dp}{dt} = -\kappa \left(\frac{p}{\alpha}\right) \frac{d\alpha}{dt}.$$

The basic state is characterized by

$$\mathbf{v}_* = 0, \qquad p_* = \text{const}, \qquad \alpha_* = \text{const}. \tag{2.2}$$

Linearizing Equations (2.1) with the aid of the perturbation method yield the form

$$\frac{\partial \mathbf{v}'}{\partial t} = -\alpha_* \nabla p', \qquad \frac{\partial p'}{\partial t} = -\kappa p_* \nabla \cdot \mathbf{v}'. \tag{2.3}$$

Eliminating $\nabla \cdot \mathbf{v}'$ we get a single equation:

$$\frac{\partial^2 p'}{\partial t^2} = a_s^2 \nabla^2 p', \qquad a_s^2 = \kappa \alpha_* p_* = \kappa R T_*, \tag{2.4}$$

where ∇^2 is the three-dimensional Laplace operator.

We have to prove that Equation (2.4) has a wave-type solution. In the case considered, the phenomenon obviously has a spherical symmetry, so it is natural to introduce spherical coordinates (r, θ, λ). Then $\nabla^2 = d^2/dr^2 + 2r^{-1}\, d/dr$ and the solution of (2.4) should be looked for in the form of a spherical wave:

$$p'(r, t) = \hat{p} r^{-1} \exp[ik(r - ct)]. \tag{2.5}$$

After substituting this for (2.4), we get the characteristic equation

$$c^2 = a_s^2, \quad \text{i.e. } c = \pm a_s = \pm (\kappa R T_*)^{1/2}. \tag{2.6}$$

This expression for the velocity of sound waves is well known.

Let us now assume that $v_* = w_* = 0$, but $u_* \neq 0$. Then instead of (2.4), the following equation is easily obtained

$$\left(\frac{\partial}{\partial t} + u_* \frac{\partial}{\partial x}\right)^2 p' = a_s^2 \nabla^2 p'. \tag{2.7}$$

Assuming $p' \sim \exp[i(k_1 x + k_2 y + k_3 z - \omega t)]$, we obtain the characteristic equation

$$\omega^2 - 2u_* k_1 \omega + u_*^2 k_1^2 - a_s^2 k^2 = 0$$

where $k^2 = k_1^2 + k_2^2 + k_3^2$. Its solution is

$$\omega = kc = k\left(u_* \frac{k_1}{k} \pm a_s\right). \tag{2.8}$$

Hence, $c = u_* k_1/k \pm a_s$. In the case of one-dimensional motion along the x-axis only, $k_2 = k_3 = 0$ and $c = u_* \pm a_s$, which represents a simple Doppler shifting of velocity due to the wind u_*. The two signs in (2.8) show that the sound waves are bidirectional in propagation.

b. Variable Basic State

In a real atmosphere α_* and p_* from (2.2) are not constant but functions of height z, and the hydrostatic condition $\alpha_*\, \mathrm{d}p_*/\mathrm{d}z = -g$ holds. For simplicity let us restrict ourselves to considering motions in the $x0z$-plane only, assuming $\partial/\partial y = 0$ in all equations. Furthermore, we assume also $v_* = w_* = 0$, $u_* = \text{const} \neq 0$. Then, the linearized set of equations will be

$$u'_t + u_* u'_x + \alpha_* p'_x = 0, \tag{2.9}$$

$$\epsilon_1 (w'_t + u_* w'_x) + \alpha_* p'_z - \frac{g}{\alpha_*}\alpha' = 0, \tag{2.10}$$

$$\alpha_*(p'_t + u_* p'_x) - gw' + \kappa p_*(\alpha'_t + u_* \alpha'_x + \alpha_{*z} w') = 0, \tag{2.11}$$

$$(u'_x + w'_z)\alpha_* - \epsilon_2(\alpha'_t + u_* \alpha'_x + \alpha_{*z} w') = 0, \tag{2.12}$$

where the letter subscripts denote differentiation and ϵ_1, ϵ_2 are two numbers introduced for convenience. They can take values of either 0 or 1. So $\epsilon_2 = 0$ means incompressibility, while $\epsilon_1 = 0$ means the quasi-staticity of the disturbed state.

The system of Equations (2.9)–(2.12) is rather complex for solving by the elimination method. If α_*, α_{*z} and p_* are further considered as constants with respect to z, then the expression

$$\psi'(x, z, t) = \hat{\psi} \exp[i(k_1 x + k_3 z - \omega t)] \tag{2.13}$$

can be directly set, where $\psi = u, w, p, \alpha$. We shall consider two particular cases:

(i) Vertically propagating sound waves, i.e., $k_1 = 0$ (or $\partial/\partial x = 0$ in the real space). Equation (2.9) drops out while the rest take the form

$$\epsilon_1 w'_t + \alpha_* p'_z - \frac{g}{\alpha_*}\alpha' = 0,$$

$$\alpha_* p'_t - gw' + \kappa p_*(\alpha'_t + \alpha_{*z} w') = 0,$$

$$\alpha_* w'_z - \epsilon_2(\alpha'_t + \alpha_{*z} w') = 0.$$

Substituting here (2.13) with $k_1 = 0$ and $k_3 = k$, we obtain the following homogeneous algebraic system of equations

$$-i\epsilon_1 \omega \hat{w} + i\alpha_* k\hat{p} - \frac{g}{\alpha_*}\hat{\alpha} = 0,$$

$$(\kappa p_* \alpha_{*z} - g)\hat{w} - i\alpha_* \omega \hat{p} - \kappa p_* i\omega\hat{\alpha} = 0, \tag{2.14}$$

$$(i\alpha_* k - \epsilon_2 \alpha_{*z})\hat{w} + i\epsilon_2 \omega\hat{\alpha} = 0.$$

The characteristic equation derived from Det (2.15) = 0 is

$$\omega[\epsilon_1 \epsilon_2 \omega^2 - \kappa RT_* k^2 + i(1-\epsilon_2)kg - \epsilon_2 g\alpha_{*z}/\alpha] = 0, \tag{2.15}$$

where $\omega = ck$. The zero root ($\omega = 0$) corresponds to the stationary solution of (2.11). For the other two roots of (2.16) at $\epsilon_1 = \epsilon_2 = 1$ we find

$$c = \pm\left(\kappa RT_* + \frac{g\alpha_{*z}}{\alpha_* k^2}\right)^{1/2}. \tag{2.16}$$

When $\alpha_{*z} = 0$, this formula coincides with (2.6). The correction in (2.16), due to the variable density $\alpha_*(z)$, is usually small but it makes the vertically-propagating sound waves dispersive – $c = c(k)$ which is an important consequence of the vertical inhomogeneity of the density field.

Much more interesting is the conclusion which can be made from (2.15) if we set ϵ_1 and ϵ_2 separately or both equal to zero. In this case, there is no solution for ω – no sound waves exist as a solution for our equations. Therefore, the *vertical* sound waves could be filtered out, not only by the incompressibility hypothesis ($\epsilon_2 = 0$), but also by the quasi-staticity hypothesis ($\epsilon_1 = 0$), which is a less restrictive condition than the first one and is a better approximation to the real atmosphere.

(ii) Horizontally propagating sound waves, i.e., $k_3 = 0$ ($\partial/\partial z = 0$ in the real space). Now Equation (2.10) should be omitted and in the other equations – $w' = 0$ assumed. Substituting (2.13) for u', α', p' with $k_3 = 0$ and $k_1 = k$, the following characteristic equation can be derived

$$\epsilon_2(u_* - c)^3 - (u_* - c)\,\kappa p_* \alpha_* = 0 \tag{2.17}$$

whose nontrivial solution at $\epsilon_2 = 1$ is

$$c = u_* \pm (\kappa RT_*)^{1/2}, \tag{2.18}$$

i.e., as for a homogeneous atmosphere. Hence, the vertical inhomogeneity of the density field ($\partial\alpha_*/\partial z \neq 0$) does not have any influence on the velocity of the horizontal sound waves and the latter are filtered out only by the incompressibility condition ($\epsilon_2 = 0$).

For an analysis of the general system (2.9)–(2.12) see [23].

3. SURFACE (EXTERNAL) GRAVITY WAVES

Surface gravity waves occur on the free top surface of a fluid in a gravitational field when this boundary is displaced from its equilibrium state. Typical and well-known examples of such waves are those on the surface of water basins (oceans, seas, lakes, etc.). They arise under the influence of winds (wind waves) or earthquakes (tsunami). The great difference between the water and air densities allows us to consider the water surface as free. There is no analogue of these waves in the real atmosphere. Nevertheless, the model of a hypothetical homogeneous atmosphere having a density, constant with height, which, if existing around the Earth, would have the height $H_S = RT_0/g \sim 8$ km where T_0 is the surface temperature, turns out to be useful. Moreover, if we are interested in the dynamics of waves whose lengths considerably exceed H_S, it is natural to refer to 'shallow water' approximation equations (Section 5 of Chapter 2).

a. Long Waves

In this approximation the equations have the form

$$\begin{aligned} &\mathbf{u}_t + (\mathbf{u} \cdot \nabla)\mathbf{u} = -g\nabla h - f\mathbf{k} \times \mathbf{u}, \\ &h_t + (\mathbf{u} \cdot \nabla h) = -h\nabla \cdot \mathbf{u}, \end{aligned} \tag{3.1}$$

where the function $h(x, y, t)$ defines the shape of the free surface. Assuming

$$\begin{aligned} &h = h_*(y) + h', &\quad u = u_* + u', \\ &v_* = 0 (v \equiv v'), &\quad u_* = -\frac{g}{f}\frac{\partial h_*}{\partial y}, \end{aligned} \tag{3.2}$$

and linearizing (3.1) we obtain

$$\begin{aligned} &\mathbf{u}'_t + u_*\mathbf{u}'_x = -g\nabla h' - f\mathbf{k} \times \mathbf{u}', \\ &h'_t + u_*h'_x + h_{*y}v' = -h_*\nabla \cdot \mathbf{u}'. \end{aligned} \tag{3.3}$$

To find the waves we are looking for in their pure and simplest possible form, let us first set in (3.3) $f = 0$ and $\partial\psi'/\partial y = 0$. Then we get $\nabla h' = h'_x$, $\nabla \cdot \mathbf{u}' = u'_x$. Besides, for simplicity, h_* and h_{*y}, i.e., u_*, are further treated as constants with respect to y.

As usual we let

$$\psi'(x, t) = \hat{\psi} \exp[i(kx - \omega t)], \quad k = k_1 \tag{3.4}$$

where $\psi = u, v, h$. Substituting (3.4) for (3.3), we obtain a system of algebraic equations

$$\begin{aligned} &(-\omega + ku_*)\hat{u} + gk\hat{h} = 0, \\ &-\omega\hat{v} + ku_*\hat{v} = 0, \\ &(-\omega + ku_*)\hat{h} + \frac{if}{g}u_*\hat{v} + h_*k\hat{u} = 0, \end{aligned} \tag{3.5}$$

where the relationship $h_{*y} = -(f/g)u_*$ has been used. Hence,

$$\mathrm{Det}(3.5) = \begin{vmatrix} -\omega + ku_* & 0 & kg \\ 0 & -\omega + ku_* & 0 \\ kh_* & iu_*f/g & -\omega + ku_* \end{vmatrix} = 0 \tag{3.6}$$

which yields the cubic equation

$$(ku_* - \omega)\,[(ku_* - \omega)^2 - k^2 gh_*] = 0.$$

Its roots are

$$\omega_1 = ku_*, \qquad \omega_{2,3} = \omega_1 \pm k(gh_*)^{1/2}, \tag{3.7}$$

or with respect to the velocity of propagation $c = \omega/k$,

$$c_1 = u_*, \qquad c_{2,3} = c_1 \pm (gh_*)^{1/2}. \tag{3.7'}$$

Since $u_* \sim 10$ m s^{-1}, the first root corresponds to very slow waves. In $c_{2,3}$ the dominant term is the second one. Actually, even at $h_* \sim 10^2$ m (lake, sea-shelf region) $(gh_*)^{1/2} \sim 30$ m s^{-1}. If $h_* \sim 4$ km which is the mean ocean depth, then $(gh_*)^{1/2} \sim$ ~ 200 m s^{-1}. The tsunami velocity of propagation, as observations show, is of this order. Finally, if $h_* \sim RT_0/g$ then $(gh_*)^{1/2} = (RT_0)^{1/2} \sim 280$ m s^{-1}, i.e., these are gravity waves on the free surface of the hypothetical homogeneous atmosphere which propagate at nearly the speed of sound.

The obtained solution (3.7) is valid for waves whose length $\lambda \gg h_*$, which corresponds to the idea of the 'shallow water' approximation. But for such long waves, the Earth's rotation should also be taken into account. This obliges us to return to the more precise system of Equations (3.3) and seek a solution of the form

$$\psi'(x, y, t) = \hat{\psi} \exp[i(k_1 x + k_2 y - \omega t)]. \tag{3.8}$$

A treatment, analogous to that previously described, yields a solution which generalizes Equations (3.7) and (3.7′)

$$\omega_1 = u_* k_1, \qquad \omega_{2,3} = \omega_1 \pm (gh_* k^2 + f^2)^{1/2}, \tag{3.9}$$

where $k^2 = k_1^2 + k_2^2$. Hence, we find

$$c_x = \frac{\omega}{k_1} = u_* \pm \left[gh_*\left(1 + \frac{k_2^2}{k_1^2}\right) + \frac{f^2}{k_1^2}\right]^{1/2}. \tag{3.10}$$

At $k_2 = 0$, this formula simplifies. The amendment in (3.10), due to the inertial factor $(f \neq 0)$, is very small and has very little influence on the value of $c = \omega/k_1$. However, it transforms the waves into dispersive waves. The reason for this are the pure inertial waves. Actually, if we set $k_2 = 0$ and $u_* = h_* = 0$ in (3.10), we obtain $c = \pm f/k_1$. It is easy to show that the equations for pure inertial motion, linearized with respect to the atmosphere's state of rest, possess such a solution. At $\lambda_1 = 2\pi/k_1 \sim 10^3$ km and $f \sim 10^{-4}$ s^{-1}, we compute $c \approx \pm 15$ m s^{-1}, i.e., the inertial waves are slow waves.

b. Short Waves

An elementary theory of these waves can be developed on the basis of the general system of Equations (2.9)–(2.12), removing in advance the sound waves by the assumption $\alpha' = 0$ (incompressibility). Then the initial equations take the form

$$\begin{aligned} &u'_t + u_* u'_x + \alpha_* p'_x = 0, \\ &\epsilon(w'_t + u_* w'_x) + \alpha_* p'_z = 0, \qquad u'_x + w'_z = 0, \end{aligned} \tag{3.11}$$

where $\epsilon = 0$ or 1, $u_* = \text{const}$, $\partial p_*/\partial z = -g\rho_*$. The following boundary conditions are imposed

$$w' = 0 \quad \text{at } z = 0 \tag{3.12}$$

$$p_t' + u_* p_x' = -w' \frac{\partial p_*}{\partial z} = w' \rho_* g \quad \text{at } z = h_* \tag{3.13}$$

the second one being a linearized version of the condition $(\mathrm{d}/\mathrm{d}t)\,(p_* + p') = 0$ at the free surface.

Leaving the derivation as an exercise (see [23]), we shall only discuss the final result:

$$c = u_* \pm (gh_*)^{1/2} \quad \text{if } \epsilon = 0, \tag{3.14}$$

$$c = u_* \pm \left(\frac{g\lambda}{2\pi} \tanh \frac{2\pi h_*}{\lambda}\right)^{1/2} \quad \text{if } \epsilon = 1. \tag{3.15}$$

(i) The external gravity waves in the general case are neutral and with dispersion – $c = c(\lambda)$.

(ii) At $\lambda \gg h_*$, formula (3.15) degenerates asymptotically into (3.14). Calculations show that $\lambda > 25\ h_*$ is enough. For $h_* \sim 8$ km, we obtain $\lambda > 200$ km. On the other hand, as far as (3.14) is obtained at $\epsilon = 0$ (i.e., according to (3.11), when the quasi-static hypothesis is assumed concerning the disturbances), the short external gravity waves can be filtered out in this way.

(iii) At $\lambda \ll h_*$ formula (3.15) yields

$$c = u_* \pm \left(\frac{g\lambda}{2\pi}\right)^{1/2}. \tag{3.16}$$

In order for (3.16) to hold, it is enough that $\lambda < 2.5\ h_*$. At $h_* \sim 8$ km this yields $\lambda < 20$ km.

c. Equatorial Atmosphere

The behaviour of the external gravity waves in an equatorial atmosphere possesses some specific peculiarities, caused by the substantial dependence of the Coriolis parameter on $y: f(y) = \beta_0 y$. Then, assuming for simplicity that $u_* = 0$ and denoting $gh' = \Phi'$, $gh_* = a_*^2$, Equations (3.3) will take the form

$$u_t' - \beta_0 y v' + \Phi_x' = 0, \qquad v_t' + \beta_0 y u' + \Phi_y' = 0, \qquad \Phi_t' + a_*^2(u_x' + v_y') = 0. \tag{3.17}$$

We seek solutions of the form

$$\psi'(x, y, t) = \hat{\psi}(y) \exp[i(kx - \omega t)], \tag{3.18}$$

where $\psi = u, v, \Phi$. Inserting into (3.17), we obtain one algebraic and two ordinary differential equations for the amplitude functions

$$\begin{aligned} &-i\omega\hat{u} + \beta_0 y\hat{v} + ik\hat{\Phi} = 0, \\ &-i\omega\hat{v} + \beta_0 y\hat{u} + \Phi' = 0, \\ &-i\omega\hat{\Phi} + a_*^2(ik\hat{u} + \hat{v}') = 0, \end{aligned} \tag{3.19}$$

where $(')$ = d/dy, while in (3.17) it denotes perturbation. From β_0 and a_* – the only dimensional parameters in (3.19) – scales of time $(a_*\beta_0)^{-1/2}$ and length $(a_*/\beta_0)^{1/2}$ can be constructed. The nondimensional analogue of (3.19) reads

$$
\begin{aligned}
&-i\omega_0\hat{u}_0 - y_0\hat{v}_0 + ik_0\hat{\Phi}_0 = 0,\\
&-i\omega_0\hat{v}_0 + y_0\hat{u}_0 + \hat{\Phi}_0' = 0,\\
&-i\omega_0\hat{\Phi}_0 + ik_0\hat{u}_0 + \hat{v}_0' = 0,
\end{aligned}
\tag{3.20}
$$

where $\omega_0, k_0, \ldots$ are dimensionless quantities. Eliminating $\hat{u}_0$ and $\hat{\Phi}_0$, one obtains a single equation for $\hat{v}_0$:

$$\hat{v}_0'' + \left(\omega_0^2 - k_0^2 - \frac{k_0}{\omega_0} - y_0^2\right)\hat{v}_0 = 0. \tag{3.21}$$

This equation is of a known type and, under natural conditions $\hat{v}_0(\pm\infty) = 0$, has the general solution

$$\hat{v}_0(y_0) = A_n H_n(y_0)\exp\left(\frac{-y_0^2}{2}\right) \tag{3.22}$$

$$\omega_0^2 - k_0^2 - \frac{k_0}{\omega_0} = 2n + 1, \quad n = 0, 1, 2, \ldots, \tag{3.23}$$

where $H_0 = 1$, $H_1 = 2y_0$, $H_2 = 4y_0^2 - 2, \ldots$ are Hermitean polynomials, while (3.23) represents a dispersion relation, defining a discrete spectrum of possible frequences. It is a cubic equation which has three roots corresponding to different types of waves. According to the physical formulation of the problem, the presence of equatorial-type inertia-gravity waves should be expected. Limiting ourselves to the approximate solutions of (3.23), we see that such are obtained if $\omega_0^2 - k_0^2 \gg k_0/\omega_0$ is assumed. Then $\omega_{0n} = \pm(2n + 1 + k_0^2)^{1/2}$ or, in dimensional form,

$$
\begin{aligned}
\omega_n &= \pm\,[gh_*k^2 + (2n+1)\beta_0(gh_*)^{1/2}]^{1/2},\\
c_n &= \pm\,[gh_* + (2n+1)\beta_0 a_*/k^2]^{1/2}.
\end{aligned}
\tag{3.24}
$$

This solution corresponds to (3.9) and (3.10) at $u_* = 0$ and $k_2 = 0$. The lowest possible frequency is at $n = 0$.

In the other asymptotic case, when $\omega_0^2 \ll k_0^2 + k_0/\omega_0$ we get from (3.23) $\omega_{0n} = -k_0/(2n + 1 + k_0^2)$ or, in dimensional form,

$$
\begin{aligned}
\omega_n &= \frac{-\beta_0 k}{k^2 + (2n+1)k_*^2},\\
c_n &= \frac{-\beta_0}{k^2 + (2n+1)k_*^2},
\end{aligned}
\tag{3.25}
$$

where $k_* = (\beta_0/a_*)^{1/2} \sim 2.10^{-7}\ \text{m}^{-1}$ is the wavenumber scale ($L_* = k_*^{-1} \sim 5 \times 10^3$ km). Obviously, the solution (3.25) represents a new type of wave, not encountered in the previous problems, which are called *Rossby waves*. Unlike inertia-gravity waves (3.24), the new waves are unidirectional – they propagate westward ($c_n < 0$).

Finally, Let us note that the condition $\hat{v}_0(\pm\infty) = 0$ could be replaced without difficulties of principle by $\hat{v}_0(\pm L) = 0$, where $L \sim 500$–700 km is the width of the equatorial belt. This makes the problem closer to reality.

The concepts and solutions considered are given by Matsuno (1966) and Blandford (1966). Since then, interest in the problem of wave motions in the equatorial atmosphere and ocean has increased considerably (Dobrishman (1977) and [10], [67], etc.).

4. INTERNAL GRAVITY WAVES

a. Waves on Internal Boundary Surfaces

As was mentioned in Section 3, free-surface gravity waves are not observed in the real atmosphere because it has no well-defined upper boundary, such as the ocean has, but nevertheless, some wave motions are similar to them. The atmosphere, however, as a rule is stratified – quasi-horizontal layers are observed; often with an abrupt transition between them. It could be considered, in an idealized way, that they are separated by a boundary surface on which some of the characteristics of the air undergo a change in value. Apparently, this is not a free boundary surface but an internal one for, on both sides, the fluid is one and the same.

Generally, if $\psi(\mathbf{x}, t)$ is an arbitrary meteorological element and discontinuity in the function $\psi(\mathbf{x}, t)$ takes place on the surface S, then S is called a *zero-order* discontinuity surface. If $\psi(\mathbf{x}, t)$ is continuous but a discontinuity in $\nabla\psi$ occurs, then S is called a *first-order* discontinuity surface and so on. Zero- and first-order discontinuity surfaces (lines) are usually encountered in the atmosphere and ocean. For example, frontal surfaces are zero-order discontinuity surfaces for temperature T, density ρ, and wind velocity $\mathbf{v}$, but are first-order for pressure p. At the same time they are boundary surfaces as well because, in practice, they consist of one and the same air particles. The empirical fact that the frontal surfaces move with a speed almost equal to that of the wind on both sides is a good enough indication. Another example of a boundary surface is the tropopause. The temperature is continuous on it, but ∇T (and more precisely $\partial T/\partial z$) undergoes a discontinuity. Internal boundary surfaces are also found at temperature inversions and around jet streams, etc. A particular kind of gravity wave, having the features of both waves on a free surface, considered in Section 3, and the true internal gravity waves to be considered below, arise and are observed on such surfaces. There are some important meteorological phenomena related to them.

The qualitative properties of this kind of gravity wave are easily demonstrated in the simplest case of a two-layer atmosphere, infinite in height, and with a uniform distribution of ρ_* and u_* (Figure 5.5). The two layers are separated by the boundary surface S, on which ρ_* and u_* change in value. A two-dimensional pattern is considered ($\partial/\partial y = 0$). A damping of the perturbations is assumed at $z \to \pm\infty$ and a boundary condition, similar to (3.13). is applied on S. Bearing in mind that $(h'_*, h''_*) \to \infty$, the theory yields a simple dispersion relation analogous to (3.16) for short waves

$$c = \frac{\omega}{k} = \frac{\rho' u' + \rho'' u''}{\rho' + \rho''} \pm \left[\frac{g\lambda}{2\pi}\,\frac{\rho' - \rho''}{\rho' + \rho''} - \rho'\rho''\,\frac{(u' - u'')^2}{(\rho' + \rho'')^2}\right]^{1/2}, \tag{4.1}$$

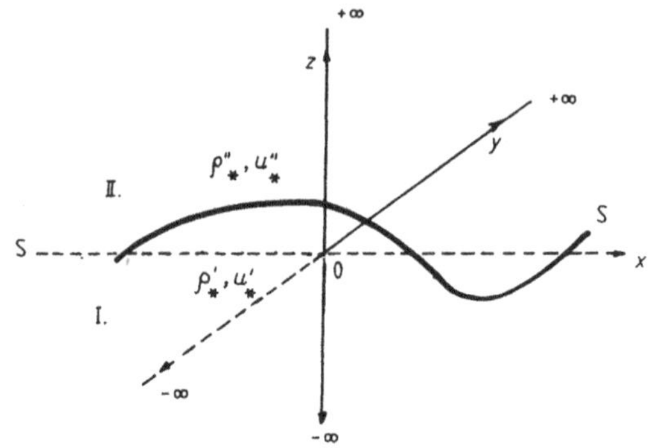

Fig. 5.5. Waves on an internal boundary surface.

where the asterisk is omitted. But $p = \rho RT$ and $p' = p''$, so in place of (4.1) we have

$$c = \frac{u'T'' + u''T'}{T' + T''} \pm \left[\frac{g\lambda}{2\pi}\,\frac{T'' - T'}{T'' + T'} - T'T''\left(\frac{u' - u''}{T' + T''}\right)^2\right]^{1/2}. \tag{4.2}$$

Two simpler cases could be mentioned:

(a) There is no wind ($u' = u'' = 0$). Then

$$c = \pm A\left(\frac{g\lambda}{2\pi}\right)^{1/2}, \qquad A^2 = \frac{T'' - T'}{T'' + T'}. \tag{4.3}$$

If $T'' = 283$ K and $T' = 273$ K, $A = 0.13$ is calculated, i.e., the wave propagation speed is about 10 times lower than that on a free surface at the same wavelength λ. At $T'' < T'$, $c = \pm ic_I$ is a pure imaginary quantity and no waves exist. As seen from (4.3), the waves are dispersive – $c = c(\lambda)$.

(b) There is no temperature discontinuity ($T' = T'' = T$) but $u' \neq u'' \neq 0$. Then

$$c = \tfrac{1}{2}\,[u' + u'' \pm i(u' - u'')] = c_R + ic_I, \tag{4.4}$$

i.e., instability holds for all wavelengths.

In the general case, the phase velocity will be complex when

$$\frac{g\lambda}{2\pi}\,\frac{T'' - T'}{T'' + T'} - T'T''\left(\frac{u' - u''}{T' + T''}\right)^2 < 0. \tag{4.5}$$

Hence, the following possibilities could be considered:

(i) $T'' < T'$ (the cold layer is above the warm one), (4.5) is satisfied for any λ and $\Delta u = u' - u''$, $c_I > 0$ and the wave motion is impossible due to the instability.

(ii) If T', T'' and λ are fixed, condition (4.5) for instability can be written in the form

$$(\Delta u)^2 > \frac{g\lambda}{2\pi}\,(T'' - T')\left(\frac{1}{T'} + \frac{1}{T''}\right) = (\Delta u)_m^2, \tag{4.6}$$

i.e., the neutrality of the waves is violated only by a velocity change greater than the minimum one $(\Delta u)_m$ according to (4.6).

(iii) If T', T'' and Δu are fixed, then (4.5) denotes

$$\lambda < \frac{2\pi}{g} \frac{(\Delta u)^2}{T''^2 - T'^2} T'T'' = \lambda_m, \tag{4.7}$$

i.e., the motion is unstable with respect to the short waves and stable in the opposite case. For $\lambda = \lambda_m$ the phase velocity of the neutral waves will be

$$c_m = \frac{u'T'' + u''T'}{T' + T''}. \tag{4.8}$$

In all the cases previously considered, $c \neq 0$ holds and the waves travel. Of particular interest in meteorology is the case of standing waves ($c = 0$). Letting $c = 0$ in (4.2), we obtain for the wavelength

$$\lambda_0 = \frac{2\pi}{g} \frac{u''^2 T' + u'^2 T''}{T'' - T'}. \tag{4.9}$$

Then, for example,

$$w'(x, z) = \hat{w}(z) \cos \frac{2\pi x}{\lambda_0}, \tag{4.10}$$

and similar expressions could be written for other wave characteristics.

The analysis made above shows that condition (4.5) can be realized for a number of combinations between the parameters, and, therefore, the waves are unstable (with growing amplitudes). It cannot be stated, however, that the observed instability will inevitably lead to their destruction, because the linear theory holds only for small perturbations. In the literature, the case described is known as the *Kelvin–Helmholtz* instability. Nevertheless, in laboratory and natural conditions (atmosphere and ocean), the destruction of unstable waves on internal boundary surfaces has been observed, due to such instability. Schematically, the process in its various stages is shown in Figure 5.6. Some of the atmospheric phenomena related to these waves are described in Section 7.

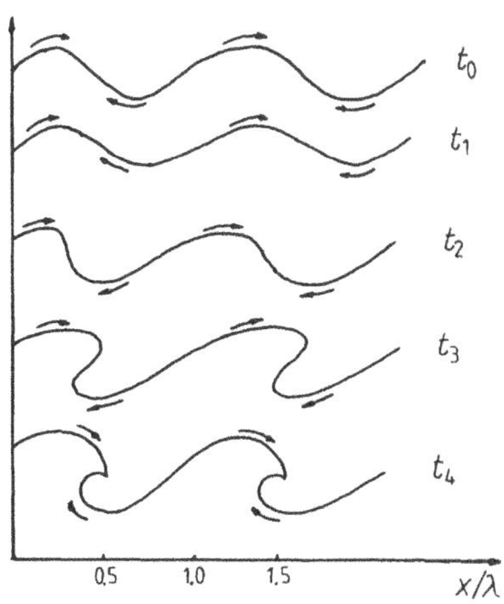

Fig. 5.6. Kelvin–Helmholtz instability and the formation of billows (after [65]).

b. Waves in a Continuously Stratified Atmosphere

The simplest example of an oscillatory motion in the field of a buoyancy force is the so-called individual air parcel. In the case of an adiabatic process, its motion along the vertical is governed by the well-known convection equation

$$\ddot{z} = \frac{d^2 z}{dt^2} = g\,\frac{T_i - T}{T} = g\,\frac{\theta_i - \theta}{\theta}, \tag{4.11}$$

where z is the vertical coordinate of the parcel, T_i and θ_i are its actual and potential temperatures, and T and θ are the same temperatures for the surrounding atmosphere. Note that the pressure $p_i = p$. Let $\theta_i(0) = \theta(0)$ at the initial level $z = 0$, i.e., the parcel does not differ from the ambient air. After adiabatically passing a short distance z along the vertical (Figure 5.7), it will have $\theta_i(z) = \theta_i(0)$ at the new level, but

$$\theta(z) \approx \theta_i(0) + z\,\frac{d\theta}{dz}.$$

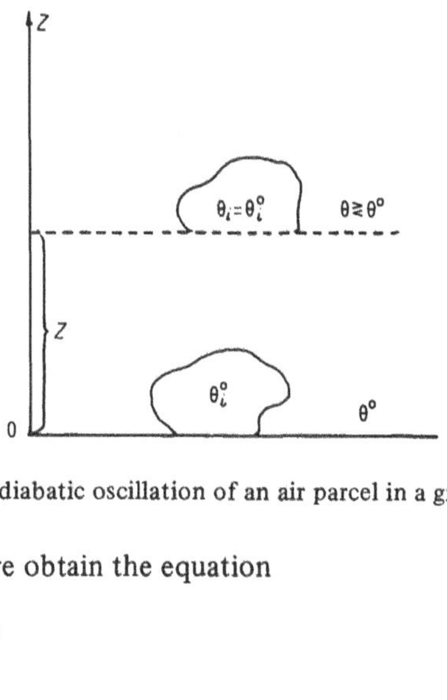

Fig. 5.7. Adiabatic oscillation of an air parcel in a gravity field.

Inserting this into (4.11), we obtain the equation

$$\ddot{z} + \omega_{BV}^2 Z = 0 \tag{4.12}$$

where

$$\omega_{BV}^2 = gB, \qquad B = \frac{\partial \ln \theta}{\partial z} = \frac{\gamma_a - \gamma}{T}, \tag{4.13}$$

$\gamma_a = g/c_p$ and $\gamma = -\partial T/\partial z$ are the dry adiabatic and local lapse rate in the atmosphere, $\omega_{BV} = (gB)^{1/2}$ is known as *Brunt–Väisälä frequency*. In isothermal atmosphere $\gamma = 0$ and $T =$ const so that $B = \gamma_a/T =$ const and $\omega_{BV} =$ const too. Since $\gamma_a = 10°\ \text{km}^{-1}$, $g \sim 10\ \text{m s}^{-2}$ and $T \sim 3 \times 10^2$ K, then $B \sim 0.3 \times 10^{-4}\ \text{m}^{-1}$ and $\omega_{BV} \sim 1.7 \times 10^{-2}\ \text{s}^{-1}$. In this case, as well as at $0 \neq \gamma < \gamma_a$ (stably-stratified atmosphere), the solution of (4.12) is of the type

$$z(t) \sim \exp[\pm i\omega_{BV} t], \tag{4.14}$$

which describes an oscillatory motion with a period $T_{BV} = 2\pi/\omega_{BV}$. As follows from the above-mentioned evaluations, $T_{BV} \sim 400$ s $\sim$ 6–7 min. In the oceanic thermocline, T_{BV} is of the same order of magnitude, but in the deep ocean regions may reach up to several hours. This example, which could also be generalized for a nonadiabatic process, (see Problem 7) is, in fact, a Lagrangian approach to the internal gravity-wave problem. Much more common and of more practical use is the Eulerian treatment of the problem in which fields instead of trajectories are studied. The example considered above, however, contributes to clarifying the physical sense of ω_{BV}, which also appears in the Eulerian variants of the problem. And there are many of them, really. The one described below is after Holton [27].

The equations in the log-pressure vertical coordinate $\zeta = -H_s \ln(p/p_0)$ derived in Section 3 of Chapter 3, will serve us for the initial equations

$$\frac{d_2\mathbf{u}}{dt} + w_\zeta \frac{\partial \mathbf{u}}{\partial \zeta} = -\nabla\Phi - f\mathbf{k} \times \mathbf{u},$$

$$\nabla \cdot \mathbf{u} + \frac{\partial w_\zeta}{\partial \zeta} - \frac{w_\zeta}{H_s} = 0, \tag{4.15}$$

$$\frac{d\Phi}{d\zeta} = \frac{RT}{H_s}, \qquad \frac{d_2 T}{dt} + w_\zeta \Gamma = 0,$$

where $H_s = RT_0/g$ and

$$\Gamma = \frac{T}{\theta}\frac{\partial\theta}{\partial\zeta} = \frac{T}{T_0}(\gamma_a - \gamma) = \frac{T^2}{gT_0}\,\omega_{BV}^2 \approx \text{const.} \tag{4.16}$$

having the value $\Gamma \approx 3°\text{C km}^{-1}$ in the troposphere.

To simplify the problem, we ignore the Coriolis force and assume $\partial/\partial y = 0$, $v = 0$. Then the linearized system has the form

$$\left(\frac{\partial}{\partial t} + u_* \frac{\partial}{\partial x}\right) u' + \frac{\partial\Phi}{\partial x} = 0,$$

$$\frac{\partial u'}{\partial x} + \frac{\partial w_\zeta}{\partial \zeta} - \frac{w_\zeta}{H_s} = 0, \tag{4.17}$$

$$\left(\frac{\partial}{\partial t} + u_* \frac{\partial}{\partial x}\right) \frac{\partial\Phi'}{\partial\zeta} + w_\zeta S = 0,$$

where $S = R\Gamma/H_s \approx$ const. Eliminating u' and w_ζ we obtain the equation

$$\left(\frac{\partial}{\partial t} + u_* \frac{\partial}{\partial x}\right)^2 \left(\frac{\partial^2\Phi'}{\partial\zeta^2} - \frac{1}{H_s}\frac{\partial\Phi'}{\partial\zeta}\right) + S\,\frac{\partial^2\Phi'}{\partial x^2} = 0. \tag{4.18}$$

Further, assuming a solution of the type

$$\Phi'(x, \zeta, t) = \hat{\Phi}(\zeta)\exp[i(k_1 x - \omega t)] \tag{4.19}$$

and substituting it for (4.18), we get an equation for the amplitude function

$$\frac{d^2\hat{\Phi}}{d\zeta^2} - \frac{1}{H_s}\frac{d\hat{\Phi}}{d\zeta} + \frac{S}{(u_* - \omega/k_1)^2}\hat{\Phi} = 0, \tag{4.20}$$

whose general solution is

$$\hat{\Phi}(\zeta) = (C_1 \exp[-ik_3\zeta] + C_2 \exp[ik_3\zeta]) \exp[\zeta/2H_s], \tag{4.21}$$

where

$$k_3 = \left[\frac{S}{(u_* - \omega/k_1)^2} - \frac{1}{4H_s^2}\right]^{1/2} \tag{4.22}$$

and C_1, C_2 are constants which might be determined by the boundary condition upon ζ.

The obtained solution (4.19)–(4.22) describes a superposition of two waves. For k_1, k_3, ω all positive, these waves propagate along the x-axis with phase velocity $c_x = \omega/k_1$ and not horizontally, although slantwise, i.e., there will also be a wave propagating along the vertical with a phase velocity $c_z = \pm\,\omega/k_3$. This is seen from the equation of the phase lines for the two waves

$$k_1 x \pm k_3\zeta - \omega t = \text{const}, \tag{4.23}$$

which lines are declined to the horizon with a slope equal to $\pm\,k_1/k_3$. If follows from (4.22) that waves are not possible at $S < 0$ (unstable atmosphere) as well as at $0 < S < (u_* - c_x)^2/4H_s^2$. In the case of standing waves along the x-axis (directed parallel to the basic flow $\mathbf{u}_*$) we have $c_x = 0$ and pure vertical gravity waves. The upward waves will propagate unlimited (no internal boundary surfaces have been assumed) whereas the downward waves will reflect on reaching the ground surface. Generally speaking, in contrast to the ocean, which is bounded at the top and bottom, the conditions in the atmosphere are more favourable for the existence of internal gravity waves, propagating vertically or having the vertical component of the phase velocity. This is confirmed by observations.

For additional information on the theory of gravity waves, see Section 3, Chapter 6.

5. THE ROSSBY WAVES

a. Two-Dimensional Pure Waves

It was mentioned in Section 3 that the linearized equations of inertial motion, namely $u'_t = fv'$ and $v'_t = -fu'$, where $f = \text{const}$, allow a wave-type solution $\psi'(x, t) = \hat{\psi}\exp[ik(x - ct)]$ where for the phase velocity of the dispersion relation $c = \pm f/k$ is obtained. This solution is analogous in Eulerian variables to the solution given in (8.7) of Chapter 3, and has no particular significance for the interpretation of large-scale motions because of $f = f(y)$. Rossby (1939) showed that the Earth's rotation around its axis in combination with its sphericity (causing the y-dependence of f), leads to the generation of a new type of long inertial waves in the atmosphere, which later received his name. It occurred that the Rossby waves had exceptionally great meteorological significance. They appeared to be no less important for the ocean's dynamics, too.

In order to isolate these waves in their purest possible form, we take the starting equations in the form of

$$\begin{aligned} &u_t + uu_x + vu_y = -\Phi_x + fv, \\ &v_t + uv_x + vv_y = -\Phi_y - fu, \qquad (5.1) \\ &u_x + v_y = 0. \end{aligned}$$

Thus, the acoustic (vertical and horizontal) and gravity waves are automatically filtered. From (5.1) immediately follows the vorticity equation $d_2(\zeta + f)/dt = 0$, which could be also taken from (5.12) in Chapter 3. In a developed form, it is

$$\zeta_t + u\zeta_x + v\zeta_y + \beta v = 0, \qquad (5.2)$$

where $\zeta = v_x - u_y$, $\beta = df/dy$. In accordance with the third equation of (5.1), we introduce the streamfunction $\psi(x, y, t)$

$$u = -\psi_y, \qquad v = \psi_x, \qquad \zeta = \nabla^2 \psi. \qquad (5.3)$$

Instead of (5.2), we shall have

$$\nabla^2 \psi_t = J(\nabla^2 \psi + f, \psi), \qquad (5.4)$$

where $J(a, b) = a_x b_y - a_y b_x$.

We linearize (5.4) assuming that

$$\psi = -u_* y + \psi', \quad u_* = \text{const.}, \qquad (5.5)$$

i.e.,

$$u = u_* - \psi'_y, \qquad v = \psi'_x.$$

The linear variant of (5.4) will be

$$\left(\frac{\partial}{\partial t} + u_* \frac{\partial}{\partial x}\right) \nabla^2 \psi' + \beta \psi'_x = 0. \qquad (5.6)$$

Assuming a solution of the kind

$$\psi'(x, y, t) = \hat{\psi} \exp[i(k_1 x + k_2 y - \omega t)],$$

we obtain the characteristic equation

$$(-i\omega + ik_1 u_*)(-k^2) + i\beta k_1 = 0,$$

where $k^2 = k_1^2 + k_2^2$. Hence, we find

$$\omega = u_* k_1 - \frac{\beta k_1}{k^2}, \qquad (5.7)$$

or

$$c_x = \frac{\omega}{k_1} = u_* - \frac{\beta}{k^2}. \qquad (5.8)$$

A simpler case than the one considered above is possible when $u_* = 0$ and $\partial/\partial y = 0$ ($k_2 = 0$). Then $c_x = c$ and

$$c = -\frac{\beta}{k^2} = -\beta\left(\frac{\lambda}{2\pi}\right)^2. \tag{5.9}$$

In this case, actually, we can speak about *pure Rossby waves*. Their outstanding property is the unidirectional propagation westward only, which is implied by the minus sign in (5.9). When a zonal flow ($u_* \neq 0$) exists, the Rossby waves propagate to the west with respect to it, but then standing waves ($c_x = 0$) are possible at $u_* > 0$ according to (5.8)

$$k_s^2 = \frac{\beta}{u_*} \quad \text{or} \quad \lambda_s = 2\pi\left(\frac{u_*}{\beta}\right)^{1/2}. \tag{5.10}$$

At $u_* \sim 10$ m s^{-1} and $\beta \sim 10^{-11}$ m^{-1} s^{-1} we get $\lambda_s \sim 6 \times 10^3$ km, i.e., these are, indeed, long waves and only 5–6 such waves can be plotted along one geographic parallel in the middle latitudes. With the aid of (5.10), formula (5.8) can be written in the form

$$c = \frac{\beta}{4\pi^2}\,(\lambda_s^2 - \lambda^2). \tag{5.11}$$

Hence, it is seen that at $\lambda < \lambda_s$ the wave will propagate to the west (referring to the Earth's surface) and vice-versa.

Another important property of the Rossby waves is that they are dispersive: $c = c(\lambda)$. For a group velocity we obtain from (5.9) and (1.14)

$$c_{gr} = \beta(\lambda/2\pi)^2 = -c \tag{5.12}$$

i.e., the wave packet propagates with the same velocity as a single wave but in the opposite direction – to the east.

If $k_2 \neq 0 (\partial/\partial y \neq 0)$, the dispersive relationships (5.7) and (5.8) also allow us to determine the meridional component of the phase velocity

$$c_y = \frac{\omega}{k_2} = \left(u_* - \frac{\beta}{k^2}\right)\frac{k_1}{k_2} = c_x\,\frac{k_1}{k_2} \tag{5.13}$$

and in contrast to c_x at $u_* = 0$, it could be both positive and negative. In this case, the analysis is simplified by introducing polar coordinates: $k_1 = k\cos\alpha$, $k_2 = k\sin\alpha$. Instead of (5.7) we shall have then

$$\omega(k, \alpha) = \left(u_* k - \frac{\beta}{k}\right)\cos\alpha$$

from where the zonal ($\partial\omega/\partial k$) and meridional ($\partial\omega/k\partial\alpha$) components of group velocity $\mathbf{c}_{gr}$ can easily be computed.

A property of the Rossby waves, interesting from a mathematical point of view, is that they are an exact solution, not only of the linearized equation (5.6) but also of the nonlinear equation (5.4). Let us write this in the form

$$\nabla^2\psi_t = J(\nabla^2\psi, \psi) - \beta\psi_x. \tag{5.14}$$

Direct substitution convinces us that

$$\psi(x, y, t) = u_* y + \hat{\psi} \exp[i(k_1 x + k_2 y - \omega t)]$$

is a solution of (5.14) when the dispersion relation (5.7) holds. The reason for this is that expressions of the form $\exp[i(\mathbf{k} \cdot \mathbf{x})]$ nullify the Jacobian $J(\nabla^2 \psi, \psi)$.

b. Two-Dimensional Mixed Waves

In the real atmosphere, the Rossby waves occur and exist simultaneously with other types of waves, and, e.g., with the long-surface gravity waves influencing each other. As a result, inertial gravity Rossby waves occur. In order to study the characteristics of these mixed waves, we start again from equations of motion in the 'shallow water' approximation:

$$\mathbf{u}_t + (\mathbf{u} \cdot \nabla)\mathbf{u} = -g\nabla h + f\mathbf{k} \times \mathbf{u} \tag{5.15}$$

$$h_t + \mathbf{u} \cdot \nabla h = -h\nabla \cdot \mathbf{u}. \tag{5.16}$$

The Rossby parameter β can be introduced in (5.15) using $f = f_0 + \beta y$. However, it is more convenient to work with the vorticity equation which follows directly from (5.15)

$$\zeta_t + \mathbf{u} \cdot \nabla\zeta + \beta v = -(\zeta + f)\, \nabla \cdot \mathbf{u}. \tag{5.17}$$

It can replace one of the equations in (5.15), say the second, so that the equations of the problem become (5.16), (5.17) and

$$u_t + uu_x + vu_y = -gh_x + fv. \tag{5.18}$$

For simplicity, let us consider a two-dimensional problem, setting $\mathbf{u}_y = 0$, $h_y = 0$ but $h_* = h_*(y)$ and $u_* = -(g/f)\,(\partial h_*/\partial y) = \text{const}$, $v_* = 0$. In this case, $\zeta = v'_x$ and linearizing Equations (5.18), (5.16) and (5.17) we get

$$\begin{aligned} &\epsilon(u'_t + u_* u'_x) - fv' + gh'_x = 0,\\ &h'_t + u_* h'_x + v' h_{*y} + h_* u'_x = 0,\\ &v'_{xt} + u_* v'_{xx} + \beta v' + fu'_x = 0, \end{aligned} \tag{5.19}$$

where $\epsilon = 1$ or 0 and $h_{*y} = -(f/g)u_*$. We look for a solution of the set (5.19) in the form $\psi'(x, t) = \hat{\psi} \exp[ik(x - ct)]$. Eventually we get

$$\begin{aligned} &\epsilon(u_* - c)ik\hat{u} - f\hat{v} + ikg\hat{h} = 0,\\ &ikh_*\hat{u} - \frac{fu_*}{g}\,\hat{v} + (u_* - c)ik\hat{h} = 0,\\ &ikf\hat{u} + [\beta - k^2(u_* - c)]\,\hat{v} = 0. \end{aligned} \tag{5.20}$$

Setting the determinant of the coefficients in front of $\hat{u}$, $\hat{v}$, $\hat{h}$ equal to zero, we obtain the following cubic characteristic equation

$$[\beta - k^2(u_* - c)]\,[gh_* - \epsilon(u_* - c)^2] - f^2\,[(u_* - c) - u_*] = 0. \tag{5.21}$$

We look for Rossby waves which have been modified by the gravity effect. They should be expected to have a low phase velocity: $gh_* \gg (u_* - c)^2$. Accounting for that, we replace (5.21) by the approximate equation

$$[\beta - k^2(u_* - c)]\, gh_* + f^2 c = 0, \tag{5.22}$$

which has the solution

$$c = u_* - \left(\beta + \frac{u_* f^2}{gh_*}\right)\left(k^2 + \frac{f^2}{gh_*}\right)^{-1}. \tag{5.23}$$

A comparison with (5.8) shows that it really corresponds to Rossby waves influenced by gravity. However, the correction for typical atmospheric values of u_*, h_*, g and f is small.

At $\epsilon = 0$, Equation (5.23) is an exact solution of (5.21). That means that the acceptance of the geostrophic relation $fv' = gh'_x$, instead of the first equation from (5.19), filters the gravity waves as far as the remaining two roots of (5.21) at $\epsilon = 1$ can be presented approximately in the form

$$c = u_* \pm \left(gh_* + \frac{f^2}{k^2}\right)^{1/2} \tag{5.24}$$

and really correspond to the fast surface-gravity waves, influenced by the Earth's rotation. The correction, however, is small.

Another, less limiting technique than the geostrophic relation $fv' = gh'_x$ exists for filtering the fast gravity waves from the solutions of our equations. It consists of assuming that $(d_2/dt)(\nabla \cdot \mathbf{u}') = 0$ which, applied to the divergence equation (5.3) of Chapter 3, means that the balance equation is supposed to hold

$$u'^2_x + v'^2_y + 2u'_y v'_x = -\nabla^2 \Phi' + f\zeta' - \beta u'. \tag{5.25}$$

It can be proved that (5.25) is a necessary and sufficient condition, as well.

Formulae (5.24) and (5.23) correspond to (3.25) and (3.24), which also describe mixed inertial-gravity waves but in an equatorial atmosphere.

c. Three-Dimensional Rossby Waves

So far we have assumed a two-dimensional structure for the Rossby solutions of the vorticity equation and, accordingly, we started from its simplified form (5.2). To find out how such waves propagate vertically we have to start from the same equation but with the vertical velocity retained on the right

$$\zeta_t + \mathbf{u} \cdot \nabla\zeta + \beta v = f\frac{\partial w_p}{\partial p}. \tag{5.26}$$

Due to the fact that all terms on the left-hand side of (5.26) have one and the same order of magnitude, without loss of accuracy we can insert the geostrophic values (quasi-geostrophic approximation, see also Chapter 7) for $\mathbf{u}$ and ζ,

$$\mathbf{u} \approx f^{-1}\mathbf{k} \times \nabla\Phi, \qquad \zeta \approx f^{-1}\nabla^2\Phi. \tag{5.27}$$

However, Equation (5.26) contains a second unknown function – $w_p = \mathrm{d}p/\mathrm{d}t$. We need one more equation. Assuming the adiabatic process for simplicity, we can use the thermodynamic energy equation (3.17) of Chapter 3

$$\left(\frac{\partial}{\partial t} + \mathbf{u} \cdot \boldsymbol{\nabla}\right) \frac{\partial \Phi}{\partial p} + m_\alpha w_p = 0. \tag{5.28}$$

Equations (5.26) and (5.28) should be linearized. In doing so we shall replace m_α with m_α^* for the basic state and write simply m:

$$m(p) = m_\alpha^* = -\frac{\alpha_*}{\theta_*}\frac{\partial \theta_*}{\partial p} = \frac{R^2 T_*}{gp^2}(\gamma_a - \gamma_*). \tag{5.29}$$

Assuming also $u_* = \text{const}$, we get

$$\left(\frac{\partial}{\partial t} + u_* \frac{\partial}{\partial x}\right) \nabla^2 \Phi' + \beta \frac{\partial \Phi'}{\partial x} = f^2 \frac{\partial w_p'}{\partial p}, \tag{5.30}$$

$$\left(\frac{\partial}{\partial t} + u_* \frac{\partial}{\partial x}\right) \frac{\partial \Phi'}{\partial p} + m w_p' = 0. \tag{5.31}$$

For further simplification of the problem, we assume that $m(p) = m = \text{const}$ (e.g., its average value for the layer). Eliminating w_p' from (5.30) and (5.31), a single equation is obtained for Φ':

$$\left(\frac{\partial}{\partial t} + u_* \frac{\partial}{\partial x}\right)\left(\nabla^2 + \frac{f^2}{m}\frac{\partial^2}{\partial p^2}\right)\Phi' + \beta \frac{\partial \Phi'}{\partial x} = 0. \tag{5.32}$$

For wave solutions

$$\Phi'(x, y, p, t) \sim \exp[i(k_1 x + k_2 y + k_3 p - \omega t)]$$

to be possible for (5.32), the dispersion relation

$$c_x = \frac{\omega}{k_1} = u_* - \beta\left[k_1^2 + k_2^2 + \left(\frac{f^2}{m}\right)k_3^2\right]^{-1} \tag{5.33}$$

must be satisfied. It modifies slightly the value of c_x compared with (5.8). However, in the case of standing waves with respect to the surface ($c_x = 0$), (5.33) yields

$$k_3^2 = \frac{m}{f^2}\left(\frac{\beta}{u_*} - k^2\right), \quad k^2 = k_1^2 + k_2^2, \tag{5.34}$$

which should be compared with (5.10). For vertical propagation to occur, k_3^2 must be positive. Since the averaged state of the troposphere is stable ($\gamma_* < \gamma_a$ and $m > 0$), the latter requirement sets the condition

$$0 < u_* < \frac{\beta}{k^2}. \tag{5.35}$$

Therefore, easterly flow ($u_* < 0$) or large westerly flows will not allow the vertical propagation of Rossby waves. Generally, this result is confirmed by observations.

In order to extract more information from the solution of the problem, we shall assume, following Wiin-Nielsen (1975), that in (5.30) and (5.31) $u_* = 0$ and $\partial/\partial_y = 0$. Then, letting

$$A'(x, y, t) = \hat{A}(p) \exp[ik(x - ct)], \quad A = \Phi, w_p,$$

and substituting into the previous equations, we get an ordinary differential equation

$$\frac{d^2 \hat{w}}{dp^2} - \frac{c + \beta/k^2}{c(f/k)^2} m(p)\hat{w} = 0. \tag{5.36}$$

It is now convenient to introduce the notations

$$c_R = \frac{\beta}{k^2}, \qquad c_I = \frac{f}{k}, \qquad \eta = \frac{p}{p_{00}}, \qquad c_m^2 = mp_{00}^2, \tag{5.37}$$

where p_{00} = 1000 mb. Evidently, $-c_R$ and $\pm c_I$ are the phase velocities of the two-dimensional pure Rossby waves and inertial waves. If $m > 0$, then c_m is a real quantity. Then Equation (5.36) becomes

$$\frac{d^2 \hat{w}}{d\eta^2} + q^2 \hat{w} = 0, \qquad q^2(\eta) = -\frac{c + c_R}{c c_I^2} c_m^2(p). \tag{5.38}$$

The behaviour of its solution depends essentially on the function $q(\eta)$. It will define the vertical profile of the waves. A simple solution is obtained at $q(\eta) = q$ = const (e.g., its average value in the layer).

Being interested in finding a wave-type solution of (5.38), we assume $q^2 > 0$. Then the general solution is

$$\hat{w}(\eta) = M \cos q\eta + N \sin q\eta. \tag{5.39}$$

Under boundary conditions

$$\hat{w} = 0 \quad \text{at } \eta = 0, 1 \tag{5.40}$$

we get $M = 0$ and

$$q_n = n\pi, \quad n = 0, 1, 2, \ldots, \tag{5.41}$$

i.e., a discrete spectrum of eigenvalues for q. Hence, and from (5.38), we find for the phase velocity

$$c_n = \frac{-c_R}{1 + n^2\pi^2(c_I^2/c_m^2)}. \tag{5.42}$$

This expression ensures the assumed inequality $q^2 > 0$. As can be seen, at $n = 0$ $c_0 = -c_R = -\beta k^{-2}$, i.e., the Rossby formula (5.9). This is the zero (normal) mode of the Rossby waves at which $\hat{w} = 0$ consistent with (5.39). At $n = 1, 2, \ldots$, higher (vertical) modes are obtained, also propagating westward ($c_n < 0$) but more slowly than for $n = 0$. For these $\hat{w}(\eta) = N \sin q_n\eta \neq 0$.

$$c_n \approx \frac{-c_R}{n^2\pi^2(c_I^2/c_m^2)} = \frac{-\beta}{n^2\pi^2(f^2/c_m^2)}, \tag{5.43}$$

i.e., c_n does not depend any more on the wavelength $\lambda = 2\pi/k$. This is known as the *ultra-long wave approximation*. As can be easily seen, it is derived from (5.38) at $c \ll c_R$, i.e., setting $c + c_R \approx c_R$ which, in its part, is equivalent to neglecting the time derivative in the vorticity equation (5.30). Then an approximate balance exists between the β-effect and the divergent factor. We shall note that the considered approximation could also be realized at small values of m when $c_m \to 0$, i.e., according to (5.29) and (5.37) at a quasi-neutrally stratified atmosphere ($\gamma_* \to \gamma_a$). Evidently, the ultra-long waves are very slowly propagating waves ($|c_n| \ll |c_0|$ at $n \gg 1$).

6. OROGRAPHIC WAVES

When an air flow encounters an obstacle such as a mountain and is forced to flow over it, then depending on the height and horizontal scale of the obstacle, different wave motions occur that have meteorological significances. Thus, the ascent of the ground surface, in accordance with the potential vorticity conservation theorem (5.15) in Chapter 3, will induce additional vorticity to the flow. In the case of a sufficient horizontal extent of the mountain range, this may be balanced by the β-effect. As a result, the so-called topographic Rossby waves occur.

In the case of an isolated mountain obstacle, the forced ascent of air masses along the upward slopes and their descent along the lee slopes generates gravity waves with a considerably smaller length than those of the Rossby waves. These are called mountain waves.

Above such obstacles, especially if these have a circumsymmetric shape, and at small Rossby numbers, another interesting phenomenon, called the Taylor column, is observed. Let us consider separately each of the phenomena mentioned above.

a. Topographic Rossby Waves

In order to discover, at least qualitatively, the physical essence of the phenomenon we shall return to the 'shallow water' approximation in the form (5.17) and (5.16),

$$\frac{d_2 \zeta_a}{dt} = -\zeta_a \nabla \cdot \mathbf{u}, \qquad \frac{d_2 h}{dt} = -h \nabla \cdot \mathbf{u}, \tag{6.1}$$

where $\zeta_a = \zeta + f$. Hence, we get

$$\frac{d_2 \zeta_a}{dt} = \left(\frac{\zeta_a}{h}\right) \frac{d_2 h}{dt}$$

or

$$\frac{d_2}{dt}\left(\frac{\zeta_a}{h}\right) = 0, \quad \text{i.e.,} \quad \frac{\zeta(x, y, t) + f(y)}{h(x, y, t)} = \zeta_* = \text{const}, \tag{6.2}$$

which is one more version of the potential vorticity conservation theorem. Using (6.2), we shall analyze the case shown in Figure 5.8.

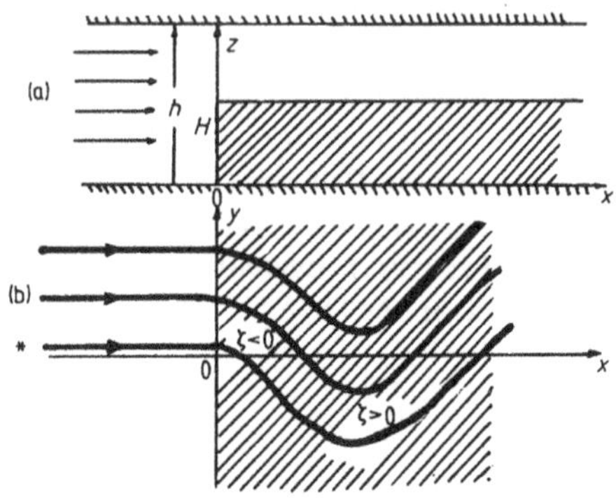

Fig. 5.8. Topographic Rossby waves above a smooth plateau: (a) vertical cross-section, (b) horizontal cross-section.

To simplify matters, we assume a flow bounded from the top by a rigid wall and with $\zeta = 0$ at $x \leqslant 0$. The plateau with height $H = \text{const}$ extends to $\pm\infty$ along the y-axis and may have an infinite or finite width l in the x-direction. According to (6.2) and Figure 5.8, we can write

$$\frac{f_0 + \beta y + \zeta}{h - H} = \frac{f_0}{H} \quad \text{or} \quad \zeta = -\frac{H}{h}\left(f_0 + \frac{h}{H}\beta y\right), \tag{6.3}$$

referring these formulae to the streamline marked by (*) in Figure 5.8. It can be seen that immediately after passing the plateau the flow receives negative (antycyclonic) vorticity (in the Northern Hemisphere) due to a decrease of the depth from h to $h - H$ (Figure 5.8a). But $\zeta < 0$ means that the streamlines (trajectories in the steady-state case) will curve to the right. At the meridional boundary of the obstacle, and westerly flow as shown in Figure 5.8, it means the generation of a northwesterly motion component and the southward moving of the air masses to a lower latitude with a lower value of $f(y)$. A tendency arises towards a change of the curvature from anticyclonic to cyclonic and to the acquisition of a southwesterly motion component. Later, a tendency towards anticyclonic deformation arises again, etc. In other words, at the westerly flow and meridional obstacle, at $x > 0$ waves may form in the velocity (and pressure) field whose lengths, as can be shown by a simple analytical description of the phenomenon, have an order of magnitude of 1000–2000 km.

Let index '1' denote $x < 0$ and '2' denote $x > 0$. We introduce the notations

$$h = h_1, \qquad u_1 = \text{const}, \qquad v_1 = 0, \qquad h - H = h_2,$$

$$u = -\psi_y, \qquad v = \psi_x, \qquad \zeta = \nabla^2\psi, \qquad f = f_0 + \beta y.$$

Obviously, at $x < 0$ $\psi = -u_1 y$. In (6.2) $\zeta_* = \zeta_*(\psi)$ is an arbitrary function. We write this equation for $x < 0$:

$$\zeta_*(\psi) = \frac{1}{h_1}(f_0 + \beta y) = \frac{1}{h_1}\left(f_0 - \frac{\beta\psi}{u_1}\right).$$

Then for $x > 0$ it follows that

$$\frac{1}{h_2}(\nabla^2\psi + f_0 + \beta y) = \frac{1}{h_1}\left(f_0 - \frac{\beta\psi}{u_1}\right),$$

or

$$\nabla^2\psi + k^2\psi = f_0\,\frac{h_2 - h_1}{h_1} - \beta y, \quad k^2 = \frac{h_2\beta}{u_1 h_1}, \tag{6.4}$$

which is a known type (Helmholtz's) nonhomogeneous partial differential equation. Evidently

$$\psi_{(1)} = \frac{1}{k^2}\left(f_0\,\frac{h_2 - h_1}{h_1} - \beta y\right)$$

is a particular solution of (6.4). Observations imply an approximate linear dependence of ψ on y. Thus, we assume the general solution of the homogeneous equation following from (6.4) to be

$$\psi_{(2)} = (c + y)\,M(x).$$

Then for M we obtain $M'' + k^2 M = 0$ and, consequently, $M(x) = A\cos kx + B\sin kx$. The general solution of (6.4) will be

$$\psi(x, y) = (y + c)(A\cos kx + B\sin kx) + \psi_{(1)}(y), \tag{6.5}$$

where A, B, c are constants to be determined.

Since at $x = 0$, $\psi_y = -u_1$ and $\psi_x = 0$ must hold, we find $B = 0$, $A = -u_1 + \beta/k^2$. Let us, for simplicity, consider a streamline such that $\psi = 0$ at $x = y = 0$. Then $c = f_0/\beta$. Thus, finally,

$$\psi(x, y) = -u_1 y - u_1\left[\frac{h_2 - h_1}{h_2}(f_0 + \beta y)\,\frac{\cos kx - 1}{\beta}\right]. \tag{6.6}$$

The obtained solution (6.6) describes an oscillating flow in the x-direction. The length of the wave is

$$L = \frac{2\pi}{k} = 2\pi\left(\frac{u_1 h_1}{\beta h_2}\right)^{1/2}. \tag{6.7}$$

If we assume $h_2 = 0.9\,h_1$, $\varphi = 45°$ and $u_1 \sim 1$ m s^{-1}, then we calculate $L \approx 1600$ km. These are the topographic Rossby waves, generated by an obstacle and developing because of the Earth's rotation and sphericity. When $l \sim L/4$, the tendency towards cyclonic deformation in the first trough will be amplified by returning to the initial layer depth h. For a very small zonal extent of the obstacle ($l \ll L$), no waves will arise. A negative vorticity will be only generated above the obstacle $\zeta = -(H/h)f$ and the motion will almost follow that expected for the same latitude.

A completely opposite picture takes place at the eastern flow. Then, an anticyclonic deformation of streamlines means their orientation in a southeast-northwest direction. But in such a motion with a northerly component, f increases and ζ, while remaining negative, should increase in magnitude in order to keep ζ_* constant. Deformation will continue, streamlines will follow a northerly direction, even with an easterly component, i.e., the flow may 'turn back'. In contrast to the previous case, there are no favourable conditions for wave generation. Often in such situations, ridges of high pressure and closed anticyclones arise.

Obviously, in this case the dynamic influence of the obstable can also extend upstream ($x < 0$), while at a westerly flow it is only felt downstream ($x > 0$).

The validity of these qualitative conclusions also remains for a complex topography of the terrain $H = H(x, y)$. On the other hand, the generation of Rossby topographic waves depends on the dynamic stability of the flow and cannot receive a comprehensive description in the framework of linear theory. For the purposes of qualitative analysis, another version of the potential vorticity conservation theorem, rather than (6.3), could be used, e.g., those in Section 5 of Chapter 3. The results are analogous. It is clear that the conclusions made above are applicable for both the atmosphere and the ocean.

b. Mountain Waves

The existence of isolated mountain obstacles changes abruptly the air flow structure. The forced lifting of the air upslope and descending downslope, which is more pronounced in the case of well elongated mountains transverse to the flow, generates downstream stationary internal gravity waves. The disturbance above the mountain itself also has a wave-type character. Waves in front of the mountain are rarely observed.

Mountain wavelengths vary from 2–3 to 40–45 km, with 10 km being the most typical. Mountain waves are vertically transverse waves with air oscillating around the equilibrium level of the undisturbed portion of the flow. The mountain is a generator of waves and the role of restoring force is played by the static stability of the atmosphere. Mountain wave amplitude (A) is of the same order of height as the mountain itself. For example, $A \sim 1$ km above the Carpathians but $A \sim 2$–3 km above the Alps and the Rocky Mountains.

The theory of this phenomenon originated in the works of Dorodnitsin, around 1940, and was further developed by Kibel, Lyra, Scorer and others. Its current variants are rather sophisticated and allow for several effects: condensation, diabaticity, nonlinearity, etc. On the other hand, it appears possible to neglect the Coriolis force because of the local character of the phenomenon. The quasi-static approximation is often applied, leading, contrary to the expectations, to results which are consistent with observations. However, baroclinicity and compressibility are taken into account.

In Figure 5.9, a schematic representation of streamlines overflowing a mountain is shown according to (a) Lyra and (b) Scorer. In the wave ridges on the leeside of the mountain, immediately above and adjacent to the ground turbulent boundary layer, vortices with horizontal axes called *rotors* are often formed with diameters of up to 0.5–1.0 km. They are overflowed by a basic flow nearly as a solid, thus increasing the effective mountain width. Periodically, the rotors are 'shed' and carried downstream by the flow.

The amplitude of the leewaves decreases with height. A phase-shift is also observed. Above the mountain itself, the streamline curvature changes several times with height.

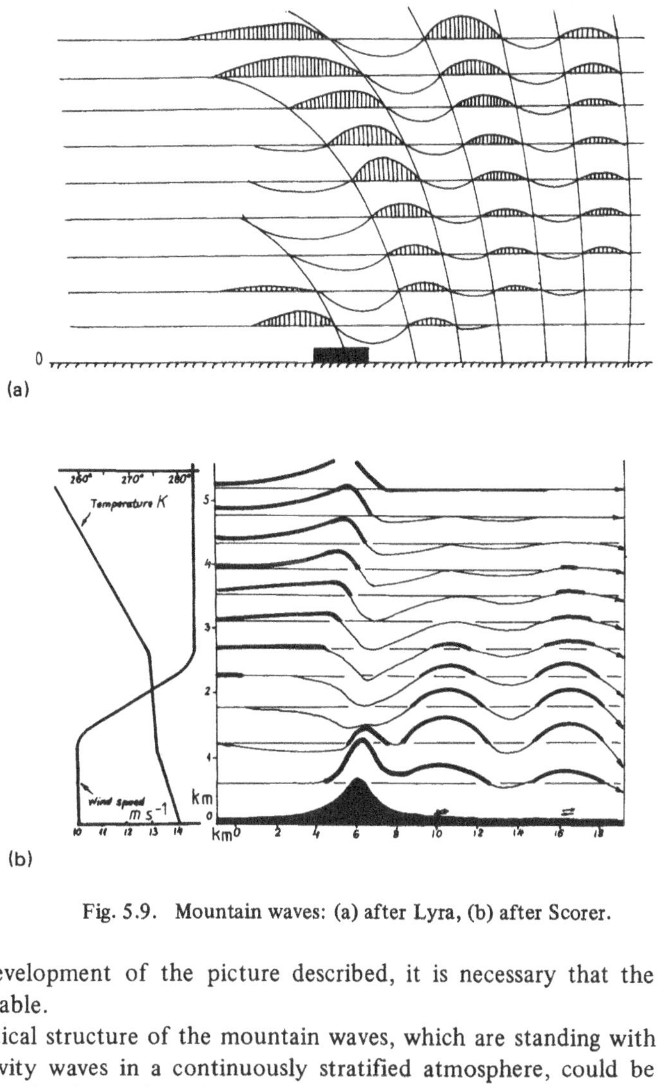

Fig. 5.9. Mountain waves: (a) after Lyra, (b) after Scorer.

For the development of the picture described, it is necessary that the atmosphere is statically stable.

The vertical structure of the mountain waves, which are standing with respect to the ground gravity waves in a continuously stratified atmosphere, could be approximately modelled by the theory from Section 4. Thus, according to (4.22), we get for the vertical wavelength

$$\lambda_3 = \frac{2\pi}{k_3} = 2\pi\left(\frac{S}{u_*^2} - \frac{1}{4H_s^2}\right)^{1/2} \approx \frac{2\pi u_*}{S^{1/2}}, \tag{6.8}$$

where, by virtue of (4.15), we have

$$S = \frac{R\Gamma}{H_s} = \frac{gT}{T_0^2}(\gamma_a - \gamma) \approx g\epsilon_0, \quad \epsilon_0 = \frac{\gamma_a - \gamma}{T_0}. \tag{6.9}$$

In Dorodnitsin's theory, the role of the static stability parameter is performed by the quantity $D^2 = g\epsilon_0/u_*^2$. Here u_* is assumed to be independent of z. At $u_* = 0$, as well as at the neutral and unstable stratification ($\gamma \geqslant \gamma_a$) obviously no waves exist, i.e., stable stratification is a necessary condition for their existence. This conclusion, however, is valid only if the Coriolis force and the dependence of u_* on z are not accounted for.

A surprising conclusion from the theory is the change of the streamline curvature above the obstacle. In the field of the vertical velocity, it appears as a change of sign of w. According to the theory

$$w(x, z) \sim \frac{dH}{dx} \sin D(H_s - z)$$

where $H(x)$ is a function describing the mountain profile and H_s is identified with the height of the tropopause. In the regions where $\sin D(H_s - z) > 0$, the signs of w and dH/dx will be the same, i.e., the streamlines will follow the shape of the terrain. At levels z_n where

$$D(H_s - z_n) = n\pi \quad \text{or} \quad z_n = H_s - \frac{n\pi}{D}, \quad n = 0, 1, \ldots \tag{6.10}$$

the sine will become zero and then will change sign. The same will happen with the function $w(z)$. This can also be seen in Figure 5.9. The layer depth between the two neighbouring nodal surfaces will be:

$$\Delta z = z_{n-1} - z_n = \pi u_* \left[\frac{T_0}{g(\gamma_a - \gamma)}\right]^{1/2} \tag{6.11}$$

For typical tropospheric conditions Δz = 2–4 km, so there will be, at most, 3–4 nodal surfaces in the limits of the troposphere.

Mountain waves are more easily identified and followed in the temperature field than in the vertical velocity field. Assuming an adiabatic vertical ascent and descent of the air particles, for temperature changes we obtain $\Delta T = -\gamma_a \Delta z$. Using the potential temperature θ instead of T and from the condition of its conservation $u\theta_x + w\theta_z = 0$, $w = -u\theta_x/\theta_z$ can be calculated. A greater accuracy of measuring the temperature often causes mountain waves to be characterized by the θ distribution above mountains.

Another interesting phenomenon connected with mountain waves, is the resonance amplification when the mountain width and wavelength are close together. Then the oscillation amplitude is maximum and can even exceed the mountain height. Waves with great amplitude have actually been observed above relatively low mountains.

As has been previously said, when the air flow encounters a long mountain at a right angle, or close to it, the mountain wave picture could be considered as two-dimensional. This is not the case, however, with isolated mountain obstacles when the air flow not only 'overflows' but also 'surrounds' the mountain. Waves arise, then, not only downstream and above the mountain, but also in a direction transverse to the flow. The phenomenon is substantially three-dimensional and cannot be described analytically. Existing theoretical models are based on the use of numerical methods.

c. Taylor's Column

When the mountain obstacle which must be overtaken by the flow is sufficiently high and steep, so that the anticyclonic vorticity (induced by topographically 'shortening' the vertical vortex tubes of the incoming flow) cannot be compensated by vorticity advection (and dissipation), a stable eddy formation may arise above the mountain which is called Taylor's column. This phenomenon can be demonstrated in a laboratory if a relative motion of liquid is created in a rotating container by heating or by stirring and if an obstacle is placed on the bottom of the tank so that the moving fluid flows around it [11]. Then the streamlines of the flow will form a column, going around the obstacle as if it extended to the top of the liquid. And this is Taylor's column. Hypotheses exist relating the sun-spots, or the famous Red Spot on Jupiter, to the existence of a stable Taylor's column. The prevailing opinion, however, accepts that it does not have any special meteorological significance in the Earth's atmosphere.

7. EMPIRICAL EVIDENCE FOR THE EXISTENCE OF WAVE MOTIONS IN THE ATMOSPHERE

It has been shown in the previous sections that weather equations possess wave-type solutions, corresponding to various wave motions.

Generation and propagation of sound waves in the atmosphere is an obvious fact. It is difficult to imagine a soundless world. It is also beyond any doubt that these waves are not related to significant weather phenomena. That is why they represent a peculiar 'noise' in weather-equation solutions. From the point of view of numerical weather prediction methods, this noise is unwanted (affecting solution stability) and it is desirable that it be removed in advance by filtering the sound waves. As we have already seen, this can be done by using the incompressibility and quasi-staticity hypotheses. Gravity and inertial waves in the atmosphere (and ocean) are related to the meteorological (oceanic) phenomena described below.

a. Gravity Waves

Some more important phenomena, associated with gravity waves, are:

(i) Relatively regularly disposed wave-type clouds, sometimes observed above plains and above mountain regions as well. In the first case, this particular cloud shape is caused by internal gravity waves which have occurred upon discontinuity surfaces (e.g., the top boundary of temperature inversions), located around the condensation level. In the wave ridges, according to (4.10), $w' > 0$ and has a maximum. Favourable conditions exist for a thickening of the cloudiness. In the troughs $w' < 0$ has a maximum too, but no cloudiness is formed there or, if there has been any, it grows thinner (due to adiabatic warming). In the case of lee mountain waves, the wave cloud formation mechanism is similar. In the case of precipitating clouds, it can lead to a regular succession of precipitation and nonprecipitation areas on the ground – an observationally well-known phenomenon which is called 'patchiness of the precipitation'. This phenomenon can be discovered, even in complex geographical conditions like those of the area around the city of Sofia [20], surrounded by mountains with different shapes and heights (Vitosha, Lyulin, the Balkan Mountains).

Another well known observational fact is the phenomenon of 'multilevel cloudiness' above the obstacle itself – an alternation of cloudy and clear layers along the vertical. It indicates the wavy character of the vertical profile of the vertical velocity $w'(z)$ above the mountain. Thus, at $u_* \sim 10$ m s^{-1}, $T \sim T_0 \sim 3 \times 10^2$ K and $\gamma \sim 6°$ km^{-1} from (6.4), we obtain a value of $\lambda_3 \sim 600$ m for the wavelength which agrees qualitatively with the observed depth of the cloudy and clear layers.

(ii) In the case of gravity-wave destruction due to Kelvin–Helmholtz-type instability, as shown in Figure 5.6, the air layer becomes extremely turbulent at the corresponding height. In a cloudless atmosphere, this phenomenon is referred to as 'clear air turbulence' (CAT). It is detected by the so-called 'angel echo' – a radar echo from the corresponding layer without any visible object to produce it. During the past 10 to 15 years, CAT has been the subject of some thorough investigations and forecasting in relation to the unpleasant and not always safe 'bumping' of aircrafts (even supersonic ones) flying in those layers.

A similar phenomenon is also observed in the ocean and leads to the formation of thin interfaces with intensive turbulence, separated by layers of quasi-laminar motion. Another suitable region for wave formation is the ocean *thermocline*. This name is given to the thin transition water layer with a sharp vertical density (temperature) gradient, separating the well-mixed surface layer from the deep and cooler homogeneous layer. Its existence has been experimentally confirmed.

Occasionally, observations and photographs have been made of clouds resembling the last stage (t_4) of billow formation due to gravity-wave instability (Figure 5.6). Along with laboratory simulation of this process, these records are evidence of the existence and realization of such a phenomenon in the atmosphere.

(iii) Gravity waves also occur upon frontal surfaces and because of their synoptic scale they are influenced by the Earth's rotation and are rather inertial gravity waves.

Finally, during powerful volcanic eruptions when, in certain locations, the atmosphere is displaced from its dynamic equilibrium in the whole characteristic depth, fast gravity waves of a free-surface type (see Section 3) occur. They propagate almost with the speed of sound. Their analogues are the tsunami waves in the ocean and related to them are tidal waves, caused by the gravitational attraction of the Moon and the Sun, existing not only in the ocean but also in the atmosphere. As far as the Sun is concerned, much more essential is the thermal wave, excited in the Earth's atmosphere because of its axial rotation. Harmonic pressure and temperature analysis allows oscillations with multiple periods of solar and lunar days, to be discovered. The semidiurnal period, acting as a conditional boundary between high- (gravity) and low-frequency (inertial) waves occupies a central place in the theory.

b. Inertial Waves

As is already known, to this category are assigned the inertial waves with period $T_i = 2\pi/f = \pi/\omega \sin\varphi$ (see Section 6 of Chapter 3, and Section 5 of this chapter) and Rossby waves. The former do not have any essential significance in the atmosphere (except probably in equatorial regions and in the ocean). Undoubtedly, Rossby waves are of greatest interest in a meteorological aspect.

Even before the fundamental work of Rossby (1939), it had been noticed that the surface pressure isobars on the map of the Northern Hemisphere were of an approximately wavy character along the latitude circles. This can be much more readily seen on the streamline charts which were drawn later for the various geopotential levels and, in particular, for the 500 mb level. Two such charts for this level, separated by a 24-hour interval, are shown in Figures 5.10a and b, borrowed from [64]. An examination of these charts shows that the wavelength along the latitude circles is of the order of 5000 km and the well-pronounced meanders propagate to the east. The wavelength along the meridians is smaller. The zonal flow is quasi-nondivergent and its velocity is approximately in an inverse proportion to the distance between the streamlines. The drift of the waves to the east shows that in the Rossby formula (5.11) $\lambda < \lambda_s = 2\pi/(u_*/\beta)^{1/2}$. It is also seen that the waves are at different stage of development – some have a regular shape, others resemble a dissipating wave with the base overtaken by the ridge, while third ones are only beginning to rise etc. This impression is confirmed after an examination of Figure 5.11, in which the constant pressure balloon trajectories, flowing at the 300 mb level in the Southern Hemisphere, are presented. Waves with a length of $\lambda \sim 6000$ km arise. These results are interpreted as qualitative evidence for the existence of nonstationary Rossby waves. On the other hand, the permanent lows familiar in meteorology (e.g., over

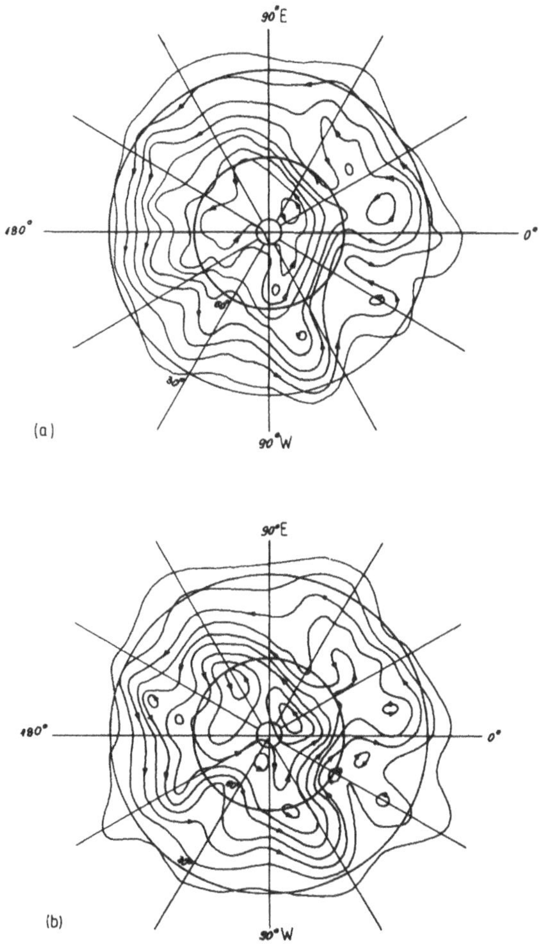

Fig. 5.10. Streamlines on a 500 mb surface with interval of 24 h (after [64]).

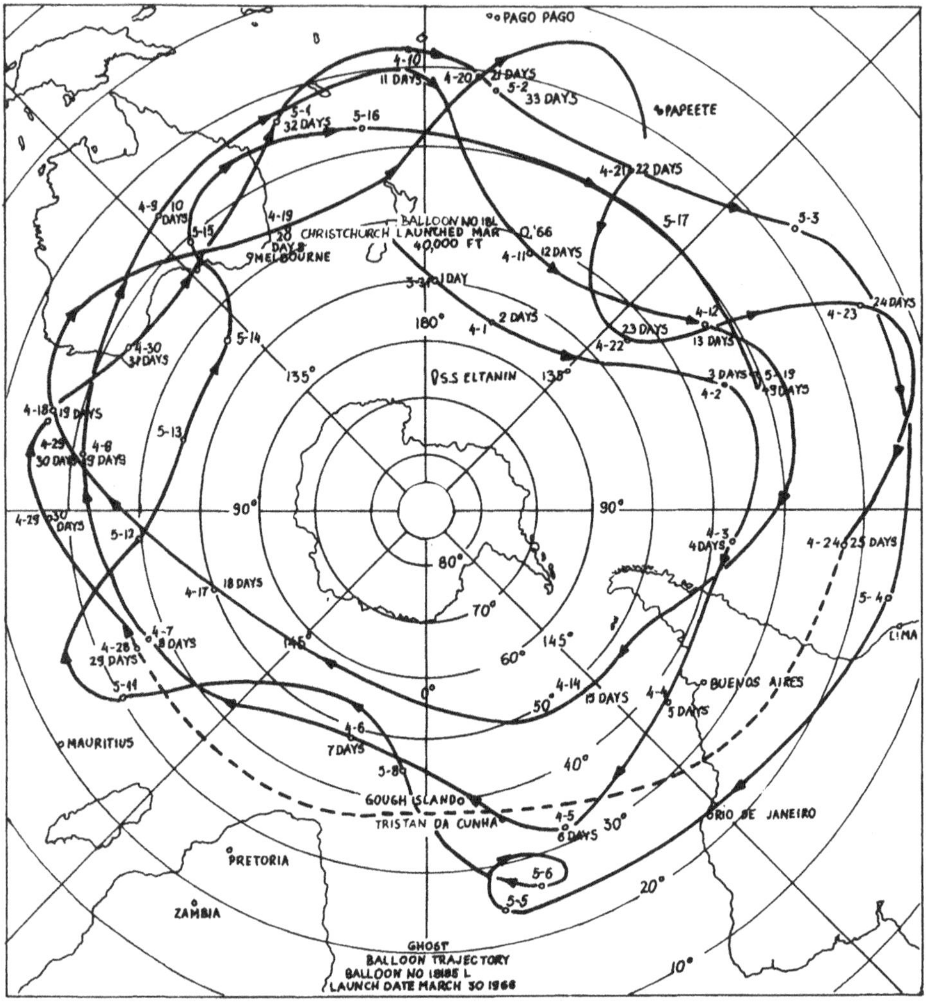

Fig. 5.11. Balloon trajectory as an indicator of global scale wave motions (after *COSPAR transactions* No 3, 1967).

Iceland) and permanent highs (e.g., over the Azores and Honolulu) which are also called atmospheric activity centers, can be considered as stationary Rossby waves because of their relatively constant geographic location. It is beyond any doubt, however, that for a more accurate description and explanation of these phenomena, it is necessary to take into consideration the large-scale baroclinicity due to continent-ocean and pole-to-equator temperature contrasts, as well as the spherical shape of the Earth. The contemporary mathematical formulation of this problem and its first solutions are associated with the name of Blinova (1943).

We shall also note here that the Rossby waves are self-sustaining in the sense that, once excited by some cause (e.g., topographic factor), they can exist without its action as far as they are sustained by the Earth's rotation. Nevertheless, if some external 'forcing' factor occurs, through specified circumstances, it can lead to an amplification of the already-existing oscillations (resonance) and the formation of well expressed cyclones and anticyclones (ridges and troughs). The role of such a forcing factor can be played by the Earth's surface with its large-scale thermal inhomogeneities, due to the existence of continents and oceans. In fact, their dimensions are of the same order as stationary Rossby wavelengths (several thousands of kilometers) and, in principle, resonance, even though for short time and over limited territory, is not excluded.

Along with forcing factors, however, dissipative ones also exist, causing a damping of the waves which have been excited by some reason. Such a dissipative factor is the friction between the atmosphere and the Earth's surface and internal eddy viscosity. Let us briefly consider the first factor; as far as the second one is concerned, this will be the subject of special attention later.

To allow for the influence of this factor in the simplest way, we shall approximate the frictional force after the Guldberg–Mon scheme: $\mathbf{R} = -r\mathbf{u}$. Then the vorticity equation (5.2) will take the form

$$\zeta_t + u\zeta_x + v\zeta_y + r\zeta + \beta v = 0, \tag{7.1}$$

which at $\partial/\partial y = 0$ ($\zeta = \partial v/\partial x = v_x$) becomes

$$v_{xt} + uv_{xx} + rv_x + \beta v = 0 \tag{7.2}$$

and after linearizing

$$v_{xt} + u_* v_{xx} + rv_x + \beta v = 0. \tag{7.3}$$

We simplify the problem further by considering stationary waves ($\partial/\partial t = 0$). Equation (7.3) becomes an ordinary equation:

$$u_* v_{xx} + rv_x + \beta v = 0. \tag{7.4}$$

Let us assume at $x = 0$, $v = 0$ and $(\partial v/\partial x)_0 = \zeta_0$ – the initial vorticity. The solution of (7.4) with these conditions is readily obtained:

$$v(x) = \frac{\zeta_0}{k'_s} \exp[-rx/2u_*] \sin k'_s x, \tag{7.5}$$

where $k'_s = (\beta/u_* - r^2/4u_*^2)^{1/2}$ is the wavenumber. It is seen that the friction acts in two ways. On the one hand, it decreases the wavenumber of the stationary Rossby waves with respect to $k_s = (\beta/u_*)^{1/2}$ from (5.10) (increasing the corresponding wavelength λ'_s). On the other hand, as could be expected from general physical considerations, friction leads to a rapid (exponential) damping of the wave amplitudes with distance to the disturbance source at point $x = 0$. Thus, at $r = 10^{-6}$ s^{-1} and $u_* = 10$ m s^{-1} the amplitude decreases twice in $x = 2\lambda'_s$. That is why the above-mentioned possibility for resonance does not occur in practice.

8. DYNAMIC INSTABILITY OF ATMOSPHERIC MOTIONS

a. General Considerations

The concept of the static stability or instability of the atmosphere is widely known in meteorology. It is associated with the vertical shifting of air masses in the field of gravitational force. Using the so-called 'parcels method' it is easy to derive the following stability criteria for dry air: $\gamma > \gamma_a$ – unstable, $\gamma < \gamma_a$ – stable, $\gamma = \gamma_a$ – neutral, where γ_a is the dry adiabatic and $\gamma = -\partial T/\partial z$ – the actual vertical temperature gradient (lapse rate) in the atmosphere.

Only this concept of stability or instability is obviously insufficient for a comprehensive analysis of real atmospheric motions. The large-scale motions in the atmosphere are quasi-horizontal and their behaviour in the presence of disturbances should be studied from the standpoint of the hydrodynamic instability (stability) theory.

The atmospheric motions we are studying are governed by a set of nonlinear differential equations for the fundamental variables u, v, w, p, ρ, T, etc. and, shortly, ψ_i. A general concept of the stability of solutions after Lyapunov exists, meaning the following [9]:

Let $\psi_i(t)$ be a solution of these nonlinear equations. One says that $\psi_i(t)$ is *stable* if given $\epsilon > 0$ and an initial condition at $t = t_0$, then there exists $\eta = \eta(\epsilon, t_0)$ such that any other solution $\varphi_i(t)$ for which $|\psi_i(t_0) - \varphi_i(t_0)| < \eta$ also satisfies $|\psi_i(t) - \varphi(t_0)| < \epsilon$ for $t \geqslant t_0$. If no such η exists, then $\psi_i(t)$ is *unstable*. If $\psi_i(t)$ is stable and, furthermore, $|\psi_i(t) - \varphi_i(t)| \to 0$ as $t \to \infty$, then we say that $\psi_i(t)$ is *asymptotically stable*.

As can be seen, this is stability (instability) with respect to disturbances in the initial conditions. It could also be named nonlinear stability (instability). A fundamental method for examining the hydrodynamic instability of fluid flows, and of geophysical fluids (atmosphere, ocean) in particular, is the linearization of the equations about small perturbations of the basic flow. For such a linearized system, however, the concept of stability is understood in another sense. Now the unstable solutions always grow infinitely while the stable ones are limited by a certain constant for any instant of time t. Thus, the latter correspond to observable states in the modelled system. It can be shown that if a motion is linearly unstable, it will be unstable in the sense of Lyapunov too. The opposite case is obvious – a nonlinear solution that is unstable will also be linearly unstable. Further, a nonlinear solution that is linearly stable will be stable only to sufficiently small perturbations. This is the case of *infinitesimal stability*. If a solution is stable to an arbitrary perturbation, we say that it is *globally stable*. This, however, cannot be determined from linearized equations. Nevertheless, a study of linearized equations can yield useful information.

The phenomenon 'loss of motion stability' or, in other words, 'occurrence of instability' is very important in geophysical fluid dynamics and in meteorology in particular. Really, atmospheric circulation arises under the influence of permanently-acting steady factors, such as solar (short-wave) radiation influx, long-wave radiation of the atmosphere and the Earth's surface and its axial rotation. As a result, some steady circulation should be established; for instance, westerly transport in the middle latitudes or close to it, influenced by the inhomogeneous Earth's surface. In reality, however, the atmospheric motions are extremely irregular. On weather maps one can identify not only damping waves, excited by some random factor, but waves with amplifying amplitudes too, which lead, at a certain stage, to the complete destruction of the zonal (west-to-east) circulation and its replacement by an eddy (cyclonic) one. Therefore, the weather phenomena are, to a considerable degree, conditioned by the zonal circulation instability in the atmosphere.

Several types of hydrodynamic instabilities exist. We shall now briefly discuss some of them.

b. Inertial Instability

For simplicity, we shall assume a purely zonal basic geostrophic motion

$$u_* = u_g = -f^{-1}\Phi_y, \qquad v_* = 0, \tag{8.1}$$

where $\Phi_y = \partial\Phi/\partial y$ and $u_g = u_g(y)$. As in Section 4, we shall again apply the parcels method.

An air parcel moves with the basic flow until the moment t_0, when some force displaces it across its equilibrium trajectory. Its motion at $t > t_0$ will be governed by the equations

$$\dot{u} = fv = f\dot{y}, \qquad \dot{v} = f(u_g - u), \tag{8.2}$$

where $(\dot{\ }) = \mathrm{d}/\mathrm{d}t$. Providing $f = \text{const}$ and $y(t_0) = 0$, we obtain from the first equation (8.2) by integrating

$$u(y) = u_g^0 + fy. \tag{8.3}$$

But at $y \to 0$

$$u_g(y) = u_g^0 + \left(\frac{\partial u_g}{\partial y}\right)_0 y. \tag{8.4}$$

Substituting both these equations into the second equation (8.2), related to $t > t_0 (y \neq 0)$, we obtain

$$\ddot{y} + \omega_f^2 y = 0, \quad \omega_f^2 = f\left(f - \frac{\partial u_g}{\partial y}\right). \tag{8.5}$$

This equation is of the same type as (4.12). Consequently, the stability criteria for the zonal motion (8.1) in the Northern Hemisphere will be

$$\left(f - \frac{\partial u_g}{\partial y}\right) \begin{cases} > 0 & \text{stable} \\ = 0 & \text{neutral} \\ < 0 & \text{unstable.} \end{cases} \tag{8.6}$$

Since $-\partial u_g/\partial y = \zeta_g$, (8.6) could be rewritten in the form

$$\zeta_a^* = (f + \zeta_g) >, =, < 0. \tag{8.7}$$

As has been commented in Section 8 of Chapter 3, according to observations, in synoptic-scale motions $\zeta_a^* > 0$, thus the zonal motions generally being dynamically stable as a whole (just as $\gamma < \gamma_a$ in the free atmosphere and the atmosphere is generally statically stable). The occurrence of areas with $\zeta_a^* < 0$ in the $x0y$-plane will be followed by unstable motion of the air masses, meridional mixing and a decreasing of the shear $\partial u_g/\partial y = -\zeta_g$ to values $\zeta_g < f$. This mechanism of disturbing and restoring the balance is called *inertial instability*.

Actually, criteria (8.6) or (8.7), similar to the hydrostatic ones ($\gamma >, =, < \gamma_a$), are rather coarse and have limited application. They do not account for the many other factors influencing the instability of the motion – such as friction, variation of f, existence of horizontal and vertical shear, etc.

A more rigorous approach to the problem of hydrodynamic instability, as has been said in the beginning, is based on the use of linearized equations of motion in Eulerian variables and the determination of the conditions for which the solution will describe disturbances, infinitely amplifying with time. This is done by looking for a wave-type solution of the form $\exp[ik(x - ct)]$ with complex phase velocity $c = c_R + ic_I$. This technique is called the *normal modes method*.

c. Barotropic Instability

At $c_I > 0$, the wave amplitude will grow exponentially with time – the motion will be unstable. Instability can occur both in a barotropic and baroclinic medium. For the wave amplitude (energy) to increase, an energy source is necessary. In the barotropic medium this can take place only at the expense of transport (advection). In the baroclinic medium, the waves grow because of the conversion of the potential energy of the baroclinic state into kinetic energy.

An important source of instability energy in both cases is the existence of basic flow shear. In a barotropic atmosphere, as we know, $\partial u_*/\partial z = 0$ and the most significant of the horizontal shears of u_* is the meridional one $\partial u_*/\partial y$ of the zonal component u_*. In a baroclinic atmosphere $\partial u_*/\partial z \neq 0$ and the thermal wind equation (4.22) of Chapter 3 represents the vertical shear of u_* with sufficient approximation, connecting it with the horizontal temperature gradient ∇T_*.

As an example, we shall consider the problem of Rossby barotropic wave stability. Since in deriving Equation (5.6) from (5.2), we assumed u_* = const, now, if considering $u_* = u_*(y)$ it would be easy to obtain in the same way the linear equation

$$\left(\frac{\partial}{\partial t} + u_* \frac{\partial}{\partial x}\right)\nabla^2 \psi' + \tilde{\beta}(y)\psi'_x = 0, \tag{8.8}$$

where

$$\tilde{\beta}(y) = \beta - \frac{d^2 u_*}{dy^2} = \frac{d}{dy}\left(f - \frac{du_*}{dy}\right) = \frac{d}{dy}\,\zeta_a^*. \tag{8.9}$$

Assuming a solution of the form

$$\psi'(x, y, t) = \hat{\psi}(y)\exp[ik(x - ct)],$$

after substitution in (8.8), we obtain the ordinary differential equation

$$(u_* - c)(\hat{\psi}_{yy} - k^2\hat{\psi}) - \tilde{\beta}(y)\hat{\psi} = 0, \tag{8.10}$$

with boundary conditions

$$\hat{\psi}(y) = 0 \quad \text{at } y = \pm d. \tag{8.11}$$

The latter means that we consider the motion in a zonally-oriented channel with rigid boundaries at $y = \pm d$ (Figure 5.12). Really, because the normal component becomes equal to zero at the boundary, we get

$$v(\pm d) = \psi'_x(\pm d) = ik\hat{\psi}(\pm d) \exp[ik(x - ct)] = 0.$$

Hence, it follows $\hat{\psi}(\pm d) = 0$, i.e., Equation (8.11). Naturally, the unrealistic boundary conditions will distort the solution close to the boundaries, but in the central parts of the jet its influence will decrease.

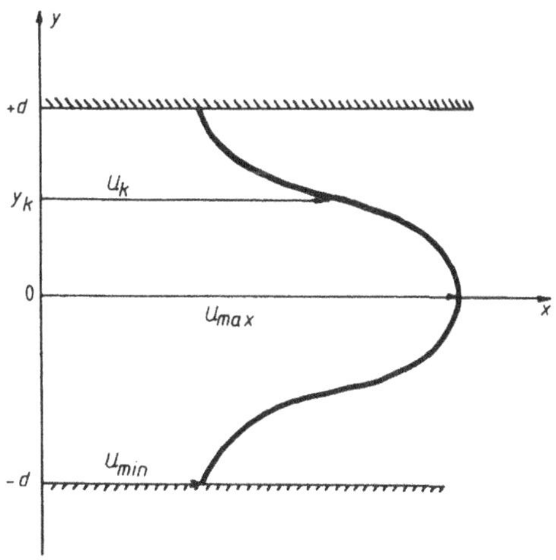

Fig. 5.12. A jet-like velocity profile in the meridional direction $u_* = U(y)$ and the problem for barotropic instability.

The so-formulated problem (8.10) and (8.11) does not possess an exact analytical solution for an arbitrary function $\tilde{\beta}(y)$, i.e., $u_*(y)$. Solutions are possible only for some special forms of $u_*(y)$ and for phase velocity values satisfying certain dispersion relation. If its solution yields $c = c_R + ic_I$ and $c_I > 0$, then the motion will be unstable. We shall not treat such particular cases.

It is possible, however, that a general form of the necessary condition can be found in order for the boundary-value problem (8.10), (8.11) to have an unstable solution in the sense of amplifying waves, without solving it. Following [23] and [29], we write $c = c_R + ic_I$, $\hat{\psi} = \hat{\psi}_R + i\hat{\psi}_I$, multiply (8.10) by the complex conjugate $\hat{\psi}_{cc} = \hat{\psi}_R - i\hat{\psi}_I$, divide by $u_* - c$, integrate over y between $\pm d$, equate the real and imaginary parts, and for the imaginary part, obtain

$$\int_{-d}^{d} \left(\hat{\psi}_I \frac{d^2}{dy^2} \hat{\psi}_R - \hat{\psi}_R \frac{d^2}{dy^2} \hat{\psi}_I\right) dy = c_I \int_{-d}^{d} \tilde{\beta}(y) \frac{|\hat{\psi}|^2}{|u_* - c|^2} dy.$$

However, $ab'' - a''b = (ab' - a'b)'$ where $(') = d/dy$. On the other hand, $\hat{\psi}_I = \hat{\psi}_R = 0$ at $y = \pm d$. Hence,

$$0 = c_I \int_{-d}^{d} \tilde{\beta}(y) \frac{|\hat{\psi}|^2}{|u_* - c|^2}\, dy.$$

If amplified (unstable) waves exist, then $c_I > 0$ and

$$\int_{-d}^{d} \tilde{\beta}(y) \frac{|\hat{\psi}|^2}{|u_* - c|^2}\, dy = 0 \tag{8.12}$$

is the required necessary condition. For (8.12) to hold, it is necessary for $\tilde{\beta}(y)$ to change sign at least once in the region $-d < y < +d$, i.e., at least one y_k should exist in the interval $(-d, d)$ for which

$$\tilde{\beta}(y_k) = 0 \quad \text{or} \quad \zeta_a^*(y_k) = \max\,(\min). \tag{8.13}$$

Conditions (8.12) and (8.13), found for barotropic instability in the general case are only necessary. However, they also prove to be sufficient for the assumed profile of $u_*(y)$, shown in Figure 5.12, which is often observed in the mid-latitude free atmosphere. The wave amplitudes will be increasing, decreasing or constant with time depending on the relation between U_k, $U_{\min}$, $U_{\max}$ and c_R (always $< U_{\max}$). For example, it appears that at $U_{\min} < c_R < U_k$ we have instability, at $c_R > U_k$ – damping, and at $c_R < U_{\min}$ the waves are neutral.

d. Baroclinic Instability

The atmosphere as a whole is a baroclinic medium. This property, however, is irregularly distributed in space. Areas exist, for example frontal regions, with a great 'concentration' of baroclinicity (sharp horizontal temperature gradients). The instability of wave disturbances upon frontal surfaces is closely related to this fact and has an exceptionally great significance with respect to the origination and development of cyclones on them.

On the other hand, regions can exist in the atmosphere with a continuous distribution of baroclinicity with height, described by the thermal wind Equation (4.22) in Chapter 3. The wave motions in this case will occur and develop against the background of a well-expressed vertical shear of the basic flow ($\partial \mathbf{u}_*/\partial z \neq 0$) that substantially influences their stability. In the general case, both the horizontal and vertical shear of the basic flow $\mathbf{u}_*$ exist and the wave motions are substantially three-dimensional.

The theory of baroclinic instability is more complicated than that of the barotropic one. Its basic aim, as in the latter case, is to determine the waves (their length λ or phase velocity c) which are baroclinically unstable.

A relatively simple theory of baroclinic instability has been proposed by Eady (1949). The theory predicts the existence of *maximum instability* for disturbances whose wavelength

$$\lambda_m = 2\pi H \omega_{\mathrm{BV}}/1.61 f \tag{8.14}$$

where H is the depth of the atmospheric layer, supposed to be confined between rigid horizontal boundaries. For typical values of the parameters in (8.14) $\lambda_m \sim 4000$ km (see also Section 3 of Chapter 7). We should expect disturbances of nearly this wavelength to be the most common ones occurring in the atmosphere. Observations in the mid-latitudes, particularly on the scales and the typical growth times of the cyclones, support this conclusion. Thus, it can be supposed that they originate as small perturbations in the baroclinically unstable basic flow. Laboratory experiments and special observations confirm the determinant role of the baroclinic instability mechanism for origination and development of the synoptic-scale wave motions in the mid-latitudes and their growth, though various other factors act as well (nonadiabatic, nonlinear, dissipative, etc.).

e. Convective Instability

Most of the atmospheric motions on all scales are of convective origin. The nonlinearity, related to the heat and vorticity (momentum) transport, is the common property of these motions. The criteria for hydrostatic stability following from (4.13) ($\gamma <, >, = \gamma_a$ or $\partial\theta/\partial z >, <, = 0$ respectively) for a given stratification to be stable, unstable or neutral are very simple and practically convenient, but they are not precise enough because of the great physical simplifications introduced. Another approach to the problem of the origination of convective motions in idealized situations, modelling real convection phenomena observed in laboratory or natural conditions (atmosphere, ocean), suggests the low-order system (9.23) of Chapter 3, derived by Lorenz. Let us write it again here:

$$\dot{X} = -\sigma X + \sigma Y, \qquad \dot{Y} = -XZ + rX - Y, \qquad \dot{Z} = XY - bZ. \tag{8.15}$$

First of all, we look for steady solutions (X_s, Y_s, Z_s) of the system (8.15). Letting $\dot{X} = \dot{Y} = \dot{Z} = 0$, we easily find

$$X_s = Y_x = Z_s = 0 \tag{8.16}$$

and

$$X_s = Y_s = \pm\sqrt{b(r-1)}, \quad Z_s = r - 1 \tag{8.17}$$

Obviously, the trivial (zero) stationary solution (8.16) corresponds to the state of rest (no convection). This solution 'meets' the next one (8.17) at the point $r_c' = 1$ – the critical value of the relative Rayleigh number. That is why it is called the *point of bifurcation* or branching point: both steady solutions (8.16) and (8.17) emanate from it in opposite directions ($r < 1$ and $r > 1$). The nontrivial solutions (there are two because of $\pm$) correspond to some circulation patterns (two-dimensional rolls or cells) in the fluid.

The next step consists of determining the stability of stationary solutions (8.16) and (8.17). For this purpose we linearize system (8.15) by letting $X = X_s + x$, $Y = Y_s + y$, $Z = Z_s + z$. Upon linearizing (8.15) with respect to the steady solution, we obtain

$$\begin{aligned} \dot{x} &= -\sigma x + \sigma y, \\ \dot{y} &= -Z_s x - X_s z + rx - y, \\ \dot{z} &= Y_s x + X_s y - bz. \end{aligned} \tag{8.18}$$

By letting $\psi(t) = \hat{\psi}\exp(\omega t)$, $\psi = x, y, z$, system (8.18) is reduced to a homogeneous algebraic one

$$\begin{aligned} &-(\sigma+\omega)\hat{x} + \sigma\hat{y} = 0,\\ &(-Z_s + r)\hat{x} - (\omega+1)\hat{y} - X_s\hat{z} = 0,\\ &Y_s\hat{x} + X_s\hat{y} - (b+\omega)\hat{z} = 0. \end{aligned} \tag{8.19}$$

Provided that Det (8.19) = 0, the system has a nontrivial solution. This yields the characteristic cubic equation

$$(b+\omega)\,[(\sigma+\omega)(1+\omega) + \sigma(Z_s - r) + X_sY_s\sigma + X_s^2(\sigma+\omega)] = 0. \tag{8.20}$$

First we apply (8.20) to (8.16). Then

$$(b+\omega)\,[\omega^2 + (\sigma+1)\omega + \sigma(1-r)] = 0, \tag{8.21}$$

so that

$$\omega_1 = -b, \qquad 2\omega_{2,3} = -(\sigma+1) \pm [(\sigma+1)^2 + 4\sigma(r-1)]^{1/2}. \tag{8.22}$$

At $r > 0$ all the roots (8.22) are real and negative provided $r < 1$, which means the asymptotic stability of the state of rest (8.16). However, at $r > 1$ one of the roots $\omega_{2,3}$ becomes positive, so that (8.16) is no longer a stable stationary state. It becomes unstable, thus giving rise to convection at $r = r_c' = 1$ and exchanging stability with the two stationary states (8.17) in the bifurcation point $r_c' = 1$.

Actually, for (8.17), the characteristic equation (8.20) takes the form

$$\begin{aligned} &\omega^3 + A\omega^2 + B\omega + C = 0,\\ &A = \sigma + b + 1, \quad B = b(1+\sigma), \quad C = 2\sigma b(r-1). \end{aligned} \tag{8.23}$$

The analytical solution of (8.23) is tedious. However, we are interested in the real parts of the roots only. It is known that the necessary and sufficient conditions for the cubic equation (8.23) to have three roots with negative real parts is that $A > 0$, $C > 0$ and $AB - C > 0$. The latter condition yields

$$r(b - \sigma + 1) > -\sigma(\sigma + b + 3). \tag{8.24}$$

Since, $\sigma, b > 0$, the first two conditions are always satisfied.

If $\sigma < b + 1$, then (8.24) is satisfied and the real parts of all the roots of (8.23) are negative and the stationary convection corresponding to (8.17) is always stable.

If $\sigma > b + 1$, then at

$$r < \frac{\sigma(\sigma + b + 3)}{\sigma - b - 1} = r_c'', \tag{8.25}$$

again the solution (8.17) is stable. But for $r > r_c''$ at least one root of (8.27) can have a positive real part. Thus, the convection becomes unstable for sufficiently high values of r. Note that $r_c'' > r_c' = 1$. For instance, for the most often used values of the parameters $\sigma = 10$, $a^2 = \frac{1}{2}$ we have $b = \frac{8}{3}$ and $r_c'' = 470/19 \approx 24.74$.

The two-dimensional rolls to which the stationary solutions (8.17) correspond are time-independent patterns of convection. When they become unstable, the motion will oscillate in intensity. What happens when the amplitudes of oscillation become large is not revealed by the linear theory. To answer this question, a numerical integration of the nonlinear system (8.15) must be performed.

9. A CONCEPT OF NONLINEAR WAVES IN THE ATMOSPHERE

a. Solitary Waves (Solitons)

Along with linear theory, there has recently been great development in the nonlinear theory of waves in the atmosphere and ocean too. One of the most interesting facts established in the course of these investigations indicates that the nonlinear equations of geophysical fluid dynamics, besides having wave solutions with a periodic character, asymptotically possess another type of solution which describes the propogation of isolated (single, solitary) disturbance with a constant form. Thus, a solitary wave is a localized perturbation propagating in a dispersive medium without altering its shape. In some cases, these solitary waves interact in such a way that a collision will not alter their velocity or shape. Such waves are called *solitons*.

In fact, the investigation of solitary waves in hydrodynamics has a history of more than 130 years. At first it was connected with the names of Scott, Russell, Boussinesq, Rayleigh and others, but after 1895, the greater parts of the theoretical works on this subject are based on the well known Korteweg–de Vries (KdV) equation which describes the evolution of long surface gravity waves with small but finite amplitudes and dispersion in the 'shallow water' approximation. This equation in its one-dimensional case reads [68]

$$h_t + (c_0 + c_1 h)h_x + \nu h_{xxx} = 0, \tag{9.1}$$

where c_0, c_1 and ν are constants, $h = h(x, t)$ is the deviation of the free surface. In the linear approximation ($c_1 = 0$), Equation (9.1) has a solution $h = a \cos\theta$ where $\theta = kx - \omega t$ and $\omega = c_0 k - \nu k^3$ is the dispersion relation.

As for the nonlinear equation (9.1), it has also a periodic solution $h = h(\theta)$ where $h(\theta)$ can be expressed by the so-called elliptic cosine $cn\theta$ (Jacobi's elliptical function). The corresponding waves are named '*cnoidal*'.

The remarkable property of the KdV Equation (9.1) is the fact that, along with the periodic solutions, it has another particular solution given by

$$\begin{gathered} h(x, t) = a \operatorname{sech}^2\left[\left(\frac{ac_1}{12\nu}\right)^{1/2} (x - ct)\right], \\ c = c_0 + \frac{ac_1}{3}, \quad \operatorname{sech} z = 2(e^z + e^{-z})^{-1}. \end{gathered} \tag{9.2}$$

It is easy to check that (9.2) solves (9.1). The factor in front of $(x - ct)$ stands for the wave number k and $c = c_0 + ac_1/3$ for the dispersion relation. Therefore, in contrast to linear waves, where $\omega = \omega(k)$, now $\omega = \omega(k, a)$ where a is the amplitude. This is the most important feature of nonlinear waves with dispersion.

In the variable $\tilde{x} = x - ct$ the shape of the function $h(\tilde{x})$ from (9.2) is shown in Figure 5.13. Obviously, $h(\pm\infty) = 0$ and $h(0) = a$. That 'hump' moves in the positive direction of x with a velocity c proportional to the amplitude a which can be arbitrary. This is exactly the so-called 'solitary wave'. Linear equations do not possess such solutions. Therefore, solitary waves are essentially nonlinear phenomena.

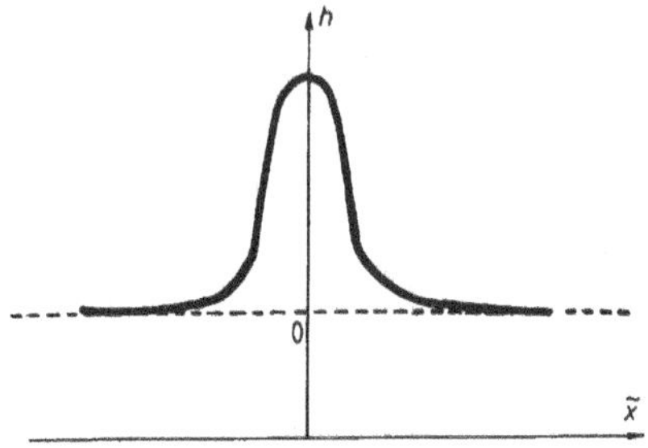

Fig. 5.13. A solitary wave (soliton).

As we know, the superposition of solutions of a linear equation is also its solution. Due to this fact, for instance, two neutral linear waves, propagating with different velocities, can pass through each other, but still retain their shape. As for the nonlinear equations, however, the superposition of their solutions does not give a new solution. Regardless of all this, most solitary waves have a 'linear' behaviour – they interact nonlinearly but after that come out unchanged in form.* This similarity to the elementary particles is the reason for them being named 'solitons'. This feature is illustrated on Figure 5.14. The higher soliton (Figure 5.14a) overtakes the lower one, interacts with it nonlinearly in conformity with Equation (9.1), and comes out 'unchanged' (Figure 5.14b), as happens in the case of a linear process. Nonlinearity is manifested only by the fact that the two solitons, after their interaction, are not to be found at those places where they would be if there was not any interaction.

The theory of nonlinear waves, in particular of the solitary waves, occupies a special place in modern physics. As well as in fluid dynamics, such waves can be found in radiophysics, plasmaphysics, geophysics and other disciplines. The very KdV equation, at first derived for waves in fluids, also appears in the theory of elasticity, plasmaphysics, etc. For a time it was thought that soliton solutions would occur for KdV equation only. In recent years it has become apparent that this is not the case. Many equations of mathematical physics have this kind of solution. Here are some examples:

(a) The Schrödinger equation with a cubic nonlinearity:

$$i\psi_t + \psi_{xx} + k|\psi|^2\psi = 0, \tag{9.3}$$

where k is a constant and $\psi(x, t)$ is a complex function, describes the two-dimensional self-focussing of light beams.

* Some solitary waves do not possess this property. Therefore, they are not solitons. For instance, the front of a bush fire.

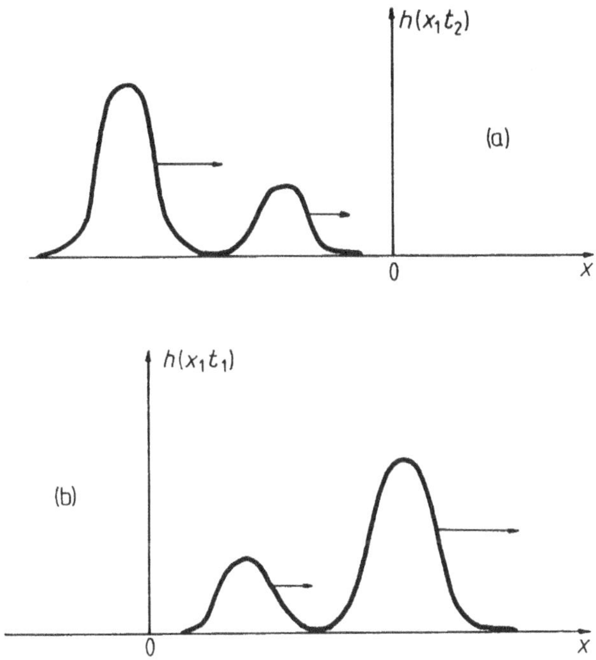

Fig. 5.14. The interaction of two solitons.

(b) The Sine–Gordon equation

$$u_{xx} - u_{tt} = \sin u, \tag{9.4}$$

describing the propagation of the so-called spin waves in superfluid He^3, of model crystal dislocations, etc.

(c) Boussinesq's equation

$$u_{xx} - u_{tt} + 6(u^2)_{xx} + u_{xxxx} = 0 \tag{9.5}$$

describes, similarly to the KdV equation, shallow-water waves.

The experimental research of solitary waves has also been developed parallel with theory – both in laboratory and natural conditions. Their detection and identification in geophysical flows (atmosphere and ocean) is of particular interest. Information about this has only recently begun to appear in meteorological literature.

b. Atmospheric Solitons

A historical survey, new data from observations and analysis of the surface atmospheric pressure, obtained from a series of highly sensitive microbarometers located in Central Australia, are presented in the paper of Christie *et al.* (1978). According to these authors, the first description of a disturbance in the field of surface pressure, interpreted as an internal gravity solitary wave, was given by Abdullah (1955).

In the early hours of 29 June 1951, the barometers in Kansas, U.S.A., registered a disturbance of +3.4 mb which was moving up to a distance of 800 km with a velocity of 18–24 m s^{-1}. It was caused by a nucleus of cold air with horizontal dimensions of about 150 km and a constant form at a height of 2 km, sliding on an inversion layer. Abdullah explains this formation as a result of an impulsive motion of the quasi-stationary cold front in the thermal inversion layer.

A series of such observations are described in the above-mentioned work of Christie *et al.* However, the solitons that they observed only had an effective width of several kilometers. Two and more than two solitons (a packet of solitons), not far away one from another, have been observed sliding along the night inversion layer. Simultaneous observations of surface humidity f_0, temperature t_0, wind velocity c_0 and direction θ_0 show that the passage of a p_0-soliton is not accompanied by a similar phenomenon in the fields of other meteorological elements. Additional investigations have shown that the passage of such a solitary wave does not exert any influence on the atmospheric turbulence. Therefore, these micro- and meso-scale waves are not only solitary in the already-accepted meaning but they are also of no obvious significance in atmospheric dynamics.

Probably, solitary waves of a synoptic and planetary scale also exist in the atmosphere. A typical example are the solitary Rossby waves (Rossby solitons). Theoretically, they are predicted as solutions of the nonlinear vorticity equation on the β-plane (Redekopp and Weidman, 1978). However, so far there are no reliable empirical data confirming the predictions of the theory. A hypothesis was proposed that the big Red Spot on Jupiter might be connected with a planetary solitary wave – a giant two-dimensional soliton.

Solitons are identified in the ocean too. A typical example is the case of a solitary tsunami wave which is usually caused by an earthquake. However, such a wave could also be caused by a deep stationary cyclone over the ocean (sea) area. Due to the low pressure in the center, the sea level increases there. The near-surface winds in the standing cyclone also contribute to this. When the cyclone starts moving, a solitary water wave also moves from the center to the periphery of the area previously occupied by the stationary cyclone.

Quite probably, the famous Leningrad floods are caused by such solitary waves. The deep cyclones, standing for several days over the Baltic Sea, are a well-known phenomenon in meteorology. Entering the Gulf of Finn, the soliton increases its height, meets the waters of the Neva river, and causes floods.

Satellite observations of solitary waves in the ocean prove to be a promising perspective in this direction.

PROBLEMS

1. Starting from the equations

$$\dot{u} + \alpha \frac{\partial p}{\partial x} = 0, \qquad \dot{w} + \alpha \frac{\partial p}{\partial z} + g = 0,$$

$$\alpha \dot{p} + \kappa p \dot{\alpha} = 0, \quad \alpha \nabla \cdot \mathbf{u} = \dot{\alpha},$$

where $\kappa = c_p/c_v$ and $\partial/\partial y = 0$, derive the linearized system (2.9)–(2.12).

2. Find the group velocity of the external gravity waves whose phase velocity is given by (3.15) and analyze the result.
3. If $\partial/\partial t = -c\,\partial/\partial x$ prove that Equation (5.4) has a first integral

$$\nabla^2 \Psi + f = F[\Psi(x - ct, y)],$$

where $\Psi = cy + \psi$, $F(\Psi)$ is an arbitrary function, find the solution if $F(\Psi) = -m^2 \Psi$.
4. Let $u_*(y) = u_0 + sy$ and $\beta = 0$. Prove that the vorticity equation

$$\left(\frac{\partial}{\partial t} + u_* \frac{\partial}{\partial x}\right)\nabla^2 \psi' = 0$$

following from (8.8), under the initial condition

$$v'(x, y, 0) = \psi'_x = -v_0 \sin kx$$

has a solution

$$\psi'(x, y, t) = \frac{v_0 \cos k(x - u_* t)}{k(1 + s^2 t^2)}.$$

Also find the velocity components $u' = u_* - \psi'_y$, $v' = \psi'_x$.
5. Consider the system of Equations (5.30) and (5.31) with $-\delta\,\partial\Phi'/\partial p$ instead of zero in the right-hand side of (5.31), thus taking into account the Newtonian heating, δ = const (Wiin-Nielsen, 1975). Repeat the derivations and the analysis described in Section 5 for the simpler (adiabatic) case. Especially show that the Newtonian heating causes an exponential damping of the type $\exp[-\delta t]$.
6. Solve Equation (5.36) with $m = m_0 p^{-2}$ which, according to (5.29), corresponds to the isothermal atmosphere.
7. The momentum equation and the heat influx equation for rising parcels with mixing due to the entrainment of surrounding air as derived in [1] read

$$\frac{\mathrm{d}X}{\mathrm{d}t} = -\alpha X^2 + \beta Y \tag{1}$$

$$\frac{\mathrm{d}Y}{\mathrm{d}t} = -aX - \alpha XY \tag{2}$$

where $X(t) = W$ is the speed of the vertical parcel rising, $Y(t) = T_p - T$ is the parcel's 'overheating' with respect to the ambient air having temperature T (T_p is the parcel's temperature), $\beta = g/\bar{T}$, α is the coefficient of entrainment, and $a = \gamma_a - \gamma$ (γ_a and γ are the dry adiabatic and the actual lapse rates).

Prove that in the case of a statically-stable stratification of the atmosphere ($\gamma < \gamma_a$ and $a > 0$) the system of Equations (1) and (2) admit periodic solutions

$$X(t) = W_* \frac{\sin \tau}{x_0 + 1 - \cos \tau}, \qquad Y(t) = \frac{a}{\alpha} \frac{\cos \tau}{x_0 + 1 - \cos \tau}, \tag{3}$$

where $\omega_{\mathrm{BV}} = (a\beta)^{1/2}$ is another expression for the Brunt–Väisälä frequency, $\tau = \omega_{\mathrm{BV}} t$,

$W_* = \omega_{\mathrm{BV}}/\alpha$, $x_0 = a/\alpha Y(0)$, $Y(0)$ is the initial (at $t = 0$) overheating of the parcel. Compare this solution with (4.12) and (4.14) for the adiabatic case ($\alpha = 0$ in (1) and (2)).

Prove also that in case of $a < 0$ (unstable stratification)

$$X(t) = -W_* \frac{1 - e^{-2\tau}}{2x_0\, e^{-\tau} - (1 - e^{-\tau})^2},$$
$$Y(t) = \frac{a}{\alpha} \frac{1 + e^{-\tau}}{2x_0\, e^{-\tau} - (1 - e^{-\tau})^2}, \qquad (4)$$

i.e., the solution is not periodic, where $(-a\beta)^{1/2}$ is now denoted by ω_{BV}.

8. Show that for $1 < r \neq 16$ the low-order system from Problem 15 in Chapter 3, has a nontrivial stationary solution

$$\theta_s = v_s - \frac{v_s w_s}{2\sigma}, \qquad \delta_s = 4w_s + \frac{v_s^2}{2\sigma},$$
$$v_s^2 = 4\sigma w_s \frac{r - 16}{8 + 2\sigma - w_s},$$

where w_s is determined from the cubic equation

$$w_s^3 + 2(5\sigma - 7)w_s^2 + 4[14\sigma^2 + \sigma(42 + r) + 8]\, w_s + 16\sigma(\sigma + 4)(r - 1) = 0.$$

Here $w_{1s} = v_s$, $w_{2s} = w_s$, $\theta_{1s} = \theta_s$, $\theta_{2s} = \delta_s$ are the stationary values of the Fourier components for which $\dot{w}_i = \dot{\theta}_i = 0$ $i = 1, 2$. Investigate the solution for $\sigma = \frac{7}{5}$.

CHAPTER 6

The Mutual Adjustment of Meteorological Elements

1. GEOSTROPHIC ADJUSTMENT: ONE-DIMENSIONAL MODEL

a. Significance of the Problem

The complete set of equations of atmospheric fluid dynamics and thermodynamics contains time derivatives of five elements: u, v, w, p, ρ (or T), i.e., it is of the fifth order in time. For initial value (Cauchy) problems, such as those for meteorological forecasting, the fields of all five elements u_0, v_0, w_0, p_0, ρ_0 (or T_0) at the initial instant t_0 should be given. These initial fields are taken from observations made and, as a rule, different elements are independently measured and are always loaded with random and systematic errors of different orders.

On the other hand, meteorological elements are related in-between by a united system of equations – the weather equations – and must satisfy it at any instant t, including the initial one (t_0). In other words, they should be initially consistent. However, due to the above-mentioned errors, the initial data are unbalanced to a certain degree. Using such a set of inconsistent initial data when solving the nonlinear system of weather partial-differential equations will be equivalent to introducing a greater or smaller perturbation in the atmospheric model, whose mathematical representation these equations are. Their solution will describe a spectrum of virtual motions which can be classified into two groups – relatively slow (synoptic) and fast (wave-like) motions. Synoptic motions have a great meteorological significance. Their investigation, however, is substantially embarrassed by the existence of fast motions. This is particularly noticeable in numerical prediction methods where the presence of fast wave-type motions is responsible for the appearance of the so-called computational instability and for a divergence of numerical solutions of the weather equations. A necessity arises for taking special measures for eliminating those effects. Two approaches are possible.

The first one consists of simplifying the weather equations in order to eliminate the unnecessary solutions describing fast (acoustic and gravity) waves, so that the simplified system would accurately describe the slow (synoptic, including wave-type) motions. As we already know, this is achieved through quasi-static and quasi-geostrophic approximations. It will be shown in Chapter 7 that their simultaneous application reduces the system to a first-order one in time, so that for a description of the evolution of the synoptic processes, an initial field of only one element – most often the pressure (geopotential) or the streamfunction – is sufficient.

When such simplification is not desirable for certain reasons, one has to turn to the second approach – the mutual adjustment of the initial data, now known as the initialization procedure – thus reducing the perturbations, introduced by them, to a minimum. The adjustment itself can be realized by different techniques which will not be considered here. Of course, the combined approach – simplification and adjustment – is possible too.

In principle, it is not possible to reach perfect mutual consistency of the initial fields. In that case, regions with larger inconsistencies will become sources of fast spreading in the remaining part of the space disturbances. After their diffusion, a balance between the model fields should be expected. In essence, it will be a process of mutual adjustment (adaptation) of the theoretically-predicted meteorological elements.

So far, our considerations have been carried out within the framework of a mathematical model of the atmosphere. Adjustment processes, however, also take place in the real atmosphere. It is an obvious observational fact that the macro-scale atmospheric motions are quasi-geostrophic and quasi-static, i.e., an approximate balance exists between pressure gradient force, Coriolis force and gravitational force respectively. Let us suppose that this balance is disturbed in some region of the atmosphere. Then fast-wave motions are generated, diffusing the perturbation energy through the whole atmosphere – the quasi-balance being restored afterwards. The fact that the atmosphere, as a rule, is in a permanent quasi-geostrophic and quasi-static balance tells of the existence of physical adaptation mechanisms. As for the physical causes disturbing this balance, they can be both internal for the atmosphere (strong fronts, hurricanes, etc.) and external (solar activity, powerful eruptions, etc.). On the other hand, the absence of an exact geostrophic and hydrostatic balance in the atmosphere shows that fundamental and secondary reasons are incessantly trying to disturb it. These two tendencies act in dialectic unity and make the inexhaustible variety of atmospheric dynamics so very interesting.

The fast wave processes in the real atmosphere have the characteristics of small oscillations around the equilibrium state and are studied by means of linear-theory methods. In linear approximation, quasi-geostrophic and quasi-static equilibrium states turn out to be steady states (explicitly independent of time t). In fact, however, the mutually-adjusted meteorological elements experience slow changes in time (they evolve) with a characteristic time scale of $T \sim 1$ day or more and this evolution is the essence of what is called the *synoptic process*. This evolution is due to nonlinear effects: advection of vorticity, entropy, etc. They cause a permanent tendency towards disturbing the existing quasi-equilibrium, at which the opposite tendency for its restoration originates. Therefore, when describing this evolution, nonlinear effects must be accounted for (using the nonlinear equations). In principle, it is impossible to describe the evolution processes within the framework of linear theory. At the same time, in order to avoid the treatment of the accompanying adjustment process, not related to significant meteorological phenomena, the meteorological elements should always be considered as mutually adjusted. The recognition of this necessity and the searching of ways for achieving compatibility of these requirements, represents the essence of the so-called quasi-static and quasi-geostrophic approximations in the theory of meteorological forecasting (see Chapter 7, Sections 2 and 3). This was a prominent achievement of dynamic meteorology of the 1950s and is of great practical importance in relation to numerical weather prediction methods and their computer-aided realization, originating at that time.

The founder of the adjustment theory in its oceanographic aspect was Rossby (1936).

He developed a one-dimensional mathematical model of the phenomenon. Later Obukhov (1949) gave rigorous mathematical formulation and a comprehensive solution of the two-dimensional problem in a barotropic atmosphere. Appreciable contributions to the study of the adjustment problem have been made by Monin, Kibel, Blumen, and Wiin-Nielsen, etc.

Consecutively, we shall consider one-, two-, and three-dimensional models of the phenomenon, above both a flat Earth and a sphere. Let us start with the one-dimensional model.

b. One-Dimensional Model

First of all, this model has a methodological value. Moreover, in the historical aspect, the first mathematical model of the geostrophic adjustment process was also a one-dimensional one. Here we shall discuss it in a contemporary formulation and solution, following [23].

Again, it is more convenient to use the equations in the shallow water approximation. After letting $\partial/\partial x = 0$ (one-dimensional space problem) and linearization the equations take the form

$$\begin{aligned} &u_t - fv = 0, \qquad h_t + h_* v_y = 0, \\ &v_t + fu + gh_y = 0. \end{aligned} \tag{1.1}$$

We specify the following initial conditions

$$\begin{aligned} &u = u_0, \quad |y| \leqslant a, \\ &u_0 = 0, \quad |y| > a, \\ &v_0 = 0, \quad (\nabla h)_0 = 0. \end{aligned} \tag{1.2}$$

Obviously, the initial fields of velocity and height are not balanced on the segment $|y| \leqslant a$, where $(\nabla h)_0 = 0$ but $u_0 \neq 0$. Thus, a disturbance is introduced into the problem.

Eliminating h and u from (1.1) we obtain

$$v_{tt} + f^2 v - gh_* v_{yy} = 0 \tag{1.3}$$

and look for a solution $v \sim \exp[i(ky - \omega t)]$. Substitution into (1.3) yields the dispersion relation

$$\omega = \pm(f^2 + gh_* k^2)^{1/2} \tag{1.4}$$

indicating that the disturbance, so introduced, generates inertial-gravity waves by means of which it will diffuse out of the segment $|y| \leqslant a$ along the whole y-axis.

The study of the adjustment process itself, however, requires a solution of the problem with given initial conditions (initial value or Cauchy problem). Thus, for Equation (1.3) at $t = 0$ we have

$$v_0 = 0, \qquad v_t(y, 0) = -fu_0 \quad \text{at } |y| \leqslant a. \tag{1.5}$$

where the second condition has been obtained with the help of (1.1) and (1.2).

We represent the solution by means of Fourier integrals

$$v(y, t) = \frac{1}{2\pi}\int_{-\infty}^{\infty} \tilde{v}(k, t) \exp[iky]\, dk,$$
$$\tilde{v}(k, t) = \int_{-\infty}^{\infty} v(y, t) \exp[-iky]\, dy. \tag{1.6}$$

Then (1.3) becomes an ordinary differential equation $\tilde{v}_{tt} + \omega^2\tilde{v} = 0$ whose general solution is

$$\tilde{v}(k, t) = A(k) \cos \omega t + B(k) \sin \omega t. \tag{1.7}$$

The initial conditions (1.5) require

$$A(k) = 0, \qquad B(k) = -\frac{2fu_0}{\omega k} \sin ak. \tag{1.8}$$

In its final form, the solution reads

$$v(y, t) = -\frac{2u_0 f}{\pi}\int_0^{\infty} \frac{1}{\omega k} \sin ak \sin \omega t \cos ky\, dk. \tag{1.9}$$

Expressions for $u(y, t)$ and $h(y, t)$ can easily be found from the first two equations (1.1).

The asymptotic behaviour of the solution at $t \to \infty$ is of special interest. It is convenient to investigate it for the central point $y = 0$. Then it follows from (1.9) that $v(0, t) \to 0$ at $t \to \infty$. In the limit at $t = \infty$, $u_t = v_t = h_t = 0$ and, for the stationary values, we obtain

$$u_s = -\frac{g}{f}\frac{\partial h_s}{\partial y}, \qquad h_s = h_s(y),\ v_s = 0 \tag{1.10}$$

i.e., geostrophic balance. In addition, we have to find the shape of the stationary functions $u_s(y)$ and $h_s(y)$. For the purpose, we shall avail of the fact that the system (1.1) possesses an invariant

$$\frac{\partial}{\partial t}\left(\frac{\partial u}{\partial y} + \frac{f}{h_*} h\right) = 0 \quad \text{or} \quad \frac{\partial u}{\partial y} + \frac{f}{h_*} h = \text{const} \tag{1.11}$$

which is a linear variant of the potential vorticity conservation theorem (6.2) in Chapter 5. Writing (1.11) for $t = \infty$ and $t = 0$, we obtain

$$\frac{du_s}{dy} + \frac{f}{h_*} h_s = \frac{du_0}{dy} + \frac{f}{h_*} h_0, \tag{1.12}$$

or

$$\frac{d^2}{dy^2} h_s - \frac{1}{L_R^2} h_s = -\frac{f}{g}\frac{du_0}{dy} - \frac{h_0}{L_R^2} \tag{1.13}$$

where

$$L_R = f^{-1}(gh_*)^{1/2} \tag{1.14}$$

is a characteristic length scale of the process introduced by Rossby and later referred to as the Rossby radius of deformation. At $h_* \sim 1$ km, $L_R \sim 10^3$ km.

Let, for instance at $t = 0$,

$$u_0 = h_0 = 0, \qquad |y| \leqslant \infty$$

$$u_0(y) = \begin{cases} 0, & |y| \geqslant \lambda \\ U\left(1 + \cos \dfrac{\pi y}{\lambda}\right), & |y| < \lambda \end{cases} \tag{1.15}$$

and also $h_s(\pm\infty) = 0$. Then Equation (1.13) becomes

$$h_s'' - \frac{h_s}{L_R^2} = \begin{cases} 0, & |y| \geqslant \lambda \\ \dfrac{\pi U f}{g\lambda} \sin \dfrac{\pi y}{\lambda}, & |y| < \lambda \end{cases} \tag{1.16}$$

and has solutions

$$h_s(y) = \begin{cases} A \exp\left(-\dfrac{y}{L_R}\right), & y \geqslant \lambda \\ B \exp\left(\dfrac{y}{L_R}\right), & y \leqslant -\lambda \end{cases} \tag{1.17}$$

$$C \exp\left(\frac{y}{L_R}\right) + D \exp\left(-\frac{y}{L_R}\right) - \frac{(\pi U f/\lambda g) \sin \pi y/\lambda}{(\pi^2/\lambda^2 + 1/L_R^2)}, \quad |y| < \lambda. \tag{1.18}$$

The constants of integration are determined from the condition of continuity of $h_s(y)$ and $h_s'(y) = \mathrm{d}h_s/\mathrm{d}y \sim u_s(y)$ at $y = \pm\lambda$:

$$A = -B = -\frac{(UfL_R/g)\,\mathrm{sh}\,(\lambda/L_R)}{1 + \lambda^2/\pi^2 L_R^2},$$

$$C = -D = \frac{(UfL_R/g)\exp[-\lambda/L_R]}{2(1 + \lambda^2/\pi^2 L_R^2)}. \tag{1.19}$$

Two asymptotic cases are of interest:

$$u_s = -\frac{g}{f}\frac{\mathrm{d}h_s}{\mathrm{d}y} = \begin{cases} 0 & \lambda \gg L_R. \\ u_0(y), & \lambda \ll L_R. \end{cases} \tag{1.20}$$

Consequently, for disturbances, localized in a limited area ($\lambda < L_R$), the final velocity field almost coincides with the initial one, while in the opposite case, it is mainly determined by the mass field. In the next sections it will be seen that this is a general feature of the adjustment phenomenon, and does not depend on the dimensions of the process.

The solution of the one-dimensional problem gives rise to at least three questions to be answered theoretically:

(1) What is the character of the process by means of which adjustment is realized and what kind of balance exists far enough from the initial instant $t = 0$?

(2) What is the activity degree of the various fields involved in the adjustment process and what does it depend on?

(3) How do the adapted fields depend on the initial ones?

We shall answer these questions by considering more realistic models than the one-dimensional model.

2. GEOSTROPHIC ADJUSTMENT: TWO-DIMENSIONAL MODEL

a. Starting Equations

In his classical work, Obukhov (1949) starts from the system of equations for a compressible, hydrostatic and barotropic atmosphere

$$u_t - fv = -\frac{p_x}{\rho}, \qquad v_t + fu = -\frac{p_y}{\rho},$$

$$0 = -g - \frac{p_z}{\rho}, \qquad \rho_t + (\rho u)_x + (\rho v)_y = 0, \tag{2.1}$$

$$p = p_{00}\left(\frac{\rho}{\rho_{00}}\right)^m, \qquad m = \text{const.}$$

In a linear approximation

$$\rho \mathbf{u}_t = (\rho \mathbf{u})_t - \mathbf{u}\rho_t = (\rho \mathbf{u})_t + \mathbf{u} \cdot \nabla\rho\mathbf{u} \approx (\rho \mathbf{u})_t \tag{2.2}$$

where the continuity equation has been used and the nonlinear terms neglected. Integrating (2.1) over z between $z = 0$ and $z = \infty$ and introducing new variables

$$\mathbf{U} = \frac{g}{p_{00}} \int_0^\infty \rho \mathbf{u}\, dz, \qquad X(x, y, t) = \frac{1}{p_{00}}\, p(x, y, 0, t) - 1, \tag{2.3}$$

Obukhov derives his fundamental system of three equations

$$U_t - fV = -a_0^2 X_x, \qquad V_t + fU = -a_0^2 X_y, \qquad X_t = -(U_x + V_y), \tag{2.4}$$

where $a_0 = (gh_0)^{1/2} = (p_{00}/\rho_{00})^{1/2} = (RT_{00})^{1/2}$ is the isothermal sound speed and p_{00}, ρ_{00}, T_{00} are standard ground values of the pressure, density and temperature of the air. In essence, (2.4) coincides with the linearized system in the shallow water approximation

$$u_t - fv = -gh_x, \qquad v_t + fu = -gh_y, \qquad h_t = -h_*(u_x + v_y). \tag{2.5}$$

Appropriate nondimensionalization makes them identical. However, the model based on (2.4) is physically more realistic.

It is more convenient, instead of $\mathbf{U} = (U, V)$ in (2.4), to introduce the streamfunction Ψ and potential Φ, letting

$$\mathbf{U} = \mathbf{k} \times \nabla\Psi + \nabla\Phi.$$

Hence, $\zeta = \nabla^2 \Psi$, $D = \nabla^2 \Phi$. Assuming f = const, standard procedures applied to (2.4) yield

$$\begin{aligned} &\text{(a)} \quad \nabla^2 \Psi_t + f \nabla^2 \Phi = 0, \\ &\text{(b)} \quad \nabla^2 \Phi_t - f \nabla^2 \Psi + a_0^2 \nabla^2 X = 0, \\ &\text{(c)} \quad X_t + \nabla^2 \Phi = 0. \end{aligned} \tag{2.6}$$

Hence

$$\begin{aligned} &\text{(a)} \quad \Psi_t = -f\Phi, \\ &\text{(b)} \quad \Phi_t = f\Psi - a_0^2 X, \\ &\text{(c)} \quad X_t = -\nabla^2 \Phi. \end{aligned} \tag{2.7}$$

A remarkable property of system (2.7) is that it possesses an invariant which follows immediately from (2.6a, c)

$$(\nabla^2 \Psi - fX)_t = 0, \tag{2.8}$$

or

$$\nabla^2 \Psi(x, y, t) - fX(x, y, t) = \nabla^2 \Psi(x, y, 0) - \\ - fX(x, y, 0) = \zeta_*(x, y, 0). \tag{2.9}$$

Another general property of system (2.7) is that it has a family of stationary solutions Ψ_s, Φ_s, X_s such that

$$\Phi_s = 0, \qquad f\Psi_s = a_0^2 X_s \tag{2.10}$$

following from (2.7) with $\Psi_t = \Phi_t = 0$. Then, if we decompose $A = \Psi, \Phi, X$ into stationary (A_s) and nonstationary (A_n) parts

$$A(x, y, t) = A_s(x, y) + A_n(x, y, t), \tag{2.11}$$

from (2.9) we easily obtain

$$\nabla^2 \Psi_n - fX_n = 0, \tag{2.12}$$

$$\nabla^2 \Psi_s - fX_s = \zeta_*(x, y, 0). \tag{2.13}$$

It is not difficult to see that Equations (2.9) and (2.12), (2.13) as well are linear expressions of the conservation of potential vorticity. Consequently, the nonstationary motion has a zero potential vorticity.

b. Character of the Adjustment Process

Since Ψ_n, Φ_n, X_n satisfy the same Equation (2.7) as do Ψ, Φ, X after differentiating (b) with respect to t and eliminating Ψ_n and X_n by means of (a) and (c), we obtain a wave-type equation for Φ_n:

$$\left(\frac{\partial^2}{\partial t^2} - a_0^2 \nabla^2 + f^2\right) \Phi_n = 0. \tag{2.14}$$

Apparently, (2.14) is a generalization of (1.3) and in a similar manner, it can be shown that it describes the propagation of inertial-gravity waves with a dispersion relation

$$c = \frac{\omega}{k} = \pm\left(a_0^2 + \frac{f^2}{k^2}\right)^{1/2}, \tag{2.15}$$

where $k^2 = k_1^2 + k_2^2$ is a wavenumber. As can be expected, the adjustment process is a wave one.

To clear up the dynamics of the relaxation process, again an initial-value problem about Equation (2.14) should be formulated and solved. For that purpose two initial conditions are needed. Concerning Φ_n they are

$$\Phi_n = \Phi_n^0 \neq 0, \qquad \frac{\partial \Phi_n}{\partial t} = f\Psi_n^0 - a_0^2 X_n^0 \neq 0 \quad \text{at } t = 0, \tag{2.16}$$

the second one being derived from (2.7b). Obviously, it does not satisfy condition (2.10) for stationarity. In other words, the initial velocity and pressure fields are not in balance. This (i.e., (2.16)) may hold in a limited area $(x, y < \infty)$ or on the unlimited plane xOy as well. Analogous initial conditions can be formulated for Ψ_n and X_n too. The solution of the so-formulated initial value problem has an asymptotic behaviour of the kind $\Phi_n, \Psi_n, X_n \to 0$ at $t \to \infty$. The time decaying is of a wave-type, i.e., qualitatively the process of adjustment proceeds like that in the hypothetical one-dimensional model (Section 1).

Theoretically at $t = \infty$, but in practice for a limited value of t after the initial instant one can consider $A_n \approx 0$ and $A \approx A_s$ in (2.11). Then relations (2.10) take place. But the second one means geostrophic relation

$$\mathbf{U}_s = \frac{1}{f\rho_{00}}\, \mathbf{k} \times \nabla p(x, y, 0). \tag{2.17}$$

Consequently, as a result of the adjustment process, a geostrophic balance between the velocity and pressure fields will be reached.

With this, the first question formulated at the end of Section 1, is answered.

c. Adjustment Activity of the Fields

To answer the second question, let us suppose that all functions in (2.7) can be represented as sums of two-dimensional Fourier components

$$A(x, y, t) = \sum_{k_1} \sum_{k_2} \hat{A}(k_1, k_2, t) \exp[i(k_1 x + k_2 y)], \tag{2.18}$$

where $A = \Psi, \Phi, X$. Then instead of (2.7), we obtain

$$\begin{aligned} &\hat{\Psi}_t = -f\hat{\Phi}, \qquad X_t = k^2 \hat{\Phi}, \\ &\hat{\Phi}_t = f\hat{\Psi} - a_0^2 \hat{X}, \end{aligned} \tag{2.19}$$

where $k^2 = k_1^2 + k_2^2$. Hence, as above, it follows that

$$\begin{aligned} &\hat{\Phi}_s = 0, \qquad f\hat{\Psi}_s = a_0^2 X_s, \\ &k^2 \hat{\Psi}_s + f\hat{X}_s = k^2 \hat{\Psi}^0 + f\hat{X}^0 = \text{const.} \end{aligned} \tag{2.20}$$

If we introduce a hypothetical streamfunction Ψ_g^0, which is in geostrophic balance with the initial pressure field

$$\hat{\Psi}_g^0 = \frac{a_0^2}{f} \hat{X}^0 \tag{2.21}$$

the last Equation (2.20) can now be written

$$\hat{\Psi}_s = m\hat{\Psi}^0 + (1 - m)\hat{\Psi}_g^0 \tag{2.22}$$

where

$$m = \frac{k^2 L_0^2}{1 + k^2 L_0^2}, \quad L_0 = \frac{a_0}{f} = \frac{(gh_0)^{1/2}}{f}, \tag{2.23}$$

L_0 being introduced by Obukhov. With $h_0 \sim 8$ km and $f \sim 10^{-4}\ \mathrm{s}^{-1}$, one obtains $L_0 \sim 2800$ km. When $h_0 = h_*$, L_0 coincides with the Rossby radius of deformation (1.13).

Instead of wavenumber k, we introduce the linear scale in the real space $L = 1/k$ interpreting it as a characteristic scale of the disturbances. Now $m = L_0^2/(L^2 + L_0^2)$. Two asymptotic cases are possible:

(i) $L \ll L_0$, $m \approx 1$, $\hat{\Psi}_s \approx \hat{\Psi}^0$ and $\hat{X}_s \approx (f/a_0^2)\hat{\Psi}^0$ following from (2.20). In case of small-scale disturbances (compared with L_0), the pressure (mass) field adjusts to the initial wind field, with the latter remaining almost unchanged during the adjustment process, i.e., a redistribution of the mass takes place. This is the case of synoptic scale motions ($L \lesssim 10^3$ km, $L/L_0 \approx 0.3$) of interest for short-range weather forecasting. Observations confirm this conclusion.

(ii) $L \gg L_0$, $m \to 0$, $\hat{\Psi}_s \approx \hat{\Psi}_g^0 = (a_0^2/f)\hat{X}^0$ and from (2.20) $\hat{X}_s \approx \hat{X}^0$ – the wind field adjusted to the initial pressure field, i.e., a redistribution of kinetic energy takes place.

Real mutual adjustment will exist only when $L \sim L_0$.

The last question concerns the shape of the adjusted fields and their dependence on the initial one. A qualitative answer can be found in Section 1. For a quantitative estimate let us turn back to Equations (2.10) and (2.13). They yield

$$\nabla^2 \Psi_s - \frac{1}{L_0^2} \Psi_s = \zeta_*(x, y, 0) = \nabla^2 \Psi^0 - fX^0. \tag{2.24}$$

Obukhov (1949) considered the next example:

$$X^0(x, y) = \text{const}, \quad \Psi^0(x, y) = A\left(2 + \frac{R^2}{L_0^2} - \frac{r^2}{R^2}\right) \exp\left[-\frac{r^2}{2R^2}\right], \tag{2.25}$$

where $r^2 = x^2 + y^2$ and R is the radius of the circular area in which the disturbance is defined. Obviously, the fields X^0 and Ψ^0 are not geostrophically balanced. This will give rise to an adjustment process. Since $\Psi^0 = \Psi^0(r)$, then Ψ_s will have a circular symmetry too: $\Psi_s = \Psi_s(r)$. Equation (2.24) reduces to an ordinary differential equation of the Bessel type:

$$\Psi_s'' + \frac{1}{r}\Psi_s' - \frac{1}{L_0^2}\Psi_s = \zeta_*(r), \tag{2.26}$$

where $(') = d/dr$. His fundamental solution reads

$$\Psi_s(x, y) = \frac{1}{2\pi} \int_{-\infty}^{\infty} \int \zeta_*(\xi, \eta) K_0\left(\frac{\rho}{L_0}\right) d\xi \, d\eta, \tag{2.27}$$

where $\rho^2 = (x - \xi)^2 + (y - \eta)^2$, $K_0(z)$ denotes the Bessel function of imaginary argument. Substituting (2.25) into (2.27) and integrating it yields

$$\Psi_s(r) = A\left(2 - \frac{r^2}{R^2}\right) \exp\left[-\frac{r^2}{2R^2}\right] \tag{2.28}$$

and $X_s(r) = (f/a_0^2)\,\Psi_s(r)$. It is easily seen that when $R \ll L_0$, the difference between $\Psi^0(r)$ (2.25) and $\Psi_s(r)$ (2.28) is inessential. However, X^0 and X_s differ drastically from one another.

3. THREE-DIMENSIONAL ADJUSTMENT MODELS

a. Geostrophic Adjustment

Obukhov's theory was generalized by Kibel [30] for a hydrostatic and adiabatic and yet baroclinic atmosphere. In a linear approximation, the fundamental equations are (3.5)–(3.7) and (3.11) of Chapter 3:

$$\begin{aligned}
&\text{(a)} \quad u_t - fv = -\Phi_x, \qquad v_t + fu = -\Phi_y,\\
&\text{(b)} \quad u_x + v_y + \left(\frac{1}{p_{00}}\right) \frac{\partial w_p}{\partial \eta} = 0,\\
&\text{(c)} \quad T_t = \left(\frac{RT}{g\eta p_{00}}\right) (\gamma_a - \gamma) w_p,\\
&\text{(d)} \quad T = -\left(\frac{\eta}{R}\right) \frac{\partial \Phi}{\partial \eta}.
\end{aligned} \tag{3.1}$$

Having in mind that $g/R = \Gamma_A = 34°\ \text{km}^{-1}$, $\gamma \sim 6°\ \text{km}^{-1}$ we estimate $\epsilon = (\gamma_a - \gamma)/\Gamma_A \approx 0.12$. Further, we introduce $a^2 = \epsilon RT$ and assume $a = \text{const}$, which implies that the temperature is replaced by some standard one – $\overline{T}$. From (3.1c–d) we determine

$$w_p = \left(\frac{Rp_{00}}{a^2}\right) \eta \frac{\partial T}{\partial t} = -\left(\frac{p_{00}}{a^2}\right) \eta^2 \frac{\partial^2 \Phi}{\partial \eta \, \partial t}, \tag{3.2}$$

and replace it in (3.1b). The result reads

$$u_x + v_y - \frac{1}{a^2} \frac{\partial}{\partial \eta} \eta^2 \frac{\partial^2 \Phi}{\partial \eta \, \partial t} = 0. \tag{3.3}$$

Henceforth, mathematically the problem is analogous to the two-dimensional one – Equations (3.1a) and (3.3) form a closed system.

Introducing streamfunction ψ and potential $\varphi(u = \varphi_x - \psi_y, v = \varphi_y + \psi_x)$, analogously, assuming f = const, we reduce the system to the new one

$$\varphi_t + \Phi - f\psi = 0, \qquad \psi_t + f\varphi = 0,$$
$$\nabla^2\varphi - \frac{1}{a^2}\frac{\partial}{\partial\eta}\left(\eta^2\frac{\partial^2\Phi}{\partial\eta\,\partial t}\right) = 0. \tag{3.4}$$

Hence, ψ and Φ can easily be eliminated to obtain Kibel's wave equation

$$a^2\nabla^2\varphi + \frac{\partial}{\partial\eta}\left[\eta^2\frac{\partial}{\partial\eta}\left(\frac{\partial^2\varphi}{\partial t^2} + f^2\varphi\right)\right] = 0. \tag{3.5}$$

The stationary solutions of system (3.4), as in the two-dimensional case, coincide with the geostrophic relations: $\varphi_s = 0$, $\psi_s = \Phi_s/f$. If the initial values φ_0, ψ_0, Φ_0 in some limited volume V of the space do not satisfy these relations, it follows from general physical considerations that a nonstationary (wave) process will arise. The volume becomes a source of waves which propagate out of it with limited speed. Since their total energy remains constant (the dissipation is neglected) and redistributes to a larger and larger volume of the space, then the characteristics of the nonstationary process, in particular the potential function φ, will tend to zero at $t \to \infty$ at an arbitrary point (x, y).

Solving the Cauchy problem for Equation (3.5), Kibel proved that under arbitrary initial conditions $\varphi \to 0$ when $t \to \infty$. In practice, this occurs 3 to 4 hours after the initial instant. The phase velocity of the waves

$$c = a = \pm\left[R\bar{T}\left(\frac{\gamma_a - \gamma}{\Gamma_A}\right)\right]^{1/2}$$

depends on the stratification of the atmosphere and is generally less than the corresponding quantity $c_0 = (RT_0)^{1/2}$ in the two-dimensional case.

Regardless of the much more sophisticated mathematics involved in Kibel's baroclinic three-dimensional model, the fundamental features of the adjustment process are quite close to those of Obukhov's two-dimensional model.

b. Geostrophic-Hydrostatic Adjustment

The formulation and solution of this most general adjustment problem in a baroclinic atmosphere was first given by Monin and Obukhov (1958). Under the assumptions of adiabaticity and no dissipation, the initial system of equations is closed

$$\rho\dot{u} = -p_x + \rho f v, \qquad \rho\dot{v} = -p_y - \rho f u,$$
$$\rho\dot{w} = -p_z - \rho g, \qquad \dot{p} = \left(\frac{\kappa p}{\rho}\right)\dot{\rho}, \tag{3.6}$$
$$\rho_t = -(\rho u)_x - (\rho v)_y - (\rho w)_z.$$

So far, all studies of the adjustment process have considered linear models in which the oscillations are superposed on a resting ($\mathbf{v}_* = 0$) hydrostatic basic state $-\rho_* = \rho_*(z)$, $p_* = p_*(z)$, $dp_*/dz = -g\rho_*$. Linearizing (3.6) and introducing the notations

$$a^2 = \kappa RT, \qquad \kappa = \frac{c_p}{c_v}, \qquad \delta = (\kappa - 1)g + \frac{da^2}{dz}, \qquad \sigma = \rho_* w,$$

one obtains

$$\begin{aligned} &\psi_t = -f\varphi, \qquad \varphi_t = f\psi - p, \\ &p_t = -a^2 \nabla^2 \varphi - \delta\sigma - a^2 \sigma_z, \\ &\mu\sigma_t = -p_z + \rho g, \qquad \rho_t = -\nabla^2 \varphi - \sigma_z, \end{aligned} \tag{3.7}$$

where ρ and p are the deviations from ρ_* and p_*, ψ and φ are the stream function and potential function, as before, and $\mu = 0, 1$ in the hydrostatic or nonhydrostatic case respectively.

System (3.7) is of fifth order with respect to t and of second order with respect to z. Consequently, it needs five initial conditions ($\psi_0, \varphi_0, \sigma_0, p_0, \rho_0$) at $t = 0$ and two boundary conditions: $\sigma \to 0$ at $z \to 0, \infty$. It can be proved that system (3.7) conserves the entropy, the potential vorticity and the total energy (kinetic plus potential energy).

By eliminating ψ, p and ρ one obtains from (3.7)

$$\left(\frac{\partial^2}{\partial t^2} + f^2\right)\varphi = \delta\sigma + a^2\left(\frac{\partial \sigma}{\partial z} + \nabla^2\varphi\right), \tag{3.8}$$

$$\mu\frac{\partial^2 \sigma}{\partial t^2} = \frac{\partial}{\partial z}\left(\delta\sigma + a^2\frac{\partial\sigma}{\partial z} + a^2\nabla^2\varphi\right) + g\left(\frac{\partial\sigma}{\partial z} + \nabla^2\varphi\right). \tag{3.9}$$

Assuming wave-type solutions

$$A(x, y, z, t) = \hat{A}(z)\exp[i(k_1 x + k_2 y - \omega t)], \tag{3.10}$$

we get

$$\begin{aligned} &\text{(a)} \quad (f^2 + k^2a^2 - \omega^2)\hat{\varphi} = \delta\hat{\sigma} + a^2\hat{\sigma}_z, \\ &\text{(b)} \quad (f^2 - \omega^2)(a^2\hat{\varphi} + g\hat{\varphi}) = (\delta g - \mu a^2\omega^2)\sigma, \end{aligned} \tag{3.11}$$

where $k^2 = k_1^2 + k_2^2$.

A particular simple solution of (3.11) can be obtained only in the case $\rho_* w = \sigma = 0$, i.e., $w = 0$. Then from (3.11a), we obtain the already-known dispersion relation $\omega^2 = f^2 + a^2k^2$ but (3.11b) becomes $a^2\partial\hat{\varphi}/\partial z + g\hat{\varphi} = 0$. Hence,

$$\hat{\varphi}(z) = \hat{\varphi}^0\exp\left(-g\int_0^z a^{-2}\,dz\right) = \hat{\varphi}^0\left(\frac{p_*}{p_*^0}\right)^{1/\kappa},$$

where $p_*^0 = p_*(z = 0)$. This solution describes waves propagating only horizontally ($w = 0$). Because of this, they are called *two-dimensional waves*. Their amplitudes decrease with height following the law $\hat{\varphi}^\kappa \sim p_*(z)$.

A further solution of (3.11) at $\sigma \neq 0$, though not attended with difficulties of principle, is mathematically quite complicated. We will only quote some final results.

(1) In the case of $\sigma \neq 0$, the waves propagate in all directions ($w \neq 0$) – they are called *internal waves*. Their amplitude decreases exponentially with z.

(2) The stationary motions (having frequency $\omega = 0$) being removed from system (3.7), its order with respect to t decreases by one, which means that the algebraic dispersion relation for the internal waves is a fourth-degree polynomial equation about ω. It turns out that two of its roots correspond to acoustic waves ($\omega_a^2 = A$) and the other two correspond to gravity waves ($\omega_g^2 = G$) where A and G are some expressions including the parameters of the problem.

(3) In an isothermal atmosphere ($dT_*/dz = 0$, i.e., $da/dz = 0$ and $\delta = (\kappa - 1)g$) the spectra of both kinds of waves are separated from each other. The calculations show that $T_a < 300$ s while $T_g > 300$ s where $T = 2\pi/\omega$ is the period of the waves. Otherwise ($da/dz \neq 0$) the spectra are overlapped.

(4) If in (3.7) we assume $\mu = 0$, (hydrostatic approximation) it results in $\omega_a = \infty$, i.e., there are no acoustic waves. The stationary solutions of (3.7) and the frequency of the two-dimensional waves, as well, remain unchanged when $\mu \to 0$. As to ω_g, its values increase slightly. Therefore, the quasi-static approximation, as we know from Chapter 5, eliminates the acoustic waves, without essentially influencing the rest of the solutions.

(5) If in (3.7) $\mu \neq 0$, we can speak about an adjustment of the atmosphere to the hydrostatic balance by means of vertically propagating the internal-acoustic waves ($\omega_a \neq 0$). Simultaneously, geostrophic adjustment holds too, and is realized by means of two-dimensional and internal-gravity waves. However, due to the small effective thickness of the atmosphere ($\sim 10^1$ km) compared to the characteristic horizontal scales of the synoptic processes ($\sim 10^3$–10^4 km), the time needed for reaching the hydrostatic balance is 10^2–10^3 times less than that for geostrophic balance. Thus, if at $t = 0$, the hydrostatic and the geostrophic balance are simultaneously violated, then the former one will be reached after a few minutes and will be needed for the wave front to cover the distance of $\sim 10^2$ km, while the latter needs a few hours. A powerful convective cloud (Cb) is a typical example of a violated hydrostatic balance in this part of the atmosphere. If we imagine that it suddenly disappears, then the balance will be practically instantaneously reestablished.

4. WAVES AND ADJUSTMENT ON A SPHERE

a. Beta Approximation

One of the essential limitations in the so-far-discussed theories of geostrophic adjustment is the assumption that the Coriolis parameter f = const. We can take into account the Earth's curvature either by introducing a β-approximation ($f = f_0 + \beta y$) or by using spherical coordinates. In the beginning, let us follow the first way.

We start from Equations (2.4) in a slightly different form

$$U_t - fV = -\pi_x, \qquad V_t + fU = -\pi_y^0,$$
$$\pi_t + a_0^2(U_x + V_y) = 0, \tag{4.1}$$

where $\pi^0 = a_0^2 X$. Letting $U = -\psi_y + \varphi_x$, $V = \psi_x + \varphi_y$, after the appropriate differentiation of the momentum equations (4.1) in order to form the vorticity ($\zeta = \nabla^2 \psi$) and divergence ($D = \nabla^2 \varphi$) equations, one obtains

$$\left.\begin{aligned} &\nabla^2 \varphi_t - f\nabla^2 \psi + \beta(\varphi_x - \psi_y) = -\nabla^2 \pi^0, \\ &\nabla^2 \psi_t + f\nabla^2 \varphi + \beta(\varphi_y + \psi_x) = 0, \end{aligned}\right\} \tag{4.2}$$

$$\pi_t + a_0^2 \nabla^2 \varphi = 0. \tag{4.3}$$

Henceforth, f and $\beta = \mathrm{d}f/\mathrm{d}y$ are considered to be constants. This is the essence (and the crudeness) of the β-approximation. The solution of the Cauchy problem for system (4.2) and (4.3) is hardly possible because of its complexity. However, some features of the fundamental wave solutions having the form $A(x, y, t) = \hat{A} \exp[i(k_1 x + k_2 y - \omega t)]$ can easily be found. Let us further simplify the problem by assuming $\partial/\partial x = 0$ and considering the one-dimensional analogue of system (4.2), (4.3). After integrating once with respect to y, we obtain

$$\varphi_{yt} - f\psi_y - \beta\psi = -\pi_y^0, \qquad \psi_{yt} + f\varphi_y + \beta\varphi = 0, \qquad \pi_t + a_0^2 \varphi_{yy} = 0. \tag{4.4}$$

It is natural to expect that system (4.2), (4.3) or (4.4) will possess two classes of solutions: (a) of a geostrophic type (eventually slowly varying with the time t) which at $\beta \to 0$ degenerates into pure geostrophic motion and (b) of an acoustic wave-type (fast waves) somehow influenced by the curvature (Rossby) parameter β. Naturally, only the first class of solutions will have a meteorological meaning.

Actually, the fast waves can be removed by setting $a_0 = \infty$ in (4.3), which is equivalent to $\varphi \equiv 0$. Then we obtain

$$\nabla^2 \pi = f\nabla^2 \psi + \beta\psi_y, \qquad \nabla^2 \psi_t + \beta\psi_x = 0. \tag{4.5}$$

The second equation coincides with Equation (5.6) of Chapter 5, at $u_* = 0$ and has a solution

$$\psi(x, y, t) = \hat{\psi} \cos \frac{2\pi}{\lambda_2} y \sin \frac{2\pi}{\lambda_1} (x - ct), \tag{4.6}$$

i.e., the Rossby waves with phase velocity

$$c = -\frac{\beta}{4\pi^2} \Big/ \left(\frac{1}{\lambda_1^2} + \frac{1}{\lambda_2^2}\right), \tag{4.7}$$

$\lambda_i = 2\pi/k_i$, $i = 1, 2$ being the wavelength along x and y axes. If $\lambda_2 = \infty$ then (4.6) coincides with (5.9) of Chapter 5. In this case, $\psi_y = 0$ and π and ψ become geostrophically related: $\pi = f\psi$ (see (4.5)). On the other hand, if $\beta \to 0$ then $c \to 0$ and we obtain from (4.7) and (4.5) the stationary fields of the geostrophic wind and pressure as it should be.

b. Spherical Earth

In the case of global scale motion and adjustment, it is necessary to use spherical coordinates. In the linear approximation, Equations (4.3) and (3.16) of Chapter 2, read

$$\text{(a)}\quad \frac{\partial v_\theta}{\partial t} - 2\omega\cos\theta\cdot v_\lambda = -\left(\frac{1}{\rho r_0}\right)\frac{\partial p}{\partial \theta},$$

$$\text{(b)}\quad \frac{\partial v_\lambda}{\partial t} + 2\omega\cos\theta\cdot v_\theta = -\left(\frac{1}{\rho r_0\sin\theta}\right)\frac{\partial p}{\partial \lambda},$$

$$\text{(c)}\quad \frac{\partial \rho}{\partial t} + \frac{1}{r_0\sin\theta}\left[\frac{\partial}{\partial\theta}(\rho v_\theta\sin\theta) + \frac{\partial}{\partial\lambda}(\rho v_\lambda)\right] = 0, \tag{4.8}$$

$$\text{(d)}\quad \frac{\partial p}{\partial z} = -\rho g,\qquad p = p(\rho)\sim\rho^m,$$

where m = const and r_0 is the Earth's radius. Obviously, the barotropic atmosphere is considered.

Yaglom (1953), whose results are summarized below, following the Obukhov approach from the beginning of Section 2, reduces system (4.8) to

$$\text{(a)}\quad \frac{\partial U}{\partial t} - 2\omega\cos\theta\cdot V = -\left(\frac{1}{r_0}\right)\frac{\partial \pi}{\partial \theta},$$

$$\text{(b)}\quad \frac{\partial V}{\partial t} + 2\omega\cos\theta\cdot U = -\left(\frac{1}{r_0\sin\theta}\right)\frac{\partial \pi}{\partial \lambda}, \tag{4.9}$$

$$\text{(c)}\quad \frac{1}{a_0^2}\frac{\partial \pi}{\partial t} + \frac{1}{r_0\sin\theta}\left[\frac{\partial}{\partial\theta}(U\sin\theta) + \frac{\partial V}{\partial\lambda}\right] = 0,$$

in which (U, V), corresponding to (v_θ, v_λ), are the wind velocity components averaged over the whole thickness of the atmosphere (see (2.3)) and $\pi(\lambda, \theta, t)$, corresponding to $X(x, y, 0, t)$ in Section 2, represents the surface pressure.

System (4.9) is well known in the theory of atmospheric and oceanic tides [6, 8]. It is a closed system and describes the motion in a two-dimensional fluid cover of the rotating Earth. By analogy with the 'plane' problem and proceeding from general physical considerations, one could expect that system (4.9) would possess solutions both of a wave and geostrophic-type. In the classical tide theory, wave solutions have been obtained, analyzed, and used for an explanation of this phenomenon (including also an explanation of the surface pressure diurnal variations). In their nature, the tide phenomena in the atmosphere and ocean are processes of breaking the dynamical balance in these geophysical fluids and subsequent adjustment. Here we shall dwell on both types of solution.

First, let us assume $\omega = 0$, i.e., the Earth does not rotate. Then, from (4.9) one obtains an equation for the pressure function

$$\left(\frac{r_0}{a_0}\right)^2\frac{\partial^2\pi}{\partial t^2} - \nabla^2\pi = 0,\qquad \nabla^2 = \frac{1}{\sin\theta}\frac{\partial}{\partial\theta}\left(\sin\theta\frac{\partial}{\partial\theta}\right) + \frac{1}{\sin^2\theta}\frac{\partial^2}{\partial\lambda^2}. \tag{4.10}$$

It admits solution of wave-type

$$\pi(\theta, \lambda, t) = \exp[i\sigma t]\,\overset{*}{\pi}(\theta, \lambda) \tag{4.11}$$

where σ in this paragraph will denote frequency and $\overset{*}{\pi}$ satisfies the equation

$$\nabla^2 \overset{*}{\pi} + \left(\frac{\sigma r_0}{a_0}\right)^2 \overset{*}{\pi} = 0. \tag{4.12}$$

It has a solution, limited on the entire sphere, only if

$$\sigma^2 \frac{r_0^2}{a_0^2} = n(n+1), \quad \text{i.e., } \sigma = \pm \frac{a_0}{r_0} [n(n+1)]^{1/2}, \tag{4.13}$$

where $n = 1, 2, 3 \ldots$ The solution is expressed in spherical functions

$$\begin{aligned} \overset{*}{\pi} &= \hat{\pi} \exp[im\lambda]\, P_n^m(\cos\theta), \\ \pi &= \hat{\pi} \exp[i(m\lambda + \sigma t)]\, P_n^m(\cos\theta), \end{aligned} \tag{4.14}$$

where $m = 0, 1, 2, \ldots n$, $P_n^m(z)$ are Legendre functions and $\hat{\pi}$, the arbitrary amplitude. The substitution of (4.14) into (4.9) at $\omega = 0$ yields the expressions

$$\begin{aligned} U &= \pm \frac{i\hat{\pi} \exp[i(m\lambda + \sigma t)]}{a_0 [n(n+1)]^{1/2}} \frac{\mathrm{d}}{\mathrm{d}\theta} P_n^m(\cos\theta), \\ V &= \mp \frac{m\hat{\pi} \exp[i(m\lambda + \sigma t)]}{a_0 [n(n+1)]^{1/2}} \frac{P_n^m(\cos\theta)}{\sin\theta}, \end{aligned} \tag{4.15}$$

describing double-directional waves:

$$c = \sigma \frac{r_0}{m} = \pm \frac{a_0}{m} [n(n+1)]^{1/2}, \tag{4.16}$$

where c is the phase velocity. Obviously, these are gravity waves which are related to those originating under similar conditions above the plane Earth.

Equation (4.10), however, also admits a trivial solution $\pi(\lambda, \theta, t) = \text{const}$. Then Equation (4.9c) implies

$$U(\theta, \lambda) = -\frac{1}{r_0 \sin\theta} \frac{\partial \psi}{\partial \lambda}, \qquad V(\theta, \lambda) = \frac{1}{r_0} \frac{\partial \psi}{\partial \theta},$$

where $\psi(\theta, \lambda)$ is an arbitrary (stationary) streamfunction. It is natural to choose

$$\psi(\theta, \lambda) = \exp[im\lambda]\, P_n^m(\cos\theta),$$

$n = 1, 2, 3, \ldots; m = 1, 2, \ldots, n$. Then $\pi = \text{const}$ and

$$U = -\frac{im \exp[im\lambda]}{r_0 \sin\theta} P_n^m(\cos\theta), \qquad V = \frac{\exp[im\lambda]}{r_0} \frac{\mathrm{d}}{\mathrm{d}\theta} P_n^m(\cos\theta). \tag{4.17}$$

Thus, in the case of the nonrotating Earth, we obtain two groups of solutions – (4.17) corresponding to the stationary velocity field which satisfies the incompressibility condition and (4.14) and (4.15) corresponding to fast gravity waves.

Let us now 'turn' the Earth round on its axis. The rotation will influence both the wave solution and the stationary one. We are interested in the second case which is of meteorological importance. Instead of (4.5) now we have

$$\begin{aligned}&\text{(a)}\quad \frac{\partial \nabla^2 \psi}{\partial t} + 2\omega \frac{\partial \psi}{\partial \lambda} = 0,\\ &\text{(b)}\quad \nabla^2 \pi = 2\omega\left(\cos\theta \cdot \nabla^2\psi - \sin\theta\,\frac{\partial \psi}{\partial \theta}\right),\end{aligned}\tag{4.18}$$

where ∇^2 is given by (4.10) and $\varphi = 0$. Assuming a solution of (4.18a) in the form

$$\psi(\theta, \lambda, t) = \sum_{n=0}^{\infty}\sum_{m=0}^{n} \psi_n^m(t)\exp[im\lambda]\,P_n^m(\cos\theta),$$

we obtain

$$\sum_n\sum_m \left[-\,n(n+1)\frac{\mathrm{d}}{\mathrm{d}t}\psi_n^m + 2i\omega m\psi_n^m\right]\exp[im\lambda]\,P_n^m(\cos\theta) = 0$$

Hence

$$[\cdots] = 0 \Rightarrow \psi_n^m(t) = \hat{\psi}\exp[i\sigma t],\tag{4.19}$$

$$\sigma = \frac{2m\omega}{n(n+1)},\tag{4.20}$$

so that

$$\psi(\theta, \lambda, t) = \hat{\psi}\exp[i(\sigma t + m\lambda)]\,P_n^m(\cos\theta),\tag{4.21}$$

while $\pi(\theta, \lambda, t)$ can be determined from (4.18b).

Therefore, in the case of a rotating Earth the geostrophic-type solution, in general, is no longer stationary. Regardless of rotation, it would be stationary only if $m = 0$, i.e., in the case of purely zonal distribution – there is no dependence on λ. In this case, as follows from (4.19), $\sigma = 0$. Consequently, the nonstationarity is related to the deviations from zonal distribution ($m \neq 0$) so that $\sigma \neq 0$. The waves propagate slowly westward with dispersion relation (4.20). Obviously, these waves represent the spherical analogue of the Rossby waves in the β-approximation and (4.20) corresponds to (4.7).

c. One-Dimensional Spectral Model

Such a model has been proposed by Wiin–Nielsen (1976). At $\partial/\partial\lambda = 0$, $v_\lambda = u$, $v_\theta = -v$, $\varphi = \pi/2 - \theta$ the corresponding equations follow from (4.8)

$$\begin{aligned}\frac{\partial u}{\partial t} &= 2\omega\sin\varphi \cdot v,\\ \frac{\partial v}{\partial t} &= -\frac{\partial \Phi}{r_0\,\partial\varphi} - 2\omega\sin\varphi\cdot u,\\ \frac{\partial \Phi}{\partial t} &= -\left(\frac{\Phi_*}{r_0\cos\varphi}\right)\frac{\partial}{\partial\varphi}(v\cos\varphi),\end{aligned}\tag{4.22}$$

where Φ_* is a standard value of Φ. Letting

$$u' = \frac{u}{2\omega r_0}, \qquad v' = \frac{v}{2\omega r_0}, \qquad \Phi' = \frac{\Phi}{4\omega^2 r_0^2}, \qquad t' = 2\omega t$$

and also

$$\mu = \sin\varphi, \qquad Q^2 = \frac{\Phi_*}{4\omega^2 r_0^2}, \tag{4.23}$$

we write (4.22) in dimensionless form (primes are omitted)

$$\begin{aligned}
\frac{\partial u}{\partial t} &= \mu v,\\
\frac{\partial v}{\partial t} &= -(1-\mu^2)^{1/2}\frac{\partial\Phi}{\partial\mu} - \mu u,\\
\frac{\partial\Phi}{\partial t} &= -Q^2\frac{\partial}{\partial\mu}\left[v(1-\mu^2)^{1/2}\right].
\end{aligned} \tag{4.24}$$

Further we write

$$\begin{aligned}
w(\mu, t) &= \sum_{n=1}^{\infty} \hat{W}(n, t) P_n^1(\mu), \qquad w = u, v,\\
\Phi(\mu, t) &= \sum_{n=1}^{\infty} \hat{\Phi}(n, t) P_n(\mu),
\end{aligned} \tag{4.25}$$

where $P_n(\mu)$ and $P_n^1(\mu) = (1-\mu^2)^{1/2}\,\mathrm{d}P_n/\mathrm{d}\mu$ are Legendre functions. As is known

$$(2n+1)\mu P_n^1 = nP_{n+1}^1 + (n+1)P_{n-1}^1,$$

$$\frac{\mathrm{d}}{\mathrm{d}\mu}\left[(1-\mu^2)^{1/2} P_n^1\right] = \frac{\mathrm{d}}{\mathrm{d}\mu}\left[(1-\mu^2)\frac{\mathrm{d}P_n}{\mathrm{d}\mu}\right] = -n(n+1)P_n.$$

Making use of these expressions during the replacement of (4.25) into (4.24) we can replace (2.4) by the spectral system

$$\begin{aligned}
\frac{\mathrm{d}}{\mathrm{d}t}\hat{U}(n, t) &= \frac{n-1}{2n-1}\hat{V}(n-1, t) + \frac{n+2}{2n+3}\hat{V}(n+1, t),\\
\frac{\mathrm{d}}{\mathrm{d}t}V(n, t) &= -\hat{\Phi}(n, t) - \frac{n-1}{2n-1}\hat{U}(n-1, t) - \frac{n+2}{2n+3}\hat{U}(n+1, t),\\
\frac{\mathrm{d}}{\mathrm{d}t}\hat{\Phi}(n, t) &= Q^2 n(n+1)\hat{V}(n, t).
\end{aligned} \tag{4.26}$$

Unlike (4.24), system (4.26) is unclosed and consists of three equations with seven unknown functions. Due to the latitudinal variations of the Coriolis parameter, the components with numbers $(n-1)$ and $(n+1)$ interact to produce a rate of change in the wind components with number n. It will be necessary to truncate the series (4.25) at some finite $n = N$ in order to get a closed set of equations. Generally, numerical methods are appropriate. Interesting information, however, can be 'extracted' by means of analytical methods too.

First of all, let us look for stationary solutions of system (4.26). With $\mathrm{d}/\mathrm{d}t = 0$ we get $\hat{V}(n) = 0$ for all n, while the middle equation yields

$$\Phi_s(n) = -\frac{n-1}{2n-1}\,\hat{U}_s(n-1) - \frac{n+2}{2n+3}\,\hat{U}_s(n+1). \tag{4.27}$$

This is a spectral analogue of the second equation of (4.22) with $\partial v/\partial t = 0$.

Numerical integration of system (4.26) performed by Wiin–Nielsen showed that, for practical applications, a system containing only $\hat{\Phi}(n)$, $\hat{V}(n)$ and $\hat{U}(n \pm 1)$ could be used, especially for large values of n. Then the first equation of (4.26) produces two approximate equations

$$\frac{\mathrm{d}}{\mathrm{d}t}\,\hat{U}(n-1,t) = \frac{n+1}{2n+1}\,\hat{V}(n,t), \qquad \frac{\mathrm{d}}{\mathrm{d}t}\hat{U}(n+1,t) = \frac{n}{2n+1}\,\hat{V}(n,t). \tag{4.28}$$

Together with the second and third equations of (4.26) they form a closed system of four equations which can be easily reduced to one equation

$$\frac{\mathrm{d}^2}{\mathrm{d}t^2}\,\hat{V}(n,t) + \nu^2(n)\,\hat{V}(n,t) = 0, \tag{4.29}$$

where

$$\nu^2(n) = n(n+1)\,Q^2 + \frac{n^2-1}{4n^2-1} + \frac{n(n+2)}{(2n+1)(2n+3)}. \tag{4.30}$$

The solution of (4.29) is

$$\hat{V}(n,t) = C_1 \cos \nu t + C_2 \sin \nu t, \tag{4.31}$$

after which (4.28) and (4.26) allow the other unknown functions to be found also.

In the general case at $t = 0$, arbitrary initial values for $\hat{V}_0(n)$, $\hat{\Phi}_0(n)$ and $\hat{U}_0(n \pm 1)$ can be chosen. For simplicity we choose

$$\hat{\Phi}_0(n) \neq 0 \quad \text{but} \quad \hat{V}_0(n) = \hat{U}_0(n \pm 1) = 0. \tag{4.23}$$

The latter means that the geostrophic relation (4.27), ensuring stationarity, doesn't hold. Under these initial conditions, the constants of integration are easily determined so that

$$\begin{aligned} \hat{\Phi}(n,t) &= \hat{\Phi}_s(n)\,(1-\cos \nu t), \\ \hat{U}(n \pm 1,t) &= \hat{U}_s(n \pm 1)\,(1-\cos \nu t), \\ \hat{V}(n,t) &= -\frac{1}{\nu}\,\hat{\Phi}_0(n) \sin \nu t, \end{aligned} \tag{4.33}$$

where

$$\begin{aligned}
\hat{\Phi}_s(n) &= \hat{\Phi}_0(n)\left[1 - \frac{n(n+1)}{\nu^2} Q^2\right],\\
U_s(n-1) &= -\frac{1}{\nu^2}\,\frac{n+1}{2n+1}\,\hat{\Phi}_0(n),\\
\hat{U}_s(n+1) &= -\frac{1}{\nu^2}\,\frac{n}{2n+1}\,\hat{\Phi}_0(n),\\
\hat{V}_s(n) &= 0,
\end{aligned} \tag{4.34}$$

is the stationary part of the solution (4.33). As might be expected, (4.34) satisfies the geostrophic condition (4.27). Consequently, the general solution (4.33) describes a periodically-oscillating process with frequency ν around the adapted spectral component (4.34) at arbitrary wavenumber n.

One-dimensional spectral models prove to be very convenient for energetic studies too.

PROBLEMS

1. Solve in details the initial value problem (1.3)–(1.5), Section 1.
   Note: See [23].
2. Derive Equation (3.5) from (3.4).
3. Prove that the substitutions

$$R = -\eta^{-1/2}\int_0^{\eta} \varphi\, d\eta, \qquad \xi = -\ln \eta$$

   reduce Equation (3.5) to the following one:

$$a^2\nabla^2 R + \left(\frac{\partial^2}{\partial \xi^2} - \frac{1}{4}\right)\left(\frac{\partial^2}{\partial t^2} + f^2\right) R = 0$$

   having a fundamental solution $\exp[i(k_1 x + k_2 y + k_3 \xi - \omega t)]$ and find the dispersion relation.
   Note: See [30].
4. Investigate system (4.2), (4.3) and (4.4) for fundamental wave-type solutions and find the corresponding dispersion relations.
5. Derive the system of Equations (4.18).
6. Show that at $f = f(y)$ U and π can be eliminated from (4.1) to obtain an equation for V:

$$V_{ttt} - a_0^2\nabla^2 V_t + f^2 V_t - a_0^2 \beta V_x = 0$$

   Assuming

$$V \sim \exp[i(k_1 x + k_2 y - \omega t)]$$

   derive the dispersion relation and solve it. Make an interpretation of the roots.

CHAPTER 7

The Theoretical Basis of Meteorological Forecasts

1. SYNOPTIC VARIATIONS OF METEOROLOGICAL ELEMENTS – EARLY THEORIES

a. Classification of the Causes

Let ψ denote an arbitrary meteorological element (scalar or vector). In the real atmosphere, ψ varies with time t in a Eulerian as well as in a Lagrangian sense, i.e., $\psi_t = \partial\psi/\partial t = 0$ and $\dot{\psi} = \mathrm{d}\psi/\mathrm{d}t = 0$. Physical intuition suggests that this can be due to a number of reasons. In order to discover and classify them, we shall try to find the conditions at which $\psi_t = \dot{\psi} = 0$. It is easily established that a motion in the atmosphere* for which the local and individual variations of all meteorological elements equal zero, should possess the following features:

(1) The motion is geostrophic and in hydrostatic balance. Actually, from the momentum equations

$$\frac{\mathrm{d}\mathbf{u}}{\mathrm{d}t} = -\frac{\nabla p}{\rho} - f\mathbf{k} \times \mathbf{u}, \qquad \frac{\mathrm{d}w}{\mathrm{d}t} = \frac{1}{\rho}\frac{\partial p}{\partial z} - g,$$

at $\dot{u} = \dot{v} = \dot{w} = 0$, we obtain the geostrophic and hydrostatic equations

$$\mathbf{u}_g = \frac{1}{\rho f}\mathbf{k} \times \nabla p, \qquad \frac{\partial p}{\partial z} = -\rho g. \tag{1.1}$$

Hence, it follows that the motion is horizontal. Really

$$\dot{p} - p_t = \mathbf{u} \cdot \nabla p + wp_z = \mathbf{u}_g \cdot \nabla p + wp_z = wp_z = 0$$

and because $p_z \neq 0$ it follows that $w = 0$.

(2) The processes are adiabatic. This conclusion immediately follows from the thermodynamic energy Equation (1.7) of Chapter 2

$$Q = \rho\left(\dot{T} - \frac{RT\dot{p}}{pc_p}\right) = 0$$

as long as $\dot{T} = \dot{p} = 0$.

* Friction is neglected, i.e., the 'ideal' atmosphere is considered.

(3) The atmosphere is horizontally barotropic. Actually,

$$\dot{T} - T_t = \mathbf{u} \cdot \nabla T + wT_z = \mathbf{u}_g \cdot \nabla T = \frac{1}{\rho f}(p_x T_y - p_y T_x) = 0. \tag{1.2}$$

(4) The motion is purely zonal, i.e., along the parallels ($v = 0$). At $\rho_t = w = 0$, the continuity equation $\rho_t + \nabla \cdot \rho\mathbf{v} = 0$ simplifies and reads $\nabla \cdot \rho\mathbf{u} = 0$. Substituting here $\rho\mathbf{u}$ by $\rho\mathbf{u}_g$ from (1.1) and taking into account that $f = f(y)$ and $\beta = \mathrm{d}f/\mathrm{d}y \neq 0$, we obtain $\beta p_x = 0$ or

$$\frac{\partial p}{\partial x} = 0 \quad \text{and} \quad v \equiv v_g = 0. \tag{1.3}$$

(5) Besides pressure, all other meteorological elements also obey zonal distribution, i.e., they do not depend on x. Actually, (1.3) and (1.2) yield $T_x = 0$ because $p_y \neq 0$ ($u_g \neq 0$ – otherwise there will be no motion at all, which is a trivial case). From the equation of state $\rho = p/RT$, it follows that $\partial\rho/\partial x = 0$, since $p_x = T_x = 0$. Finally, from the equality

$$\dot{u} - u_t = uu_x + vu_y + wu_z = uu_x = 0$$

and $u = u_g \neq 0$, it follows that $u_x = 0$.

Thus, the only possible motion of the atmosphere at which $\dot{\psi} = \psi_t = 0$ turns out to be purely zonal with a constant velocity of the air particles which move along the geographical parallels. All the meteorological elements have a zonal distribution ($\psi_x = 0$). Violation of any of the above-established properties of the motion, or of all of them, would cause time variations of ψ. As we already know, these variations are of two kinds: fast, wave-like, and slow, synoptic. The motions of the first kind are of no interest for the theory of the synoptic processes. More precisely, they are interesting but only from the point of view of their elimination as solutions of the weather equations. This problem was discussed in Chapter 5. Let us recall the obtained results:

(i) The acoustic waves are generally eliminated by the incompressibility condition $\nabla \cdot \mathbf{v} = 0$ or $\mathrm{d}\rho/\mathrm{d}t = 0$ and the vertical acoustic waves can be removed by means of the less-restrictive quasi-static equation. The latter also eliminates the short surface gravity waves.

(ii) The gravity waves can be eliminated with the help of a quasi-geostrophic hypothesis or balance equation (5.25) of Chapter 5.

(iii) The neglecting of the Earth's sphericity (the condition $\beta = \mathrm{d}f/\mathrm{d}y = 0$) eliminates the Rossby waves from the solutions of the equations.

These methods can be applied independently and in combinations as well, in accordance with the particular problem.

On the basis of the considerations made, a classification of the factors causing variations of the meteorological elements in time can be proposed:

(1) The *Diabatic* factor is connected with the breaking down of the adiabatic condition $Q \neq 0$ that leads to $\dot{T} \neq 0$ and $\dot{p} \neq 0$. Contributions to the value of Q in some subvolume of the atmosphere may come from water phase transitions, radiation, thermal effects of the underlying surface, etc. In the free atmosphere* and for time periods of not longer

* I.e., above the planetary boundary layer (see Chapter 10), with a thickness of 1–1.5 km.

than 24 hours, this factor can be neglected (quasi-adiabaticity hypothesis). This simplifying assumption lies in the basis of recent short-range weather prediction theories but it is not applicable, however, for longer periods. Accounting for the diabatic heating Q is one of the main difficulties in the theory of climate and long-range weather prediction.

(2) The *inertial* factor is connected with the breaking down of the condition for the zonal distribution of all elements and the presence of a meridional component of motion at distances of 10^3 km and greater. Synoptic and subsynoptic scale motions are not greatly influenced by this factor.

(3) The *dynamic* factor is connected with the breaking down of the geostrophic condition, i.e., with the presence of air particle accelerations (local, advective and convective). Accounting for this factor makes the problems essentially nonlinear.

(4) The *Baroclinic* (*advective*) factor is connected with the breaking down of barotropic condition (1.2). Due to the crossing of isobars and isotherms, there will be nonzero advection of temperature and, consequently, variation in time at a fixed point.

It is clear that a comprehensive weather prediction theory should take all these factors into account. In principle, they are included in the full system of atmospheric fluid and thermodynamic equations. However, because of the difficulty of solution, or due to the specific features of the particular problems, the creation of such a theory has been historically proceeded by way of approximations and special simplifications.

An unsuccessful attempt at numerical weather prediction on the basis of a complete system of weather equations, was made during the Twenties by L. Richardson, but it was not until 1940 that the first successful solution was obtained by the prominent Russian theoretician and geophysical fluid dynamist, I. A. Kibel. Despite its shortcomings, Kibel's theory satisfactorily endured the verification of practice and, for many years, was the working tool of synopticians and prognosticators. Even now, it has not completely lost its significance for the operative process, although recent short-range weather prediction theories have been developed on a rather different basis. Here we shall briefly examine the general principles and results of Kibel's theory.

b. Kibel's Theory

This is a theory of temperature and pressure prediction at the lower boundary of the atmosphere for a period of up to 24 hours. The lower boundary is identified with the Earth's surface and the diabatic heating is ignored. Consequently, the results would be applicable for the free atmosphere and would need corrections for the boundary layer. The limitation to the period (up to 24 hours) yields a limitation to the scales ($L \lesssim 10^3$ km) of the processes under consideration, so that the inertial factor can also be neglected. As for the dynamic factor, Kibel also neglects it in the first approximation, assuming the wind to be geostrophic, and then introduces corrections for the ageostrophic deviations (second approximation).

And so the starting equations in Kibel's theory are

$$\mathbf{u} = \frac{1}{\rho f}\mathbf{k} \times \nabla p \tag{1.4}$$

$$T_t + uT_x + vT_y = \frac{RT}{pc_p}\,p_t. \tag{1.5}$$

Hence, with the help of some additional suppositions, Kibel derives the following two fundamental equations of his theory

$$\frac{\partial p_0}{\partial t} = -aJ(T_0, p_0), \qquad \frac{\partial T_0}{\partial t} = bJ(T_0, p_0), \tag{1.6}$$

where a, b are two coefficients, considered approximately constant, $J(\ ,\ )$ denotes the Jacobian on x, y and p_0 and T_0 are the surface values (at the lower boundary) of these quantities.

The two Equations (1.6) are nonlinear and could not have been solved analytically or numerically some 40 years ago. Despite this, much information has been extracted from them by means of graphical methods based on the following transformations of the equations.

Equations (1.6) imply that

$$a\frac{\partial T_0}{\partial t} + b\frac{\partial p_0}{\partial t} = 0 \quad \text{or} \quad \frac{\partial \theta}{\partial t} = 0,$$

where

$$\theta(x, y) = aT_0(x, y, t) + bp_0(x, y, t). \tag{1.7}$$

Consequently, linear combination (1.7) is time-invariant in a local sense ($\partial\theta/\partial t = 0$). Then

$$\begin{aligned} J(p_0, \theta) &= J(p_0, aT_0 + bp_0) = -aJ(T_0, p_0), \\ J(T_0, \theta) &= bJ(T_0, p_0), \end{aligned} \tag{1.8}$$

because $J(A, A) \equiv 0$, $J(A, B) = -J(B, A)$. Therefore, instead of (1.6) we can write

$$\frac{\partial T_0}{\partial t} = J(T_0, \theta), \qquad \frac{\partial p_0}{\partial t} = J(p_0, \theta). \tag{1.9}$$

However, the Jacobian $J(A, \theta)$, as it is known, expresses the advection of the quantity A along the isolines of θ with velocity $c_\theta = \theta_x^2 + \theta_y^2$. In other words,

$$\frac{\partial p_0}{\partial t} = -(\mathbf{u}_\theta \cdot \nabla p_0), \qquad \frac{\partial T_0}{\partial t} = -(\mathbf{u}_\theta \cdot \nabla T_0), \tag{1.10}$$

where $\mathbf{u}_\theta = \mathbf{k} \times \nabla\theta$. In this way, calculating the tendencies $\partial p_0/\partial t$ and $\partial T_0/\partial t$ is simply reduced to an ordinary transport of the quantities p_0 and T_0 along the isolines of θ with velocity $\mathbf{u}_\theta$, which is easily achieved graphically.

It follows from (1.7) that the θ-field can be constructed by using the initial data at the discrete points of observation

$$\theta(x_i, y_i) = aT_0(x_i, y_i, 0) + bp_0(x_i, y_i, 0), \quad i = 1, 2, \ldots$$

Later on, it was established that the isoline configuration of θ practically coincides with that of the 500 mb isohypses or of the isobars at level $z \approx 5.6$ km, which is the same. Such a level is called the 'steering' level and the wind there is the 'steering' current. Consequently, having analyzed maps of this level, the surface pressure and temperature can be predicted following Kibel's method.

Finally, we shall point out that according to (1.9)

$$\frac{\partial}{\partial t}\left(\frac{\partial A_0}{\partial t}\right) = J\left(\frac{\partial A_0}{\partial t}, \theta\right), \quad A_0 = T_0, p_0, \tag{1.11}$$

i.e., the tendencies $\partial T_0/\partial t$ and $\partial p_0/\partial t$ are also advected.

In the first approximation, Kibel's theory does not describe the evolution of baric and thermal formations. The corrections made, proposed by Kibel himself and by others later on, partially remove this defect and improve the prognostic features of the scheme. The intransitive advantage of Kibel's theory, however, is the fact that it has been shown for the first time how the weather equations can be worked up, selecting only that which is typical for the synoptic atmospheric processes and eliminating the unnecessary from a meteorological point of view. So the beginning of a new stage in the development of dynamic meteorology and numerical weather prediction commenced.

2. BAROTROPIC PROGNOSTIC MODELS

a. Quasi-Geostrophic Approximation

After Kibel's work, the second most important point in the recent history of dynamic meteorology was the construction of the so-called barotropic prognostic model and, on that basis, the first successful experiment on numerical weather prediction (virtually, only of pressure), was realized.

In a barotropic atmosphere without friction and in adiabatic approximation, according to the classification from Section 1, only dynamic and inertial factors are significant.

In this case, the vorticity equation, after the appropriate simplification (see Section 5 of Chapter 3), takes the form

$$\zeta_t = -\mathbf{u}\cdot\nabla\zeta - \beta v + f\left(w_z \quad \text{or} \quad \frac{\partial w_p}{\partial p}\right), \tag{2.1}$$

where $w_z = \partial w/\partial z$, $\zeta_t = \partial\zeta/\partial t$. As we know, a prominent feature of this equation is that its left-hand side has the order of the terms on the right. In principle, this gives us an opportunity to intentionally modify the terms on the right by means of equation simplification without the fear that the error introduced in ζ_t would be inadmissibly large.

First of all, if we accept that at the upper and lower boundaries of the atmosphere w, $w_p = 0$, then there would exist at least one level at which $\partial w/\partial z$ or $\partial w_p/\partial p$ would equal zero. As far as $\partial w_p/\partial p = -\nabla_p\cdot\mathbf{u}$ and $\partial w/\partial z = -\nabla\cdot\mathbf{u}$, this would be a level of nondivergent motion. With the same degree of accuracy with which the vorticity equation (2.1) is obtained, from the divergence equation follows (see Chapter 3, Section 5, formulae (5.31)–(5.38))

$$\zeta \approx \zeta_g = \frac{\nabla^2\Phi}{f}, \quad \text{i.e., } \mathbf{u} \approx \mathbf{u}_g = f^{-1}\mathbf{k}\times\nabla\Phi. \tag{2.2}$$

Let us now introduce (2.2) into (2.1). We obtain

$$\frac{\partial\zeta_g}{\partial t} = -\mathbf{u}_g\cdot\nabla\zeta_g - \beta v_g, \tag{2.3}$$

or

$$\nabla^2 \left(\frac{\partial \Phi}{\partial t}\right) = J(f^{-1} \nabla^2 \Phi + f, \Phi), \tag{2.4}$$

which is one equation for one unknown function – the absolute geopotential Φ of the so-called *middle* level at which $\partial w_p/\partial p = 0$. It is usually identified with the 500 mb level (about 5 km). Obviously, the remarkable nonlinear Equation (2.4) is a direct consequence of the assumption that the absolute vorticity conservation on the middle level is a fundamental property of the synoptic processes.

In the real atmosphere, however, due to baroclinic and dissipative effects (frictional forces), the absolute vorticity is not a conservative quantity. This property can be ascribed to it only for periods of not longer than 1–2 days. Namely, the most successful forecasts of a geopotential field using Equation (2.4) have been recorded for such periods and under conditions corresponding to those theoretically proposed.

A more general approach can be applied to derive Equation (2.4). Let us write Equation (2.1) in the form

$$\zeta_t + \mathbf{u} \cdot \nabla(\zeta + f) = f \frac{\partial w_p}{\partial p} \tag{2.5}$$

and assume that the wind velocity vector $\mathbf{u}$ does not change its direction with height but may change its magnitude, i.e.,

$$\mathbf{u} = B(p)\overline{\mathbf{u}} \quad \text{and} \quad \zeta = B(p)\overline{\zeta} \tag{2.6}$$

where

$$\overline{(\ )} = \frac{1}{p_{00}} \int_0^{p_{00}} (\ )\, dp, \qquad \overline{B}(p) = 1, \quad p_{00} = 1000 \text{ mb}. \tag{2.7}$$

$B(p)$ is an empirical function of the pressure. We substitute (2.6) into (2.5)

$$B\overline{\zeta}_t + B^2\overline{\mathbf{u}} \cdot \nabla\overline{\zeta} + \beta B\overline{v} = f \frac{\partial w_p}{\partial p}, \tag{2.8}$$

then apply the averaging operator (2.7) and obtain

$$\overline{\zeta}_t + \overline{B^2}\overline{\mathbf{u}} \cdot \nabla\overline{\zeta} + \beta\overline{v} = \frac{w_p^0 f}{p_{00}}, \quad w_p^0 = w_p(p_{00}). \tag{2.9}$$

In the case of a flat surface (no mountains), it is admissible to let $w_p^0 = 0$. Then (2.9) and (2.6) yield

$$\zeta_t + \left[\frac{\overline{B^2}}{B(p)}\right] \mathbf{u} \cdot \nabla\zeta + \beta v = 0. \tag{2.10}$$

If one or more levels p^*, on which $B(p^*) = \overline{B^2}$ exist, then Equation (2.10) simplifies to

$$\zeta_t^* + \mathbf{u}^* \cdot \nabla\zeta^* + \beta v^* = 0 \quad \text{at} \quad p = p^*, \tag{2.11}$$

and after expressing u^* and ζ^* by Φ^*, i.e., after the introduction of the quasi-geostrophic approximation in (2.11), we again reach an equation of the type (2.4) for Φ^*

$$\nabla^2 \Phi_t^* = J(f^{-1} \nabla^2 \Phi^* + f, \Phi^*). \tag{2.12}$$

The model which depends on this equation is called *equivalent-barotropic*. Analysis of climatic empirical data leads to the conclusion that the equation $B(p^*) = \overline{B^2}$ has two roots: $p_1^* \approx 200$ mb and $p_2^* \approx 600$ mb. In such a way, the prognostic value of (2.12) is higher than that of Equation (2.4), since a prognosis can be given for two levels instead of for one, using the same technique.

Let us now discuss the assumption $w_p^0 = 0$. If we turn to the exact formula (3.16) of Chapter 3

$$w_p = \rho(\Phi_t + \mathbf{u} \cdot \nabla\Phi - gw), \tag{2.13}$$

we see that at $p = p_{00} = 1000$ mb none of the terms in the right-hand side equals zero. Really, the Earth's surface $z = 0$ does not coincide with the isobaric surface $p_{00} = 1000$ mb, so that $w \neq 0$. The latter is much more evident in the case of a mountainous surface. In an atmosphere without friction, such as we treat here, the geostrophic relation can be extended to the lower boundary so that at $p = p_{00}$ we can consider $\mathbf{u} \cdot \nabla\Phi = 0$. However, in the real atmosphere this is inadmissible. Finally, $\Phi_t(p_{00}) \neq 0$ too. If on the basis of these considerations $w_p^0 \neq 0$ is assumed, then instead of (2.12) one obtains

$$\nabla^2 \Phi_t - k_0^2 \Phi_t = J(f^{-1} \nabla^2 \Phi + f, \Phi), \tag{2.14}$$

where $L_0 = 1/k_0 = a_0/f$ is the scale (2.23) of Chapter 6, and the asterisk is omitted.

Equation (2.14) can also be derived by starting from the equations of motion in the shallow-water approximation (5.19) and (5.20) of Chapter 2, appropriately transformed

$$\zeta_t + \mathbf{u} \cdot \nabla(\zeta + f) = -f\nabla \cdot \mathbf{u}, \tag{2.15}$$

$$\nabla \cdot \mathbf{u} = -h^{-1}(h_t + \mathbf{u} \cdot \nabla h) \approx -h_*^{-1} \frac{\mathrm{d}h}{\mathrm{d}t}. \tag{2.16}$$

By inserting $\nabla \cdot \mathbf{u}$ in (2.15), introducing the quasi-geostrophic approximation, and letting $gh = \Phi$, we obtain Equation (2.14), where $k_0^2 = f^2/gh_*$. Hence, one can conclude that the condition $w_p^0 \neq 0$ is equivalent to $\nabla \cdot \mathbf{u} \neq 0$. For this reason, (2.14) is called a *divergent* barotropic equation while (2.12) or (2.4) are *nondivergent*. Thus, two prognostic equations for the geopotential field have been obtained:

$$\nabla^2 \Phi_t = A_\Phi \quad \text{and} \quad (\nabla^2 - k_0^2)\Phi_t = A_\Phi, \tag{2.17}$$

where $A_\Phi = J(f^{-1} \nabla^2 \Phi + f, \Phi)$. With respect to Φ, A_Φ is essentially a nonlinear term and the equations are insolvable by means of analytical methods. If, however, at $t = t_0$ A_Φ, which contains derivatives with respect to x and y only, is considered as a known function $A_\Phi(x, y, t_0)$ (e.g., from the observations, though in discrete points), then (2.17) are correspondingly Poisson's and Helmholtz's equations with respect to Φ_t which are well known in mathematical physics. Their solution under appropriate boundary conditions will provide us with the tendency Φ_t at the instant t_0. This will permit us to calculate

$$\Phi(t_0 + \Delta t) = \Phi(t_0) + \Phi_t(t_0)\Delta t$$

for a time step Δt ahead at any point (x, y). Then $A_\phi(t_0 + \Delta t)$ can be calculated, and so on. In detail, this iterative procedure of a numerical solution is the object of special courses in numerical weather prediction [3, 16, 23].

Here we shall only point out that the appearance of the term $k_0^2 \Phi_t$ on the left-hand side of (2.14) considerably influences the properties of the solution, despite the fact that this term is small in magnitude compared to the first one. This is demonstrated in the simplest way in the linear approximation of the Helmholtz Equation (2.17), namely

$$\left(\frac{\partial}{\partial t} + u_* \frac{\partial}{\partial x}\right) \nabla^2 \Phi' - k_0^2 \frac{\partial \Phi'}{\partial t} + \beta \frac{\partial \Phi'}{\partial x} = 0. \tag{2.18}$$

At $k_0 = 0$, it coincides with Equation (5.6) of Chapter 5, for nondivergent Rossby waves. Repeating the procedure of seeking wave-type solutions of (2.18), one obtains the dispersion relation

$$c = \left(u_* - \beta \frac{\lambda^2}{4\pi^2}\right) \Big/ \left(1 + \frac{\lambda^2}{4\pi^2 L_0^2}\right). \tag{2.19}$$

Hence, it follows that for $\lambda \ll L_0(\sim 2\text{–}3000 \text{ km})$, the correction to the classical formula (i.e., the denominator in (2.19)) is negligible. But for $\lambda \sim L_0$, the phase velocity of the waves computed by using (2.19), is in better agreement with the observations. It is curious that for $L_0 < \infty$ and $\lambda \to \infty$, $c(\lambda)$ is limited: $c_\infty = -\beta L_0^2$.

More generally, the equivalent-barotropic model also possesses the advantage that it permits an equation for the vertical velocity w_p at the level p^* to be easily obtained. In reality, if (2.9) is multiplied by B and then subtracted from (2.8), one would obtain

$$f \frac{\partial w_p}{\partial p} = (B^2 - B\overline{B^2})\overline{\mathbf{u}} \cdot \nabla \overline{\zeta} + \left(\frac{fB}{p_{00}}\right) w_p^0. \tag{2.20}$$

This is a simple diagnostic equation. After integration between p_{00} and p^*, $w_p(p^*)$ is obtained. The latter can then be used for cloudiness and precipitation forecasting at p^* level (see Section 5, below).

On the basis of the so-described barotropic models in quasi-geostrophic approximation, the following prognostic scheme has been proposed and realized in practice:

(i) At given initial observational data and appropriate boundary conditions Equation (2.17) are solved to determine $\Phi_t(t_0)$ and then $\Phi(t_0 + \Delta t)$.

(ii) Through the geostrophic relations (2.2) $\mathbf{u}(t_0 + \Delta t)$ is calculated and then $A_\phi(t_0 + \Delta t)$ too and so on.

(iii) At each time step, using (2.20), w_p can be calculated ($B(p)$ is determined in advance on climatic data, $\overline{\zeta}$ and $\overline{\mathbf{u}}$ are expressed through Φ, $w_p^0 = 0$ or is a priori given).

b. Quasi-Solenoidal Approximation

Some specification of the above-described model and prognostic scheme can be achieved if we reject the quasi-geostrophic relations (2.2) and assume that

$$\mathbf{u} = \mathbf{k} \times \nabla\psi + \nabla\varphi = \mathbf{u}_\psi + \mathbf{u}_\varphi,$$

where $\nabla \cdot \mathbf{u}_\psi = 0$ and $\nabla \times \mathbf{u}_\varphi = 0$. In practice, in the middle latitudes at synoptic scale motions $|\mathbf{u}_\psi| \gg |\mathbf{u}_\varphi|$ so that $\mathbf{u} \approx \mathbf{k} \times \nabla\psi$ and consequently $\zeta \approx \nabla^2\psi$. Inserting this in (2.1) (without the last term on the right, since $\nabla \cdot \mathbf{u}_\psi = 0$), we obtain a prognostic equation for the streamfunction

$$\nabla^2 \psi_t = J(\nabla^2 \psi + f, \psi). \tag{2.21}$$

The geopotential field can be found from the balance Equation (5.38) in Chapter 3:

$$\nabla^2 \Phi = f\nabla^2 \psi + 2(\psi_{xx}\psi_{yy} - \psi_{xy}^2) + \beta\psi_y = A_\psi .$$

At a given $A_\psi (x, y, t_0)$, Equation (2.22) is a Poisson's equation for Φ. We know how to solve it. Obviously, in quasi-geostrophic approximation $A_\psi = f\nabla^2\psi$.

Similarly, instead of (2.14) we now have

$$(\nabla^2 - k_0^2)\psi_t = J(\nabla^2 \psi + f, \psi). \tag{2.22}$$

After introducing the potential vorticity $\zeta_* = f + \nabla^2\psi - k_0^2\psi$, we write (2.22) in the form

$$\frac{\partial \zeta_*}{\partial t} + J(\psi, \zeta_*) = 0. \tag{2.22$'$}$$

The fault with this approximation is that the term $f\nabla \cdot \mathbf{u}_\varphi$ has been neglected in the right-hand side of Equation (2.21). If it is retained and replaced with $f\partial w_p/\partial p$, then the identical procedure to that in the previous case allows the construction of an equivalent-barotropic model in a quasi-solenoidal approximation which, in principle, should have better prognostic features than the previous models and, in practice, its application is quite good confirmation of this. The baroclinic prognostic models to be discussed below in Sections 3 and 4 suggest a better solution to this problem but at the expense of an increased number of equations. At a certain stage of the development of the theory and its application in prognostic practice, the relative mathematical simplicity of the equivalent barotropic scheme was of considerable importance.

c. Energetics of the Model

Let us consider the more general Equation (2.22)

$$(\nabla^2 - k_0^2)\psi_t + J(\psi, \nabla^2\psi) + \beta\psi_x = 0. \tag{2.23}$$

The appropriate scaling of the time and length variables yields a simpler dimensionless form of this equation

$$(\nabla^2 - 1)\psi_t + J(\psi, \nabla^2\psi) + \psi_x = 0. \tag{2.24}$$

Hence, three integrals of motion can be derived:

$$\bar{E}_{kp} = \iint [\tfrac{1}{2}(\nabla\psi)^2 + \tfrac{1}{2}\psi^2]\, dx\, dy = \text{const} \tag{2.25}$$

or

$$\bar{E}_{kp} = \bar{E}_k + \bar{E}_p = \tfrac{1}{2}\overline{(\nabla\psi)^2} + \tfrac{1}{2}\overline{\psi^2} = \text{const} \tag{2.25$'$}$$

expressing the conservation of the total (kinetic plus potential) energy;

$$\bar{\zeta}_* = \iint (\nabla^2 \psi - \psi)^2 \, dx \, dy = \text{const} \tag{2.26}$$

expressing the conservation of the potential vorticity; and

$$\bar{M} = \iint \psi \, dx \, dy = \text{const} \tag{2.27}$$

expressing the mass conservation.

In the simplest case, when in (2.23) $k_0 = \beta = 0$, the kinetic energy $\bar{E}_k = b$ and the so-called *enstrophy* $e = \overline{\zeta^2}/2$ are conserved

$$b = \tfrac{1}{2}\overline{(\nabla\psi)^2} = \text{const}, \qquad e = \tfrac{1}{2}\overline{(\nabla^2\psi)^2} = \text{const}. \tag{2.28}$$

We now write

$$\psi = \sum_n a_n \psi_n \tag{2.29}$$

where ψ_n are eigenfunctions of the Laplace operator ∇^2, i.e., $\nabla^2 \psi_n = -k_n^2 \psi_n$, k_n are wavenumbers. As is known, these eigenfunctions are orthogonal in the sense that

$$\overline{\psi_m \psi_n} = \delta_{mn} = \begin{cases} 0, & m \neq n \\ 1, & m = n. \end{cases} \tag{2.30}$$

Inserting (2.29) into (2.28), after some calculations one obtains

$$\begin{aligned} b &= \sum_n \tfrac{1}{2} a_n^2 k_n^2 = \sum_n b_n = \text{const}, \\ e &= \tfrac{1}{2} \sum_n k_n^4 a_n^2 = \sum_n k_n^2 b_n = \text{const}. \end{aligned} \tag{2.31}$$

Consequently

$$k = \left(\frac{e}{b}\right)^{1/2} = \left(\frac{\sum_n k_n^2 b_n}{\sum_n b_n}\right)^{1/2} = \text{const} \tag{2.32}$$

can be interpreted as a mean wavenumber, while b_n is the kinetic energy spectral density. Then

$$k^2 b = \sum_n b_n k_n^2 = \text{const}, \tag{2.33}$$

i.e., $k(t) = \text{const}$. Hence, it follows that a systematic transport of kinetic energy from spectral components with small wavenumbers k_n to such with larger k_n is impossible (Fjortoft's theorem, 1953). To clarify this conclusion Charney (see [49]) proposed the following mechanical interpretation of the relationship (2.33):

Let us imagine that we have a weightless rod on which we suspend weights b_1, b_2, etc., at distances k_1^2, k_2^2, etc., from the point 0 and that they are balanced by the weight $\bar{E}_k$, suspended at distance k^2 on the left from the point 0, Figure 7.1. Then formula

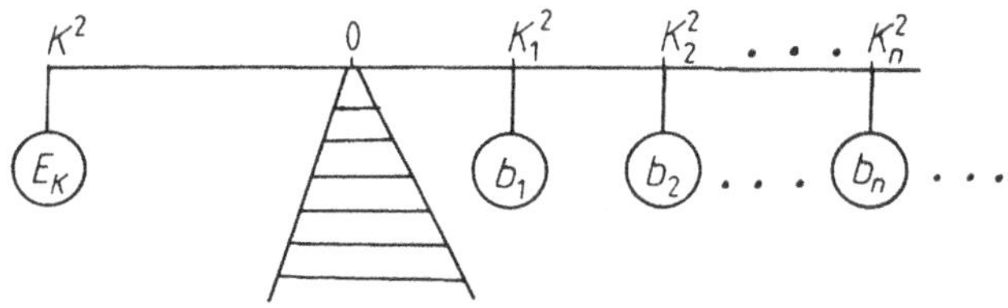

Fig. 7.1. Illustration to Fjortoft's Theorem (after Charney, [49, 45]).

(2.33) manifests that the mass b together with the moment of inertia, has to be preserved. It is clear that one cannot change the places of two weights, no matter which, e.g., b_i and b_k, without disturbing the balance. At least three weights should be involved in such a transposition.

Consequently, the transport of kinetic energy toward the higher wavenumbers is clearly limited. The higher the wavenumber, the more it is limited. This feature sharply distinguishes the two-dimensional flows from the three-dimensional ones and we shall return to it later on. It appears to be a direct consequence of the simultaneous existence of both invariants (2.31).

d. Nonlinear Interactions

(i) *Low-Order Barotropic Model*. The equation

$$\nabla^2 \psi_t + J(\psi, \nabla^2 \psi) = 0 \tag{2.34}$$

which follows from (2.23) at $k_0 = \beta = 0$ is, perhaps, the most elegant equation of dynamic meteorology. Besides being used for the construction of prognostic schemes, it can also be used for studying various atmospheric phenomena which are intrinsically nonlinear, but because of its quadratic nonlinearity, this equation cannot be solved accurately by analytical methods. Approximate analytical solutions are possible, however, in particular, the spectral representation of (2.34) with its subsequent appropriate truncation, first proposed by Lorenz (1960), permits us to obtain a low-order system corresponding to (2.34), which is solvable in special functions.

Actually, let us assume that ψ is a doubly periodic function at all times, i.e.,

$$\psi\left(x + \frac{2\pi}{m}, y + \frac{2\pi}{n}, t\right) = \psi(x, y, t),$$

where m, n are fixed wavenumbers. As is shown by Lorenz (1960, 1982), maximum simplification can be achieved if one seeks a solution of (2.34) in the form

$$\psi(x, y, t) = A(t) \cos mx + B(t) \cos ny + C(t) \sin mx \sin ny. \tag{2.35}$$

Then Equation (2.34) reduces to the following low-order system of three nonlinear ordinary differential equations

$$\frac{dA}{dt} = \alpha BC, \qquad \frac{dB}{dt} = \beta CA, \qquad \frac{dC}{dt} = \gamma AB, \tag{2.36}$$

where α, β, γ are constants depending on m, n and satisfying the relations

$$m^2\alpha + n^2\beta + \tfrac{1}{2}(m^2 + n^2)\gamma = 0,$$
$$m^4\alpha + n^4\beta + \tfrac{1}{2}(m^2 + n^2)^2\gamma = 0. \tag{2.37}$$

Physically, B represents the strength of westerly and easterly currents at alternating latitudes, while A and C together define the amplitude and phase of superposed waves. System (2.36) has two important properties:

(a) It conserves the total kinetic energy b and enstrophy e as (2.24) does:

$$4b = m^2A^2 + n^2B^2 + \tfrac{1}{2}(m^2 + n^2)C^2,$$
$$4e = m^4A^2 + n^4B^2 + \tfrac{1}{2}(m^2 + n^2)^2C^2, \tag{2.38}$$

i.e., b and e are quadratic invariants of the motion – $db/dt = de/dt = 0$.

(b) Due to the latter fact, Equations (2.36) are easily solved analytically in elliptic (i.e., periodic) functions by eliminating the two variables through (2.38). Besides, Equations (2.36) are much more easily numerically solved than Equation (2.34) itself. On the basis of these solutions, the nonlinear interaction between the zonal current and the superposed waves can be studied.

(ii) *Resonance Rossby-Wave Interaction.* Let us go back again to Equation (2.24). As was pointed out in Section 5 of Chapter 5, this nonlinear equation admits a wave solution of the type

$$\psi(\mathbf{x}, t) = \hat{\psi}\exp[i(\mathbf{k}\cdot\mathbf{x} - \omega t)],$$
$$\mathbf{x} = (x, y), \qquad \mathbf{k} = (m, n), \tag{2.39}$$

with dispersion relation

$$\omega(m^2 + n^2 + 1) + m = 0, \tag{2.40}$$

as the linear analogue of (2.24) does. However, a superposition of the two Rossby waves

$$\psi_j(\mathbf{x}, t) = \hat{\psi}_j\exp[i(\mathbf{k}_j\cdot\mathbf{x} - \omega_j t)], \quad j = 1, 2,$$
$$\omega_j(m_j^2 + n_j^2 + 1) + m_j = 0, \quad j = 1, 2, \tag{2.41}$$

will no longer be a solution of the nonlinear Equation (2.24). Due to the quadratic nonlinearity of the latter, a term

$$\hat{\psi}_1\hat{\psi}_2\exp\{i[(\mathbf{k}_1 + \mathbf{k}_2)\cdot\mathbf{x} - (\omega_1 + \omega_2)t]\} \tag{2.42}$$

will appear. It could be interpreted as an external force, acting on the linear system, described by (2.39). Then, if

$$\mathbf{k} = \mathbf{k}_1 + \mathbf{k}_2, \qquad \omega = \omega_1 + \omega_2 \tag{2.43}$$

resonance will appear. Thus, in the case of the resonance interaction of a triad of Rossby waves, the nine quantities (m, m_1, m_2), (n, n_1, n_2) and $(\omega, \omega_1, \omega_2)$ must satisfy five equations (2.40), (2.41) and (2.43). Therefore, considerable freedom for satisfying these conditions exists.

3. BAROCLINIC PROGNOSTIC MODELS

a. Quasi-Geostrophic Approximation

We start from the vorticity equation (4.28) of Chapter 3

$$\zeta_t = -\mathbf{u} \cdot \nabla(\zeta + f) - f\nabla \cdot \mathbf{u} \tag{3.1}$$

and the thermodynamic energy equation (3.17) of Chapter 3

$$\frac{\partial \Phi_t}{\partial p} = -\mathbf{u} \cdot \nabla\left(\frac{\partial \Phi}{\partial p}\right) - mw_p, \tag{3.2}$$

where, according to (3.13) of Chapter 3, $m = m_\alpha = -(\alpha/\theta)\, \partial\theta/\partial p$ is a parameter of the static stability of the atmosphere.

The introduction of the quasi-geostrophic approximation into system (3.1) and (3.2) consists of letting

$$\mathbf{u} \approx \mathbf{u}_g = f^{-1}\mathbf{k} \times \nabla\Phi, \qquad \zeta \approx \zeta_g = f^{-1}\nabla^2\Phi \tag{3.3}$$

except in the divergent term in (3.1) which is transformed with the help of the exact equation $\nabla \cdot \mathbf{u} = -\partial w_p/\partial p$. We obtain

$$\nabla^2\Phi_t = -f\mathbf{u}_g \cdot \nabla(f^{-1}\nabla^2\Phi + f) + f^2\,\frac{\partial w_p}{\partial p}, \tag{3.4}$$

$$\frac{\partial \Phi_t}{\partial p} = -\mathbf{u}_g \cdot \nabla\left(\frac{\partial \Phi}{\partial p}\right) - mw_p. \tag{3.5}$$

From these equations, either w_p or Φ can be excluded to obtain a single equation. For simplicity we shall assume m = const.*

Let us apply the operator $(f^2/m)\,\partial/\partial p$ to both sides of Equation (3.5) and add the result to (3.4). We obtain

$$\left(\nabla^2 + \frac{f^2}{m}\,\frac{\partial^2}{\partial p^2}\right)\Phi_t = -f\mathbf{u}_g \cdot \nabla\left(\frac{1}{f}\,\nabla^2\Phi + f\right) + \frac{f^2}{m}\,\frac{\partial}{\partial p}\left(-\mathbf{u}_g \cdot \nabla\,\frac{\partial\Phi}{\partial p}\right). \tag{3.6}$$

This is a prognostic equation for the geopotential tendency Φ_t the right-hand side of which is essentially nonlinear but does not contain derivatives with respect to time t. If we denote it by A_ϕ and look on it as being known from the initial data of observations, then (3.6) is a linear equation with respect to Φ_t which is valid for each level in the atmosphere. The equation

$$\left[\nabla^2 + \frac{f^2}{m}\,\frac{\partial^2}{\partial p^2}\right]\Phi_t = A_\phi = A'_\phi + A''_\phi \tag{3.6'}$$

is of second order with respect to p and it can be solved numerically, or even analytically

---

* According to (3.13) in Chapter 3, $m = a^2p^{-2}$ where $a^2 = RT(\gamma_a - \gamma)/\Gamma_A$, $\Gamma_A = g/R \approx 3.42°\ \mathrm{km}^{-1}$, but with greater accuracy, we can assume that $a \approx$ const, then $m(p) \sim p^{-2}$. In an isothermal atmosphere ($\gamma = 0$, T = const), a = const.

under two appropriate boundary conditions. When this solution is carried out over a limited area of the plane xOy, a boundary condition on x, y will also be needed.

In addition to being prognostic, Equation (3.6′), or its solution, also serves for a theoretical analysis of the contribution of both factors: the advection of vorticity (A'_Φ) and the so-called *differential advection* of temperature (A''_Φ) for the formation of geopotential tendency Φ_t. Examples of such analysis over idealized synoptic situations in the middle latitudes are considered in detail by Holton [27].

Together with the geopotential field Φ, the knowledge of the vertical velocity w_p, which is directly connected with the cloudiness and precipitation formation (see Section 5, below), is of considerable importance for weather forecasting. As we know, w_p cannot be determined from the continuity equation $\partial w_p/\partial p = -\nabla \cdot \mathbf{u}$ because the empirical value of $\nabla \cdot \mathbf{u}$ is burdened with a tremendous error. Equations (3.4) and (3.5) are also not suitable because they contain Φ_t, which is also the subject of calculation. It is desirable to find a direct relationship between w_p and Φ.

The form of Equations (3.4) and (3.5) implies that Φ_t can easily be eliminated to obtain a single equation for w_p. For this purpose, we apply the differential operators $\partial/\partial p$ and $\nabla^2 = \partial^2/\partial x^2 + \partial^2/\partial y^2$ correspondingly to Equations (3.4) and (3.5) and subtract and bearing in mind once more that $m = \text{const}$ and $\nabla^2 \partial/\partial p = \partial \nabla^2/\partial p$ we find

$$\left[\nabla^2 + \left(\frac{f^2}{m}\right)\frac{\partial^2}{\partial p^2}\right] w_p = A_w = A'_w + A''_w,$$
$$A_w = \frac{f}{m}\frac{\partial}{\partial p}\left[\mathbf{u}_g \cdot \nabla\left(\frac{1}{f}\nabla^2\Phi + f\right)\right] - \frac{1}{m}\nabla^2\left(\mathbf{u}_g \cdot \nabla\frac{\partial\Phi}{\partial p}\right), \tag{3.7}$$

where the term A'_w is called *differential vorticity advection.*

Note that Equation (3.7) for w_p is a diagnostic one – it does not contain derivatives with respect to t. It is of the same type as Equation (3.6) and allows w_p to be determined in an arbitrary instant t in which Φ is known (observed or prognosed). Similarly to (3.6), Equation (3.7) can be used for a theoretical analysis of the contribution (in magnitude and sign) of both factors A'_w and A''_w [27] in idealized synoptic situations.

b. Quasi-Solenoidal Approximation

Again, the starting equations are (3.1) and (3.2) where now $\mathbf{u} = \mathbf{u}_\psi + \mathbf{u}_\varphi$, $\nabla \cdot \mathbf{u}_\psi = 0$, $\nabla \times \mathbf{u}_\varphi = 0$, $\mathbf{u}_\psi = \mathbf{k} \times \nabla\psi$. Then the continuity equation takes the form $\nabla \cdot \mathbf{u} = \nabla \cdot \mathbf{u}_\varphi = -\partial w_p/\partial p$. Since $\zeta = \nabla^2\psi$, then from (3.1) we obtain

$$\frac{\partial \nabla^2 \psi}{\partial t} = -\mathbf{u}_\psi \cdot \nabla(\nabla^2\psi + f) + f\frac{\partial w_p}{\partial p}. \tag{3.8}$$

In the heat-influx Equation (3.2) we let $\Phi = f\psi$ ($f = \text{const}$):

$$\frac{\partial^2 \psi}{\partial t\,\partial p} = -\mathbf{u}_\psi \cdot \nabla\left(\frac{\partial\psi}{\partial p}\right) - \left(\frac{m}{f}\right) w_p, \tag{3.9}$$

where the fact that $|\mathbf{u}_\varphi| \ll |\mathbf{u}_\psi|$ has been used.

Equations (3.8) and (3.9) correspond to Equations (3.4) and (3.5) and could be treated in the same way. Here we shall apply some other transformations.

We multiply (3.9) by f^2/m where $m = m(p)$, differentiate with respect to p and add to (3.8). We obtain

$$\frac{\partial \zeta_*}{\partial t} = -\mathbf{u}_\psi \cdot \nabla \zeta_* = -J(\psi, \zeta_*), \tag{3.10}$$

where

$$\zeta_* = \nabla^2 \psi + f + f^2 \frac{\partial}{\partial p} \frac{1}{m} \frac{\partial \psi}{\partial p}, \tag{3.11}$$

i.e., $d\zeta_*/dt = 0$, where ζ_* is the potential vorticity. At $m = \text{const}$ (3.11) simplifies. Consequently, ζ_* is a conservative quantity and is transported along the streamlines. An alternative form of Equation (3.10) looks like (3.6)

$$\left(\nabla^2 + f^2 \frac{\partial}{\partial p} \frac{1}{m} \frac{\partial}{\partial p}\right) \psi_t = -J(\psi, \zeta_*). \tag{3.12}$$

Almost identical to (3.7) will be the equation for w_p which can be obtained from (3.8) and (3.9).

The basic equations of the baroclinic prognostic model (3.6), (3.7) and (3.12) are applicable for an arbitrary level in the atmosphere, including the standard ones for which observational data are available to be used as initial conditions. Since computers are available, these equations, in principle, can be solved numerically for each of these levels. However, in practice, this is hardly possible, even for the standard levels because this calls for an enormous memory and speed of the computers. That is why, on the basis of the above-mentioned equations, some multi-layer baroclinic models have been developed, of which the two-layer one is the simplest.

c. The Two-Layer Baroclinic Model

The geometrical structure of the model is presented in Figure 7.2. The whole atmosphere is divided into two layers, bounded by the continuous lines: 4 – lower boundary, 0 –

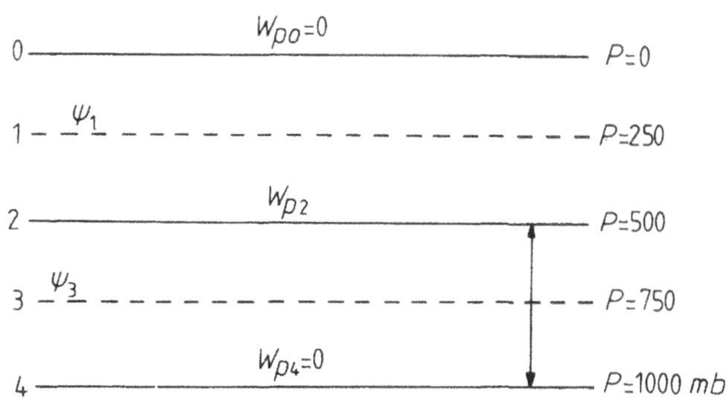

Fig. 7.2. Two-layer baroclinic model.

upper boundary, 2 – internal boundary. For simplicity we let $w_p(0) = w_p(1000 \text{ mb}) = 0$. Two intermediate levels are also introduced: 1 and 3. Then all derivatives with respect to p in Equations (3.8) and (3.9) or (3.4) and (3.5) can be replaced by finite difference ratios as follows:

$$\left(\frac{\partial w_p}{\partial p}\right)_1 = \frac{1}{\Delta p}(w_{p_2} - w_{p_0}), \quad \left(\frac{\partial w_p}{\partial p}\right)_3 = \frac{1}{\Delta p}(w_{p_4} - w_{p_2}) = -\frac{1}{\Delta p} w_{p_2},$$

$$\left(\frac{\partial \psi}{\partial p}\right)_2 \approx \frac{1}{\Delta p}(\psi_3 - \psi_1).$$

Equation (3.8) is written for levels 1 and 3

$$\frac{\partial}{\partial t}\nabla^2\psi_1 = -(\mathbf{k} \times \nabla\psi_1) \cdot \nabla(\nabla^2\psi_1 + f) + \frac{f}{\Delta p} w_{p_2}, \tag{3.13}$$

$$\frac{\partial}{\partial t}\nabla^2\psi_3 = -(\mathbf{k} \times \nabla\psi_3) \cdot \nabla(\nabla^2\psi_3 + f) - \frac{f}{\Delta p} w_{p_2}, \tag{3.14}$$

and Equation (3.9) is written for level 2:

$$\frac{\partial}{\partial t}(\psi_1 - \psi_3) = -(\mathbf{k} \times \nabla\psi_2) \cdot \nabla(\psi_1 - \psi_3) + \frac{m\Delta p}{f} w_{p_2}. \tag{3.15}$$

The system formed from the last three equations, however, is unclosed – the unknown functions are four in number: ψ_1, ψ_3, w_{p_2} and ψ_2, while there is no equation for ψ_2. The simplest way to close it is to assume

$$\psi_2 = \tfrac{1}{2}(\psi_1 + \psi_3). \tag{3.16}$$

Further transformations and the solution of the derived system of equations are the subject of courses in numerical weather prediction [3, 16, 23]. Here we point out that even this simple operative prognostic baroclinic model has an important advantage over the barotropic or equivalent-barotropic model because it accounts for temperature advection as well. This fact makes it capable of predicting the origin of new baric systems, while the barotropic models predict (in the middle latitudes well enough for a period of 1–2 days) only the evolution due to the vorticity advection of systems already existing. In other words, the problem of baroclinic instability and cyclogenesis should be studied on the basis of the baroclinic models of the atmosphere, particularly the two-layer one described here. An example of such application can be found in detail in Holton [28]. One of the most interesting results finds expression in the conclusion about the existence of baroclinic waves of maximum instability having the wavenumber $k_m^2 = \sqrt{2}\,\nu^2$ where $\nu^2 = f^2/m\Delta p^2$. In the troposphere of the middle latitudes $\nu \sim 1.4 \times 10^{-6}$ m^{-1}. Under normal conditions of static stability, the wavelength $\lambda_m = 2\pi/k_m$ of the maximum baroclinic instability is about 3.7×10^3 km, which is close to the average wavelength for midlatitude synoptic systems (see also Section 8 of Chatper 5).

4. PROGNOSTIC MODELS WITH PRIMITIVE EQUATIONS

a. General Characteristics

The various prognostic models discussed in the previous sections are based on the 'filtered' equations in quasi-geostrophic or quasi-solenoidal approximation which do not possess solutions of the fast (first of all, gravity) type waves. For the geopotential (pressure) these equations are of first order with respect to time t and, consequently, need initial conditions only for this element. The equation for the vertical velocity was of the diagnostic type. The mathematical form of these equations allows an application of analytical and even graphical methods of solution. They were solved with the help of the first computers – having small speed and memory – by using not very precise numerical schemes.

The main shortcoming of the filtered equations is the inaccurate description of the wind field by the quasi-geostrophic or quasi-solenoidal approximation. This defect is very apparent during the processes of origination of new disturbances in the velocity and pressure fields and their mutual adjustment. Then the largest deviations of the real wind from the geostrophic (or more generally, from the nondivergent) one are observed and the filtered equations become 'insensitive' to these processes. The attempts to describe the wind field by means of ageostrophic schemes make the equations so complex that they lose one of their important advantages. On the other hand, the quasi-geostrophic approximation is not applicable for the equatorial atmosphere while the quasi-solenoidal one requires a solution of balance equation (5.38) in Chapter 3, for determining the streamfunction when the geopotential Φ is known, which is accompanied by great mathematical difficulties.

Due to these reasons, and as early as the Fifties when an intensive development of the prognostic models with filtered equations had been observed, the search began for some other ways of solving the weather prediction problem based on a fundamental system of equations. Thus, a new direction in the theory of meteorological forecasting emerged in which no approximation regarding the real wind is used, except the quasi-static one which is much more precise than the others. The latter means that the wind velocity is determined from the nontransformed equations of the horizontal motion. Written down most often in isobaric coordinates, together with the third prognostic equation, that of the First Principle, and the three diagnostic equations (of statics, state, and continuity), they form a system of six equations, called *primitive*, about the unknown functions u, v, w_p (or w_σ) Φ, α, θ (or T). In the p-system they are (see Section 3 in Chapter 3)

$$\frac{\mathrm{d}\mathbf{u}}{\mathrm{d}t} = -\nabla\Phi - f\mathbf{k}\times\mathbf{u}, \qquad \nabla\cdot\mathbf{u} + \frac{\partial w_p}{\partial p} = 0,$$

$$\frac{\partial\Phi}{\partial p} = -\frac{RT}{p} = -\alpha, \qquad \frac{\partial T}{\partial t} + \mathbf{u}\cdot\nabla T = -m_T w_p, \tag{4.1}$$

where

$$m_T = \frac{RT(\gamma_a - \gamma)}{gp}, \qquad \frac{\mathrm{d}}{\mathrm{d}t} = \frac{\mathrm{d}_2}{\mathrm{d}t} + w_p\frac{\partial}{\partial p}.$$

As is seen, the return towards primitive equations (4.1) increases the order of the system with respect to t compared with that of the filtered equations and preserves the gravity waves in the solutions of the prognostic equations. The former factor requires initial conditions to be given not only for the geopotential, as was the case with the filtered equations, but also for the wind velocity, i.e.,

$$u = u^0, \qquad v = v^0, \qquad \Phi = \Phi^0 \quad \text{at } t = t_0 \tag{4.2}$$

On the other hand, for practical purposes, the primitive equations are integrated into an area, limited in the horizontal direction. Well-formulated boundary conditions are necessary because if there are insufficient or too many, the numerical solutions of the equations become unstable. The errors, originating at each step of integration with respect to time, would spread still deeper from the boundary into the area and would completely 'discredit' the solution. Quite often, for simplicity of analysis, rigid boundaries are assumed. For instance, a problem of numerical prediction related to the entire Northern Hemisphere admits that a hard boundary (wall) is to be found at the equator which does not allow a mass exchange between both hemispheres. Mathematically this is expressed by the equalities

$$u_n \bigg|_B = 0, \qquad \frac{\partial u_t}{\partial n}\bigg|_B = 0, \qquad \frac{\partial \Phi}{\partial n}\bigg|_B = 0 \tag{4.3}$$

where u_n and u_t are the normal and tangential components of the velocity $\mathbf{u}$ on the boundary B.

Great contributions to the development of this approach to the theory of meteorological forecasting have been made by Kibel, Marchuk, Smagorinsky, Phillips, etc. Here we are going to briefly discuss two methods of solving primitive equations, which, together with other equations, are studied in detail in special courses in numerical prediction.

(i) Direct integration of prognostic Equations (4.1), which can be written in the form

$$\frac{\partial F}{\partial t} = M_F, \qquad F = u, v, T,$$

where M_F denotes the sum of all terms which do not contain derivatives with respect to t. Having observational data at $t = t_0$ about u, v and Φ, one can determine T and w_p and then M_F as well, i.e., $F_t = \partial F/\partial t$. Then

$$F(t_0 + \Delta t) = F(t_0) + \left(\frac{\partial F}{\partial t}\right)_0 \Delta t.$$

This method can hardly be realized in practice because of the fact that $\partial F/\partial t$ is determined as a small difference of great quantities burdened with otherwise admissible errors of observations. Special differential approximations of the exact derivatives, of the same type as those developed by Marchuk [42, 43], for instance, and a special preliminary processing of the initial data (initialization), described below, are needed in order that the error with which $\partial F/\partial t$ is calculated is decreased.

(ii) Another way for overcoming this difficulty was proposed by Kibel (1958, 1960, 1962). The primitive equations are written in the form

$$\mathbf{u}_t + f\mathbf{k} \times \mathbf{u} + \nabla\Phi = -\mathbf{N},$$

$$\nabla \cdot \mathbf{u} + \left(\frac{1}{p_{00}}\right) \frac{\partial w_p}{\partial \eta} = 0, \tag{4.4}$$

$$\eta^2 \frac{\partial \Phi_t}{\partial \eta} + \left(\frac{a^2}{p_{00}}\right) w_p = R\eta N_T,$$

where $a^2 = \epsilon RT$, $\epsilon = (\gamma_a - \gamma)/\Gamma_A$, $\Gamma_A = g/R$. The third equation results from the equations of heat-influx and static

$$T_t - \left(\frac{a^2}{p_{00}}\right) w_p = -N_T, \qquad T = -\left(\frac{\eta}{R}\right) \frac{\partial \Phi}{\partial \eta},$$

where $\eta = p/p_{00}$, $p_{00} = 1000$ mb – standard pressure. Besides,

$$\mathbf{N} = (\mathbf{u} \cdot \nabla)\mathbf{u} + \left(\frac{w_p}{p_{00}}\right) \frac{\partial \mathbf{u}}{\partial p},$$

$$N_T = (\mathbf{u} \cdot \nabla T) = -\left(\frac{\eta}{R}\right)\left(\mathbf{u} \cdot \nabla \frac{\partial \Phi}{\partial \eta}\right). \tag{4.5}$$

From the equations of motion (4.4), following the procedures known from Section 5 in Chapter 3, we derive the vorticity ($\zeta = v_x - u_y$) and divergence ($D = u_x + v_y$) equations. The second and third equations in (4.4) allow the elimination of w_p. As a result, we obtain three equations

$$\frac{\partial \zeta}{\partial t} + fD = -\mathbf{k} \cdot (\nabla \times \mathbf{N}) = A_\zeta,$$

$$\frac{\partial D}{\partial t} + \nabla^2 \Phi - f\zeta = -\nabla \cdot \mathbf{N} = A_D, \tag{4.6}$$

$$\frac{\partial}{\partial \eta}\left(\eta^2 \frac{\partial^2 \phi}{\partial \eta\, \partial t}\right) - a^2 D = R \frac{\partial}{\partial \eta} \eta N_T = A_T,$$

where f = const has been assumed.

At the given A_ζ, A_D and A_T at a fixed instant, e.g., at $t = t_0$ calculated from observational data, Equations (4.6) represent a linear system for ζ, D and Φ, which can be solved numerically as well as analytically.

Obviously, the system (4.6) consists of nonfiltered equations and would describe the propagation and evolution of the wave disturbances in the model atmosphere. The fast ones (i.e., the gravity waves), will participate in mutual adjustment of the velocity and pressure fields. The existence of free (nonlinear) terms A_ζ, A_D, A_T in system (4.6), distinguishing it from the linear system (3.4) of Chapter 6, reflects the process of a continuous generation of new waves. That is, A_ζ, A_D and A_T play the role of sources. Consequently, system (4.6) would describe motions, oscillating near the geostrophic one. It allows the forecasting of synoptic motions, while at the same time taking into account the proximity of the wind to the geostrophic one, as well as the process of adjustment. By using this approach to the problem, the difficulties in directly integrating $\partial F/\partial t = M_F$, discussed in point (i), immediately disappear when primitive Equations (4.4) are written down.

The development of numerical weather prediction methods, based on primitive equations, has been determined by physical, mathematical, and technical achievements, which have been directly stimulated by the present problem. A typical example is the numerical method of 'splitting', proposed and developed by Marchuk [42, 43] in connection with solving primitive weather equations. On the other hand, current progress in computer techniques makes a computer realization of prognostic schemes by means of primitive equations not much more difficult than solving filtered equations, as it was at the beginning.

All this, however, does not necessarily mean that we should rely entirely on the utilization of primitive equations only, and moreover, this is not even inevitable because the application of prognostic schemes, based on the filtered equations, are far from exhausted. This is due to the unsuitable approximations of the differential operators in the equations by finite differences, and also to unreliable boundary conditions, to errors in the initial data, etc.

b. Initialization

As was mentioned above, the initial data contain errors, as well as local inhomogeneities of independent meteorological elements that are of no importance in predicting the evolution of the synoptic processes. It is desirable to filter them because it is almost certain that the initial data would not satisfy the prognostic system of equations which we have chosen for a mathematical model of the atmosphere. In general, none of the previously-discussed systems is adequate, nor should we expect them to be such or to reflect even the smallest details in the behaviour of the atmosphere, because individual descriptions are of no interest. But, after all, if we still think that our mathematical model is in some degree adequate, it is quite natural that we should expect the initial data to satisfy, with the same degree of accuracy, the corresponding equations in order to avoid some unwanted effect (intense adjustment, computational instability, etc.). The procedures by means of which the initial data are made consistent with a definite set of model equations is known as *initialization*. In the very process of initialization, some errors in the initial data can be found and excluded.

Different initialization methods have been proposed. Here we are going to briefly discuss the method based on the variational principle, and proposed by Sasaki (1958).

Let $\mathbf{u}^0$, Φ^0 and T^0 denote the observed values of the wind velocity, geopotential and temperature, while $\mathbf{u}$, Φ, T denote their adjusted values at the same points. As an example let us consider the case when our elements satisfy the geostrophic and thermal wind relationship as well as the hydrostatic equation

$$\mathbf{u} = f^{-1}\mathbf{k} \times \nabla\Phi, \qquad \frac{\partial \mathbf{u}}{\partial p_*} = f^{-1}\mathbf{k} \times \nabla T, \quad T = \frac{\partial \Phi}{\partial p_*} \tag{4.7}$$

where $p_* = -R \ln(p/p_{00})$, $p_{00} = 1000$ mb. In general, the observed values $\mathbf{u}^0$, Φ^0 and T^0 will not satisfy these constraints. We let

$$\mathbf{u} = \mathbf{u}^0 + \mathbf{u}', \qquad \Phi = \Phi^0 + \Phi', \qquad T = T^0 + T', \tag{4.8}$$

and construct the quadratic form

$$I = \int_G [\alpha_u^2 |\mathbf{u}'|^2 + \alpha_\phi^2 \Phi'^2 + \alpha_T^2 T'^2]\, \mathrm{d}x\, \mathrm{d}y\, \mathrm{d}p_*, \tag{4.9}$$

where, according to (4.7)

$$\mathbf{u}' = f^{-1}\mathbf{k} \times \nabla(\Phi^0 + \Phi') - \mathbf{u}^0, \qquad T' = \left(\frac{\partial}{\partial p_*}\right)(\Phi^0 + \Phi') - T^0, \tag{4.10}$$

and α_u^2, α_ϕ^2, α_T^2 are the empirical weights attached to the corresponding information.

Obviously, $I = I(\Phi')$. We now require that the functional $I(\Phi') = \min$, i.e., $\delta I = 0$ in the sense of a calculus of variation. Following the corresponding rules, one can obtain the partial differential equation

$$\nabla_*^2 \Phi' - q_\phi^2 \Phi' = f\zeta^0 + q_T^2 \frac{\partial T^0}{\partial p_*} - \nabla_*^2 \Phi^0, \tag{4.11}$$

where $\zeta^0 = v_x^0 - u_y^0$ is the observed vorticity, $q_\phi = f\alpha_\phi/\alpha_u$, $q_T = f\alpha_T/\alpha_u$ and

$$\nabla_*^2 = \nabla^2 + q_T \frac{\partial^2}{\partial p_*^2}, \qquad \nabla^2 = \frac{\partial^2}{\partial x^2} + \frac{\partial^2}{\partial y^2}.$$

The right-hand side of (4.11) can be calculated from observational data. If it vanishes, this means that they are all satisfying (4.7) and can be directly used. If, however, this does not happen, then the elliptic equation (4.11) has to be solved for Φ'. After this $\mathbf{u}'$ and T' can be found from (4.10).

If $q_T = 0 (\alpha_T = 0)$, Equation (4.11) simplifies to

$$\nabla^2 \Phi' - q_\phi^2 \Phi' = f\zeta^0 - \nabla^2 \Phi^0. \tag{4.12}$$

This is only the case for a geostrophic constraint imposed on the velocity and geopotential field. Alternatively, instead of (4.12) we can write

$$\nabla^2 \Phi - q_\phi^2 \Phi = f\zeta^0 - q_\phi^2 \Phi^0. \tag{4.13}$$

On the other hand, Equations (2.4) or (2.5) from Chapter 6, describing the process of geostrophic adjustment in a barotropic atmosphere, are of the type

$$u_t - fv = -\Phi_x, \qquad v_t + fu = -\Phi_y, \qquad \Phi_t + a_*^2(u_x + v_y) = 0.$$

As we know, this system admits an invariant $\zeta_* = a_*^2\zeta - f\Phi$ which can also be written as

$$\nabla^2 \Phi_s - \left(\frac{f}{a_*}\right)^2 \Phi_s = f\zeta^0 - \left(\frac{f}{a_*}\right)^2 \Phi^0. \tag{4.14}$$

Equations (4.13) and (4.14) completely coincide if we let $q_\Phi = f/a_*$, i.e., $\alpha_\Phi/\alpha_u = 1/a_*$. Then the results of both processes – the geostrophic adjustment and initialization, following Sasaki's method – will also coincide.

Equations (4.7), used to illustrate Sasaki's method, were diagnostic. The application of the variational principle, however, is not restricted to diagnostic constraints only. In principle, the complete equations of the forecasting model can be used as a constraint on the observations. Then, the initialization scheme will be much more complex.

5. METHODS OF CLOUDINESS AND PRECIPITATION FORECASTING

The forecast of the pressure or height of the isobaric surfaces, of the temperature and the wind, although of great importance, is not sufficient for yielding the complete characteristics of the weather. Among other meteorological elements necessary for this purpose, cloudiness and precipitation are of special interest. Sometimes they stand as the only characteristics of the weather, expressed in the words 'the weather is cloudy or rainy'. Their forecasting is one of the most challenging problems of present-day dynamic meteorology and its related disciplines, synoptic meteorology and numerical weather prediction. The progress achieved, though modest, is quite promising indeed.

a. Basic Equations

Clouds and precipitation are formed as a result of water vapour condensation and sublimation in the atmosphere. The processes of heat and moisture transfer, resulting in the formation and subsequent evolution of clouds, are described by the following equations

$$\frac{\mathrm{d}q}{\mathrm{d}t} = \mathrm{Diff}(q) - \frac{m}{\rho}, \tag{5.1}$$

$$\frac{\mathrm{d}\theta}{\mathrm{d}t} = \mathrm{Diff}(\theta) + \frac{Lm}{\rho c_p}, \tag{5.2}$$

where q is the specific air humidity, θ is the potential temperature, m is the absolute rate of water vapour condensation, i.e., the amount of water and ice crystals formed in unit volume in unit time, L is the latent heat of condensation, and ρ is the air density,

$$\frac{\mathrm{d}}{\mathrm{d}t} = \frac{\partial}{\partial t} + (\mathbf{u}\cdot\boldsymbol{\nabla}) + \left(w\frac{\partial}{\partial z} \quad \text{or} \quad w_p\frac{\partial}{\partial p}\right). \tag{5.3}$$

The symbol 'Diff' stands for the contribution of the diffusion mechanism (molecular and turbulent). As is known from general courses in physics and meteorology (see also Section 3 of Chapter 8), 'Diff' is a linear differential operator, such that, for instance

$$\mathrm{Diff}(a) + \mathrm{Diff}(b) \approx \mathrm{Diff}(a+b). \tag{5.4}$$

Assuming that the cloud elements (drops and crystals) have negligible inertia, so that they follow the air motion, we can write, for the specific water content Δ, the equation

$$\frac{\mathrm{d}\Delta}{\mathrm{d}t} = \mathrm{Diff}(\Delta) + \frac{m}{\rho}. \tag{5.5}$$

Then, for the total water content $s = q + \Delta$, Equations (5.1) and (5.5) yields

$$\frac{ds}{dt} = \mathrm{Diff}(s), \tag{5.6}$$

where (5.4) has been used. One more equation which does not contain the rate of condensation m can be obtained from (5.1) and (5.2) by introducing the function

$$M = \theta + \left(\frac{L}{c_p}\right) q. \tag{5.7}$$

Actually

$$\frac{dM}{dt} = \mathrm{Diff}(M). \tag{5.8}$$

For a cloud, where the water vapour is in a saturated condition, the system of Equations (5.1) and (5.2) is supplemented by the well-known formula

$$q = q_m = \frac{0.622\, E(T)}{p}, \tag{5.9}$$

where q_m is the maximum specific humidity, and $E(T)$ is the maximum water vapour pressure (outside of the cloud $m = 0$).

The quantities s and M demonstrate an important property – they are invariant with respect to the cloud-formation process since differential equations (5.6) and (5.8) have one and the same form before and after the beginning of condensation – they do not contain the m parameter.

A general simplification of the problem under consideration can be achieved if we assume that the diffusion can be neglected Diff( ) = 0. With this assumption hereafter introduced, we obtain that

$$\frac{ds}{dt} = 0, \qquad \frac{dM}{dt} = 0, \tag{5.10}$$

i.e., s = const and M = const regardless of the condensation.

b. Semiempirical Method

This method is based on the empirically established correlations between the vertical velocity $w_p = dp/dt$ and the dew-point deficit (also called the hydroscopic difference)

$$S = T - t_d \tag{5.11}$$

on the one hand and the cloudiness and precipitations on the other. An example of such a correlation in the form of a nomogram is shown in Figure 7.3 taken from [39]. It has been produced from about 500 cases of available data on computed (diagnostic) values of w_p and actual values of S on the one hand, and observation of the character of the weather on the other: precipitation; cloudy (7–10 tenths); slight cloudiness (4–6 tenths); clear (0–4 tenths). The demarcation lines on the nomogram separate the areas of precipitation

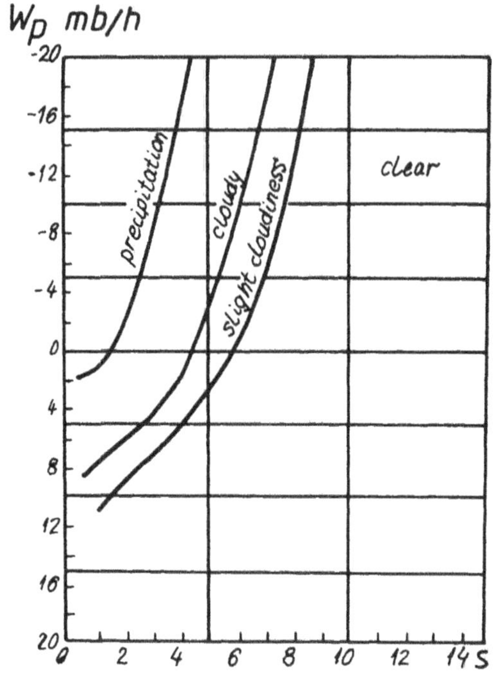

Fig. 7.3. Nomogram of the correlation between computed vertical motion (w_p), measured hygroscopic difference (S) at 700 mb and lower levels and the weather phenomena: precipitation, cloudy (7–10 tenths), slight cloudiness (4–6 tenths), clear (less than 4 tenths) (after [39], p. 490).

and considerable amount of clouds with about 80% confidence and an intermediate area of slight cloudiness with about 50% confidence. For the 850 mb level, according to [39] the demarcation lines are well approximated by

$$
\begin{aligned}
(w_p)_{\text{pr}} &= -1.1522S^2 - 0.1805S + 2.4152,\\
(w_p)_{\text{cl}} &= -0.6182S^2 + 1.0090S + 6.9140, \qquad (5.12)\\
(w_p)_{\text{clear}} &= -0.4691S^2 + 1.1885S + 9.0277.
\end{aligned}
$$

Having such empirical formulae or nomograms and forecasted values of w_p and S, a forecast of cloudiness and precipitations can be made in the above-described aspect. The accuracy of forecasts made by using this method is approximately 80%.

Thus, the problem is reduced to forecasting the vertical velocity w_p and the hygroscopic difference S. Assuming that we know how to predict w_p from the previous sections, let us concentrate our attention on the pre-calculation of $S = T - t_d$, t_d is the dew-point temperature in °C. When condensation does not occur, $m = 0$. Then equation (5.1) becomes

$$\frac{\mathrm{d}}{\mathrm{d}t} q = \frac{\mathrm{d}}{\mathrm{d}t} \ln q = \frac{\mathrm{d}_2}{\mathrm{d}t} \ln q + w_p \frac{\partial}{\partial p} \ln q = 0, \qquad (5.13)$$

where $d_2/dt = \partial/\partial t + \mathbf{u} \cdot \nabla$ is a two-dimensional material derivative. On the other hand, from the well-known formulae

$$q = 0.622\frac{e}{p}, \quad e = e_0 \times 10^{at_d/(b+t_d)}, \tag{5.14}$$

where $a = 7.63$, $b = 241.9$ for water and $a = 9.5$, $b = 265.5$ for ice, one can easily derive the expression

$$\ln q = \ln 0.622 + \ln e_0 + \frac{\nu a t_d}{b + t_d} - \ln p,$$

where $\nu = \ln 10$, e is the water-vapour pressure. Hence $\partial \ln q/\partial p$ and $d_2 \ln q/dt$ can be determined and substituted in (5.13). The final result is

$$\frac{d_2 t_d}{dt} = \frac{(b + t_d)^2}{\nu ab} \frac{w_p}{p} - w_p \frac{\partial t_d}{\partial p}. \tag{5.15}$$

In order to derive an equation for S, we write Equation (5.2) in the form of (3.11) in Chapter 3, for the adiabatic case

$$\frac{d_2}{dt} T = \frac{\kappa - 1}{\kappa} \frac{T}{p} w_p - w_p \frac{\partial T}{\partial p} \tag{5.16}$$

and subtract (5.15) from (5.16). We obtain

$$\frac{d_2}{dt} S = \alpha \frac{w_p}{p} - w_p \frac{\partial S}{\partial p}, \quad \alpha = \frac{\kappa - 1}{\kappa} T - \frac{(b + t_d)^2}{\nu ab}. \tag{5.17}$$

This is the prognostic equation for S. To make a forecast of S based on this equation, one needs to have a forecast for $\mathbf{u}$ and T, as well as to give the proper boundary condition for S.

However, empirical investigations have shown that within the range of possible variations of T and t_d in the lower and middle troposphere, α from (5.17) can be considered as approximately constant at levels of 850, 700 and 500 mb. For instance, over the European part of the U.S.S.R., $\alpha_{850} = 63.92°$, $\alpha_{700} = 62.73°$, $\alpha_{500} = 60.53°$. Moreover, statistical investigations have also shown that without any practical loss of accuracy, the second term on the right-hand side of Equation (5.17) can be neglected – $d_2 S/dt \approx \alpha w_p/p$. Hence

$$\frac{\partial S}{\partial t} = -\left(u \frac{\partial S}{\partial x} + v \frac{\partial S}{\partial y}\right) + \frac{\alpha}{p} w_p. \tag{5.18}$$

After introducing the quasi-geostrophic approximation, the prognostic equation for S takes the final form

$$\frac{\partial S}{\partial t} = f^{-1} J(S, \Phi) + \frac{\alpha}{p} w_p. \tag{5.19}$$

Having predicted w_p and Φ, this equation allows us to calculate $S_t = \partial S/\partial t$ and, consequently, $S(t + \Delta t) = S(t) + \Delta t S_t(t)$, i.e., to precalculate (numerically predict) S by the stepwise method. In practice, Δt is chosen to be equal to one hour.

c. Method of Invariants

We start from Equations (5.1) and (5.2) with $m = 0$ and Diff( ) = 0:

$$\frac{\mathrm{d}q}{\mathrm{d}t} = \frac{\mathrm{d}\ln q}{\mathrm{d}t} = 0, \tag{5.20}$$

$$\frac{\mathrm{d}T}{\mathrm{d}t} = \left(\frac{1}{\rho c_p}\right)\frac{\mathrm{d}p}{\mathrm{d}t}. \tag{5.21}$$

We also use the equations $q = 0.622\ e/p$, $e = rE(T)$, where r is the relative humidity of the air. Hence, it follows that

$$\frac{\mathrm{d}\ln e}{\mathrm{d}t} = \frac{\mathrm{d}\ln p}{\mathrm{d}t}, \tag{5.22}$$

$$\frac{\mathrm{d}\ln r}{\mathrm{d}t} + \frac{\mathrm{d}\ln E}{\mathrm{d}T}\frac{\mathrm{d}T}{\mathrm{d}t} - \frac{\mathrm{d}\ln p}{\mathrm{d}t} = 0. \tag{5.23}$$

Inserting $\mathrm{d}T/\mathrm{d}t$ from (5.21), we obtain

$$\frac{\mathrm{d}\ln r}{\mathrm{d}t} + \delta\frac{\mathrm{d}\ln p}{\mathrm{d}t} = 0, \tag{5.24}$$

where

$$\delta = \frac{\kappa - 1}{\kappa}T\frac{\mathrm{d}\ln E}{\mathrm{d}T} - 1, \quad \kappa = \frac{c_p}{c_v}. \tag{5.25}$$

Further, one may use the Clausius–Clapeyron formula $\mathrm{d}\ln E/\mathrm{d}t = L/R_w T^2$, where R_w is the gas constant of water vapour. Thus, $\delta(T)$ is a function of the temperature only. But, approximately, it may be considered as constant. Then Equation (5.24) can be written as an invariant

$$\frac{\mathrm{d}}{\mathrm{d}t}[\ln(rp^\delta)] = 0 \quad \text{or} \quad rp^\delta = \text{const} = r^0(p^0)^\delta \tag{5.26}$$

where r^0, p^0, r and p are, respectively, the relative humidity and pressure at the initial and final points of the air parcels trajectories. However, in adiabatic approximation $\mathrm{d}\theta/\mathrm{d}t = 0$, i.e., $Tp^{-\lambda} = \text{const}$ alongside of the trajectories, $\lambda = R/c_p = (\kappa - 1)/\kappa$. Then (5.26) also reads

$$rT^{\kappa\delta/(\kappa - 1)} = \text{const}. \tag{5.27}$$

Having predicted the temperature, one can now predict the relative humidity $r(t + \Delta t)$, and the spatial trajectory of the air parcel should be constructed for this purpose, since

d/dt in (5.26) denotes the three-dimensional material derivative. To do this one must compute the vertical velocity. At the cost of some simplification of the above invariants (5.26) and (5.27), this rather complex procedure can be avoided.

Equation (5.24) can be rewritten in the form

$$\frac{d_2}{dt} \ln r + \left(p \frac{\partial \ln r}{\partial p} + \delta\right) \frac{w_p}{p} = 0. \tag{5.28}$$

Again, the empirical data allow us to conclude that $p\, \partial \ln r/\partial p \ll \delta$, so that we can write

$$\frac{d_2 \ln r}{dt} + \frac{\delta}{p} w_p = 0 \tag{5.29}$$

instead of (5.24). To eliminate w_p, we make use of the vorticity Equation (4.7) in Chapter 3

$$\frac{d_2 \ln \zeta_a}{dt} = \frac{\partial w_p}{\partial p}, \quad \zeta_a = \zeta + f. \tag{5.30}$$

Here an assumption is made that

$$w_p(x, y, p, t) = \bar{w}_p(x, y, t)\, F(p). \tag{5.31}$$

Then Equations (5.29) and (5.30) are combined to yield

$$\frac{d_2}{dt} (r^k \zeta_a) = 0 \quad \text{or} \quad r^k \zeta_a = \text{const} \tag{5.32}$$

along the two-dimensional trajectories. Here $k = pF'/\delta F$, $F' = dF/dp$. Generally $k = k(p)$. However, it has been shown that k slightly depends on the form of the function $F(p)$ (solve Problem 7). Note that in order to derive (5.32) from (5.29)–(5.31) it is not necessary to assume nondependency of k on p as was the case with δ in (5.24), since now we consider plane trajectories of the air parcels.

A similar invariant to (5.32) can be derived under more general (adiabatic) conditions in the atmosphere –

$$e^{k_1} \zeta_a = \text{const} \tag{5.33}$$

along a horizontal trajectory, where

$$k_1 = \frac{pF'}{\delta_1 F}, \quad \delta_1 = \frac{\gamma}{\gamma_a} (\delta + 1) - 1, \tag{5.34}$$

where $\gamma_a = g/c_p$ and $\gamma = -\partial T/\partial z$ are the dry adiabatic and actual lapse rates. For this purpose, Equation (5.22) is used

$$\left(\frac{d_2}{dt} + w_p \frac{\partial}{\partial p}\right) \ln e - \frac{w_p}{p} = 0. \tag{5.35}$$

Making use of the equations $e = rE$ and $dp = -\rho g\, dz$, (5.35) is reduced to

$$\frac{d_2 \ln e}{dt} + \left(p\, \frac{\partial \ln r}{\partial p} + \delta_1\right) \frac{w_p}{p} = 0 \tag{5.36}$$

corresponding to (5.28) where

$$\delta_1 = \frac{\gamma R}{g}\, \frac{\partial \ln E}{\partial \ln T} - 1. \tag{5.37}$$

Further transformations reduce (5.37) to (5.34). Since $\gamma \sim \gamma_a$ in the troposphere, then $\delta_1 \sim \delta \gg p\, \partial \ln r/\partial p$ and (5.36) can be simplified. The remaining transformations are identical to (5.29)–(5.31).

In conclusion let us mention that, by definition, $e = E(T_d)$ so that (5.33) can also be used in the form

$$E(T_d)\zeta_a^{1/k_1} = \text{const}. \tag{5.38}$$

When humidity is predicted (either r or e with the use of the corresponding invariants (5.32) and (5.33)) together with the velocity field for different pressure levels, again empirical correlations (nomograms, formulae) are used to predict the cloudiness and precipitations. Though the accuracy of the prognoses made by this method increases by about 10% for some levels compared with the previous method, both have been shown to produce approximately the same results. It suggests that the processes of cloud formation and precipitation cannot be described accurately by means of empirical correlations between the humidity characteristics and vertical velocity of the air motion, on the one hand, and the weather phenomena under consideration on the other. In recent years, much more complex prognostic schemes have been developed, based on the primitive equations of the type (5.1)–(5.8) supplemented with the equations of motion, continuity, radiation transfer, microphysics and kinetics of the processes in the clouds, etc., in which the diffusion mechanisms are taken into consideration. Their discussion, however, is beyond the scope of this book.

6. PREDICTABILITY OF THE METEOROLOGICAL ELEMENTS

a. The Nature of the Problem

The methods of weather forecasting, currently in use at present, can be classified in three groups:

(i) *Synoptic method* – historically the first and most popular one, which until 40 years ago was the only known weather prediction method. It consists of an ensemble of rules derived on the basis of the physical laws, applied to the atmosphere. In combination with the knowledge, experience and even physical intuition of the meteorologist-forecaster, the method allows a qualitative prediction of the future behaviour of the atmosphere above a given area. Thus, the subjective element plays an important role in this method when analyzing the synoptic situation by means of a synoptic map and when making forecasts as well. This encourages the meteorologists to seek methods which, once developed, will not rely upon human judgement, i.e., to be more or less objective.

(ii) *Statistical method* – based only on experience but expressed in the form of a set of numbers representing the characteristics of various atmospheric properties. These numbers are substituted into mathematical (regression) formulae derived from past observations. The output of these formulae is a similar set of numbers which expresses the predicted weather features. Thus, the method is objective as long as it is based on an algorithm. The development of large computers stimulated the application of this method, but in the meantime, however, statistical forecasting has not overcome the shortcomings of synoptic forecasting in everyday practice.

(iii) The *Dynamical (numerical) method* – based on systems of equations like those discussed in Sections 1–5 of this chapter and their numerical solution under appropriate initial and boundary conditions. As far as these equations are more or less adequate for the real atmosphere and their solution is realized with the help of prechosen schemes, then this method can also be considered objective.

There also exists a mixed (stochastic-dynamic) method of prediction which is a combination of methods (ii) and (iii), but with advantages and shortcomings of its own.

The common consumer of meteorological prognoses usually thinks that the weather is predictable and that if there are some failures, they are due only to imperfect methods of forecasting. This is really true but only to a certain extent. If we confine ourselves to only considering the dynamical method, it is clearly seen how the more perfect prediction models lead to higher-quality prognoses. In a limit, if we dispose of an absolutely perfect prognostic technique (equations, method and computers for solution) and the necessary initial conditions which fix exactly the state of the atmosphere at each point at $t = 0$, then, in principle, it could be possible to predict its behaviour up to the smallest details at an arbitrary future $t > 0$. This statement is valid for weather forecasting, as well as for predicting Sun (Moon) eclipses and many other astronomical phenomena. However, in both cases, the range of prediction and accuracy differs considerably. At least three reasons can be noted for this:

(1) The equations of celestial mechanics describe considerably better the real motion of the planets and the Sun in space. What is more, it is admissible to consider them as material points moving under the action of a central force – the Newtonian gravitational force, depending only, in a simple way, on the coordinates. Such kind of simplification is impossible in the case of the atmosphere – it is a continuous medium. A number of forces depending, in a complex and implicit way on coordinates and time, act upon the air particles. The atmospheric motions are of many spatial scales (between 10^{-2} and 10^7 m) and practically unlimited from the above oscillation periods. In addition, the dynamics of the atmosphere cannot be separated from its thermodynamics, including the radiation processes. However, a system of equations, accurately describing this complex of factors, has not yet been constructed. The existing mathematical models are rather crude approximations of reality. Finally, the weather equations, though incomplete, are rather complex, i.e., they are nonlinear partial differential equations. It now becomes clear that the problems of celestial mechanics and dynamic meteorology are of different categories of difficulty.

(2) The initial information of the problems in celestial mechanics is incomparably more precise and complete than in meteorology. Actually, only three numbers (coordinates) are needed to specify the location of the Earth's gravity centre and three more (components) will specify the velocity. The same is true of the Sun and Moon. Thus, a

total of only 18 numbers describes the present Sun-Earth-Moon configuration from which the future one may be predicted. They are easily measured and quite precise.

At the same time it is impossible to describe the initial state of a continuous medium like the atmosphere by a finite set of quantities. In practice, this is done by a finite though very large set due to the fact that the meteorological observation stations are located at discrete points along the Earth's surface. Their proximity, as well as the shortening of the time interval between observations in order to increase the volume of initial information, become economically unprofitable beyond some limits. Consequently, the conventional meteorological world network of stations for surface and height observations in principle, does not allow for the registration of phenomena and processes of spatial and time scales which are correspondingly smaller than the distances between the stations and the time intervals between the observations. On the other hand, the numerical solution of the equations by finite difference methods, in principle, also does not allow the prediction of phenomena of arbitrarily small scales.

(3) Let us assume that the initial meteorological information is as full and precise as the astronomical information is and that the equations governing the behaviour of both physical systems (atmosphere and Earth-Sun-Moon) are adequate in equal degrees of reality. Even in this case, the astronomical prognoses would be much more precise than the meteorological ones. This is so because an almost perfect forecast of the future of a system, at an almost perfectly known present is possible, provided that the system is stable. Here, possibly, the main reason for the success of the astronomical prognoses can be found; the system Earth-Sun-Moon, for instance, is much more stable than the atmosphere of our planet. It would be more true to say that the atmosphere is an unstable dynamical system.

The strict periodicity in the motion of these celestial bodies makes the motion predictable for a long time ahead and with great precision too. On the other hand, observations of the atmosphere indicate that it is not a periodically-varying system. It does have some periodic components, notably the diurnal and annual cycles and their overtones, caused by strictly periodic astronomical factors and are, thus, predictable (after winter inevitably comes summer!). Some other periodicities are also suspected and used, though with a great amount of uncertainty, for medium- and long-range weather and climate forecasting. When all these periodic components are subtracted from the total 'signal', a large residual is still present which behaves aperiodically and, thus, is unpredictable in the sufficiently distant future.

It was mentioned above that some nontrivial periodicities or regimes are suspected to exist or have been established. A typical example of periodic phenomenon is the so-called quasi-biennial oscillation. This is an oscillation between persistent easterly and westerly winds in the equatorial stratosphere, generally taking, somewhat, over two years (about 26 months) to complete a cycle. A typical example for regime phenomenon is the so-called 'blocking'. It consists of the continuous presence of troughs and ridges in the pressure field at preferred longitudes over extended periods of time, even though the long-term average circulation may not possess such troughs and ridges. There is no doubt that some extended-range predictability exists in both cases.

From all that is said so far, we may conclude that predictability is a property of the atmosphere and not of a prognostic model, and that there exists a range of predictability depending on: (a) the nature of the process, (b) the initial data, (c) the prediction method.

It is clear that an individual (deterministic) forecast of a particular meteorological element would make sense only for such a period ahead within the limits of which the prognostic error (the difference between the predicted value and the true one) is smaller than the mean climatic variations of the element. The determination of the range of predictability represents the content of the predictability problem.

b. Range of Predictability

The deciding factor in predictability is the instability of the system under consideration (in our case, the atmosphere). As a result small initial errors will amplify as time progresses and the determination of the amplification rate is one of the main tasks of the predictability theory. We shall demonstrate one possible solution with the example of the barotropic prognostic model described by Equation (2.21) at $f = \text{const}$

$$\nabla^2 \psi_t = -J(\psi, \nabla^2 \psi). \tag{6.1}$$

Let $\psi_0(x, y) = \psi(x, y, 0)$ be the true field of the streamfunction at $t = 0$ and $\tilde{\psi}_0 = \psi_0 + \delta_0$ be the calculated one with some error $\delta_0(x, y)$. Due to the initial error δ_0, the predicted field $\tilde{\psi}$ will differ from the true one: $\tilde{\psi}(x, y, t) = \psi(x, y, t) + \delta(x, y, t)$. Inserting this into (6.1), we obtain an equation governing δ

$$\nabla^2 \delta_t = -J(\psi, \nabla^2 \delta) - J(\delta, \nabla^2 \psi) - J(\delta, \nabla^2 \delta). \tag{6.2}$$

If the error δ is small enough compared to ψ, Equation (6.2) can be linearized as

$$\nabla^2 \delta_t = -J(\psi, \nabla^2 \delta) - J(\delta, \nabla^2 \psi). \tag{6.3}$$

Let us introduce

$$E_k = \tfrac{1}{2}\overline{(\nabla\psi)^2}, \qquad e_k = \tfrac{1}{2}\overline{(\nabla\delta)^2}. \tag{6.4}$$

Obviously, E_k and e_k are measures for the mean kinetic energies of the true and 'erroneous' motions. The problem reduces to the determination of the function $e_k(t)$. With some additional assumptions, Thompson (1957) and Novikov (1959) first solved this problem by starting from Equation (6.2). Their solution was in the form of the infinite series

$$\frac{e_k(t)}{e_k(0)} = 1 + at^2 + \cdots \tag{6.5}$$

but they succeeded only in showing that $a > 0$, i.e., at least in the beginning $\mathrm{d}e_k/\mathrm{d}t > 0$. According to their calculations, variations in the velocity field with a timescale $T \sim 1$ day are unpredictable for a period larger than a few days. Another example of an analytical approach to the problem will be discussed in Section 3 of Chapter 11. It seems to us that the numerical experiments are more promising for the future.

c. Numerical Experiments of Predictability

With the numerical method of weather prediction, the initial conditions represent the real data from the observations. But they could be also calculated and changed within certain limits under some consideration, while the solution could be extended for an arbitrary interval of time (weeks, months and even years). Then the results from the solution can be regarded as an ensemble of data which can be further treated statistically, as is done with real data. This procedure is known as *numerical experimentation* or *simulation*. Its purpose is not to produce good weather forecasts, but the realistic overall behaviour of a simulated atmosphere which should be compared (quantitatively) with the behaviour of the real atmosphere in the past or present or should be used for predicting the future. Thus, for instance, we are not able to change the influx of Sun radiation or the dislocation of the continents and oceans on the Earth in order to see how the real atmosphere would react to these new conditions. But in the numerical experiments related to climatic theory, this combination of conditions (as well as many others) is possible and the results will allow us to obtain important conclusions about the past and future of the Earth's climate. It is a simple matter to demonstrate the degree of instability, as measured by the rate at which typical small perturbations will grow, for a simulated atmosphere. For this purpose, one simply performs the same numerical experiments twice, with slightly differing initial conditions, and observes how rapidly the two solutions diverge from each another. It is clear that this would be an experiment on predictability which could be repeated with other prognostic models. During the past two decades, such experiments have been performed by Charney, Smagorinsky, Leith, Mintz and others and Lorenz (1965, 1969, 1981) has also made important contributions to the theory of the problem.

The time at which a given initial error doubles is called *doubling time*. Evaluated at the beginning to be of order of 10 days and later reduced to 5 days, the generally accepted value is now about 3 days. If we, for instance, consider the temperature, a typical observational error may be as low as 1°C. In 3 days it would grow to 2°C, in 6 days – to 4°C and in 9 days – to 8°C, etc. Consequently, reasonably good forecasts of the temperature should be possible a week in advance.

In conclusion, we shall point out that the more general the characteristics of the weather we are interested in are, the longer the range of their predictability is. So if the typical individual synoptic systems – cyclones and anticyclones – are unpredictable for a month ahead, then such an important generalized characteristic of their ensemble manifestation as the monthly sum of precipitation appears to be predictable for this period. Generally speaking, the existence of a limit beyond which the details (in a previously defined sense) are individually unpredictable, is not of particular importance in practice because, at least for the time being, most often they are not necessary. When needed, their effect could be statistically evaluated in a definite way for each particular case.

PROBLEMS

1. Derive an expression for the group velocity c_{gr} of the divergent Rossby waves (2.19) and investigate the difference $\Delta c = c_{gr} - c$ as a function of L.
2. On the basis of Equations (2.6) and (2.7), prove that $\overline{B}(p) = 1$.
3. Prove that (2.28) follows from (2.27) and also that an arbitrary function of $\nabla^2 \psi$ is invariant too, i.e., $\overline{F(\nabla^2 \psi)} = \text{const}$.
4. Starting from (3.4) and (3.5), derive equations for w_p and Φ for the case when $T(p) = \text{const}$ and $m \sim p^{-2}$.

5. Starting from Equations (5.1) and (5.5), prove that the quantity $M^* = \theta - (L/c_p)q = M - (L/c_p)s$ satisfies Equation (5.8) with M replaced by M^*.
6. Derive with details Equation (5.15).
7. Consider the invariant (5.32): $r^k \zeta_a = \text{const}$, $k = pF'/\delta F$. Assume the following three examples for vertical profiles of w_p from (5.31), namely

   (i) $F(p) = \sqrt{2} \sin(\pi p^*/2)$

   (ii) $F(p) = \frac{1}{2}(2p^* + p^{*2})$

   (iii) $F(p) = 2p^*$

   where $p^* = (p_0 - p)/(p_0 - p_e)$, $p_0 = 1000$ mb, p_e – the pressure at the nondivergent level (500 mb).

   Prove by computations that k slightly depends on the vertical profile of w_p, at least for $700 \leqslant p \leqslant 900$ mb. Assume $\delta \approx 5$ (some mean value of δ from (5.25)).
8. Prove that expressions (5.34) and (5.37) are identical for δ_1.
9. Derive the low-order system (2.37) by substituting (2.35) into Equation (2.27).

   Determine α, β, γ as functions of m and n.

   Prove that system (2.37) can be solved analytically in an elliptic function.

   Prove that in the case of quadratic area ($m = n$) $C(t) = \text{const}$ and the solution for $A(t)$ and $B(t)$ is expressed in trigonometric functions.

PART II

THE DYNAMICS OF A REAL (WITH FRICTION) ATMOSPHERE

Geophysical fluids – the atmosphere and ocean (i.e., the air and water) – are viscous fluids. Moreover, along with molecular viscosity, which is a physical property of the fluid, they also possess turbulent viscosity, which is no longer a property of the fluid itself but of the flow, in so far as the motions in the atmosphere and ocean are nearly always turbulent. Neglecting the viscous effects in the theory of atmospheric motions undoubtedly simplifies and idealizes in some way the so-created models. That is why the determination of the problems, where this is possible, is of particular importance. As we know, in motions with a Reynolds number $Re \to \infty$, molecular friction force can be neglected. In the atmosphere this condition is nearly always realized. Attention will be paid to the exceptions in this and the following chapters.

The analogous force of turbulent friction cannot always be neglected, even in the free atmosphere and this kind of force should be taken into account in the boundary layers. On the whole, the effects connected with the turbulent character of atmospheric motions seem to be extremely important in understanding the atmospheric processes, including those responsible for the weather and this necessitates the systematic study of the very phenomenon 'atmospheric turbulence' to which the second part of the book is dedicated.

CHAPTER 8

The General Theory of Atmospheric Turbulence

1. TURBULENT MOTIONS: GENERAL INFORMATION

When the general theory of turbulent motions is set out, we can speak about some kind of abstract fluid. In particular cases this may be air, water, oil, blood, and so on, but for definiteness here, when we speak about a fluid in turbulent motion, we shall be referring to atmospheric air, i.e., we shall be investigating the turbulence in natural conditions.

a. Definition of Turbulence

Despite the 100-year history of turbulence research, there is still not a generally accepted and wholely satisfactory definition of this phenomenon. Proceeding from [63], we shall remark on a number of its distinguishing features and, in combination, we shall accept them as a definition. Seven of the most important features are given below.

(1) *Turbulent flows are highly irregular and chaotic* (*random*). Their characteristics are random functions of spatial coordinates and time. However, every motion, possessing this feature, should not be classified as turbulent; for instance, a wave motion with a random amplitude and phase or a superposition of such waves is not turbulent. It shows some degree of regularity of its elementary components – predominant direction of displacement of the material particles with regard to the direction of the wave vector and the existence of dispersion correlation that represents a simple connection between wave-frequency and wave-vector, etc. Such simple correspondence between the time (frequency ω) and spatial (wave vector $\mathbf{k}$) characteristics does not exist in pure turbulent motion. A wide frequency region can correspond to a Fourier component with a fixed $\mathbf{k}$. In this connection, it is also said that the turbulent motion has an infinitely large number of excited degrees of freedom.

Turbulent motion is often compared with laminar motion – a smooth, regular motion of viscous fluids and gases, governed by Navier–Stokes equations – in order that its 'randomness' can be stressed. At fixed initial and boundary conditions, in principle, its characteristics can be determined at an arbitrary future instant from the solutions of these equations, i.e., the laminar motion is predictable in its details without any limitations on the time period.

(2) *Turbulent flows possess high diffusion capabilities*. Due to the chaotic movement of fluid parcels, called *turbulent eddies*, an intensive mixing and transporting of different substances (heat, momentum, water vapour and other admixtures) is realized. This kind

of mechanism is specified as *turbulent diffusion* and is analogous to the mechanism of molecular diffusion, but is much more intensive. From an application point of view, this is possibly the most important feature of the turbulent flows. It is enough to mention its role in spreading detrimental admixtures in the atmospheric surface layer which have been emitted by ground industrial and daily waste sources.

A motion which has (or seems to have) a random character but does not show diffusion abilities, cannot be regarded as turbulent.

(3) *Turbulence always occurs at very high Reynolds number values.* $\mathrm{Re} = UL/\nu$, where U and L are typical velocity and length scales and ν is a kinematic coefficient of molecular viscosity. The role of the nondimensional combination UL/ν as a turbulence criterion was ascertained by Reynolds in his classical works from 1883 and 1895. He observed the transition of a laminar flow in tubes into a turbulent one at a somewhat critical value $\mathrm{Re}_{\mathrm{cr}} \sim 10^3$. From a dynamical point of view, Re can be interpreted as a ratio of the inertial forces to the forces of molecular friction acting in the fluid: $U^2L^{-1}/\nu UL^{-2} = UL/\nu = \mathrm{Re}$. The meaning of $\mathrm{Re} \to \infty$ is that the inertial forces predominate over the frictional forces. The so-originated flow disturbances in the velocity field cannot be 'extinguished' by the viscosity and the flow becomes turbulent. At $\mathrm{Re} \to 0$, it is quite the contrary. But the inertial forces are represented by the nonlinear terms of the Navier–Stokes equations. Therefore, the 'turbulence' phenomenon is essentially nonlinear and, for its description by means of these equations, the nonlinear terms in them should be compulsorily preserved. Ignoring them means depriving the equations of their ability of describing the turbulent motion of the fluid.

High values of Re always correspond to the atmospheric motion. For instance, if we assume that at breeze circulation on a lake shore $L_1 \sim 10^3$ m, $U_1 \sim 1.5$ m s^{-1}; at monsoon circulation $L_2 \sim 5 \times 10^2$ km, $U_2 \sim 10$ m s^{-1}; and at general atmospheric circulation $L_3 \sim 10^4$ km, $U_3 \sim 10$ m s^{-1}, then at $\nu = 0.13 \times 10^{-4}$ m^2 s^{-1}, we calculate accordingly $\mathrm{Re}_1 \sim 10^8$, $\mathrm{Re}_2 \sim 10^{11}$, $\mathrm{Re}_3 \sim 10^{13}$, i.e., $\mathrm{Re} \gg \mathrm{Re}_{\mathrm{cr}}$. This fact indirectly shows that the motions in the atmosphere are turbulent, which is also confirmed by direct Eulerian and Lagrangian observations.

(4) *Turbulent flows are rotational, with a high level of vorticity* $\mathbf{\Omega} = \nabla \times \mathbf{v}$ *and intensive fluctuations of* $\mathbf{\Omega}$. Of course, $\mathbf{\Omega}$ is a random vector, too, and what is more, it is a three-dimensional one, as is $\mathbf{v}$. In the opposite case (in two-dimensional flow), the vorticity-maintenance mechanism, by means of vortex stretching, should be missing, while in a three-dimensional flow it is crucial. Hence, random and yet potential flows should not be regarded as turbulent. For instance, random waves on the ocean surface are not turbulent since they are irrotational.

(5) *Turbulent flows are always dissipative.* The kinetic energy of the turbulent motion transforms into internal energy (heat) due to molecular friction (viscosity). The rate of this process, according to Equation (1.15) in Chapter 2, is

$$Q_d = \frac{\nu}{2}\left(\frac{\partial v_i}{\partial x_k} + \frac{\partial v_k}{\partial x_i}\right)^2, \quad \nu = \frac{\eta}{\rho}. \tag{1.1}$$

Therefore, in order to exist, a turbulent flow needs a continuous supply of energy which should compensate for the viscous losses. Otherwise the turbulence would decay with time.

(6) *Turbulent flows are continuous* in the sense that the scales of even the smallest turbulent elements (eddies) are far greater than any molecular length scale. In other words, turbulence is a typical form of motion of the continuous media, governed by the fluid mechanics (Navier–Stokes) equations.

(7) *Turbulence is a feature of fluid flows but not of the fluid itself.* The motion of different fluids (liquids and gases) can be turbulent with different molecular viscosity and heat conductivity. As we shall see, they may be subjected to universal laws if only Re is large enough.

In common, the so-enumerated seven fundamental features can be accepted as a definition for turbulence. A fluid flow that possesses all these features should be called turbulent. This does not exclude speaking about turbulence as an inviscid (ideal) fluid flow with $\nu = 0$, or about two-dimensional turbulence as an idealized model of geophysical importance as well.

b. Methods of Description

The random (stochastic) character of turbulent motions as one of their most distinguishing features, presupposes the use of statistical methods of description. The individual (deterministic) description of the trajectory of a fixed air 'particle' (parcel) into a turbulent flow is impossible and is not necessary because of its extremely complicated form or because of the random changes of any element $f(t)$ at a fixed point. Some kind of average (statistical) characteristics are necessary that should be easily estimated on the basis of empirical data.

The mathematical apparatus of contemporary theory of turbulent motions, i.e., the theory of random functions, was created in the Thirties. Here we shall give only the most necessary information.

Let $f(t)$ be a random function of time t (an ensemble of its 'realizations', each one having an extremely complicated form, obtained from separate experiments). For simplicity, we shall accept that $\overline{f(t)} = 0$, where the overbar denotes the average value over the ensemble (statistical averaging). Then, a very useful and widely used statistical characteristic is the correlation function

$$B_{ff}(\tau) = \overline{f(t)f(t+\tau)}, \quad \tau >, < 0. \tag{1.2}$$

When B_{ff} depends on τ, but not on t, then $f(t)$ is called a *stationary random function*. Evidently, $B_{ff}(\tau) = B_{ff}(-\tau)$ and $B_{ff}(0) = \overline{f^2} = \sigma_f^2 = \text{const}$ is the variance of $f(t)$.

Together with $B_{ff}(\tau)$, great application has also obtained the structure function

$$D_{ff}(\tau) = \overline{[f(t+\tau) - f(t)]^2}. \tag{1.3}$$

It can be easily proved that

$$D_{ff}(\tau) = 2[B_{ff}(0) - B_{ff}(\tau)]. \tag{1.4}$$

Evidently, $D_{ff}(0) = 0$ and $D_{ff}(\infty) = 2B_{ff}(0)$, since $B_{ff}(\infty) = 0$ (at $\tau \to \infty$, $f(t)$ and $f(t+\tau)$ become statistically independent).

The correlation function $B_{ff}(\tau)$ has a very prominent feature: its Fourier transformation is nonnegative

$$0 \leqslant E_{ff}(\omega) = \frac{2}{\pi} \int_0^\infty B_{ff}(\tau) \cos \omega\tau \, \mathrm{d}\tau. \tag{1.5}$$

To make clear the meaning of $E_{ff}(\omega)$, we use the opposite transformation

$$B_{ff}(\tau) = \int_0^\infty E_{ff}(\omega) \cos \omega\tau \, \mathrm{d}\omega. \tag{1.6}$$

Hence,

$$B_{ff}(0) = \overline{f^2} = \sigma_f^2 = \int_0^\infty E_{ff}(\omega) \, \mathrm{d}\omega. \tag{1.7}$$

Since ω has a meaning of frequency, $E_{ff}(\omega)$ should represent the *spectral density* giving the distribution of energy σ_f^2 over the frequency spectrum.

Now let $f(\mathbf{x})$ be a scalar random function of the coordinates (random field) where $\mathbf{x}$ is a radius vector of a point in the space or on the plane. If $\mathbf{x}'$ and $\mathbf{x}''$ are two points and $r = |\mathbf{x}'' - \mathbf{x}'|$ is the scalar distance between them and $\overline{f} = 0$, then, similarly to (1.2) and (1.3), the spatial correlation and structure function are introduced

$$B_{ff}(r) = \overline{f(\mathbf{x}')f(\mathbf{x}'')}, \qquad D_{ff}(r) = \overline{[f(\mathbf{x}'') - f(\mathbf{x}')]^2}, \tag{1.8}$$

and a relation, analogous to (1.4), exists between $B_{ff}(r)$ and $D_{ff}(r)$. When B_{ff} and D_{ff} depends only on r, the field $f(\mathbf{x})$ is called *homogeneous* and *isotropic*.

In the case of a random vector field (i.e., the velocity $\mathbf{v}(\mathbf{x})$), one distinguishes *longitudinal, transverse* and *total* correlation and the structure functions

$$B_{ll}(r) = \overline{v_l' v_l''}, \qquad B_{nn}(r) = \overline{v_n' v_n''}, \qquad B_{cc}(r) = \overline{\mathbf{v}' \cdot \mathbf{v}''} \tag{1.9}$$

and, similarly for $D(r)$, where v_l and v_n are projections of $\mathbf{v}$ on the straight line determined by $\mathbf{r} = \mathbf{x}'' - \mathbf{x}'$ and on its norm, i.e., longitudinal and transverse components of $\mathbf{v}$.

It can be proved [53] that

$$B_{cc}(r) = B_{ll}(r) + 2B_{nn}(r) \tag{1.10}$$

for a three-dimensional vector field $\mathbf{v}$ and

$$B_{cc}(r) = B_{ll}(r) + B_{nn}(r) \tag{1.11}$$

in the two-dimensional case, i.e., on the plane. It can also be proved that if $\nabla \cdot \mathbf{v} = 0$, then

$$B_{nn}(r) = \left(\frac{r}{2} \frac{\mathrm{d}}{\mathrm{d}t} + 1\right) B_{ll}(r) \quad \text{or} \quad B_{nn}(r) = \left(r \frac{\mathrm{d}}{\mathrm{d}r} + 1\right) B_{ll}(r), \tag{1.12}$$

correspondingly. In this case $B_{cc}(r)$ can be expressed by $B_{ll}(r)$ only. Analogous relations are valid for $D(r)$ too.

The correlation and structure functions of the random fields also allow a representation in terms of the Fourier integral. Thus, for the three-dimensional case

$$\frac{1}{2} B_{cc}(r) = \int_0^\infty E_{cc}(k) \frac{\sin kr}{kr} \, \mathrm{d}k \tag{1.13}$$

or

$$\frac{1}{2} B_{cc}(0) = \frac{1}{2} \sigma_c^2 = \frac{1}{2} \overline{\mathbf{v} \cdot \mathbf{v}} = \int_0^\infty E_{cc}(k) \, \mathrm{d}k. \tag{1.14}$$

Since k has a wavenumber meaning, then $E_{cc}(k)$ should represent the distribution of the 'kinetic energy' $\frac{1}{2} \sigma_c^2$ over the wavenumber spectrum. With a random vector field on the plane instead of (1.13), we have

$$B_{cc}(r) = \int_0^\infty E_{cc}(k) J_0(kr) \, \mathrm{d}k \tag{1.15}$$

where $J_0(z)$ is the Bessel function. The same formula is also valid for a scalar field $f(\mathbf{x})$ on the plane.

So far, we have been dealing with pure mathematical (probability) methods of describing the 'turbulence' phenomenon. In building up the theory, however, a number of other methods (dimensional analysis, similarity method, analogies with phenomena from other spheres of physics, etc.) are used. Many of them will be discussed later.

c. On the Averaging Procedure

Reynolds suggested the fruitful idea that a turbulent flow can be regarded as a superposition of an average (laminar-like) motion and a fluctuating one. In accordance with this, any quantity may be decomposed into two parts

$$f = \bar{f} + f', \tag{1.16}$$

where the overbar denotes averaging.

Now a question appears about what should be understood by the operation 'averaging'. In probability theory, there is no ambiguity on this question – it is statistical. Besides this, however, there may be an averaging of $f(\mathbf{x}, t)$ on time over a finite or unlimited time interval, on space over a finite or unlimited volume, or on space and time together which we shall call empirical averaging.

On the other hand, having decomposed $f(\mathbf{x}, t)$ after (1.16) and assuming *à priori* that $f(\mathbf{x}, t)$ satisfies the fundamental equations of fluid- and thermodynamics, new equations for $\bar{f}$ should be derived from them. This appears to be a central task of the theory. It is quite natural to use such a kind of averaging procedure that should lead to comparatively simple equations for $\bar{f}$. The importance of this question was also understood by Reynolds and that is why he formulated the following five conditions to which the averaging operations should be subjected [48]:

$$
\begin{aligned}
&(1) \quad \overline{f+g} = \bar{f} + \bar{g}, \\
&(2,3) \quad \overline{af} = a\bar{f}, \qquad \bar{a} = a \quad \text{if } a = \text{const}, \\
&(4) \quad \overline{\frac{\partial f}{\partial s}} = \frac{\partial \bar{f}}{\partial s}, \qquad s = x, y, z, t, \\
&(5) \quad \overline{\bar{f}g} = \bar{f} \cdot \bar{g}.
\end{aligned}
\tag{1.17}
$$

Consecutively assuming that $g = 1, g = \bar{h}$ and $g = h' = h - \bar{h}$ from condition (5) we get the following important equalities

$$
\bar{\bar{f}} = \bar{f}, \qquad \overline{f'} = 0, \qquad \overline{\bar{f}\bar{h}} = \bar{f}\bar{h}, \qquad \overline{\bar{f}h'} = \bar{f}\overline{h'} = 0. \tag{1.18}
$$

It is not difficult to see that the first four conditions (1.17) are valid at any of the above-mentioned kinds of averaging. The fifth condition, however, is fulfilled at probability averaging only. However, it can be realized only in theory. In actual fact, in experimental researches we never have an unlimited ensemble of realizations at our disposal, but very often only one and, what is more, over finite time interval T and spatial volume V. That is why during working up the data, we necessarily take the average procedure on time or space. But now the fifth condition (1.17) cannot be satisfied exactly. However, with a suitable choice of T or V the error can be kept within admissible boundaries. It is enough, for instance, for T to be much greater compared to the characteristic periods of the fluctuating field f' and smaller compared to the periods of the average field $\bar{f}$.

The problem of averaging in turbulence investigations has another aspect too, which is of practical importance for the verification of the theory by means of experimental data. As we have already said, probability averaging is used in theory while in practice, the empirical one is used. The results of the two kinds of averaging do not generally coincide. But under some not very restricting conditions that define the *ergodic* property of the quantity $f(\mathbf{x}, t)$, the results coincide. We shall assume that the random fields characterizing the turbulent flow possess this property (*ergodic hypothesis*). In such a way, we may justifiably compare the predictions of the theory with the data from the experiments and the observations.

2. THE REYNOLDS EQUATIONS

a. Derivation of the Equations

For simplicity, we consider incompressible fluid, i.e.,

$$
\nabla \cdot \mathbf{v} = \frac{\partial v_k}{\partial x_k} = 0 \tag{2.1}
$$

Then expression (2.3) in Chapter 2 for the molecular stress tensor takes the form

$$
\sigma_{ik} = \rho\nu\left(\frac{\partial v_i}{\partial x_k} + \frac{\partial v_k}{\partial x_i}\right). \tag{2.2}
$$

The equations of motion

$$\rho \frac{dv_i}{dt} = -\frac{\partial p}{\partial x_i} + \rho\nu\nabla^2 v_i + \rho F_i \tag{2.3}$$

can be written down in the following way

$$\rho \frac{dv_i}{dt} = -\frac{\partial p}{\partial x_i} + \frac{\partial \sigma_{ik}}{\partial x_k} + \rho F_i, \tag{2.4}$$

where

$$\frac{dv_i}{dt} = \frac{\partial v_i}{\partial t} + \frac{\partial v_i v_k}{\partial x_k} \tag{2.5}$$

and the external force F_i, according to (2.9) and (4.18) in Chapter 2, is given by the expression

$$F_i = -2\epsilon_{ijk}\omega_j v_k + \delta_{3i} g(1 \text{ or } \beta_0 T) \tag{2.6}$$

depending on whether the flow is thermally stratified or not. Here T is the temperature deviation from the standard value according to the requirement of the Boussinesq approximation.

We copy down Equation (2.4) as follows

$$\rho \frac{\partial v_i}{\partial t} = -\frac{\partial p}{\partial x_i} + \frac{\partial}{\partial x_k}(\sigma_{ik} - \rho v_i v_k) + \rho F_i \tag{2.7}$$

and in accordance with (1.16) we decompose p, v_α and σ_{ik} into mean and fluctuating parts

$$p = \bar{p} + p', \qquad v_\alpha = \bar{v}_\alpha + v'_\alpha, \qquad \sigma_{ik} = \bar{\sigma}_{ik} + \sigma'_{ik}. \tag{2.8}$$

With the help of (1.18), it can be proved that

$$\overline{v_i v_k} = \bar{v}_i \bar{v}_k + \overline{v'_i v'_k}. \tag{2.9}$$

Under the condition that ρ = const and after the substitution of (2.8) into (2.7), averaging and using (2.9), we obtain the Reynolds equations for the mean motion

$$\rho \frac{\partial \bar{v}_i}{\partial t} = -\frac{\partial \bar{p}}{\partial x_i} + \frac{\partial}{\partial x_k}(\bar{\sigma}_{ik} - \rho \bar{v}_i \bar{v}_k - \rho\overline{v'_i v'_k}) + \rho \bar{F}_i \tag{2.10}$$

or

$$\rho \frac{d\bar{v}_i}{dt} = -\frac{\partial \bar{p}}{\partial x_i} + \frac{\partial}{\partial x_k}(\bar{\sigma}_{ik} - \rho\overline{v'_i v'_k}) + \rho \bar{F}_i, \tag{2.11}$$

where

$$\frac{d\bar{v}_i}{dt} = \frac{\partial \bar{v}_i}{\partial t} + \bar{v}_k \frac{\partial \bar{v}_i}{\partial x_k},$$

$$\bar{\sigma}_{ik} = \rho\nu\left(\frac{\partial \bar{v}_i}{\partial x_k} + \frac{\partial v_k}{\partial x_i}\right), \tag{2.12}$$

$$\bar{F}_i = -2\epsilon_{ijk}\omega_j \bar{v}_k + \delta_{3i} g(1 \text{ or } \beta_0 \bar{T}).$$

In addition, the equation

$$\frac{\partial \bar{v}_k}{\partial x_k} = 0 \tag{2.13}$$

results from (2.1), which allows us to write down $\bar{v}_k$ in (2.12) out of the operator $\partial/\partial x_k$. An alternative form of the Reynolds equations is obtained when $\bar{\sigma}_{ik}$ from (2.12) is substituted in (2.11)

$$\frac{\partial \bar{v}_i}{\partial t} + \bar{v}_k \frac{\partial \bar{v}_i}{\partial x_k} = -\frac{1}{\rho}\frac{\partial \bar{p}}{\partial x_i} + \frac{\partial}{\partial x_k}\left(\nu \frac{\partial \bar{v}_i}{\partial x_k} - \overline{v_i' v_k'}\right) + \bar{F}_i. \tag{2.14}$$

Analogous equations can be derived for any scalar passive substance θ, which satisfies equations of the same type as (1.19) in Chapter 2

$$\frac{\partial \theta}{\partial t} + \frac{\partial v_k \theta}{\partial x_k} = \kappa_m \nabla^2 \theta, \tag{2.15}$$

where κ_m is a molecular diffusion coefficient. Hence, after letting $\theta = \bar{\theta} + \theta'$ and $v_k = \bar{v}_k + v_k'$, substitution into (2.15) and averaging, we obtain

$$\frac{\partial \bar{\theta}}{\partial t} + \frac{\partial}{\partial x_k}\bar{v}_k\bar{\theta} = \kappa_m \nabla^2 \bar{\theta} - \frac{\partial}{\partial x_k}\overline{v_k'\theta'} = -\frac{\partial}{\partial x_k}\left(-\kappa_m \frac{\partial \bar{\theta}}{\partial x_k} + \overline{v_k'\theta'}\right). \tag{2.16}$$

In particular, θ may denote temperature T (see Equation (1.18) of Chapter 2).

b. Analysis and Interpretation of the Results

The four starting Equations (2.1) and (2.3) for the unaveraged quantities, make a system for the same number of unknowns (v_1, v_2, v_3, p). The density ρ = const is assumed to be a given physical characteristic of the fluid, as well as ν. Therefore, from a mathematical point of view, the system is closed.

The Reynolds system of Equations (2.13) and (2.14) also consists of four equations, but it has 10 unknown functions: $\bar{v}_1$, $\bar{v}_2$, $\bar{v}_3$, $\bar{p}$ and the six components of the symmetrical second order tensor $\overline{v_i' v_k'}$. Therefore, the system is not closed and this is the main obstacle in the theory of turbulence. Evidently, the reason for this is the nonlinearity of the Navier–Stokes Equation (2.7) from where, according to (2.9), the extra unknowns appear.

Analogous is the situation with Equations (2.15) and (2.16) too. The second one contains three unknowns more: the vector components $\overline{v_k'\theta'}$. Again, the reason is the nonlinear term in the left-hand side of (2.15).

The average quantities $\bar{v}_k$, $\bar{p}$ and $\bar{\theta}$ are the simplest statistical characteristics of the random fields v_k (velocity), p (pressure) and θ (scalar substance, more often temperature). They are connected with one another through a number of differential Equations (2.13), (2.14) and (2.16). It is worth mentioning that in the last two, along with $\bar{v}_k$ and $\bar{p}$, characteristics of the fluctuating motion $\overline{v_i' v_k'}$ and $\overline{v_k'\theta'}$ are also present. This means that the mean and fluctuating motions, into which, according to Reynolds, the real turbulent motion can be 'split up', are not dynamically independent. The fluctuating motion interacts with the mean one, which is expressed mathematically by the presence of additional unknowns in the Reynolds equations. Let us examine their physical meaning more precisely.

In order to do this we shall consider the average quantity

$$\overline{\rho v_i v_k + p\delta_{ik} - \sigma_{ik}} = \rho \bar{v}_i \bar{v}_k + \bar{p}\delta_{ik} - T_{ik}, \tag{2.17}$$

where

$$T_{ik} = \bar{\sigma}_{ik} - \rho\overline{v_i' v_k'} = \bar{\sigma}_{ik} + \tau_{ik}, \quad \tau_{ik} = -\rho\overline{v_i' v_k'}. \tag{2.18}$$

It is seen from (2.17) that T_{ik} plays the role of viscous stress tensor with regard to the mean motion. But (2.18) shows that T_{ik} consists of two terms, the first one of which, i.e., $\bar{\sigma}_{ik}$, has a clear physical meaning. This is a tensor of the molecular stresses for the mean motion. Naturally, then we shall interpret τ_{ik} as an additional tensor of the turbulent (Reynolds) stresses that originate into the flow due to the chaotic velocity fluctuations.

In terms of the symbols (x, y, z) and (u, v, w), we shall write τ_{ik} in the form

$$\tau_{ik} = \begin{pmatrix} \tau_{xx} & \tau_{xy} & \tau_{xz} \\ \tau_{yx} & \tau_{yy} & \tau_{yz} \\ \tau_{zx} & \tau_{zy} & \tau_{zz} \end{pmatrix} = -\rho \begin{pmatrix} \overline{u'^2} & \overline{u'v'} & \overline{u'w'} \\ \overline{v'u'} & \overline{v'^2} & \overline{v'w'} \\ \overline{w'u'} & \overline{w'v'} & \overline{w'^2} \end{pmatrix}. \tag{2.19}$$

The diagonal terms are called *normal stresses* and the others, *tangential stresses*. What is more, due to the symmetry of the tensor τ_{ik}, the corresponding components from the two sides of the main diagonal are equal. The denominations are made clear by the geometrical interpretation in Figure 8.1.

We can imagine that on the walls of an elementary volume, normal to the coordinate axes, the stresses τ_x, τ_y and τ_z act respectively. Each one of them can be decomposed

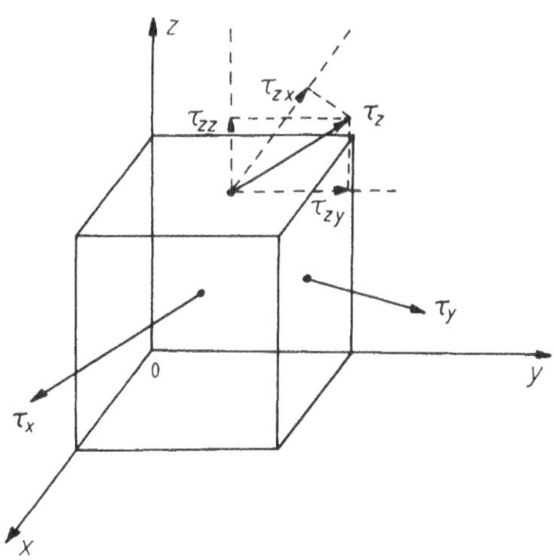

Fig. 8.1. Geometrical interpretation of Reynolds stresses.

into three components as is shown with τ_z. It is seen that τ_{xx}, τ_{yy}, τ_{zz} represent the additional dynamic pressure caused by the velocity fluctuations which acts normally to the walls. The tangential stresses $\tau_{xy} = \tau_{yx}$, $\tau_{xz} = \tau_{zx}$ and $\tau_{yz} = \tau_{zy}$ are generally not equal to zero. This indicates the existence of a certain correlation between the coupled components of the velocity as far as

$$\tau_{ik} = -\rho\overline{v_i'v_k'} = -\rho R_{ik}\,(\overline{v_i'^2} \cdot \overline{v_k'^2})$$

where R_{ik} is a correlation coefficient and the variances $\overline{v_\alpha'^2}$ differ from zero. It appears that in a developed turbulent flow with a large value of Re, $\tau_{ik} \gg \bar{\sigma}_{ik}$, so that $T_{ik} \approx \tau_{ik} = -\rho\overline{v_i'v_k'}$.

An analogous interpretation can be given to Equation (2.16). If θ is a temperature (potential or absolute – T), then

$$q_k^t = \rho c_p \overline{v_k'\theta'} \tag{2.20}$$

will be a turbulent heat flux in the direction of the coordinate axes to be summed up with the molecular flux $q_k^m = \rho c_p \kappa_m \cdot \partial\bar{\theta}/\partial x_k$. As in the case of momentum flux, however, $q_k^t \gg q_k^m$ so that $q_k^m + q_k^t \approx q_k^t \equiv q_k$ which notation we shall use further on.

In the general case, $\tau_{ik} \neq 0$ and $q_k \neq 0$. There is, however, an extremely simple hypothetical case of *isotropic* turbulence at which $\overline{v_k'\theta'} = 0$ and

$$\tau_{ik} = \begin{cases} 0, & \text{at } i \neq k \\ \tau, & \text{at } i = k \end{cases} \tag{2.21}$$

where $\tau = \tau_{11} = \tau_{22} = \tau_{33} = \sigma^2$, σ^2 is the variance of any velocity component. But even in this, probably the most simple case, the number of unknown functions exceeds by one that of the equations and the main difficulty in the theory remains. Further on, we shall deal with the different ways of partially overcoming it. First, let us write down the Reynolds equations for the case of atmospheric turbulence in a standard coordinate system with the usual simplifications

$$\begin{aligned}
\rho\frac{\mathrm{d}\bar{u}}{\mathrm{d}t} &= -\frac{\partial\bar{p}}{\partial x} + \rho f\bar{v} - \left(\frac{\partial\tau_{xx}}{\partial x} + \frac{\partial\tau_{xy}}{\partial y} + \frac{\partial\tau_{xz}}{\partial z}\right)\\
\rho\frac{\mathrm{d}\bar{v}}{\mathrm{d}t} &= -\frac{\partial\bar{p}}{\partial y} - \rho f\bar{u} - \left(\frac{\partial\tau_{yx}}{\partial x} + \frac{\partial\tau_{yy}}{\partial y} + \frac{\partial\tau_{yz}}{\partial z}\right)\\
\rho\frac{\mathrm{d}\bar{w}}{\mathrm{d}t} &= -\frac{\partial\bar{p}}{\partial z} - \rho g - \left(\frac{\partial\tau_{zx}}{\partial x} + \frac{\partial\tau_{zy}}{\partial y} + \frac{\partial\tau_{zz}}{\partial z}\right)
\end{aligned} \tag{2.22}$$

$$\begin{aligned}
&\frac{\partial\bar{u}}{\partial x} + \frac{\partial\bar{v}}{\partial y} + \frac{\partial\bar{w}}{\partial z} = 0\\
&\frac{\mathrm{d}\bar{\theta}}{\mathrm{d}t} = -\left(\frac{\partial\overline{u'\theta'}}{\partial x} + \frac{\partial\overline{v'\theta'}}{\partial y} + \frac{\partial\overline{w'\theta'}}{\partial z}\right)
\end{aligned} \tag{2.23}$$

where

$$\frac{\mathrm{d}}{\mathrm{d}t} = \frac{\partial}{\partial t} + \bar{u}\frac{\partial}{\partial x} + \bar{v}\frac{\partial}{\partial y} + \bar{w}\frac{\partial}{\partial z}.$$

These equations serve as a basis for solving many problems of planetary boundary-layer physics and dynamics.

3. FUNDAMENTALS OF THE SEMIEMPIRICAL THEORY OF TURBULENCE

a. Equations for the Reynolds Stresses

Let us, for simplicity, consider a thermally-nonstratified flow, i.e., in (2.6) $\beta_0 = 0$. Then, as we have already shown in the previous paragraphs, the system of Equations (2.10) and (2.13) contains six more unknowns, i.e., the Reynolds stresses $\overline{v_i' v_k'}$. These equations are an exact consequence of the initial ones.

It is quite natural that we should also try to derive such kind of equations for τ_{ik}. For this purpose, we term-by-term subtract the Reynolds Equations (2.10) from the initial ones (2.4) and obtain equations for the fluctuating motion

$$\rho \frac{\partial v_i'}{\partial t} + \frac{\partial}{\partial x_\alpha} [\rho(\overline{v}_i v_\alpha' + \overline{v}_\alpha v_i' + v_i' v_\alpha' - \overline{v_i' v_\alpha'}) - p'\delta_{i\alpha} - \sigma_{i\alpha}'] = \rho F_i'. \tag{3.1}$$

Hence, also using the identity

$$\frac{\partial}{\partial t} v_i' v_k' = v_i' \frac{\partial v_k'}{\partial t} + v_k' \frac{\partial v_i'}{\partial t} \tag{3.2}$$

we derive an equation for the Reynolds stress tensor

$$\begin{aligned} \rho \frac{\partial}{\partial t} \overline{v_i' v_k'} &+ \frac{\partial}{\partial x_\alpha} [\rho \overline{v}_\alpha \cdot \overline{v_i' v_k'} + \rho \overline{v_i' v_k' v_\alpha'} + \\ &+ (\overline{p' v_i'} \, \delta_{k\alpha} + \overline{p' v_k'} \, \delta_{i\alpha}) - (\overline{v_i' \sigma_{k\alpha}'} + \overline{v_k' \sigma_{i\alpha}'})] \\ &= \rho(\overline{v_i' F_k'} + \overline{v_k' F_i'}) + \overline{p'\left(\frac{\partial v_i'}{\partial x_k} + \frac{\partial v_k'}{\partial x_i}\right)} - \\ &- \left(\overline{\sigma_{i\alpha}' \frac{\partial v_k'}{\partial x_\alpha}} + \overline{\sigma_{k\alpha}' \frac{\partial v_i'}{\partial x_\alpha}}\right) - \left(\rho \overline{v_i' v_\alpha'} \frac{\partial \overline{v}_k}{\partial x_\alpha} + \rho \overline{v_k' v_\alpha'} \frac{\partial \overline{v}_i}{\partial x_\alpha}\right). \end{aligned} \tag{3.3}$$

Unfortunately, however, the so-obtained Equation (3.3), along with the 'old' unknowns $\overline{v}_\alpha$ and $\overline{v_\alpha' v_\beta'}$ also contains some new unknowns that cannot be expressed by the old ones. These are, in the first place, the third statistical moments $\overline{v_\alpha' v_\beta' v_\gamma'}$. If we try to compose equations for them in the same manner, the fourth moments $\overline{v_\alpha' v_\beta' v_\gamma' v_\delta'}$ would appear, and so on. The system will always remain unclosed while the number of new unknowns will rise catastrophically. In other words, this strict approach to the problem does not yield its solution. That is why it is necessary to stop at a definite 'level' and the missing equations to be postulated (constructed) on the basis of additional physical requirements, ideas and hypotheses related to the turbulent motion. In such a way, the strictness of the theory is broken and the semiempirical approach is introduced. It is clear that by its nature this approach will contain, to a rather large extent, an element of speculation, as far as an unrestricted number of more or less physically-grounded hypotheses and postulates can be proposed. Nevertheless, it was the semiempirical theory of turbulence which gained great success in fluid dynamics related to the names of Karman, Taylor, Prandtl and others and, later, was successfully applied to the case of atmospheric turbulence.

At the present stage of development of the theory, attempts to close the system at the level of Equations (3.3) for the Reynolds stresses are quite fashionable. Several, so-called 'second-order closure schemes' have been proposed. They are based on proposals for analytical relations between the surplus unknowns in (3.3) and the basic unknowns $\overline{v}_\alpha$, $\overline{v'_\alpha v'_\beta}$ and $\overline{p}$ which contain a lot of unknown numerical constants. Since these schemes have not yet been properly worked out, we shall confine ourselves to the classical closure schemes, at the level of the Reynolds Equations (2.10), when postulating new equations for $\overline{v'_i v'_k}$. They are comparatively simple and easy to test in practice. However, we shall first use (3.3) to obtain the so-called turbulence kinetic energy balance equation.

b. Energy Balance Equation

We denote by $\overline{E'} = \frac{1}{2}\rho\overline{v'_i v'_i}$ the kinetic energy of the fluctuating motion (a sum should be taken on i from 1 to 3). For this quantity from (3.3) at $i = k$, we obtain the equation

$$\frac{\partial \overline{E'}}{\partial t} + \frac{\partial}{\partial x_\alpha}(\overline{v}_\alpha \overline{E'} + \overline{v'_\alpha E'} + \overline{p' v'_\alpha} - \overline{v'_j \sigma'_{\alpha j}}) = \rho\overline{v'_\alpha F'_\alpha} - \rho\overline{\epsilon'} - \rho\overline{v'_\alpha v'_j}\frac{\partial}{\partial x_\alpha}\overline{v}_j \tag{3.4}$$

where

$$\overline{\epsilon'} = \epsilon = \frac{1}{\rho}\overline{\sigma'_{\alpha j}\frac{\partial}{\partial x_\alpha}v'_j} = \frac{\nu}{2}\overline{\left(\frac{\partial}{\partial x_j}v'_i + \frac{\partial}{\partial x_i}v'_j\right)^2} \tag{3.5}$$

obviously is the rate of dissipation of the kinetic energy into heat under the influence of viscosity. Equation (3.4) is extremely interesting. Each one of its terms has a proper physical meaning. Thus, the vector term in parentheses on the left-hand side is the density of the turbulent energy flux connected with its transport by the mean flow ($\overline{v}_\alpha$), by the fluctuating flow (v'_α) (i.e., turbulent flux of turbulent energy $\overline{v'_\alpha E'}$), by the pressure fluctuations, and by the viscous (molecular) forces. The divergence $\partial(\;)/\partial x_\alpha$ of this vector gives the corresponding influx, which is one of the factors determining $\partial\overline{E'}/\partial t$. Upon integrating over a closed volume of zero energy flux through the surface, according to the Ostrogradsky–Gauss theorem, this term vanishes.

The first term on the right-hand side of (3.4) describes the work done by the force F'_i and the second one describes the dissipation. Special attention should be paid to the last term

$$\mathrm{Tr}^* = \rho\,\mathrm{Tr} = -\rho\overline{v'_\alpha v'_j}\frac{\partial \overline{v}_j}{\partial x_\alpha}. \tag{3.6}$$

It is worth mentioning that if, in a similar way, we try to derive from (2.10) a balance equation for the kinetic energy of the mean motion $\overline{E} = (\rho/2)\overline{v}_i^2$, we shall obtain

$$\frac{\partial \overline{E}}{\partial t} = -\rho\overline{\epsilon} + \rho\overline{v'_\alpha v'_j}\frac{\partial \overline{v}_j}{\partial x_\alpha} + \cdots \tag{3.7}$$

(the unwritten terms do not concern us). Here $\rho\bar{\epsilon} = \overline{\sigma_{\alpha j} \cdot \partial \bar{v}_j/\partial x_\alpha}$ is the dissipation of the mean motion and, like $\overline{\epsilon'}$, is a positive quantity. While $\overline{\epsilon'}$ and $\bar{\epsilon}$ enter in (3.4) and (3.7) with a minus sign, which is absolutely clear, the sign in front of Tr in both equations is opposite so that at summing up, these terms are cancelled. This means that in both equations they describe one and the same mechanism of converting (transforming) the kinetic energy of the one motion into an energy of the other. If at a point of the space occupied by the flow Tr* > 0, then according to (3.4) $\overline{E'}$ at this point increases with time, i.e., the fluctuating motion receives energy from the mean one. This is the reason why Tr* is also called the 'energy production term'. It is possible, however, for Tr to be negative? At a first glance, it seems impossible. Later we shall return to this important question.

If in (3.4) $F'_\alpha = 0$, then the only source of energy for the turbulent motion in the closed volume V with zero flux of energy through its surface is the transformation. And what is more, developing or stationary turbulence, i.e.,

$$0 \leqslant \int_V \frac{\partial \overline{E'}}{\partial t}\, \mathrm{d}V = -\int_V \epsilon\, \mathrm{d}V + \int_V \mathrm{Tr}^*\, \mathrm{d}V$$

is possible only if

$$\int_V \mathrm{Tr}^*\, \mathrm{d}V > 0, \tag{3.8}$$

which means that a part of the kinetic energy of the mean motion is being transformed into turbulence energy. Modern technical devices allow the accurate measurement of both factors $\overline{v'_\alpha v'_j}$ and $\partial \bar{v}_j/\partial x_\alpha$ in (3.6). It appears that for incompressible fluids at ordinary turbulent flows in tubes and boundary layers, condition (3.8) is fulfilled. This fact is also reflected in some formulae of the semiempirical theory (see (3.19)).

Let us now assume that $F'_\alpha \neq 0$ and $\overline{v'_\alpha F'_\alpha} \neq 0$, i.e., the turbulent flow has an external source of energy. A typical example in the atmosphere is the energy of instability, generated by the Archimedean force. Then, in some finite subarea of the flow, it may happen that Tr < 0, i.e., an energy flux from the fluctuating to the mean motion. As we shall see in Chapter 11, we meet such a case with the general atmospheric circulation (GAC), considered as an average current on which the irregular motion of the cyclones and anticyclones (eddies of the so-called *macroturbulence*) is superimposed. These eddies may appear as a result of local heat influxes and further on they transmit their energy to the planetary motions in the GAC system. Direct estimates of Tr* from (3.6), in this case by using the empirical data for $\bar{v}_\alpha$ and v'_α, have confirmed the supposition that Tr* < 0 (see [22]).

If we make use of expression (2.6), then

$$\overline{v'_i F'_i} = g\beta_0 \overline{v'_3 T'} = \beta \overline{w'T'}, \quad \beta = g\beta_0. \tag{3.9}$$

Although Equation (3.4) has been derived for an incompressible fluid, in the Boussinesq approximation which is valid for the atmosphere, we may substitute (3.9) into (3.4) and obtain an equation of the energy balance for the more interesting and important case of thermally-stratified turbulent flow. Henceforth, instead of E' we shall introduce the quantity $b' = E'/\rho$, so that

$$b = \overline{b'} = \tfrac{1}{2}\overline{v_i'^2} = \tfrac{1}{2}(\overline{u'^2} + \overline{v'^2} + \overline{w'^2}). \tag{3.10}$$

Besides, in (3.4) we neglect the term $\overline{v_j' \sigma_{\alpha j}'}$ as small. At a more rough approximation (but sufficiently exact for atmospheric applications) the term $\overline{p' v_\alpha'}$ may also neglected. Thus, we obtain the approximate equation

$$\frac{db}{dt} = -\overline{v_\alpha' v_j'}\frac{\partial \bar{v}_j}{\partial x_\alpha} - \frac{\partial \overline{v_\alpha' b'}}{\partial x_\alpha} + \beta \overline{w' T'} - \epsilon \tag{3.11}$$

which is most often used in the theory of atmospheric turbulence, being added to Reynolds Equations (2.10), (2.13) and (2.16) at $\theta = T$. In such a way, their number is being increased by one equation and three new unknowns are introduced: b, $\overline{v_\alpha' b'}$ and ϵ.

The idea for the use of (3.11) together with the Reynolds equations was put forward by Kolmogorov in 1942. It is a kind of a compromise between the latest models, based on the six equations of system (3.3) and the previous ones in which (3.3) is completely ignored. The idea turns out to be extremely fruitful because, in contrast to (3.3), the 'surplus' unknowns in (3.11) have a very clear physical meaning which, in fact, facilitates the construction of additional links between them and the main unknowns.

c. Coefficients of Turbulence

Again following [48] we shall consider the case when $\mathrm{Tr} > 0$, i.e., the fluctuating motion receives kinetic energy from the mean one. We shall also rely on the formal analogy existing between the chaotic movement of the turbulent eddies and the similar motion of the molecules in the kinetic theory of gases and fluids. On the basis of these assumptions, a proposition may be made that

$$\tau_{ik} = \varphi(\Phi_{ik}), \qquad \Phi_{ik} = \frac{\partial \bar{v}_i}{\partial x_k} + \frac{\partial \bar{v}_k}{\partial x_i}, \tag{3.12}$$

i.e., the symmetrical tensor of Reynolds stresses τ_{ik} is a function of the deformation tensor Φ_{ik} which is also symmetrical. If by analogy with (2.2) we assume that this dependence is linear, then the most general form of such a relationship will be

$$\tau_{ik} = -\rho\overline{v_i' v_k'} = \rho K_{iklm}\Phi_{lm} - \tfrac{2}{3}\rho b\,\delta_{ik}, \tag{3.13}$$

where K_{iklm} is a fourth-order tensor, symmetrical with respect to the coupled indexes i, k and l, m and satisfying the condition

$$K_{iilm}\Phi_{lm} = 0. \tag{3.14}$$

This condition, as well as the second term in (3.13), ensures its transformation into an identity at $i = k$ when $\Phi_{ii} = 0$ (incompressible flow) and $\delta_{ii} = 3$. The newly-introduced tensor K_{iklm} is called the *coefficient of turbulence* and $A_{iklm} = \rho K_{iklm}$ the *turbulent* or *eddy viscosity*. They have the same dimensions, with ν and $\eta = \rho\nu$ respectively.

A particular case of (3.13) is the classical Boussinesq formula

$$\tau_{ik} = \rho K \Phi_{ik} - \tfrac{2}{3}\rho b\, \delta_{ik}, \tag{3.15}$$

where K is a scalar quantity, a full analogue of ν.

Among these two extreme cases is the formula proposed by Monin [48]

$$\tau_{ik} = \tfrac{1}{2}\rho b^{1/2}(l_{i\alpha}\Phi_{\alpha k} + l_{k\alpha}\Phi_{\alpha i}) - \tfrac{2}{3}\rho b\, \delta_{ik}, \tag{3.16}$$

where l_{ij} is a symmetrical tensor of linear scales ('mixing length') which is an analogue to the mean free path of the molecules l_m. The introduction of this geometrical characteristic allows us to take into account the fact that turbulence is not generally isotropic and different directions may have different characteristic scales. Obviously,

$$K_{ij} = \sqrt{b}\, l_{ij} \tag{3.17}$$

has the dimension of a turbulence coefficient. Expression (3.17) is written by analogy with $\nu \sim u_m l_m$, where u_m is the average velocity of molecules. At $l_{ij} = l\, \delta_{ij}$ (isotropy), (3.16) coincides with (3.15) as

$$K = l\sqrt{b}. \tag{3.18}$$

Monin's formula (3.16) is in agreement with (3.13) and imposes additional restrictions on the tensor K_{iklm}.

If τ_{ik} from (3.15) is introduced into (3.6) and taking into account that $\delta_{ij}\, \partial\bar{v}_j/\partial x_i = \partial\bar{v}_i/\partial x_i = 0$, we obtain the formula

$$\mathrm{Tr} = K\Phi_{ij}\frac{\partial \bar{v}_j}{\partial x_i} = \tfrac{1}{2}K\Phi_{ij}^2 \tag{3.19}$$

coinciding with (3.5) in structure. It is seen that $\mathrm{Tr} > 0$ is equivalent to the condition $K > 0$. The application of (3.15) and (3.16) in cases when $\mathrm{Tr} < 0$ (for instance, atmospheric macroturbulence), would mean the introduction of a negative eddy viscosity coefficient ($K < 0$) [60].

By analogy with (3.13), for the first-order tensor (the vector) $\overline{v_i'\theta'}$ which appears in (2.16), may be written down as

$$\overline{v_i'\theta'} = -\left(K_{i\alpha}^{\theta}\frac{\partial\bar{\theta}}{\partial x_\alpha} \quad \text{or} \quad K^{\theta}\frac{\partial\bar{\theta}}{\partial x_i}\right). \tag{3.20}$$

The introduction of a coefficient of turbulent diffusion evidently means that the process of diffusion is assumed to be isotropic. As can be seen, the dimensions of K and K^θ are one and the same. Their values, however, differ in general ($K \neq K^\theta$). But often it is accepted that $K^\theta = \alpha_\theta K$ (α_θ being a dimensionless constant and empirically determined). Besides, formula (3.18) is used for K. In (3.20), θ' means any passive substance. In particular, it may be $\theta' \equiv b' = \tfrac{1}{2}v_i'v_i'$ and then for the diffusive term in the balance Equation (3.11) we would have

$$\overline{v_i'b'} = -\alpha_b K\frac{\partial b}{\partial x_i}, \quad (b = \overline{b'}). \tag{3.21}$$

At last, if we assume, similarly to (3.18) that $\epsilon = \epsilon(b, l)$, then dimensional arguments yield

$$\epsilon = c_1 b^{3/2} l^{-1} = c_1 K^3 l^{-4} = c_1 b^2 K^{-1}, \quad c_1 \sim 1 \tag{3.22}$$

or

$$K = c_2 \epsilon^{1/3} l^{4/3}, \quad c_2 = c_1^{-1/3}. \tag{3.23}$$

Formulae (3.18) and (3.22) are proposed by Kolmogorov. In such a way we have at hand a number of new (postulated) equations for the 'surplus' unknowns in the Reynolds equations and the energy Equation (3.11) that bring us close to the correct mathematical formulation of the problem: the number of equations should be equal to the number of unknowns.

4. FUNDAMENTALS OF THE STATISTICAL THEORY OF TURBULENCE

a. Homogeneous and Isotropic Turbulence

We consider an idealized case of a turbulent flow, occupying the whole space in which there are no privileged directions (isotropy). Such a flow would be without a mean velocity ($\overline{v}_\alpha = 0$) and without rigid boundaries and energy sources, for otherwise they would cause anisotrophy. In a real (viscous) fluid the turbulence should decay with time due to a continuous dissipation of the kinetic energy, which is initially introduced into the flow for its turbulization. Regardless of the great idealization, this case permits one to establish and interpret some fundamental physical regularities which, to a large extent, are also preserved in real and particularly geophysical turbulent flows.

The Navier–Stokes equations for incompressible fluid ($\partial v_k/\partial x_k = 0$) would serve as a starting point:

$$\frac{\partial v_i}{\partial t} + \frac{\partial v_i v_k}{\partial x_k} = -\alpha \frac{\partial p}{\partial x_i} + \nu \nabla^2 v_i \tag{4.1}$$

where $\alpha = 1/\rho$, p and v_i stands for the fluctuations of these quantities ($\overline{v}_i = 0$) at an instant t in a point $\mathbf{x}$. If we denote

$$p(\mathbf{x}', t) = p' \quad \text{and} \quad v'_\alpha = v_\alpha(\mathbf{x}', t),$$

we may write

$$\frac{\partial v'_j}{\partial t} + \frac{\partial v'_j v'_k}{\partial x'_k} = -\alpha \frac{\partial p'}{\partial x'_j} + \nu \nabla'^2 v'_j \tag{4.2}$$

Since $\partial v'_\alpha/\partial x_\alpha = \partial v_\alpha/\partial x'_\alpha = 0$, after multiplying (4.1) by v'_j and (4.2) by v_i, summing both equations, taking the average procedure, and having in mind that due to the homogeneity condition $\partial/\partial x'_\alpha = \partial/\partial r_\alpha$, $\partial/\partial x_\alpha = -\partial/\partial r_\alpha$, where $r_\alpha = x'_\alpha - x_\alpha$, we obtain the tensor equation

$$\left(\frac{\partial}{\partial t} - 2\nu \nabla_r^2\right) B_{ij} = W_{ij} + P_{ij} \tag{4.3}$$

where $\nabla_r^2 = \partial^2/\partial r_\alpha \partial r_\alpha$,

$$W_{ij}(\mathbf{r}, t) = \frac{\partial}{\partial r_k}(\overline{v_i v_k v_j'} - \overline{v_i v_k' v_j'}),$$

$$P_{ij}(\mathbf{r}, t) = -\alpha\left(\frac{\partial}{\partial r_j}\overline{p'v_i} - \frac{\partial}{\partial r_i}\overline{pv_j'}\right), \tag{4.4}$$

$$B_{ij}(\mathbf{r}, t) = \overline{v_i v_j'} = \overline{v_i(\mathbf{x}, t)\, v_j(\mathbf{x}', t)},$$

and $\mathbf{x}' = \mathbf{x} + \mathbf{r}$. At $i = j$ and $r = 0$, (4.3) turns into a turbulent energy balance equation as far as $B_{ii}(0, t) = 2b(t)$. According to (4.4) $W_{ii}(0, t) = 0$, but $W_{ii}(\mathbf{r}, t) \neq 0$. Physically this means that the inertial forces do not change the total kinetic energy of the turbulent flow but only redistribute it between the fluctuations. As for P_{ij}, with the help of the continuity equation $\partial v_k/\partial x_k = 0$ it can easily be proved that $P_{ii}(\mathbf{r}, t) = P_{ii}(0, t) \equiv 0$ in a homogeneous flow. Therefore, this term in the balance equation is neither a source nor a consumer of kinetic energy. The pressure fluctuations only redistribute the kinetic energy among different velocity components as far as at $r \neq 0$ $P_{11}, P_{22}, P_{33} \neq 0$. It is to be expected that by this mechanism the higher energetic velocity components would transmit energy to the lower energetic ones and would determine a tendency toward isotropy ($\overline{u'^2} = \overline{v'^2} = \overline{w'^2}$) in the velocity field. Further on we shall consider that the state of isotropy has been reached, so that $\overline{pv_\alpha'} \equiv 0$.

The above conclusions become physically clearer if we convey our discussion from the real to the phase space by way of Fourier transformation

$$B_{ij}(\mathbf{r}, t) = \int E_{ij}(\mathbf{k}, t)\exp[i(\mathbf{k}\cdot\mathbf{r})]\, d\mathbf{k}. \tag{4.5}$$

Then instead of (4.3), we shall have

$$\left(\frac{\partial}{\partial t} + 2\nu k^2\right) E_{ij}(\mathbf{k}, t) = S_{ij}(\mathbf{k}, t), \tag{4.6}$$

where S_{ij} is the spectral analogue to W_{ij}. The relation between them is such that

$$\int S_{ii}(\mathbf{k}, t)\, d\mathbf{k} = W_{ii}(0, t) = 0. \tag{4.7}$$

This formula also allows us to add to the above-mentioned interpretation of the equality $W_{ii}(0, t) = 0$ the fact that the inertial forces redistribute the kinetic energy among the fluctuations with a different wave vector, i.e., among different eddies. On the other hand, due to the factor k^2 in (4.6), the role of the dissipation increases considerably at $k = |\mathbf{k}| \to \infty$. Thus, if the turbulent flow has gained its kinetic energy within the range of the small wavenumbers ($k \to 0$ – large scales) and loses it at $k \to \infty$ (small scales), then the inertial flux of energy that has been described by the term S_{ij} in (4.6) would go from the small to large wavenumbers (i.e., from the larger turbulent eddies to the smaller ones). This fundamental result is of great importance for the theory.

We shall also mention that at $i = j$ and summing up in (4.6), after some rearrangements the scalar spectral equation is obtained

$$\frac{\partial E(k, t)}{\partial t} = W(k, t) - 2\nu k^2 E(k, t) \tag{4.8}$$

where $E(k, t)$ is the kinetic energy spectral density, describing its distribution across the spectrum of wavenumbers, i.e.,

$$b(t) = \frac{1}{2}\overline{v_i^2} = \int_0^\infty E(k, t)\, \mathrm{d}k. \tag{4.9}$$

As can be seen, (4.8) is one equation for two unknown functions: $E(k, t)$ and $W(k, t)$ (energy flux across the spectrum), i.e., again we come upon the well-known difficulty, due to the nonlinearity of the initial equations. Scientists such as A. M. Obukhov and W. Heisenberg are associated with a new branch of the theory related to attempts at constructing a second equation to (4.8). More information on these problems can be found in the literature on turbulence [48, 53, 26].

b. Locally Homogeneous and Isotropic Turbulence

Homogeneous and isotropic turbulent flows do not exist in nature. Thus, the atmosphere has a rigid boundary (the Earth's surface), privileged direction (vertical), mean sheared velocity of the air currents ($\bar{v}_\alpha \neq 0$, $\nabla\bar{v}_\alpha \neq 0$), and so on. Consequently, the atmospheric turbulence would not be homogeneous and isotropic in the sense described above. It may be presumed, however, that far away from the rigid boundaries and in a rather small subregion of the area occupied by the flow as a whole, the influence of the factors leading to nonisotropy can be neglected and the turbulence can be regarded as *locally-homogeneous* and *isotropic*. This supposition, as well as the above terminology, was first introduced by A.N. Kolmogorov in 1941. Some call it his first hypothesis. It is well known that this idea, complemented by two other physically-substantiated postulates, lead to the greatest discovery in the modern theory of turbulence: the so-called '$\frac{2}{3}$ law' or its spectral analogue 'the $-\frac{5}{3}$ law'. Proceeding in historical succession, we shall briefly present here Kolmogorov's theory of the locally-isotropic turbulence, assuming that $\mathrm{Re} \to \infty$ for the whole flow.

We have to begin, however, with Richardson's notion from 1926 about the micro-structure of a developed turbulent flow at $\mathrm{Re} \to \infty$. According to this notion, at every instant the flow consists of an ensemble (hierarchy) of eddies (disturbances, nonhomogeneities) of a different order (scale). Eddies of a given order originate as a result of the instability of larger ones, thus receiving their energy and becoming unstable, they break down to still smaller eddies and transmit their own kinetic energy and so forth. This 'cascade' process of 'splitting' the eddies and transmitting the kinetic energy from large to very small goes on until stable eddies appear. It is physically clear that the fluid molecular viscosity determines this lower boundary but does not permit a splitting down to molecular scales. The dissipation mechanism would turn the kinetic energy into heat.

Richardson expressed these considerations in a qualitative form (even poetic*) which were not assessed at that time. It was Kolmogorov who breathed new life into these ideas, developed them and expressed them in the form of clear physical postulates. On this basis he derived the fundamental laws and, what is more, he derived them by means of elementary mathematical tools.

Meanwhile, in 1935, J. Taylor introduced the notion of homogeneous and isotropic turbulence. Kolmogorov's concept about local homogeneity and isotropy had lead him to bring the theoretical model nearer to reality while the theory of random functions was used to generalize the mathematical description.

Thus, some of the largest turbulent eddies having scale Λ, determined by the geometry of the flow as a whole and originated because of the instability of the mean motion, would exist in the turbulent flow together with some of the smallest eddies having scale λ_0, determined by the molecular viscosity. They are respectively called the 'outer' (or integral) scale (Λ) and 'inner' (or Kolmogorov's micro-) scale (λ_0). Among them a cascade process is realized which we shall schematically mark like this:

$$\begin{array}{ccccccccc} \Lambda & \longrightarrow & \lambda_1 & \longrightarrow & \lambda_2 & \longrightarrow & \cdots & \longrightarrow & \lambda_0 \\ u_\Lambda & & u_{\lambda_1} & & u_{\lambda_2} & & \cdots & & u_{\lambda_0} \\ \mathrm{Re}_\Lambda & & \mathrm{Re}_{\lambda_1} & & \mathrm{Re}_{\lambda_2} & & \cdots & & \mathrm{Re}_{\lambda_0} \end{array} \tag{4.10}$$

where $\mathrm{Re}_\lambda = u_\lambda \lambda/\nu$ is the spectral Reynolds number for the eddies having scale λ and velocity fluctuations u_λ.

According to Kolmogorov, the influence of the factors leading to anistropy can be ignored for eddies of $\lambda \ll \Lambda$. Then, the statistical regime of the small scale eddies must be stationary and subjected to some universal laws. Kolmogorov postulated that within the range $\lambda \ll \Lambda$, called the *equilibrium* range, this regime is determined only by two-dimensional physical parameters: the dissipation rate ϵ and the molecular viscosity ν. Really, under the stationary regime, the rate of dissipation in the region around λ_0 must be equal to the kinetic energy flux through the cascade of eddies in the region $\lambda \ll \Lambda$. The only length and velocity scales that can be constructed from ϵ and ν are $(\nu^3/\epsilon)^{1/4}$ and $(\epsilon\nu)^{1/4}$. Kolmogorov interpreted them as

$$\lambda_0 = \left(\frac{\nu^3}{\epsilon}\right)^{1/4}, \qquad u_{\lambda_0} = (\epsilon\nu)^{1/4}, \tag{4.11}$$

so that $\mathrm{Re}_{\lambda_0} = \lambda_0 u_{\lambda_0}/\nu \sim 1$.

As a fundamental characteristic of the microstructure of the turbulent flows in the equilibrium range $\lambda \ll \Lambda$, Kolmogorov introduces the structure function (1.8), i.e., the longitudinal one $D_{ll}(r)$. Then, if the distance r between the points $\mathbf{x}'$ and $\mathbf{x}''$ is much

---

* Big whorls have little whorls,
Which feed on their velocity;
Little whorls have smaller whorls,
And so on unto viscosity.

smaller than the outer scale $\Lambda (r \ll \Lambda)$, on the basis of the previous considerations we may write

$$D_{ll}(r) = u_{\lambda_0}^2 \varphi\left(\frac{r}{\lambda_0}\right), \tag{4.12}$$

where $\varphi(z)$ is a universal function of $z = r/\lambda_0$. As can be seen, (4.12) expresses a condition for the similarity of D_{ll}.

The next Kolmogorov hypothesis says that at Re $\to \infty$ a subrange $\lambda_0 \ll \lambda \ll \Lambda$ exists (the *inertial* subrange), where still $\mathrm{Re}_\lambda \to \infty$ and, consequently, the influence of the viscosity can be neglected. In other words, in this subrange the splitting of eddies under the action of inertial forces and the flux of energy through the spectrum are realized almost without dissipation. This means that at $\lambda_0 \ll r \ll \Lambda$, the structure function must not depend on ν. As can be easily seen, this is possible only if in (4.12) $\varphi(z) = \alpha_r z^{2/3}$. Then

$$D_{ll}(r) = \alpha_r \epsilon^{2/3} r^{2/3}, \qquad \lambda_0 \ll r \ll \Lambda. \tag{4.13}$$

This is Kolmogorov's so-called '$\frac{2}{3}$ law', where $\alpha_r \sim 1$ is a universal dimensionless constant and its exact value is estimated experimentally ($\alpha_r \approx 1.9$).

The considerations and conclusions of Kolmogorov's theory, presented here, can easily be preformulated in terms of spectral language. Instead of the structure function, the basic characteristic would now be the spectral density $E(k)$. Since $k \sim 1/r$, the width of the inertial subrange would now be determined by the inequality $k_\Lambda \ll k \ll k_0$ where $k_0 = \lambda_0^{-1} = (\epsilon/\nu^3)^{1/4}$, $k_\Lambda = \Lambda^{-1}$. In this subrange $E = E(\epsilon, k)$, a dimensional analysis leads to the spectral analogue of (4.13)

$$E(k) = \alpha_k \epsilon^{2/3} k^{-5/3}, \quad k_\Lambda \ll k \ll k_0, \tag{4.14}$$

where $\alpha_k \sim 1$ is the universal constant that is experimentally estimated. By the way, it is connected with the constant α_r from (4.13) in so far as (4.14) is the Fourier transformation of (4.13): $\alpha_k \approx 0.76\ \alpha_r$. Therefore, (4.13) and (4.14) reflect one and the same regularity of the velocity field microstructure in locally-isotropic turbulence. In spite of the apparent complexity of the spectral method of investigation, it is preferred not only in theory but also in experimental investigations, because of the clear physical meaning of the basic characteristics and various technical reasons.

Having estimated ϵ, both expressions (4.13) and (4.14) can be used in the applications. By definition $\epsilon = -\mathrm{d}b/\mathrm{d}t$ so that we may write down $\tau = -b/(\mathrm{d}b/\mathrm{d}t) = b/\epsilon$ for the 'time of relaxation (degeneration)' of the turbulence. Further, if l_E, b_E and τ_E are scales of energy-containing (largest) eddies, kinetic energy and characteristic life-time, then $\tau_E = |b_E/(\mathrm{d}b_E/\mathrm{d}t)| \sim l_E/b_E^{1/2} \approx l_E/b^{1/2}$ because practically $b_E \approx b$. Under a stationary regime $\tau_E \sim \tau$. Hence, we obtain $\epsilon \sim b^{3/2}/l_E$, i.e., formula (3.22). Proceeding further, we can also consider that $\epsilon = \mathrm{Tr}$, where Tr is given by (3.19). This means that, under a stationary regime, the dissipation ϵ equals the transformation, i.e., the quantity of kinetic energy supplied from the mean motion to the fluctuating one.

c. Microstructure of Scalar Fields

Velocity field turbulence also generates turbulent fluctuations of various scalar quantities in the flow: temperature T (or θ), humidity (q), pressure (p), and so on. Let us, for definiteness, speak about temperature which satisfies the equation

$$\frac{\partial T}{\partial t} + \frac{\partial v_k T}{\partial x_k} = \kappa \nabla^2 T. \tag{4.15}$$

The temperature field microstructure problem can be treated in the approximation of the isotropy or local isotropy of the field. According to the above presentation of the velocity field, the second case is of direct interest for atmospheric applications. As an example, we shall examine only the question about the dependency of the structure function $D_{TT}(r)$ on the distance r when it pertains to the inertial subrange.

By analogy with the quantity $b = \overline{v_i'^2}/2$ we introduce $b_T = \overline{T'^2}$ (the 'energy' of temperature fluctuations). The chaotic motion of turbulent eddies would lead to a local strengthening of the temperature gradients due to the possibility of a collision of eddies with different temperatures. The mechanism of the molecular heat diffusion, however, would act to smooth these gradients. In such a way, a scattering ('dissipation') of the temperature fluctuation 'energy' would be observed. Similarly to $\epsilon = -\mathrm{d}b/\mathrm{d}t$, the rate of this process may be determined by $\epsilon_T = -\mathrm{d}b_T/\mathrm{d}t$, a quantity introduced by Obukhov. Conditionally, ϵ_T could be called the 'temperature dissipation'.

If we wish only to determine the structure function $D_{TT}(r)$ in the inertial subrange,* besides on r, it should also depend on ϵ and ϵ_T (but not on ν and κ), i.e., $D_{TT} = \varphi_T(\epsilon, \epsilon_T, r)$. A dimensional analysis immediately yields

$$D_{TT}(r) = \beta_r \epsilon_T \epsilon^{-1/3} r^{2/3}, \quad \lambda_0 \ll r \ll \Lambda, \tag{4.16}$$

i.e., again the '$\frac{2}{3}$ law' derived by Obukhov in 1949. Independently, in the same way in 1951, Corrsin derived the spectral analogue of (4.16), i.e., the '$-\frac{5}{3}$ law'

$$E_{TT}(k) = \beta_k \epsilon_T \epsilon^{-1/3} k^{-5/3}, \quad k_\Lambda \ll k \ll k_0. \tag{4.17}$$

Naturally, (4.17) can be derived purely mathematically as a Fourier transformation of (4.16) and a link between the universal constants is to be found: $\beta_k \approx 0.4\,\beta_r$.

The fundamental laws (4.13), (4.14) and (4.16), (4.17) were substantially and experimentally verified through measurements in natural geophysical turbulent flows (the atmosphere and the ocean) and also in laboratory conditions. They are considered as classical results in the modern theory of turbulence.

PROBLEMS

1. Prove formulae (1.4) and (1.10)–(1.13).
2. Derive Equation (3.7) in a complete form.
3. Similarly to (3.4), derive an equation for $\overline{v'_\alpha \theta}$ and interpret its terms.

* Strictly speaking, in the 'inertial-convective' subrange $\lambda_0^T \ll r \ll \Lambda$, but if $\kappa \sim \nu$, then $\lambda_0^T = (\kappa^3/\epsilon)^{1/4} \sim \lambda_0 = (\nu^3/\epsilon)^{1/4}$. That is the case, considered here for simplicity.

4. If $\partial v_k/\partial x_k = 0$ and the random fields $\mathbf{v}(\mathbf{x})$ and $p(\mathbf{x})$ are homogeneous and isotropic, prove that

$$\overline{p'(\mathbf{x})v'(\mathbf{x}+\mathbf{r})} \equiv 0, \qquad \frac{\partial B_{ij}}{\partial r_j} = \frac{\partial B_{ij}}{\partial r_i} = 0.$$

Note: See [53].

5. Making use of (1.16), prove that at $r \to 0$

$$B_{ff}(r) = \sigma_f^2 - \tfrac{1}{4}\sigma_{\nabla f}^2 r^2 + \tfrac{1}{64}\sigma_{\nabla^2 f}^2 r^4 - \cdots \qquad \text{(P1)}$$

where $\sigma_\varphi^2 = \overline{\varphi^2}$ is the variance of $\varphi = f, \nabla f, \nabla^2 f, \ldots$

Solution: At $x \to 0$ for the Bessel function $J_0(x)$ we can write

$$J_0(x) = 1 - \frac{x^2}{4} + \frac{x^4}{64} - \cdots$$

By the definition of correlation functions it follows

$$B_{\nabla f \nabla f} = -\nabla_r^2 B_{ff}(r), \qquad B_{\nabla^2 f \nabla^2 f}(r) = \nabla_r^4 B_{ff}(r), \cdots$$

Substituting these expressions in (1.16) we obtain (P1).

CHAPTER 9

The Dynamics of the Atmospheric Surface Layer

1. TURBULENT SURFACE LAYER: GENERAL PROPERTIES

a. Definition of Surface Layer (SL)

The Earth's surface is a rigid lower boundary of the atmosphere. Due to this, a boundary layer exists in the wind velocity field or in other meteorological elements. Because the air flow is generally turbulent, this is a turbulent boundary layer rather than molecular. Naturally, the influence of the underlying surface is greatest in the air layer immediately adjacent to it, and depends on the distribution of its properties in a horizontal direction. The latter indicates the existence of horizontal gradients of the averaged quantities, which only exceptionally are greater than vertical ones (see below Section 6). That is why in all other sections of this chapter we shall refer only to the case of a horizontally-homogeneous underlying surface – e.g., a vast, even steppe area (bare or evenly covered with one and the same vegetation or a similar type of agricultural tract). This presupposes a uniform distribution of soil characteristics (temperature, moisture, heat conductivity, etc.) which may exert some influence on the atmospheric processes.

With this fundamental supposition, the derivatives with respect to x and y in Reynolds equations will vanish except $\partial\bar{p}/\partial x$ and $\partial\bar{p}/\partial y$ which depend on synoptic factors of a much larger scale. In such a way we come to the initial system of equations

$$\frac{d\bar{u}_i}{dt} = -\alpha\frac{\partial\bar{p}}{\partial x_i} - f\epsilon_{i3k}\bar{u}_k - \frac{\partial\overline{u_i'w'}}{\partial z},$$
$$\frac{d\bar{\theta}}{dt} = -\frac{\partial\overline{w'\theta'}}{\partial z}, \quad \frac{\partial\bar{w}}{\partial z} = 0. \tag{1.1}$$

Since at $z = 0$, $\bar{w} = 0$, then (1.1) yields $\bar{w}(z) = 0$, i.e., under the assumption of horizontal homogeneity, the mean motion is also horizontal (but $w' \neq 0$). Then $d/dt = \partial/\partial t$ and Equations (1.1) read

$$-\frac{\partial\overline{u_i'w'}}{\partial z} = \frac{\partial\bar{u}_i}{\partial t} + F_i = M_i, \quad i = 1, 2$$
$$-\frac{\partial\overline{w'\theta'}}{\partial z} = \frac{\partial\bar{\theta}}{\partial t} = N, \tag{1.2}$$

where $F_i = \alpha\partial\bar{p}/\partial x_i + f\epsilon_{i3k}\bar{u}_k$.

After integration between 0 and z, we obtain

$$(\overline{u_i'w'})_0 - (\overline{u_i'w'})_z = \int_0^z M_i \, dz,$$
$$(\overline{w'\theta'})_0 - (\overline{w'\theta'})_z = \int_0^z N \, dz. \tag{1.3}$$

Since M_i and N are continuous functions of z, and with a suitable choice of z, the integrals can be made arbitrarily small, so that at $z \leqslant h$, we can write

$$(\overline{u_i'w'})_z \approx (\overline{u_i'w'})_0, \qquad (\overline{w'\theta'})_z \approx (\overline{w'\theta'})_0, \tag{1.4}$$

i.e., at $z \leqslant h$ the vertical turbulent fluxes of different 'substances' (momentum, temperature, humidity, etc.) vary little from its surface values and can be considered as constant with height. It appears that in the atmosphere $h \sim 30$–50 m. It is this constant flux layer that is called the *surface layer* (SL). It is also clear that, as well as horizontal homogeneity, another condition for its existence and finite thickness is the quasi-stationarity of the averaged fields ($\partial/\partial t = 0$). Strictly speaking, the inference about constancy with height refers to the total fluxes (molecular plus turbulent), the dominant components of which, however, are the turbulent ones.

Turning towards Figure 8.1, we see that the quantities (1.4) $-\rho\overline{u'w'}$ and $-\rho\overline{v'w'}$ are x- and y-components of the vector τ_z, acting on the horizontal wall. For simplicity we shall write $\tau = \mathbf{i}\tau_x + \mathbf{j}\tau_y$. Then conditions τ_x, τ_y = const mean that at $z \leqslant h$ the turbulent stress vector τ is constant with height in magnitude as well as in direction and, therefore, removes one privileged direction in the SL.

On the other hand, a well-known observational fact is that in the first several decameters above the ground, the mean wind almost preserves its direction (does not rotate). This is a measure for the negligibly small influence of the Coriolis force in this layer. We may come to this conclusion by also evaluating the order of the last two terms in Equation (1.1). Mathematically it means that $\overline{\mathbf{u}} \times \partial\overline{\mathbf{u}}/\partial z = 0$. Therefore, the mean wind velocity vector $\overline{\mathbf{u}}$ determines another new privileged direction in the coordinate plane $x0y$. Henceforth, it is convenient to choose axis the $0x \| \overline{\mathbf{u}}$ so that $\overline{v} = 0$. Then from considerations of symmetry with respect to the plane $x0z$ follows $\overline{v'w'} = 0$.

Summing up the additional features of the SL together with (1.4), we may also write,

$$\overline{u} = \overline{u}(z), \qquad \overline{v} = \overline{w} = 0, \qquad \overline{u'v'} = \overline{v'w'} = 0,$$
$$-\overline{u'w'} = \frac{\tau}{\rho} = v_*^2 = \text{const}_z, \quad z \leqslant h. \tag{1.5}$$

The quantity v_*, having a dimension of velocity, is often called the *friction* (or *dynamic*) *velocity*. Note that $\overline{u'w'}$ is negative, corresponding to a flux of horizontal momentum *down* the mean velocity gradient or in the $-z$ direction. Similarly, for the 'normalized' heat flux

$$q_* = \frac{q}{\rho c_p} = \overline{w'\theta'} = \text{const}_z, \quad z \leqslant h \tag{1.6}$$

The quantities v_* and q_* represent the important scaling inner physical parameters of the SL which do not depend on z at $z \leqslant h$.

Let us also note the great vertical temperature gradients, particularly close to the ground, as another characteristic feature of the SL. Due to this $\partial \bar{T}/\partial z \approx \partial \bar{\theta}/\partial z$ where T and θ are real and potential temperature.

The surfaces layer could also be introduced as a sublayer of the entire boundary layer generated by the rigid lower boundary – the Earth's surface. Figure 9.1 shows some simplified distribution of the mean wind velocity $|\mathbf{u}|$, horizontal stress $|\boldsymbol{\tau}|$ and vertical heat flux $|q|$ in the boundary layer over a uniform flat surface. The wind velocity increases smoothly from zero to its free-stream value at $z = h_g$ – the boundary-layer thickness.* Then, as a next approximation instead of (1.4)–(1.6), one can write

$$\tau(z) = \rho v_*^2 \left(1 - \frac{z}{h_g}\right), \qquad q(z) = \rho c_p q_* \left(1 - \frac{z}{h_g}\right), \quad z < h_g.$$

Therefore, at heights $z < 0.1\, h_g$ the stress is within 10% of its surface value. But h_g is evaluated to be about 10^3 m, so that for the thickness of the 'constant' flux layer we have $h \sim 10^2$ m.

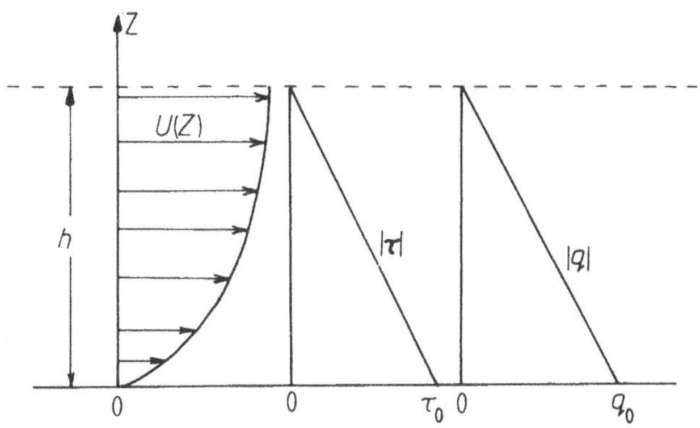

Fig. 9.1. Idealized wind, horizontal shear stress and heat flux distribution in the boundary layer (after [25]).

b. Energetics of the SL

Let us turn again to the Reynolds stress Equations (3.3) in Chapter 8, and let $i = k$ (without summing!), $i = 1, 2, 3$ consecutively. For simplicity, we shall consider $F'_\alpha = 0$. We denote $E'_i = \frac{1}{2}\rho v_i'^2$ and write down the result in the form [48]

$$\frac{\partial \overline{E'_i}}{\partial t} + \frac{\partial}{\partial x_\alpha}\left(\bar{v}_\alpha \overline{E'_i} + \overline{E'_i v'_\alpha} + \overline{p' v'_i}\,\delta_{i\alpha} - \nu\,\frac{\partial \overline{E'_i}}{\partial x_\alpha}\right)$$

$$= \overline{p'\,\frac{\partial v'_i}{\partial x_i}} - \rho\overline{\epsilon'_i} - \rho\overline{v'_i v'_\alpha}\,\frac{\partial \bar{v}_i}{\partial x_\alpha}, \tag{1.7}$$

* See also Section 1 in Chapter 10.

where $\overline{\epsilon_i'} = \epsilon_i$ can be interpreted as the dissipation of energy E_i' of the ith velocity component. The molecular-diffusion term (the last on the left-hand side) in (1.7) can be neglected, compared to the turbulent one (the second in the parentheses).

We write (1.7) for a horizontally homogeneous SL. Having in mind the equation $\partial v_\alpha'/\partial x_\alpha = 0$ and the features of the surface layer previously discussed, after a change of notation we obtain

$$\frac{\partial \overline{E_x'}}{\partial t} + \left(\frac{\partial}{\partial z}\right)\overline{w'E_x'} = \overline{p'\frac{\partial u'}{\partial x}} - \rho\epsilon_x + \tau\frac{\partial \bar{u}}{\partial z}, \tag{1.8}$$

$$\frac{\partial \overline{E_y'}}{\partial t} + \left(\frac{\partial}{\partial z}\right)\overline{w'F_y'} = \overline{p'\frac{\partial v'}{\partial y}} - \rho\epsilon_y, \tag{1.9}$$

$$\frac{\partial \overline{E_z'}}{\partial t} + \left(\frac{\partial}{\partial z}\right)(\overline{w'E_z'} + \overline{p'w'}) = \overline{p'\frac{\partial w'}{\partial z}} - \rho\epsilon_z, \tag{1.10}$$

where $\tau = -\rho\overline{u'w'}$. It can be seen that the mean motion directly 'feeds up' with energy only the longitudinal (u') fluctuating component which is testified by the presence of the production term $\tau\ \partial\bar{u}/\partial z$ in Equation (1.8). The lateral components (v' and w') receive energy from u' by way of pressure fluctuations p' (the first term on the right-hand sides of the Equations (1.8)–(1.10)). The mechanism determines a tendency towards isotropy, since it 'transfers' energy from u' to v' and w' components and does not permit a 'storing up'. On the whole, however, energy has not been lost or generated. Really, at a stationary regime ($\partial\overline{E'}/\partial t = 0$) if we sum up (1.8)–(1.10) and integrate over the volume V with zero energy flux through the boundaries, we obtain

$$\int_V \tau\frac{\partial \bar{u}}{\partial z}\,\mathrm{d}V = \rho\int_V \epsilon\,\mathrm{d}V$$

where $\epsilon = \epsilon_x + \epsilon_y + \epsilon_z$ is the total dissipation.

Under the specific conditions of the SL, Equation (3.11) of Chapter 8 for the total kinetic energy considerably simplifies:

$$\frac{\partial b}{\partial t} = -\overline{u'w'}\frac{\partial \bar{u}}{\partial z} - \frac{\partial}{\partial z}\overline{w'b'} + \beta\overline{w'T'} - \epsilon. \tag{1.11}$$

This equation finds wide application in the theory.

c. Semiempirical Equations

The fundamental equations of the semiempirical theory of turbulence (Section 3, Chapter 8) also simplify in the SL. Thus, based on (3.15) and (3.20) of Chapter 8, one can write

$$\begin{aligned} \tau &= -\rho\overline{u'w'} = \rho K\frac{\mathrm{d}\bar{u}}{\mathrm{d}z}, \\ q &= \rho c_p\overline{w'\theta'} = -\rho c_p K^\theta\frac{\mathrm{d}\bar{\theta}}{\mathrm{d}z}, \end{aligned} \tag{1.12}$$

where τ and q are vertical turbulent fluxes of momentum and heat, K and K^θ are turbulent (eddy) coefficients. The problem is not settled, with the introduction of K and K^θ, however, because they remain unknown – only a substitution of τ and q with K and K^θ has been made. An approach in atmospheric turbulence theory exists, in which K and K^θ are considered as known a priori ('a priori' or 'K' models). However, the principles that should be observed in choosing K and K^θ which, to a certain extent, are even more abstract characteristics of turbulence than τ and q, are not clear. In this sense there is considerable power in the idea of introducing a more primitive and geometrical-meaning characteristic of turbulence – the 'mixing length' – by means of which K and K^θ should be expressed. This is a classical idea and belongs to one of the founders of semiempirical turbulence theory in fluid dynamics, L. Prandtl. It is expressed in a new version by Monin's formula (3.16) in Chapter 8. Specified for the SL, this approach leads to simple and, we shall see, extremely useful equations. To derive them we need a clear schematization of the mechanism of turbulent exchange in the SL. It consists of the following suppositions:

(1) In a developed turbulent flow with nonuniformly distributed properties in the space, the mean (averaged) motion permanently generates turbulent eddies that move chaotically in all directions.

(2) The eddies travel quasi-statically ($p_i = p_a$, i.e., the inner pressure equals the outer) without any exchange of features with the surroundings. After passing some finite distance l, called the 'mixing length', they suddenly 'dissolve' into the flow, thus delivering the features which they carry.

(3) The characteristics of the eddies (velocity, temperature, etc.) in the place of their origination are equal to the averaged characteristics on this level. Successively passing through the point of observation, the eddies cause fluctuations equal to the difference between the 'brought up' feature and the averaged one on the level of observation.

(4) The features (substances) that are transported by the eddies are conservative, i.e., on the whole way l, they neither disappear nor originate. If this condition is not observed, the individual changes during the motion should be determined.

(5) Because of the horizontal homogeneity of the SL, the turbulent transport of the features should be observed in the vertical direction only. As a result, a turbulent flux of substances from regions with a higher mean concentration to such with a lower concentration will exist. The initially nonuniform distribution along the vertical axis will tend to smooth. Correspondingly, l should be considered as the vertical mixing length.

Now let θ be an arbitrary characteristic of the flow carried by the eddies. We consider a unit-horizontal area at level z above the ground. According to the above-described scheme, the fluctuation θ' on this level should be

$$\theta' = \overline{\theta}(z - l'_\theta) - \overline{\theta}(z) = -l'_\theta \frac{\mathrm{d}\overline{\theta}}{\mathrm{d}z} + \frac{1}{2} l'^2_\theta \frac{\mathrm{d}^2\overline{\theta}}{\mathrm{d}z^2} + \cdots \tag{1.13}$$

where l'_θ is the mixing length for θ; l'_θ is a random quantity which changes from eddy to eddy and may be different from different characteristics (momentum, temperature, humidity, etc.). In (1.13) $z - l'_\theta$ is the level at which the eddy has originated. For eddies crossing the chosen area from below $l'_\theta > 0$ and $w' > 0$ while for eddies moving downwards, $l'_\theta < 0$ and $w' < 0$. Keeping the linear term in the series (1.13), for the total flux through the area we find

$$\overline{w'\theta'} = -\overline{w'l'_\theta}\,\frac{d\bar{\theta}}{dz}, \tag{1.14}$$

where the overbar means ensembles averaging: If θ is not a conservative property and the change of $\bar{\theta}$ following the motion of the eddy is characterized by the gradient $\delta\bar{\theta}/\delta z$, then instead of (1.13) we shall have

$$\theta' = \bar{\theta}(z - l'_\theta) + l'_\theta\,\frac{\delta\bar{\theta}}{\delta z} - \bar{\theta}(z) = -l'_\theta\left(\frac{d\bar{\theta}}{dz} - \frac{\delta\bar{\theta}}{\delta z}\right), \tag{1.15}$$

so that

$$\overline{w'\theta'} = -\overline{w'l'_\theta}\left(\frac{d\bar{\theta}}{dz} - \frac{\delta\bar{\theta}}{\delta z}\right). \tag{1.16}$$

Comparing (1.14) and (1.16) with (1.12) we obtain

$$K^\theta = \overline{w'l'_\theta}. \tag{1.17}$$

The temperature T is an example of a nonconservative characteristic. With the dry-adiabatic motion of the eddies, as is already known, $-\delta\bar{T}/\delta z = \gamma_a = 10^{-2}$ °C m^{-1}, but, in contrast, the potential temperature θ is conservative $-\delta\theta/\delta z = 0$. In the general case $-\delta\bar{\theta}/\delta z = \gamma_e$ can be interpreted as an equilibrium gradient, so that at $d\bar{\theta}/dz = -\gamma_e$ (1.16) yields $\overline{w'\theta'} = 0$. Obviously, in the case of the adiabatic process $\gamma_e = \gamma_a$. However, the assumption for adiabaticity of the eddies' motion is, to a large extent, groundless, so that generally $\gamma_e \neq \gamma_a$. Having in mind the well-known formula for the potential temperature gradient

$$\frac{d\bar{\theta}}{dz} = \left(\frac{\bar{\theta}}{\bar{T}}\right)\left(\frac{d\bar{T}}{dz} + \gamma_a\right) \tag{1.18}$$

we see that both expressions (1.14) and (1.16) do not contradict each other.

An analogous approach can be applied to the transport of momentum $\rho u'$. Neglecting the influence of viscosity, this quantity can be considered as a conservative one. Thus, on a level z we shall have

$$\begin{gathered} u' = \bar{u}(z - l) - \bar{u}(z) = -l'\,\frac{d\bar{u}}{dz}, \\ \overline{u'w'} = -\overline{w'l'}\,\frac{d\bar{u}}{dz}, \qquad K = \overline{w'l'}. \end{gathered} \tag{1.19}$$

It is physically acceptable to have $l'_\theta \neq l'$. This means that the turbulent eddy may cease to exist as an isolated body in the field of θ (for instance, the temperature) but with respect to the momentum which it carries, this may not be so yet. Consequently, $\overline{w'l'} \neq \overline{w'l'_\theta}$, i.e., $K \neq K^\theta$. In many problems of surface-layer physics and dynamics, such a refinement is not necessary. In other cases it is indispensable.

The next step, which is in agreement with the continuity equation $\nabla \cdot \mathbf{v}' = 0$, is the assumption that

$$w' \sim -u' \sim l'\,\frac{d\bar{u}}{dz}. \tag{1.20}$$

Substituting in (1.19) we obtain

$$K = l^2 \frac{d\overline{u}}{dz}, \quad l = (\overline{l'^2})^{1/2},$$
$$\tau = -\rho\overline{u'w'} = \rho l^2 \left(\frac{d\overline{u}}{dz}\right)^2 = \rho K \frac{d\overline{u}}{dz}, \tag{1.21}$$

or

$$\left.\begin{aligned} l \frac{d\overline{u}}{dz} &= v_* = \text{const}_z \\ K \frac{d\overline{u}}{dz} &= v_*^2 = \text{const}_z \end{aligned}\right\}, \quad z \leqslant h, \tag{1.22}$$

where const_z denotes independence on z. The mixing length, as defined in (1.21), is an averaged quantity, while l' was a random one.

The general expression for the energy production term also simplifies considerably in the surface layer

$$\text{Tr} = -\overline{v_i'v_j'} \frac{\partial}{\partial x_i} \overline{v}_j = -\overline{u'w'} \frac{d\overline{u}}{dz} = K \left(\frac{d\overline{u}}{dz}\right)^2. \tag{1.23}$$

Eliminating $d\overline{u}/dz$ from here and from (1.21), we obtain an interesting expression

$$K = \text{Tr}^{1/3} l^{4/3} \tag{1.24}$$

which coincides with (3.23) in Chapter 8, at $\epsilon = \text{Tr}$.

In addition to (1.22), the equation for θ may be supplemented

$$K^\theta \frac{d\overline{\theta}}{dz} = -q_* = \text{const}_z, \quad z \leqslant h. \tag{1.25}$$

In the approximation $K^\theta = K$, formula (1.21) can be used.

In conclusion, let us make the following stipulation: henceforth θ will denote potential temperature only. Observations indicate that $|p'/\overline{p}| \ll |T'/\overline{T}|$. On the other hand, by the definition $\theta = T(p_{00}/p)^\lambda$, $\lambda = R/c_p$, so that $T'/\overline{T} \approx \theta'/\overline{\theta}$. In the surface layer $\overline{\theta} \approx \overline{T}$ and $T' \approx \theta'$, $\overline{w'\theta'} \approx \overline{w'T'}$. This fact permits us to replace T' by θ' in Equations (3.9) and (3.11) of Chapter 8, as well as in (1.11). This very stipulation is observed in all subsequent equations.

2. VERTICAL PROFILES OF THE WIND AND OTHER METEOROLOGICAL ELEMENTS IN THE SURFACE LAYER

a. Neutral Stratification: Logarithmic Law

In the general case, the turbulent regime in the SL is determined by two factors: dynamic and thermal. To the first belong the ground surface with its roughness and the wind, while temperature stratification belongs to the second. That is why the following three cases are possible:

(i) Pure mechanical turbulence – only the first factor plays a role. This case is realized under neutral conditions in the SL which practically means isothermy.

(ii) Pure thermal turbulence (free convection). This is realized with superadiabatic gradients and windless conditions.

(iii) Mixed regime of turbulence (arbitrary stratification).

Under neutral stratification ($d\bar{\theta}/dz = 0$) according to (1.25) $q_* = 0$. Thus, the only inner scaling velocity parameter remains v_* (= const_z at $z \leqslant h$). Then either of the Equations (1.22) solved for $d\bar{u}/dz$:

$$\frac{d\bar{u}}{dz} = \frac{v_*}{l} = \frac{v_*^2}{K} \tag{2.1}$$

may be used for determining $\bar{u}(z)$ if $l = l(z, v_*)$ or $K = K(v_*, z)$ are known in explicit form. Simple dimensional arguments lead to the only possible dependences, $l \sim z$, $K \sim v_* z$. Following Prandtl let us write

$$l = \kappa z, \tag{2.2}$$

where $\kappa = 0.4$ is Kármán's constant. Substituting (2.2) into (2.1) and solving this simple differential equation for $\bar{u}(z)$, we obtain

$$\bar{u}(z) = \frac{v_*}{\kappa} \ln z + C. \tag{2.3}$$

The underlying surface is aerodynamically rough. To characterize this property the roughness parameter z_0 is introduced, such that $\bar{u}(z_0) = 0$. With this as a boundary condition, (2.3) yields

$$\bar{u}(z) = \frac{v_*}{\kappa} \ln \frac{z}{z_0}. \tag{2.4}$$

Then

$$K(z) = \kappa v_* z. \tag{2.5}$$

The question about the interpretation of z_0 is not so simple as it seems at a first glance. Let d_0 be the averaged height of the geometrical roughness on the Earth's surface. In its immediate proximity, the molecular viscosity ν of the air would also influence the motion and would determine the existence of a very thin molecular sublayer. If $d_0 \leqslant z_* = \nu/v_*$ the surface is called *aerodynamically smooth*. For the air $\nu = 0.15\ \text{cm}^2\ \text{s}^{-1}$ and $10 \leqslant v_* \leqslant 100\ \text{cm s}^{-1}$, so that $1.5 \leqslant 10^3 z_* \leqslant 15$ cm. In real conditions, however, even at low vegetation (grass or something else) $d_0 \gg z_*$, so that the underlying surface is *aerodynamically rough*. What is the ratio between z_0 and d_0? It is natural to suppose that $z_0 = a d_0$, where $a < 1$. Experimental data indicate that $a = 0.1$–0.2, i.e., $z_0 \sim z_*$. Strictly speaking, the logarithmic law (2.4) can be correct only at $z \gg z_0$.

Two modifications of the logarithmic law (simply 'log-law') (2.4) are of special interest.

(i) For vegetative covers (canopies) that are not too closely spaced, field observations and laboratory experiments indicate that instead of (2.4), the relation

$$\bar{u}(z) = \frac{v_*}{\kappa} \ln \frac{z - d_0}{z_0}, \quad z \geqslant d_0 + z_0 \tag{2.6}$$

holds, where d_0 is the so-called *zero-plane displacement*. Some theoretical support to (2.6) could be given if one investigates the more general expression for this case

$$\frac{\mathrm{d}\bar{u}}{\mathrm{d}z} = \frac{v_*}{\kappa z}\,\varphi\left(\frac{d_0}{z}\right), \tag{2.7}$$

where $\varphi(\eta)$ is a dimensionless function to be determined experimentally or theoretically on the basis of some additional assumptions.

(ii) The log law (2.4) also remains valid for an air flow above the water surface. Due to waves, however, the roughness is not constant but may depend on v_* and g (gravity acceleration), i.e., $z_0 = z_0(g, v_*)$. Dimensional arguments lead to Charnok's formula

$$z_0 = \frac{m v_*^2}{g}, \qquad m = 0.02\text{–}0.08,$$

where m is an empirical constant.

Very often the dynamical regime in the surface layer, both above land and water, is characterized by the so-called *drag coefficient*

$$C_d = \left[\frac{v_*}{\bar{u}(z)}\right]^2 \tag{2.8}$$

With the log law (2.4) we obtain

$$C_d(z) = \frac{\kappa^2}{(\ln z/z_0)^2}\,. \tag{2.8$'$}$$

So far, the log law (2.4) has obtained undisputable support from innumerable field and laboratory measurements in appropriate conditions consistent with the physical requirements for its validity. It is worth noting, however, that this, to a large extent universal law, is not 'sensitive' to the method of its derivation.

Actually, if as a supplement to the above-described way of obtaining (2.4), we use Kármán's formula for the mixing length

$$l = -\kappa \frac{\mathrm{d}\bar{u}}{\mathrm{d}z}\bigg/\frac{\mathrm{d}}{\mathrm{d}z}\left(\frac{\mathrm{d}\bar{u}}{\mathrm{d}z}\right) \tag{2.9}$$

and replace it in (2.1), we obtain the equation

$$\left(\frac{\mathrm{d}\bar{u}}{\mathrm{d}z}\right)^2\bigg/\frac{\mathrm{d}^2\bar{u}}{\mathrm{d}z^2} = -\frac{v_*}{\kappa} \tag{2.10}$$

whose solution is (2.4). On the other hand, (2.9) is a necessary condition for self-similarity of the mean wind velocity [48]

$$\frac{\bar{u}(z) - \bar{u}(z_0)}{\bar{u}(z_0 + l) - \bar{u}(z_0)} = f\left(\frac{z - z_0}{l}\right). \tag{2.11}$$

Other ways for deriving (2.4) are discussed in [48].

Having determined explicit expressions for $\bar{u}(z)$, $K(z)$ and $l(z)$, one can also find the vertical profile $b(z)$ or the turbulent kinetic energy in a steady regime when dissipation equals production: $\epsilon = \mathrm{Tr}$. First we substitute (2.4) and (2.5) into (1.23)

$$\mathrm{Tr} = K\left(\frac{d\bar{u}}{dz}\right)^2 = \frac{v_*^3}{\kappa z} \tag{2.12}$$

and make use of formula (3.22) in Chapter 8: $\epsilon = c_1 b^2/K$, $c_1 = \text{const} \sim 1$. Hence,

$$b(z) = \frac{v_*^2}{c_1^{1/2}} = \text{const}_z, \tag{2.13}$$

i.e., b does not depend on z and consequently there is no energy flux through the ground surface $db/dz = 0$ at $z = z_0$.

Let now s be some passive scalar substance in the SL (for instance, specific humidity) with a coefficient of turbulent diffusion $K^s = \alpha_s K$. Analogous to (1.25)

$$\alpha_s K \frac{d\bar{s}}{dz} = -\frac{q_s}{\rho} = -q_s^* = \text{const}_z, \quad z \leqslant h. \tag{2.14}$$

Substituting K from (2.5), we obtain

$$\frac{d\bar{s}}{dz} = -\frac{q_s^*}{\kappa \alpha_s v_* z}.$$

Hence the logarithmic law follows

$$\bar{s}(z) = \bar{s}(z_0) - s_* \ln \frac{z}{z_0}, \tag{2.15}$$

where $s_* = q_s^*/\kappa\alpha_s v_*$.

The question about the temperature profile is more difficult. At neutral stratification $d\bar{\theta}/dz = 0$ and from (1.12) $q = 0$, so that it should be $\bar{T}(z) = \bar{T}(z_0) - \gamma_a(z - z_0)$. If the heat flux $q \neq 0$, then at very small z it is physically acceptable to neglect the influence of the stratification. Consequently, $\alpha_\theta \kappa v_* z \, d\bar{\theta}/dz = -q_*$. Hence, the log law again follows

$$\bar{\theta}(z) = \bar{\theta}(z_0) - \theta_* \ln \frac{z}{z_0}, \tag{2.16}$$

where $\theta_* = q_*/\kappa\alpha_\theta v_*$. This question is discussed in more detail in Section 3.

b. Arbitrary Stratification: Power Model

From a qualitative point of view, the influence of the surface-layer temperature stratification on turbulence is obvious. At other equal conditions the turbulence would be weaker at stable stratification and stronger at unstable. In the first case, the stability 'extinguishes' the so-originated turbulent fluctuations, while, in the second, the instability contributes to their development further on. One can suggest that in this case of a mixed turbulent regime, when dynamical and thermal factors act simultaneously, the log law and the other regularities established above, will not be valid.

On the other hand, the neutral equilibrium in the surface layer is realized for a very short period of time through the day or the year. Actually, in daytime or summer season we usually have $\gamma > \gamma_a$, while through the night or winter season $\gamma < \gamma_a$ including $\gamma < 0$ (temperature inversions), so that $\gamma \approx \gamma_a$, i.e., neutral equilibrium, can be observed only in the evening and morning or spring and autumn.

As far as the neutral stratification is an intermediate case between the inversion and superadiabatic conditions, it can be assumed that the averaged-over-day or year-values of the meteorological elements would satisfy the simple laws established in the previous section. However, this is not the case. The reason for this is that, at stable stratification, the heat flux, which is directed downwards ($q_* < 0$), is small in magnitude due to the small value of the turbulent exchange coefficient (K is a measure for turbulence intensity), while in the opposite case $q_* > 0$ and is large in magnitude because K is large. Hence, the averaged-over-all situations during the day or the year value proves to be positive: $\overline{q}_* > 0$. Therefore, a kind of surface-layer ventilation is realized.

The necessity for a generalization of the theory is obvious. In solving this problem, the turbulent kinetic energy balance equation (1.11), specified for the surface layer, is used. After introducing the fundamental formulae of the semi-empirical theory, it takes the form

$$\frac{\partial b}{\partial t} = K\left(\frac{\mathrm{d}\overline{u}}{\mathrm{d}z}\right)^2 - \alpha_\theta \beta K(\gamma_a - \gamma) + \alpha_b \frac{\partial}{\partial z} K \frac{\partial b}{\partial z} - \epsilon, \tag{2.17}$$

where $\beta = g/\overline{T} \approx g/\overline{\theta}$, $\gamma = -\mathrm{d}\overline{T}/\mathrm{d}z$ and for the dissipation some of the expressions (3.22) in Chapter 8, can be used. Usually, the diffusion term is neglected as being very small. Then (2.17) can be written in the form

$$\frac{\partial b}{\partial t} = \mathrm{Tr}(1 - \mathrm{Rf}) - \epsilon, \quad \mathrm{Tr} = K\left(\frac{\mathrm{d}\overline{u}}{\mathrm{d}z}\right)^2, \tag{2.18}$$

where

$$\mathrm{Rf} = \beta \frac{\overline{w'\theta'}}{\overline{u'w'}\,(\mathrm{d}\overline{u}/\mathrm{d}z)} = \alpha_\theta \mathrm{Ri} \tag{2.19}$$

is called the *flux Richardson number* and

$$\mathrm{Ri} = \beta(\gamma_a - \gamma)\Big/\left(\frac{\mathrm{d}\overline{u}}{\mathrm{d}z}\right)^2 = \beta \frac{\mathrm{d}\overline{\theta}}{\mathrm{d}z}\Big/\left(\frac{\mathrm{d}\overline{u}}{\mathrm{d}z}\right)^2 \tag{2.20}$$

is the *gradient Richardson number* and $\alpha_\theta = K^\theta/K$. If $\alpha_\theta = 1$, then $\mathrm{Rf} \equiv \mathrm{Ri}$.

It follows from (2.18) that the Richardson number is a criterion for turbulence. As a ratio of buoyant production to stress production of turbulent kinetic energy, its sign will depend on the sign of $\overline{w'\theta'}$ since in the SL $\mathrm{d}\overline{u}/\mathrm{d}z > 0$ and $\overline{u'w'} < 0$ always. If the heat flux is upward ($\overline{w'\theta'} > 0$), then $\mathrm{Rf} < 0$. But $\overline{w'\theta'} > 0$ corresponds generally to $\mathrm{d}\overline{\theta}/\mathrm{d}z < 0$. This is the case of unstable atmosphere. In the opposite case of stable stratification $\overline{w'\theta'} < 0$, $\mathrm{d}\overline{\theta}/\mathrm{d}z > 0$ and $\mathrm{Rf} > 0$. If the positive Rf is large enough ($\mathrm{Rf} > 1$), as follows from (2.18), $\partial b/\partial t < 0$ – this leads to the complete suppression of the turbulence. Observations have shown, however, that turbulence cannot be maintained if $\mathrm{Rf} > 0.2$, approximately. Generally, some critical Richardson number $\mathrm{Rf}_{\mathrm{cr}}$ exists, whose value is to be determined more accurately, such that $\partial b/\partial t = 0$ at $\mathrm{Rf} = \mathrm{Rf}_{\mathrm{cr}}$.

In the case of a stationary turbulence regime in the SL $\partial b/\partial t = 0$ and Equation (2.18) with the help of Equation (3.22) in Chapter 8, and (1.22) can be written in the form

$$v_*^4(1 - \text{Rf}) = \frac{K^4}{c^4 l^4}, \quad c = c_1^{-1/4}.$$

Hence,

$$K = v_* cl(1 - \text{Rf})^{1/4}.$$

As we know, at neutral stratification (Rf = 0) $cl = \kappa z$, so that we may properly write $cl = \kappa z \lambda_*(\text{Rf})$. Then we finally obtain that, at an arbitrary stratification,

$$K = \kappa v_* z \lambda_*(\text{Rf})\,(1 - \text{Rf})^{1/4} = \kappa v_* z \lambda(\text{Rf}), \tag{2.21}$$

where $\lambda(\text{Rf})$ would be a dimensionless function – a correction to (2.5) concerning the influence of the stratification on the eddy exchange coefficient. Substituting (2.21) into (2.1), we find a similar correction for the shear:

$$\frac{d\bar{u}}{dz} = \frac{v_*}{\kappa z}\,\Phi(\text{Rf}) \tag{2.22}$$

where $\Phi = \lambda^{-1}$.

The so-introduced functions $\lambda(\text{Rf})$ and $\Phi(\text{Rf})$ can be experimentally or theoretically determined on the basis of additional hypotheses and physical considerations. Since $\text{Rf} = \text{Rf}(z)$, and this dependence is generally unknown, then $K(z)$ and $u(z)$ cannot be obtained in an explicit form while, for the practical applications (for instance, in the theory of turbulent diffusion, Section 5), this is namely the case we need. An elegant and generally-accepted solution to this problem has been given by the similarity theory (Section 3). However, there are empirical models which, for better or for worse, approximate the real profiles of K, $\bar{u}$, $\bar{\theta}$, etc., in the SL and are simple and convenient for application. One of them is Laikhtman's model.

Laikhtman proposed, instead of $l = \kappa z$, another expression for the mixing length at an arbitrary stratification:

$$l(z) = A(\epsilon) z^{1-\epsilon}, \quad A(\epsilon) = \frac{\kappa z_0^{\epsilon}}{(1-\epsilon)} \tag{2.23}$$

where $\epsilon(\text{Ri})$ is a stability parameter, such that

$\epsilon = 0$ at $\text{Ri} = 0$ (neutral stratification),

$\epsilon > 0$ at $\text{Ri} > 0$ (stable stratification),

$\epsilon < 0$ at $\text{Ri} < 0$ (unstable stratification),

which is tabulated as a function of Ri on the basis of observational data. Equation (2.23), introduced in (2.1), yields

$$\bar{u}(z) = \frac{v_*}{\epsilon A}\,(z^{\epsilon} - z_0^{\epsilon}), \tag{2.24}$$

where the boundary condition $\bar{u}(z_0) = 0$ has been used. To avoid the uncertainty related to the factor $A(\epsilon)$, one can use (2.24) in the form

$$\bar{u}(z) = \bar{u}_1 \frac{z^\epsilon - z_0^\epsilon}{z_1^\epsilon - z_0^\epsilon}, \tag{2.25}$$

where $\bar{u}_1 = \bar{u}(z_1)$ and $z_1 < h$ can be, for instance, the height at which the anemometer is being settled.

From (2.1) and (2.23) it also follows that

$$K(z) = A v_* z^{1-\epsilon} = K_1 \left(\frac{z}{z_1}\right)^{1-\epsilon}, \tag{2.26}$$

i.e., at $\epsilon < 0$ (> 0) K increases faster (slower) than linearly ($\epsilon = 0$). The results from the observations of the wind profile in the SL confirm the good approximative qualities of the model (2.25). Besides, it appears that $-\frac{1}{2} \leqslant \epsilon \leqslant \frac{1}{2}$. The similarity theory, however (see Section 3) specifies the left inequality: $\epsilon \geqslant -\frac{1}{3}$. It can be easily proved that at $\epsilon \to 0$ log law (2.4) follows from (2.25).

An expression of type (2.25) can also be obtained for the temperature profile too

$$\frac{\bar{\theta} - \bar{\theta}_1}{\bar{\theta}_2 - \bar{\theta}_1} = \frac{z^\epsilon - z_1^\epsilon}{z_2^\epsilon - z_1^\epsilon}, \qquad z_1, z_2 < h.$$

Analogous to Laikhtman's model, Deacon has proposed

$$\frac{d\bar{u}}{dz} = \frac{v_*}{\kappa z_0} \left(\frac{z}{z_0}\right)^{-\beta} \tag{2.27}$$

where $\beta = \beta(\mathrm{Ri})$, $\beta(0) = 1$. If (2.27) is integrated from $z = z_0$ (where $\bar{u}(z_0) = 0$) to z we obtain

$$\bar{u}(z) = \frac{v_*}{\kappa(1-\beta)} \left[\left(\frac{z}{z_0}\right)^{1-\beta} - 1\right]. \tag{2.28}$$

The Laikhtman and Deacon formulae represent binomial expressions. Very often, however, in theoretical investigations (for instance, in turbulent diffusion theory), $\bar{u}(z)$ enters as a coefficient into the differential equation. Then such a form of $\bar{u}(z)$ appears to be useless for carrying out the solution in an analytical form. That is why in such cases $\bar{u}(z)$ is approximated to the less exact, but simpler power expressions

$$\bar{u}(z) = \bar{u}_1 \left(\frac{z}{z_1}\right)^n, \tag{2.29}$$

$$K(z) = K_1 \left(\frac{z}{z_1}\right)^{1-n}, \tag{2.30}$$

where $0 < n < 1$.

The power law (2.29) efficiently describes the wind profile $\bar{u}(z)$ only under stable conditions at $n = \frac{1}{4}$ and under neutral conditions at $n = \frac{1}{7}$ ($\ln z$ is well approximated by $z^{1/7}$). In the case of unstable stratification, Equation (2.29) is a poor approximation to the real profile $\bar{u}(z)$.

There are many other empirical models for the wind profile $\bar{u}(z)$ in the surface layer.

3. THE SIMILARITY THEORY FOR THE STRUCTURE OF THE SURFACE LAYER

a. Fundamental Suppositions and Formulae

The power-function empirical models discussed in Section 2 more or less approximate satisfactorily the vertical profiles of the meteorological elements and, particularly, the wind in the surface layer at arbitrary stratification but, on the other hand, they are lacking in physical content and do not explain the stratification influence. The most general theory, free of the above-mentioned shortcomings, was proposed by the Soviet researchers Monin and Obukhov about 30 years ago. This theory gained universal recognition and firm experimental confirmation under field and laboratory conditions and became a basis for all future works in surface-layer physics and dynamics.

The following hypothesis lies at the basis of Monin–Obukhov theory: At developed turbulence in the SL, the influence of the molecular diffusion and viscosity is negligible and at heights considerably exceeding the roughness scale z_0 of the underlying surface, its feature do not directly influence the vertical profiles $\bar{u}(z)$, $\bar{\theta}(z)$, etc., nor other statistical characteristics ($\overline{u'^2}$, $\overline{v'^2}$, . . .). The only dimensional physical parameters they can depend on are dynamic velocity v_*, heat flux q_* and convection parameter $\beta = g/\bar{T} \approx g/\bar{\theta}$. All of them do not depend on z if $z < h$. From the three governing parameters (v_*, q_*, β) one can form the scales of velocity, length and temperature as follows:

$$v_*,\ L = -\frac{v_*^3}{\kappa\beta q_*}, \qquad \theta_* = -\frac{q_*}{\kappa v_*}, \tag{3.1}$$

where $\kappa \approx 0.4$ is Kármán's constant. Since $v_* > 0$, then the sign of q_* will determine the signs of L and θ_*:

$$\begin{array}{ll} \text{unstable stratification} & q_* > 0,\ L < 0,\ \theta_* < 0, \\ \text{stable stratification} & q_* < 0,\ L > 0,\ \theta_* > 0, \\ \text{neutral stratification} & q_* \to 0,\ |L| \to \infty,\ |\theta_*| \to 0. \end{array} \tag{3.2}$$

Consequently, the scales L (Monin–Obukhov length) and θ_* can serve as characteristics of the surface-layer stratification. Thus, we come to the fundamental part of the Monin–Obukhov hypothesis: the assumption of self-similarity of the vertical profiles of $\bar{u}$ and $\bar{\theta}$:

$$\frac{d(\bar{u}/v_*)}{d(z/L)} = g\left(\frac{z}{L}\right), \qquad \frac{d(\bar{\theta}/\theta_*)}{d(z/L)} = g_\theta\left(\frac{z}{L}\right). \tag{3.3}$$

It is more convenient to rewrite these expressions in the form

$$\frac{\kappa z}{v_*}\frac{d\bar{u}}{dz} = \varphi\left(\frac{z}{L}\right), \qquad \frac{z}{\theta_*}\frac{d\bar{\theta}}{dz} = \varphi_\theta\left(\frac{z}{L}\right), \tag{3.4}$$

where $\varphi(\zeta)$ and $\varphi_\theta(\zeta)$ are dimensionless profiles – universal functions of the dimensionless variable

$$\zeta = \frac{z}{L}. \tag{3.5}$$

Evidently, through (3.2) ζ can also serve as a stability parameter.

A combination of (3.4) with the fundamental formulae of semiempirical theory

$$\frac{d\bar{u}}{dz} = \frac{v_*^2}{K}, \qquad \frac{d\bar{\theta}}{dz} = -\frac{q_*}{K^\theta}$$

yields

$$K = \frac{\kappa v_* z}{\varphi(\zeta)}, \qquad K^\theta = \frac{\kappa v_* z}{\varphi_\theta(\zeta)} \tag{3.6}$$

i.e.,

$$\alpha_\theta(\zeta) = \frac{K^\theta}{K} = \frac{\varphi(\zeta)}{\varphi_\theta(\zeta)}, \tag{3.7}$$

and $\varphi(\zeta) \equiv \varphi_\theta(\zeta)$ at the approximation $\alpha_\theta = 1$. On the other hand,

$$\mathrm{Ri} = \beta \frac{d\bar{\theta}}{dz} \bigg/ \left(\frac{d\bar{u}}{dz}\right)^2 = \zeta \varphi_\theta \varphi^{-2} = \frac{\zeta}{\alpha_\theta \varphi(\zeta)}, \tag{3.8}$$

i.e.,

$$\mathrm{Rf} = \alpha_\theta \mathrm{Ri} = \frac{\zeta}{\varphi(\zeta)}. \tag{3.9}$$

If Equations (3.4) are integrated, we obtain

$$\begin{aligned} \bar{u}(z) - \bar{u}(z_1) &= \frac{v_*}{\kappa} [f(\zeta) - f(\zeta_1)], \\ \bar{\theta}(z) - \bar{\theta}(z_1) &= \theta_* [f_\theta(\zeta) - f_\theta(\zeta_1)], \end{aligned} \tag{3.10}$$

where

$$\varphi, \varphi_\theta = \zeta \cdot \left(\frac{df}{d\zeta}, \frac{df_\theta}{dz}\right), \qquad z_0 \ll z, z_1 < h,$$

h is the thickness of the SL. As it was with the log law (2.4), however, the boundary condition could be set on the level z_0, i.e., $z_1 \equiv z_0$ where $\bar{u}(z_0) = 0$. Then from (3.10) it follows that

$$\bar{u}(z) = \frac{v_*}{\kappa} [f(\zeta) - f(\zeta_0)]. \tag{3.11}$$

b. Asymptotic Cases

The similarity and dimensional arguments have enabled us to write down formulae (3.3)–(3.11), but it is not possible to find the explicit form of the universal functions $\varphi(\zeta)$ and $\varphi_\theta(\zeta)$ for an arbitrary ζ (arbitrary stratification) only on such a basis. However, their asymptotic behaviour can be studied both in cases of pure mechanical and pure thermal turbulence (free convection) without additional assumptions.

(i) Suppose that $|\zeta| \to 0$. We can reach this in two ways: (a) by letting $|L| \to \infty$ ($q_* \to 0$) at fixed z, i.e., by approaching neutral stratification. Thus $\varphi(0) = \varphi_\theta(0) = 1$; (b) by letting $z \to 0$ at fixed $|L|$, i.e., fixed $q_* \neq 0$. From a dynamical point of view, this case is equivalent to the first one. For any q_*, close enough to the surface (but keeping $z \gg z_0$), the stability is effectively neutral and the logarithmic law holds. This conclusion is a direct consequence of similarity hypothesis (3.3) and explains the experimentally-observed validity of the log law close to the surface under arbitrary stratification. This fact increases the universality of this law. Consequently, the Monin–Obukhov scale L can be interpreted as determining the thickness of the sublayer of dynamical turbulence since, at $z \ll |L|$, the thermal factor could be ignored. The observations show that this can be done at $z < 0.05\,|L|$. Besides, at $|\zeta| \to 0$ (3.9) yields $\mathrm{Rf} \approx \zeta$.

At small but finite values of $|\zeta|$ (stratification close to the neutral) $\varphi(\zeta)$ can be presented as

$$\varphi(\zeta) = 1 + \beta_1\zeta + \beta_2\zeta^2 + \cdots \tag{3.12}$$

so that

$$\begin{aligned} f(\zeta) &= \text{const} + \ln|\zeta| + \beta_1\zeta + \cdots, \\ \mathrm{Rf}(\zeta) &= \zeta - \beta_1\zeta^2 - \cdots \end{aligned} \tag{3.13}$$

The constants $\beta_1, \beta_2, \ldots$ have to be determined empirically. Usually only the linear term is retained in (3.12). Then, for $\bar{u}(z)$ and $K(z)$ one obtains

$$\bar{u}(z) = \frac{v_*}{\kappa}\left(\ln\frac{z}{z_0} + \beta_1\frac{z - z_0}{L}\right), \tag{3.14}$$

$$K(z) = \frac{\kappa v_* z}{1 + \beta_1 (z/L)}. \tag{3.15}$$

Formula (3.14) is known as a log-linear law for the wind profile. As to the constant β_1, it is positive ($\beta_1 > 0$) which follows from elementary considerations of the influence of stability on $\bar{u}(z)$. Thus, at $L < 0$ (instability), due to the intensive vertical mixing, $\bar{u}(z)$ must increase slower than $\ln z$, which is possible if $\beta_1 > 0$. The opposite case yields the same conclusion. Though the experimental data of various authors show considerable scatter, $\beta_1 \approx 5$ can be recommended.

(ii) Let us now suppose that $\zeta \to -\infty$. We can reach this in two ways: Firstly, by letting $q_* \to +\infty$ and $v_* \to 0$ at fixed z, i.e., by approaching the free convection-like state under very unstable conditions – a windless state where the turbulent motion is only caused by the buoyant force due to strong surface heating. Such an extreme state never exists in the atmosphere since we always have some wind and, hence, surface stress $v_* \neq 0$. However, it could be thought of as a pure asymptotic case. Thus, the eddy exchange coefficient $K = v_*^2/\mathrm{d}\bar{u}/\mathrm{d}z$ is nonsensical in this case, but $K^\theta = q_*/\mathrm{d}\bar{\theta}/\mathrm{d}z$ does exist. Since $K^\theta = \psi(\beta, q_*, z)$ dimensional analysis yields

$$K^\theta = \frac{3}{c}(q_*\beta)^{1/3} z^{4/3}, \tag{3.16}$$

where the coefficient $3/c$ is chosen for convenience. Then

$$\frac{\mathrm{d}\bar{\theta}}{\mathrm{d}z} = -\frac{c}{3}\, q_*^{2/3}\beta^{-1/3} z^{-4/3}.$$

Hence,

$$\bar{\theta}(z) \approx \bar{T}(z) = \bar{T}_\infty + c q_*^{2/3}\beta^{-1/3} z^{-1/3}. \tag{3.17}$$

A similar analysis concerning the velocity and temperature variances yields

$$\sigma_u^2 \sim \sigma_v^2 \sim \sigma_w^2 \sim (q_*\beta z)^{2/3},$$

$$\sigma_\theta^2 \sim q_*^{4/3}(\beta z)^{-2/3},$$

where $\sigma_f^2 = \overline{f'^2}$.

The second way is by letting z increase (but still keeping $z \leqslant h$) at fixed $L < 0$ ($q_* > 0$ and $v_* \neq 0$). From a dynamical point of view, this mixed regime of turbulence, as the similarity theory predicts, is equivalent to the local free convection; v_*, though nonzero, should no longer be important and the governing parameters remain the same as in the previous case: β, q_*, z. Then from (3.16) and the relation $\alpha_\theta K = K^\theta$ we find

$$K(z) = \frac{3}{c'}(q_*\beta)^{1/3} z^{4/3}. \tag{3.16'}$$

The equation $d\overline{u}/dz = v_*^2/K$, after integration, yields

$$\overline{u}(z) = \overline{u}(z_1) - c' v_*^2 (\beta q_*)^{-1/3} (z^{-1/3} - z_1^{-1/3}), \tag{3.18}$$

where $z, z_1 \gg |L|$, $c' = c\alpha_\theta(-\infty)$. Thus, at $\zeta \to -\infty$, the following asymptotes are found

$$\varphi(\zeta) = -\frac{c'}{3}\kappa^{4/3}\zeta^{-1/3}, \qquad f(\zeta) = \text{const} + c'\kappa^{4/3}\zeta^{-1/3}, \tag{3.19}$$

$$\text{Rf}(\zeta) = -\left(\frac{3}{c'\kappa^{4/3}}\right)|\zeta|^{4/3}. \tag{3.20}$$

Let us summarize the predictions of the unstable stratification theory. Suppose we have a well-developed surface layer with a large enough thickness h, so that $|L| < h$ holds. Then, as was mentioned above, at $z_0 \ll z \ll |L|$ a sublayer of dynamical turbulence exists, while at $|L| \ll z \leqslant h$ the corresponding sublayer is of thermal turbulence. When $q_* \to 0$, sublayer No. 1 grows and may occupy the whole surface layer. Similarly, when $v_* \to 0$, sublayer No. 2 presses the lower sublayers down and occupies the whole surface layer. Intermediate sublayer No. 3 has a real mixed regime. This structure of the surface layer is shown in Figure 9.2.

A third asymptotic regime could occur at $\zeta \gg 1$, i.e., $L > 0$ ($q_* < 0$) – very stable stratification and $z \gg L$. For example, at a strong inversion in the surface layer (an almost windless state), the turbulence degenerates, becomes local and intermittent. Some characteristics behave as if z is no longer important, for instance $\text{Rf} \to R = \text{const}_z$, $K \to K_\infty = \text{const}_z$. Then

$$\varphi(\zeta) = \frac{\zeta}{\text{Rf}} \to \frac{\zeta}{R}, \qquad f(\zeta) = \text{const} + \frac{\zeta}{R}, \tag{3.21}$$

so that

$$\overline{u}(z) = \frac{v_* z}{\kappa R L} + \text{const}, \qquad \overline{\theta}(z) = \frac{\theta_* z}{RL} + \text{const}.$$

The last conclusions are, however, uncertain because at a very strong stability in the surface layer, a considerable role may also be played by the radiative factor which is not included in the similarity theory.

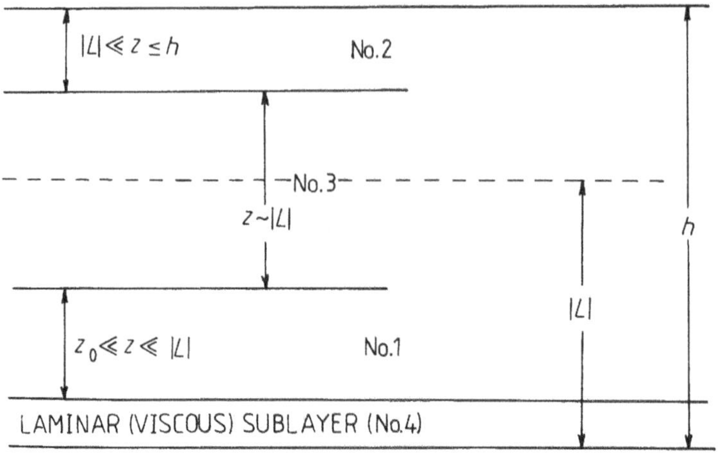

Fig. 9.2. Structure of a developed surface layer as predicted by the similarity theory at unstable stratification.

c. Universal Functions

The similarity theory leads to interesting predictions and asymptotic expressions for universal functions, but it gives no opportunity for finding their explicit form for arbitrary ζ. This may be done experimentally or theoretically, but on the basis of additional assumptions. A third way also exists – constructing formal interpolation formulae for the intermediate values of $z \sim |L|$ (sublayer No. 3 in Figure 9.2) on the basis of the theoretical asymptotes for sublayers Nos. 1 and 2. Qualitatively, the behaviour of $\varphi(\zeta)$ and $f(\zeta)$ is presented in Figure 9.3; $\varphi_\theta(\zeta)$ and $f_\theta(\zeta)$ behave similarly.

The empirical formulae constructed from experimental data collected in the State of Kansas (USA) by Businger *et al.* [25] have recently been in regular use in the applications:

$$\varphi(\zeta) = (1 - 15\zeta)^{-1/4}, \quad \zeta < 0, \qquad \varphi(\zeta) = 1 + 4.7\zeta, \quad \zeta > 0 \tag{3.23}$$

$$\varphi_\theta(\zeta) = 0.74\,(1 - 9\zeta)^{-1/2}, \quad \zeta < 0, \qquad \varphi_\theta(\zeta) = 0.74 + 4.7\zeta, \quad \zeta > 0 \tag{3.24}$$

where, through (3.7), $\varphi_\theta(0) = 0.74$ means that $\alpha_\theta(0) \approx 1.35$.

Having $\varphi(\zeta)$ and $\varphi_\theta(\zeta)$ specified and with the help of (3.7)–(3.9), it is not difficult to find the explicit expressions for $\alpha_\theta(\zeta)$, $\mathrm{Ri}(\zeta)$ and $\mathrm{Rf}(\zeta)$, necessary for a number of practical applications. Moreover, if we integrate Equations (3.4) – and this proves to be possible in the elementary functions for $\varphi(\zeta)$ and $\varphi_\theta(\zeta)$ given by (3.23) and (3.24) – $\bar{u}(z)$ and $\bar{\theta}(z)$ will be obtained for arbitrary ζ.

It may be noted that at large negative $\zeta(\zeta \to -\infty)$, (3.23) and (3.24) yield $\varphi \sim \zeta^{-1/4}$ and $\varphi_\theta \sim \zeta^{-1/2}$, which is not in agreement with (3.19). Other experimental data have also shown such a contradiction. The reason for this is not yet clear.

From the theoretical model formulae proposed for $\varphi(\zeta)$ we shall mention only the one which appears to be a solution of the algebraic equation

$$\varphi^4 - \sigma\zeta\varphi^3 - 1 = 0, \quad \zeta < 0, \tag{3.25}$$

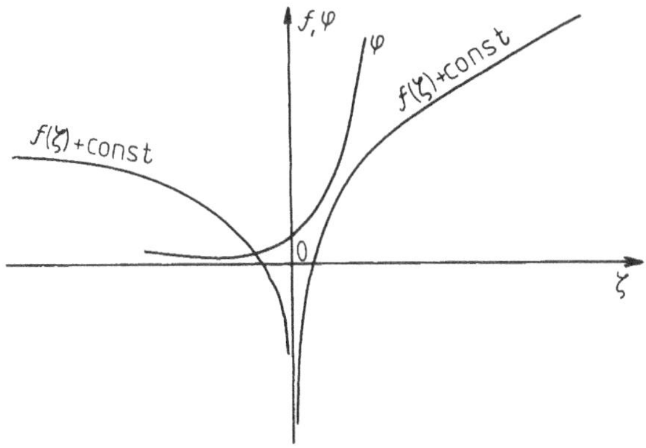

Fig. 9.3. General behaviour of the universal functions $\varphi(\zeta)$ and $f(\zeta)$ (after [48]).

which, by itself, follows from the turbulent energy balance equation and some additional suppositions; σ is a numerical constant. In the formal approach to the problem, intermediate sublayer No. 3 (Figure 9.2) could be 'shared' between Nos. 1 and 2. For instance

$$K(z) \sim \begin{cases} z, & z \leqslant |L| \\ z^{4/3}, & z \geqslant |L| \end{cases} \tag{3.26}$$

which implies $\varphi(\zeta) = 1$ at $|\zeta| \leqslant 1$ and $\varphi(\zeta) \sim \zeta^{-1/3}$ at $|\zeta| \geqslant 1$.

More details can be found in the literature [41, 48, 65, 72], etc.

4. MICROSTRUCTURE OF ATMOSPHERIC TURBULENCE IN THE SURFACE LAYER

a. Spatial Microstructure

Atmospheric turbulence (particularly in the surface layer) has turned out to be very suitable for the examination and application of nearly all the results of the semiempirical and statistical theory of turbulence. Moreover, a number of theoretical problems have been set and settled in connection with their application to micrometeorology (SL physics). The reason for this is the fact that in the atmosphere as a whole, and even in the SL, the spatial and time scales of turbulence are considerably greater than those in aerodynamic wind tunnels. That is why the values of the Reynolds number (Re) are very large, which usually ensures the existence of an inertial subrange in the spectrum, i.e, the implementation of the '$\frac{2}{3}$ and $-\frac{5}{3}$ laws' (Equations (4.13)–(4.17) in Chapter 8). Here we shall define these laws in relation to the surface layer.

$$D_{ll}(r) = \alpha_r(\epsilon r)^{2/3}, \qquad E(k) = \alpha_k \epsilon^{2/3} k^{-5/3}, \tag{4.1}$$

$$D_{\theta\theta}(r) = \beta_r \epsilon_\theta \epsilon^{-1/3} r^{2/3}, \qquad E_{\theta\theta}(k) = \beta_k \epsilon_\theta \epsilon^{-1/3} k^{-5/3}. \tag{4.2}$$

The variable quantities here are the dissipation rates ϵ and ϵ_θ. At a stationary regime ($\partial b/\partial t = 0$) and a neutral stratification (Rf = 0), as follows from (2.12),

$$\epsilon = \mathrm{Tr} = K(z)\left(\frac{d\bar{u}}{dz}\right)^2 = \frac{v_*^3}{\kappa z}. \tag{4.3}$$

A similar expression can be derived for ϵ_θ (see Problem 5)

$$\epsilon_\theta = \alpha_\theta K(z)\left(\frac{d\bar{\theta}}{dz}\right)^2 = \frac{\alpha_\theta \kappa^2 v_* \theta_*^2}{\kappa z}, \tag{4.4}$$

where (3.4) has been used. Substituting (4.3) and (4.4) into (4.1) and (4.2), we obtain

$$\begin{aligned} D_{ll}(r) &= \alpha_r v_*^2 \left(\frac{r}{l}\right)^{2/3}, \\ D_{\theta\theta}(r) &= \beta_r \alpha_\theta (\kappa\theta_*)^2 \left(\frac{r}{l}\right)^{2/3}, \end{aligned} \tag{4.5}$$

at $r \ll l = \kappa z$. Consequently, it is enough to know v_* and θ_* (and, of course, the universal constants α_r, β_r) in order to calculate the structure functions (4.5). Similar expressions can be obtained for the spectra $E(k)$ and $E_{\theta\theta}(k)$.

The influence of the stratification on the structure characteristics is two fold. On the one hand, due to the Archimedean force, acting in a vertical direction, the equilibrium range (of the local isotropy) and the inertial subrange becomes smaller, and may even completely disappear. On the other hand, formulae (4.3) and (4.4) are no longer valid. They should be replaced by more general ones, based on the similarity theory

$$\epsilon = \frac{v_*^3}{\kappa z}\varphi_\epsilon\left(\frac{z}{L}\right), \qquad \epsilon_\theta = \frac{\kappa^2 v_* \theta_*^2}{\kappa z}\varphi_\epsilon^\theta\left(\frac{z}{L}\right) \tag{4.6}$$

where $\varphi_\epsilon(\zeta)$ and $\varphi_\epsilon^\theta(\zeta)$ are new universal functions which have to be determined. For this purpose, the kinetic energy balance equation will be used. For instance from (2.18) at $\partial b/\partial t = 0$ we have

$$\epsilon = K(1 - \mathrm{Rf})\left(\frac{d\bar{u}}{dz}\right)^2.$$

Substituting here K, Rf and $d\bar{u}/dz$ from (3.6), (3.9) and (3.4), we obtain

$$\epsilon = \frac{v_*^3}{\kappa z}\,[\varphi(\zeta) - \zeta],$$

i.e.,

$$\varphi_\epsilon(\zeta) = \varphi(\zeta) - \zeta = \zeta\left(\frac{df}{d\zeta} - 1\right), \tag{4.7}$$

where $\varphi(\zeta)$ is considered as a known function (see Section 3).

Much more difficult and so far unsatisfactorily resolved is the problem concerning the influence of stratification on the form of the spectra and structure functions. Physically, it is clear that with a stable stratification, the kinetic energy from the turbulent flow can be 'taken away' and transformed into potential energy, while with unstable conditions, it is quite the reverse, the kinetic energy is 'injected' at the expense of the potential energy of instability. Consequently, in that region of the spectrum, in which, at neutral stratification, only the cascade inertial transport of energy from larger to smaller scales (smaller to larger wave numbers k) is realized, there would be a divergence from this regime under stable or unstable conditions. From a dimensional analysis point of view, this means that a parameter, characterizing the Archimedean force, should be added to the governing parameters ϵ and ϵ_θ. Two alternative possibilities exist, namely

$$\beta = \frac{g}{\overline{\theta}} \approx \frac{g}{\overline{T}} \quad \text{or} \quad \omega_{\mathrm{BV}}^2 = \frac{g}{\overline{\theta}} \frac{\mathrm{d}\overline{\theta}}{\mathrm{d}z} = \beta(\gamma_a - \gamma) = N^2 \tag{4.8}$$

where ω_{BV} is the Brunt–Väisälä frequency, here traditionally denoted by N. The appearance of a new parameter (β or N), however, breaks the 'automodelity' of the regime and permits the construction of a new length scale.

Actually, from each group of parameters (ϵ, ϵ_θ, β) or (ϵ, ϵ_θ, N) one can construct only one length scale

$$L_\beta = \epsilon^{5/4} \epsilon_\theta^{-3/4} \beta^{-3/2} \quad \text{or} \quad L_N = \epsilon^{1/2} N^{-3/2} \tag{4.9}$$

proposed by Obukhov (1959) and Ozmidov [51]. Then, instead of (4.1) and (4.2), we can write

$$D_{ll}(r) = \alpha_r (\epsilon r)^{2/3} \, \mathrm{d}\left(\frac{r}{L_i}\right), \qquad E(k) = \alpha_k \epsilon^{2/3} k^{-5/3} e(kL_i), \tag{4.10}$$

$$D_{\theta\theta}(r) = \beta_r \epsilon_\theta \epsilon^{-1/3} r^{2/3} \mathrm{d}_\theta\left(\frac{r}{L_i}\right), \qquad E_{\theta\theta}(k) = \beta_k \epsilon_\theta \epsilon^{-1/3} k^{-5/3} e_\theta(kL_i), \tag{4.11}$$

where $L_i = L_\beta$ or L_N, while d, d_θ, e, e_θ are dimensionless universal functions to be determined experimentally or theoretically but on the basis of additional assumptions. For instance, Bolgiano (1959) set forth a hypothesis assuming that at very strong stability, the loss of kinetic energy for overcoming static stability is very great, so that it transforms into potential energy long before it reaches the dissipation interval. Consequently, the dissipation parameter ϵ becomes unimportant and should be dropped. Then (4.10) and (4.11) yield different predictions, correspondingly

$$E(k) \sim \epsilon_\theta^{2/5} \beta^{4/5} k^{-11/5}, \qquad E_{\theta\theta}(k) \sim \epsilon_\theta^{4/5} \beta^{-2/5} k^{-7/5}, \tag{4.12}$$

$$E(k) \sim N^2 k^{-3}, \qquad E_{\theta\theta}(k) \sim \epsilon_\theta N^{-1} k^{-1}. \tag{4.13}$$

In a log-log plot ($\ln E$, $\ln k$) power dependences of type k^{-n} are expressed by a straight line with a slope equal to the power ($-n$). Thus, as follows from (4.12) and (4.13), the strong stability makes the slope of the energy spectrum steeper (Figure 9.4) and that of the temperature spectrum less steep compared with the '$-\frac{5}{3}$ slope'. The spectral subrange, where (4.12) and (4.13) are valid, is called the *buoyant subrange*. It is located on the left-hand side of the inertial subrange which becomes narrower in this case, or may even totally disappear. Unfortunately, so far no experimental data exists which agrees with (4.12) or (4.13). Analogous expressions can also be obtained for the structure functions.

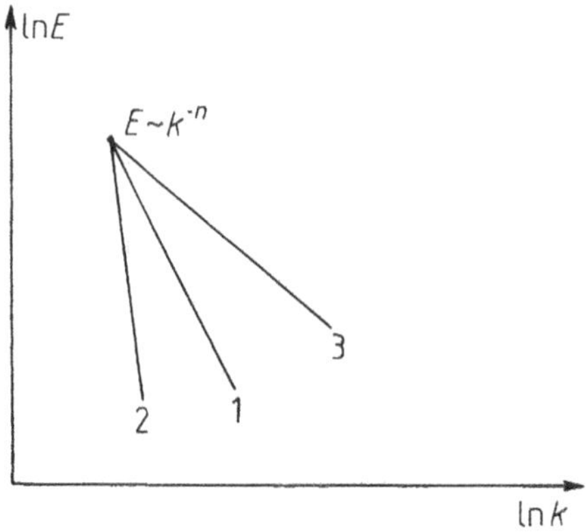

Fig. 9.4. Influence of the stratification on the slope of the energy spectrum: (1) $n = \frac{5}{3}$ – neutral; (2) $n > \frac{5}{3}$ – stable; $n < \frac{5}{3}$ – unstable.

b. Time Microstructure

For experimental determination of the structure function $D_{ff}(r)$, one needs synchronous (in time) measurements of the fluctuations f' at two points at distance r. However, it is easier and more reliable to measure $f'(t)$ at a fixed point and determine $D_{ff}(\tau)$. The same also applies to the spectral characteristics. On the other hand, the theory of the locally-isotropic turbulence predicts the form of the spatial microstructure characteristics $D_{ff}(r)$ and $E_{ff}(k)$ but not of the time ones. A question arises whether it would be possible, as is the case with wave motions on the basis of a dispersion relation, to relate the spatial with the time structural characteristics of the turbulent flow. The proper answer to this question is negative [48]. Approximately, however, but with sufficient accuracy, such a transition from spatial to time characteristics and vice-versa is realized by means of Taylor's hypothesis of 'frozen turbulence'.

At $|\mathbf{v}'| \ll |\bar{\mathbf{v}}|$ (which is fulfilled in the SL) the turbulent fluctuations $f'(\mathbf{x}, t)$ in a fixed point $\mathbf{x}$ for a rather small interval of time τ are observed as a result of the transport of turbulence disturbances (eddies), situated upward in the flow at a distance $r = \bar{u}\tau$, at the mean velocity of the flow and without any changes (as 'frozen'). Hence, the name of the hypothesis.

Two shortcomings of this hypothesis are obvious: the transport is considered not at the mean velocity but at the instantaneous one and, second, with changes during the transportation. However, physical intuition tells us that at large $\bar{u}$ and small τ, this hypothesis would be exact enough. This offers an opportunity to write down the relationship

$$\widetilde{D}_{ff}(\tau) = D_{ff}(\bar{u}\tau), \qquad r = \bar{u}\tau, \tag{4.14}$$

where time structure function is denoted by ~ in order to be distinguished from the spatial one. Thus, for instance, from (4.1) and (4.14) we have

$$\begin{aligned} \widetilde{D}_{ll}(\tau) &= \alpha_r(\epsilon\bar{u})^{2/3}\tau^{2/3} = a\tau^{2/3}, \\ \widetilde{D}_{\theta\theta}(\tau) &= \beta_r\epsilon_\theta\epsilon^{-1/3}(\bar{u}\tau)^{2/3} = a_\theta\tau^{2/3}. \end{aligned} \tag{4.15}$$

Similarly, for the spectra

$$\tilde{E}(k) = \left(\frac{1}{\bar{u}}\right) E\left(\frac{\omega}{\bar{u}}\right), \quad \frac{\omega}{\bar{u}} = k. \tag{4.16}$$

Evidently, $k = \omega/\bar{u}$ plays the role of a dispersion relation. Hence, it follows that

$$\tilde{E}(\omega) \sim \omega^{-5/3} \quad \text{when } E(k) \sim k^{-5/3}.$$

Taylor's hypothesis for 'frozen turbulence' occupies an extremely important place in turbulence investigations because it enables the theory to 'meet with' the experiment. Time statistical characteristics (frequency spectra and structure functions) are determined, then transformed into spatial ones and compared with the theory. In this way the most convincing experimental proofs can be obtained concerning the '$\frac{2}{3}$ and $-\frac{5}{3}$ laws' in the surface layer, as well as in many other natural (the ocean) and laboratory (the wind tunnel) turbulent flows.

c. Practical Applications

The theoretical and experimental results related to the statistical microstructure of meteorological elements have not only a cognitive significance but find application in solving many practical problems too. One problem relates to the inertia of meteorological devices (instruments). Let us, for simplicity, consider 'linear' devices with the equation of inertia

$$\frac{\mathrm{d}F}{\mathrm{d}t} + \frac{F}{T} = \frac{f}{T} \tag{4.17}$$

where $f(t)$ is the real quantity to be measured, $F(t)$ is what the device shows, T is the inertial constant – a characteristic of the device (anemometer, thermometer, etc.). If $f(t)$ and $F(t)$ are stationary random functions, then (4.17) yields $\bar{f} = \bar{F}$, i.e., the inertia does not influence the averaged value. The same equation (4.17) is valid for the fluctuations $f'(t)$ and $F'(t)$. It can be directly verified that its solution at $t \gg T$ is

$$F'(t) = \frac{1}{T}\int_0^\infty \exp\left[\frac{-\tau'}{T}\right] f'(t - \tau')\,\mathrm{d}\tau'. \tag{4.18}$$

Squaring, averaging and integrating once we obtain

$$\sigma_F^2 = \sigma_f^2 - \frac{1}{2T}\int_0^\infty \exp\left[\frac{-\tau}{T}\right] D_{ff}(\tau)\,\mathrm{d}\tau = \sigma_f^2 - \Delta^2. \tag{4.19}$$

Obviously, Δ^2 is a systematic error of the variance due to the inertia, so that its true value $\sigma_f^2 = \sigma_F^2 + \Delta^2$. To calculate Δ^2, we need to know the structure function $D_{ff}(\tau)$. If $f(t)$ is a temperature or wind velocity, we may use (4.15). Then

$$\Delta^2 = \frac{a_f}{2T}\int_0^\infty \tau^{2/3} \exp \frac{-\tau}{T}\,\mathrm{d}\tau = \frac{1}{2}\Gamma\left(\frac{5}{3}\right) a_f T^{2/3} = 0.45\, D_{ff}(T),$$

where $\Gamma(x)$ is the gamma-function.

Another example of an important application is the investigation of sound-, light- and radio-wave propagation in a turbulent atmosphere. The characteristics of the waves (phases, amplitudes) also fluctuate. The structural characteristics of these fluctuations depend entirely on the corresponding characteristics of the atmosphere and, particularly, of its surface layer.

5. TURBULENT DIFFUSION OF ADMIXTURES IN THE SURFACE LAYER

a. Semiempirical Equation of Diffusion

Let $s(\mathbf{x}, t)$ be a specific concentration of some conservative and passive admixture in a turbulent flow, satisfying the equation

$$\frac{\partial s}{\partial t} + \frac{\partial v_i s}{\partial x_i} = \kappa_s \nabla^2 s, \tag{5.1}$$

where κ_s is the coefficient of molecular diffusion. Letting $s = S + s'$, $v_i = \overline{v}_i + v_i'$, we easily derive the Reynolds-like equation for the averaged concentration S:

$$\frac{\partial S}{\partial t} + \frac{\partial \overline{v}_i S}{\partial x_i} = \frac{\partial}{\partial x_i}\left(\kappa_s \frac{\partial S}{\partial x_i} - \overline{v_i' s'}\right). \tag{5.2}$$

As is usual with this kind of equation, the dominant term on the right-hand side of Equation (5.2) is the second one, so that, without loss of accuracy, we can write

$$\frac{\partial S}{\partial t} + \frac{\partial \overline{v}_i S}{\partial x_i} = -\frac{\partial Q_i}{\partial x_i}, \tag{5.3}$$

where $Q_i = \overline{v_i' s'}$. Further, we introduce the eddy diffusion coefficient $K_{ij} = K_{ij}(\mathbf{x}, t)$ by letting

$$Q_i = -K_{ij} \frac{\partial S}{\partial x_j}. \tag{5.4}$$

Then

$$\frac{\partial S}{\partial t} + \overline{v}_i \frac{\partial S}{\partial x_i} = \frac{\partial}{\partial x_i} K_{ij} \frac{\partial S}{\partial x_j} \tag{5.5}$$

represents the so-called semiempirical equation of turbulent diffusion. It is rather complicated and difficult for handling but it can be significantly simplified through assumptions concerning $\overline{v}_i$ and K_{ij}. In the surface layer, for instance,

$$\overline{v}_1 = \overline{u}(z), \qquad \overline{v}_2 = \overline{v}_3 = 0, \qquad K_{ij} = K_{ij}(z). \tag{5.6}$$

As to the tensor K_{ij}, we assume that its main axes coincide with the coordinate axes $0x$, $0y$, $0z$, so that

$$K_{ij} = \begin{pmatrix} K_x & 0 & 0 \\ 0 & K_y & 0 \\ 0 & 0 & K_z \end{pmatrix}. \tag{5.7}$$

Thus, for the surface layer, Equation (5.5) becomes

$$\frac{\partial S}{\partial t} + \bar{u}(z)\frac{\partial S}{\partial x} = K_x \frac{\partial^2 S}{\partial x^2} + K_y \frac{\partial^2 S}{\partial y^2} + \frac{\partial}{\partial z} K_z \frac{\partial S}{\partial z}. \tag{5.8}$$

If the horizontal eddy diffusion coefficients are equal ($K_x = K_y = K_H$), the first two terms on the right-hand side can be written in the form $K_H \nabla^2 S$, where

$$S = S(x, y, z, t). \tag{5.9}$$

Even in a horizontally-homogeneous surface layer the field of averaged concentration S may depend on (x, y) since it is determined by the geometry of the sources and their location in space.

If the admixture whose concentration is denoted by S consists of heavy particles, having a sedimentation velocity w_s, then a new term $(-w_s \partial S/\partial z)$ should be added on the left-hand side of Equation (5.8). In the case of a nonconservative admixture, the term αS should be added on the right-hand side of Equation (5.8), where α is a constant, characterizing the rate of creation or annihilation of the admixture. In the general case, an arbitrary function $F(S, \mathbf{x}, t)$ can appear in the diffusion equation, to be specified in the particular cases.

The diffusion equation (5.8) is a linear parabolic partial differential equation and appropriate boundary conditions must be formulated. In the case of diffusion in an unlimited space, the natural boundary and initial conditions are

$$\begin{gathered} S(\mathbf{x}, 0) = 0, \qquad S(\infty, t) = 0, \\ \int S(\mathbf{x}, t)\, \mathrm{d}\mathbf{x} = M, \end{gathered} \tag{5.10}$$

where $\mathrm{d}\mathbf{x} = \mathrm{d}x\, \mathrm{d}y\, \mathrm{d}z$ and the integral condition means conservation of the substance during the diffusion (M-power of the source).

When a rigid boundary exists, for instance at $z = 0$, as in the surface layer,

$$K_z \frac{\partial S}{\partial z} + w_s S = \beta S \quad \text{at } z = 0, \tag{5.11}$$

is a sufficiently general boundary condition, describing practically all situations of interest. In (5.11), $\beta = 0$ corresponds to the 'reflection' of the admixture from the boundary, $\beta = \infty$ to the 'absorption' and $0 < \beta < \infty$ to the intermediate cases.

b. Particular Solutions and Analysis

Let us consider the case of a permanently-working stationary source of the admixture with a power M, located on the ground ($z = 0$) or at a height h_s above it. Then in Equation (5.8) $\partial S/\partial t = 0$. Usually $|K_x\, \partial^2 S/\partial x^2| \ll |\bar{u}\, \partial S/\partial x|$. In the case of an unlimited linear source perpendicular to the wind (i.e., parallel to the axis Oy), we have $\partial S/\partial y = 0$.

Thus, from (5.8) we obtain the two-dimensional semiempirical equation of turbulent diffusion

$$\bar{u}(z)\,\frac{\partial S}{\partial x} = \frac{\partial}{\partial z}\,K_z(z)\,\frac{\partial S}{\partial z}, \tag{5.12}$$

where $S = S(x, z)$ and, in the case of neutral stratification, in the surface layer as we already know

$$K_z = \kappa v_* z, \qquad \bar{u}(z) = \frac{v_*}{\kappa}\ln\frac{z}{z_0}. \tag{5.13}$$

The presence of $\ln z$, however, is a great obstacle against obtaining the exact analytical solution of (5.12). That is why, instead of (5.13), we shall use the power dependences (2.29) and (2.30), written for $z_1 = 1$ m

$$K_z(z) = k_1 z^{1-n}, \qquad \bar{u}(z) = u_1 z^n \tag{5.14}$$

At $n = \frac{1}{7}$ the logarithmic law (5.13) is well approximated.

So we are going to solve Equation (5.12) with coefficient (5.14) under boundary conditions

$$S \to 0 \quad \text{at } x, z \to \infty; \qquad K_z \frac{\partial S}{\partial z}\bigg|_{z=0} = 0 \tag{5.15}$$

$$S = 0 \quad \text{at } x = 0,\ z \neq h_s; \qquad \int \bar{u} S\,\mathrm{d}z = M, \quad x \geqslant 0. \tag{5.16}$$

Instead of (5.16), one can write

$$\bar{u}(h_s)\,S(0, z) = M\delta(z - h_s)$$

where $\delta(a)$ is a delta function.

Under these conditions and $h_s = 0$, the solution is

$$S(x, z) = \frac{(2n+1)M}{u_1\Gamma[(n+1)/(2n+1)]}\left[\frac{u_1}{(2n+1)^2 k_1 x}\right]^{(n+1)/(2n+1)} \times$$

$$\times \exp\left[-\frac{u_1 z^{2n+1}}{(2n+1)^2 k_1 x}\right], \tag{5.17}$$

where $\Gamma(a)$ is a gamma function. One can account approximately for the influence of the stratification on the field of concentration S by varying n. The ground concentration $S(x, 0)$ is of special interest. From (5.17)

$$S(x, 0) \sim x^{-(n+1)/(2n+1)} \quad (= x^{-8/9} \text{ at } n = \tfrac{1}{7}). \tag{5.18}$$

Hence, for the height z_s of the surface of the constant relative concentration $S_r(x, z) = S(x, z)/S(x, 0) = \text{const}$, we find

$$z_s(x) \sim x^{1/(2n+1)} \quad (= x^{7/9} \text{ at } n = \tfrac{1}{7}). \tag{5.19}$$

These results (Figure 9.5) agree well with the observations.

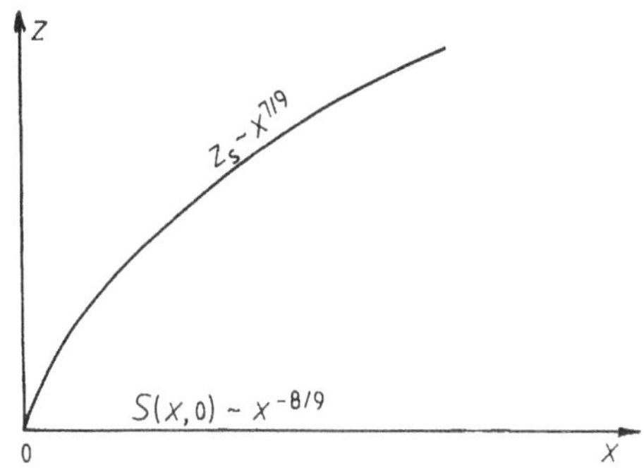

Fig. 9.5. Some characteristics of the diffusion process according to (5.17) at $n = \frac{1}{7}$.

If we accept $K_z = k_1 z$ but consider $\bar{u}(z) = \bar{u} = \text{const}$ (for instance, an averaged wind velocity for the surface layer) then Equation (5.12) has a very simple solution

$$S(x, z) = \left(\frac{M}{k_1 x}\right) \exp\left[-\frac{\bar{u}z}{k_1 x}\right] \tag{5.20}$$

obtained by Bosanquet and Pearson. Hence, it follows that

$$S(x, 0) \sim x^{-1}, \qquad z_s(x) \sim x^1, \tag{5.21}$$

i.e., results close to (5.18) and (5.19).

In our example, the source was supposed to be linear and permanent in time. However, various other sources are possible: point, plane, volume, instantaneous or temporal. Accordingly, one or another form of the diffusion equation and initial and boundary conditions is taken. Moreover, because of the linearity of the diffusion equation (5.8), the *principle of superposition* is used. Having obtained a fundamental solution of this equation for an instantaneous point source, solutions for various kinds of sources could be constructed by integration on time and space variables.

More details can be found in the literature [7, 48].

6. HORIZONTALLY NONHOMOGENEOUS SURFACE LAYER

Together with the quasi-stationarity ($\partial/\partial t = 0$), the requirement for the horizontal homogeneity of the underlying surface and, hence, of the surface layer itself, was the second fundamental condition for existence and finite thickness of a constant flux layer. Present-day investigations show that there would be no serious nonstationary effects in the surface layer, except possibly near sunrise and sunset. This is not the case with the second factor, however. The so-far exploited assumption for horizontal homogeneity is much more restrictive than the first one. A strictly homogeneous underlying surface is never encountered in the real word. Its properties change continuously or suddenly. The latter case occurs, for instance, near the seashore. The surface temperature, humidity, roughness, etc., change considerably. The air flow above responds to these changes. A process of adjustment to the new surface takes place. First, let us consider some scalar characteristics of the flow.

a. Adjustment of Scalar Characteristics

Let S denote the humidity of the air and an absolutely dry air flow crosses the seashore perpendicularly, in a land-sea direction. The seashore line is supposed to be straight and parallel to the $0y$ axis, Figure 9.6. The half-plane $x > 0$ evaporates and the water vapour

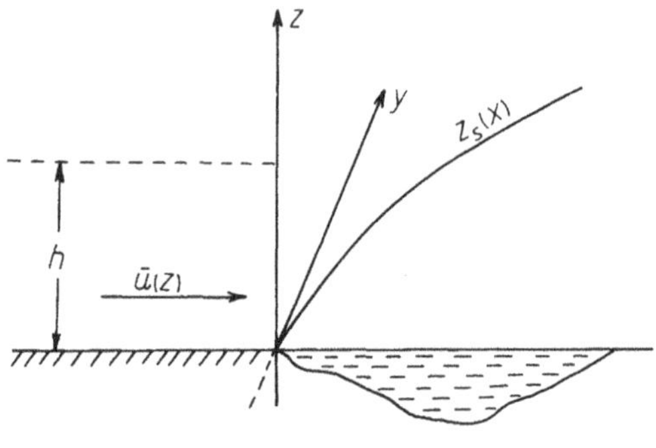

Fig. 9.6. Water vapour propagation as an advective and diffusion process.

propagates in the surface layer, thus changing the initial humidity of the air flow. The process is diffusive in its nature and is governed by the equation*

$$\bar{u}(x, z)\frac{\partial S}{\partial x} = \frac{\partial}{\partial z} K_z(x, z)\frac{\partial S}{\partial z}. \tag{6.1}$$

Besides $S(x, z)$, this equation contains two more unknown functions $\bar{u}(x, z)$ and $K_z(x, z)$, since the wind velocity and the eddy diffusion coefficient also respond to the new conditions at $x > 0$. As a crude approximation, we shall neglect the latter fact by assuming that at $x > 0$

$$\bar{u} = u_1 z^n, \qquad K_z = k_1 z^{1-n}. \tag{6.2}$$

Equation (6.1) takes the form

$$a^2 z^n \frac{\partial S}{\partial x} = \frac{\partial}{\partial z}\left(z^{1-n}\frac{\partial S}{\partial z}\right) \tag{6.3}$$

where $a^2 = u_1/k_1$. In agreement with Figure 9.6, the boundary conditions are

$$\begin{aligned} &S(0, z) = S(x, \infty) = 0, \\ &S(x, 0) = S_0 \neq 0 \quad \text{at } x > 0. \end{aligned} \tag{6.4}$$

Following [35] we introduce a new variable $\zeta = x/a^2 z^{1+2n}$ and obtain

$$\begin{aligned} &\zeta^2 S'' + (p\zeta - q)S' = 0, \\ &S(0) = 0, \qquad S(\infty) = S_0, \end{aligned} \tag{6.5}$$

* Compare with Equation (5.12).

where $(') = d/d\zeta$, $p = (1 + 3n)/(1 + 2n)$, $q = (1 + 2n)^{-2}$. The ordinary differential equation (6.5) can easily be integrated twice to obtain its general solution

$$S(\zeta) = A_1 + A_2\Gamma\left(p-1, \frac{q}{\zeta}\right) \tag{6.6}$$

where

$$\Gamma(\alpha, x) = \int_0^x y^{\alpha-1} \exp[-y]\, dy.$$

The boundary conditions yield $A_1 = S_0$, $A_2 = -S_0/\Gamma(p-1)$. Finally

$$S(x, z) = S_0\left[1 - \frac{1}{\Gamma(p-1)}\,\Gamma\left(p-1, \frac{u_1 q z^{1+2n}}{k_1 x}\right)\right]. \tag{6.7}$$

Having in mind that $\Gamma(\alpha, 0) = 0$ and $\Gamma(\alpha, \infty) = \Gamma(\alpha)$, the dependence of S on the problem's parameters can be investigated. Thus, intensive turbulent exchange (large k_1) intensifies the adjustment process, while strong wind (large u_1) restricts it to the lower layers. The latter is better seen from the equation of the surfaces of constant concentration $S(x, z_s) = \text{const}$. Equation (6.7) yields

$$z_s(x) \sim \left(\frac{k_1 x}{u_1 q}\right)^{1/(1+2n)} \tag{6.8}$$

Evidently $z_s(0) = 0$. One of these surfaces – say, that corresponding to $S/S_0 = 0.01$ – could define the upper level of what is called the *internal boundary layer* above which (at $z \geqslant z_s(x)$) in practice, the influence of the new underlying surface is negligible. Mathematically, this is expressed by the conditions

$$S(x, z_s) = 0, \qquad \left.\frac{\partial S}{\partial z}\right|_{z_s} = 0, \tag{6.9}$$

the second meaning zero flux of S through the surface z_s. On this basis, a mathematical method has been created for approximately solving this kind of problem when the exact solution is impossible. It is known as 'the finite boundary-layer' method.* We shall demonstrate this method using Equation (6.3).

We seek a solution in the form

$$S(x, z) = \sum_i S_i(x, z), \tag{6.10}$$

where the functions S_i satisfy the equations

$$\frac{\partial}{\partial z} z^{1-n} \frac{\partial S_1}{\partial z} = 0, \qquad \frac{\partial}{\partial z} z^{1-n} \frac{\partial S_i}{\partial z} = a^2 z^n \frac{\partial}{\partial x} S_{i-1}, \quad i = 2, 3, \ldots \tag{6.11}$$

* In Russian literature it is referred to as Shwetz's method.

and the boundary conditions

$$S_1(x, 0) = S_0, \qquad S_1(x, z_s) = 0$$

$$S_i(x, 0) = S_i(x, z_s) = 0, \quad i = 2, 3, \ldots$$

The unknown function $z_s(x)$ can be determined from the second condition (6.9) and the requirement that $z_s(0) = 0$. If series (6.10) converges fast enough, we can truncate it and write $S \approx S_1 + S_2$. Another reason for doing this may be the complexity of the next approximations in more realistic models.

b. Adjustment of the Wind

Assumption (6.2) at $x > 0$ makes the formulation of the problem internally contradictory since $\bar{u}$ and K_z also change and may depend on x in the transitional area. This particularly concerns the wind velocity $\bar{u}$ which is extremely sensitive to abrupt changes of surface roughness z_0 (Figure 9.7). Let us consider a neutral surface layer.

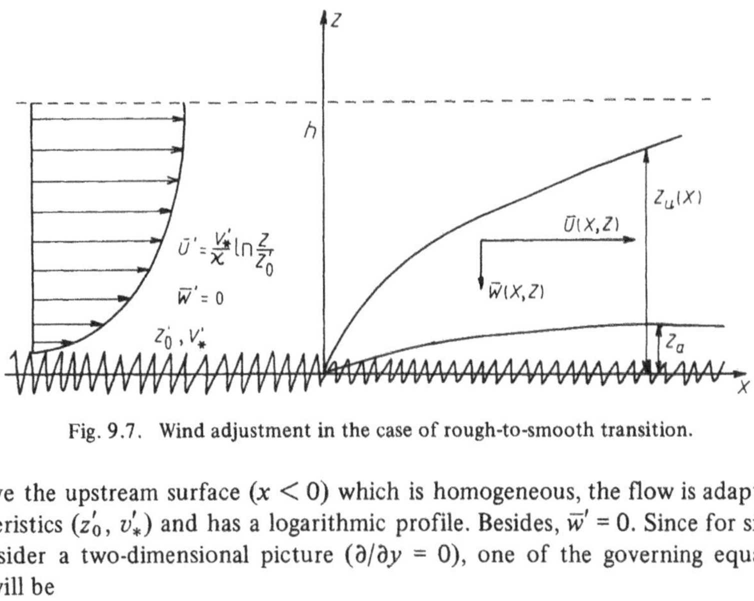

Fig. 9.7. Wind adjustment in the case of rough-to-smooth transition.

Above the upstream surface ($x < 0$) which is homogeneous, the flow is adapted to its characteristics (z_0', v_*') and has a logarithmic profile. Besides, $\bar{w}' = 0$. Since for simplicity we consider a two-dimensional picture ($\partial/\partial y = 0$), one of the governing equations at $x > 0$ will be

$$\frac{\partial \bar{u}}{\partial x} + \frac{\partial \bar{w}}{\partial z} = 0. \tag{6.12}$$

Hence, we conclude that if $z_0 < z_0'$ (the case on Figure 9.7), then $\bar{w} < 0$ and at one and the same height z, $\bar{u} > \bar{u}'$. In the opposite case ($z_0 > z_0'$) $\bar{w} > 0$ and $\bar{u} < \bar{u}'$. If the half-plane $x > 0$ has uniform roughness z_0, then at $x \to \infty$ $\bar{w}(x, z) \to 0$ and $\bar{u}(x, z) \to (v_*/\kappa)$ ln (z/z_0). What happens in between? This is a response region and we can imagine an interface $z_u(x)$ growing out of $x = 0$, as shown in Figure 9.7. It separates the internal boundary layer ($z < z_u$) which develops over the new roughness from the undisturbed flow ($z > z_u$). At $z < z_u$ the flow adjusts to the new v_* and z_0. If $z_u < h$ the phenomenon is of micrometeorological scales and the Coriolis force and pressure-gradient force can be neglected. Then, for the stationary flow, the equation of motion will be

$$\bar{u}\,\frac{\partial \bar{u}}{\partial x} + \bar{w}\,\frac{\partial \bar{u}}{\partial z} = \frac{\partial}{\partial z}\,v_*^2, \quad v_*^2(x, z) = \frac{\tau}{\rho} = -\overline{u'w'} \tag{6.13}$$

at $x > 0$. System (6.12) and (6.13) contains three unknown functions – $\bar{u}$, $\bar{w}$, v_*. If we write

$$v_*^2(x, z) = K(x, z)\,\frac{\partial}{\partial z}\,\bar{u}(x, z) \tag{6.14}$$

and assume $K(x, z) = K(z) = \kappa v_* z$, where $v_* = v_*(\infty, z) = \text{const}_z$, the system will close and may be solved. However, one may go further by adding the equations

$$\bar{u}\,\frac{\partial b}{\partial x} + \bar{w}\,\frac{\partial b}{\partial z} = K\left(\frac{\partial \bar{u}}{\partial z}\right)^2 - \frac{cb^2}{K}, \tag{6.15}$$

$$K = l\sqrt{b}, \quad l = -\kappa\,\frac{\partial \bar{u}}{\partial z}\Big/\frac{\partial^2 \bar{u}}{\partial z^2}. \tag{6.16}$$

In such a way, we have six equations for six functions – $\bar{u}$, $\bar{w}$, v_*, k, l, b. By the way, there is one more unknown function – the height of the internal boundary layer $z_u(x)$, an equation for which can be found from the boundary conditions formulated at $z = 0$ and $z = z_u$. For instance, at $x > 0$ and $z = z_u$

$$\bar{u} = \frac{v'_*}{\kappa}\,\ln\left(\frac{z_u}{z'_0}\right), \qquad K\,\frac{\partial \bar{u}}{\partial z} = v_*'^2 \tag{6.17}$$

while at $z = z_0$ and $x > 0$

$$\bar{u} = \bar{w} = 0, \qquad K = \kappa v_*(x) z_0, \qquad K\,\frac{\partial \bar{u}}{\partial z} = v_*^2(x). \tag{6.18}$$

Even for this simplest type of inhomogeneity and neutral stratification, the mathematical model of the phenomenon (Equations (6.12)–(6.18)) can only be realized numerically. Numerical calculations and some experiments have shown that within the internal layer $z_u(x)$ a sublayer ($z_a(x)$) exists (Figure 9.7) which has reached full equilibrium with the new surface. However, the equilibrium sublayer grows very slowly – z_a is about 1% of the downstream distance. Therefore, $\Delta z(x) = z_u(x) - z_a(x)$ is essential and the details of the response within the internal layer Δz will be important in real applications. Above it (at $z > z_u$), the flow still 'keeps in mind' the properties of the 'old' surface ($x < 0$).

A single change of the roughness, though a great idealization, could be met in reality. Theoretical models for double changes of this parameter ($z_0 \neq z'_0 \neq z''_0$, for instance, wind blowing perpendicularly to an aerodrome runway) have also been proposed. Though some observations under real and laboratory conditions have been made during the past years, the most prospective tool for studying this very complicated phenomenon seems to be the numerical turbulence modelling experiments.

PROBLEMS

1. Prove that (2.4) is a solution of (2.10) and that (2.9) follows from (2.11).
   Note: see [48].
2. Prove that at $\epsilon \to 0$ (2.25) is consistent with (2.4).
3. Construct models for $\bar{u}(z)$ in the surface layer if a priori is given that: (a) $K(z) = K_\infty(1 - e^{-mz})$; (b) $l(z) = \kappa z\, e^{mz}$; (c) $l(z) = \kappa z/(1 + mz)$.
4. Using (3.23) and (3.24), integrate Equations (3.4).
   Note: Make substitutions

$$x = (1 - 15\zeta)^{1/4} = \frac{1}{\varphi}, \qquad y = (1 - 9\zeta)^{1/2} = \frac{0.74}{\varphi_\theta}.$$

5. Starting from the equations

$$\frac{\partial \theta}{\partial t} + \frac{\partial v_k \theta}{\partial x_k} = \kappa_m \nabla^2 \theta$$

and (2.16) in Chapter 8, for $\bar{\theta}$, derive equations for θ' and hence for $b_\theta = \overline{\theta'^2}/2$ in the surface layer:

$$\frac{\partial b_\theta}{\partial t} = -\overline{w'\theta'}\,\frac{d\bar{\theta}}{dz} - \epsilon_\theta + \text{Diff}_\theta$$

where

$$\epsilon_\theta = \kappa_m \overline{(\nabla\theta')^2}, \qquad \text{Diff}_\theta = \kappa_m \frac{\partial^2 b_\theta}{\partial z^2} - \frac{\partial}{\partial z}\frac{\overline{w'\theta'^2}}{2}, \qquad \overline{w'\theta'} = -K^\theta \frac{d\bar{\theta}}{dz}.$$

6. Obtain expressions for the structure functions (4.10) and (4.11) when the dissipation parameter ϵ is unessential.
7. Derive (4.19) from (4.18).
   Note: see [53], Chapter 8, Section 5.
8. Prove that (6.7) is a solution of Equation (6.5).
9. Substituting (6.14) with $K = \kappa v_* z$ and (6.12) into (6.13), an integro-differential equation can be derived

$$\bar{u}\,\frac{\partial \bar{u}}{\partial x} - \frac{\partial \bar{u}}{\partial z}\int_{z_0}^{z} \frac{\partial \bar{u}}{\partial x}\,dz = \kappa v_* \frac{\partial}{\partial z}\, z\, \frac{\partial \bar{u}}{\partial z}.$$

Find an approximate solution to this equation (say $\bar{u} \approx \bar{u}_1 + \bar{u}_2$) by applying the method of the finite boundary layer from Section 6.

CHAPTER 10

The Dynamics of the Atmospheric Planetary Boundary Layer

1. TURBULENT PLANETARY BOUNDARY LAYER (PBL): GENERAL PROPERTIES

a. Definition for PBL: Equations

The boundary-layer concept is well known in fluid dynamics. It is always related to the presence of a rigid boundary in the flow which can be either laminar or turbulent. The most characteristic feature of the boundary layer is the existence of a mean-flow shear in a direction perpendicular to the boundary. In this sense, the surface layer ($h \sim 30$–50 m) of the atmosphere, studied in Chapter 9, is a boundary layer. However, the dynamic influence of the Earth's surface can be found much more higher – up to $h_g = 1$–2 km in the middle latitudes. To this height the wind velocity increases (Figure 10.1) and approaches the geostrophic wind $\mathbf{u}_g$, i.e., $z \to h_g$ $\mathbf{u} \to \mathbf{u}_g$. In the *free atmosphere* ($z > h_g$), the wind is nearly geostrophic. This layer with a thickness h_g is called the *planetary boundary layer*. It is clear that the transition between the PBL and the free atmosphere is smooth and that the surface layer, as defined in Chapter 9, is a sublayer of the PBL – its lowest part. On the Earth's surface $u = v = w = 0$.

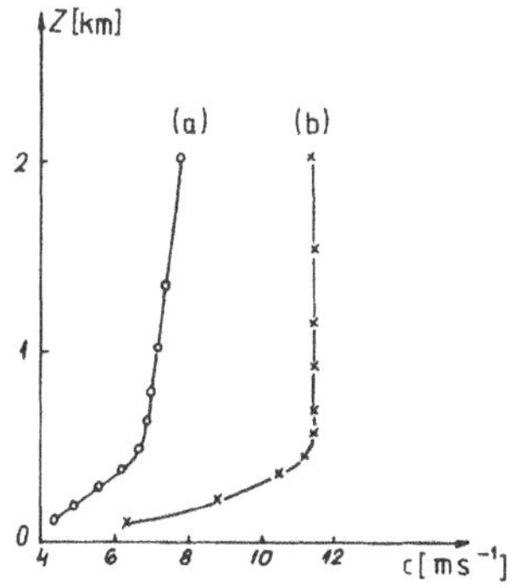

Fig. 10.1. Vertical profile of the monthly mean wind (a – summer, b – winter) at Voeykovo Station, U.S.S.R. (after [35]).

The above definition of the PBL is, to a large extent, incomplete and qualitative. According to Wipperman [69], the boundary layer of the atmosphere is called PBL if

(1) the boundary layer is turbulent;

(2) the mean flow and the turbulence properties are stationary and horizontally homogeneous;
(3) the molecular transport of momentum, heat or moisture can be neglected as being small compared with the corresponding turbulent transport.

Therefore, the **PBL** is a turbulent, well-mixed layer. In fair weather (Figure 10.2a), its upper limit can be defined as

(a) the height h_g at which $c(h_g) = c_g$, where $c = |\mathbf{u}|$ and $c_g = |\mathbf{u}_g|$;
(b) the height h_g at which $dc/dz = 0$;
(c) the height h_g at which the wind $\mathbf{u}$ becomes, for the first time, parallel to $\mathbf{u}_g$, i.e.,

$$\left(\frac{v}{u}\right)_{z=h_g} = \frac{v_g}{u_g}. \tag{1.1}$$

Usually the last definition of h_g is used.

The question of stationarity and horizontal homogeneity is more complicated. In the case of a flat surface and cumulus or stratocumulus clouds (Figure 10.2b), it might still

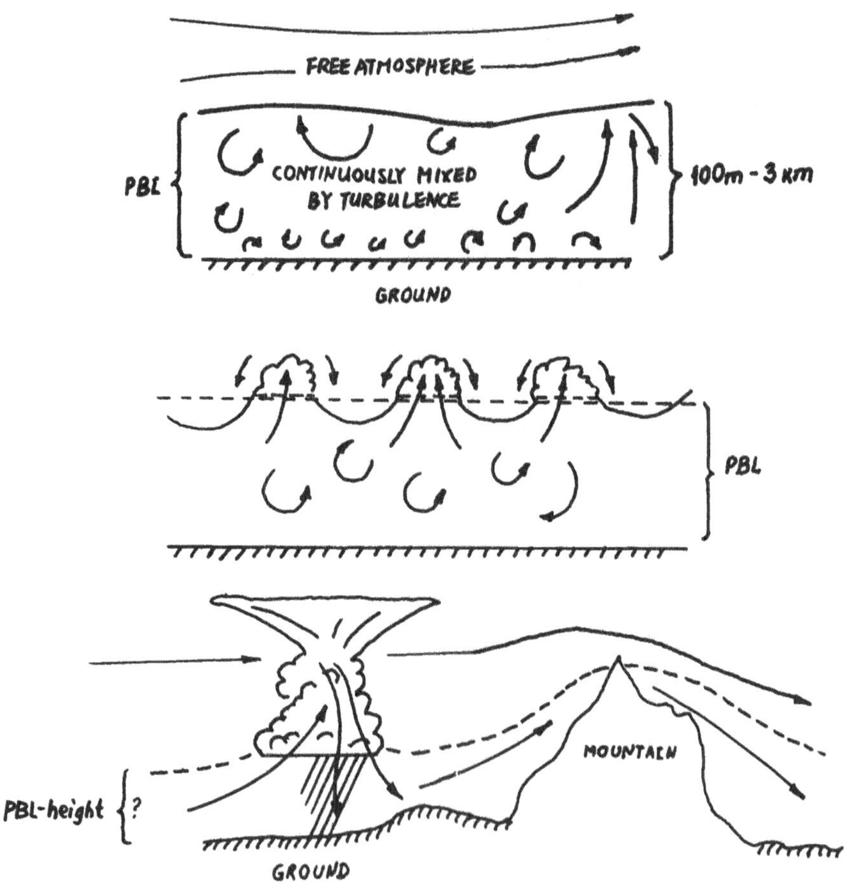

Fig. 10.2. Definition of the planetary boundary layer (after S. Bodin, 1978).

be possible to define an upper limit of the PBL, although it will not be continuous as in the first case. But in the case of active Cb clouds or a flow over sharp mountains where the layers close to the ground interact strongly with the free atmosphere, we can hardly speak about PBL in the above sense. The dynamics of the subcloud layer in this case is, to a large extent, governed by the dynamics of the cloud. From a practical point of view, most applied models have disregarded interactions of these kinds and in the case of smooth and not very high mountains, it has been generally assumed that a well-defined PBL follows the terrain height (Bodin, 1978). Its upper limit $h_g(x, y)$ does not necessarily repeat the Earth's topography $H(x, y)$ (for details see Section 6).

We shall only be interested in the vertical profiles of the mean velocity $\bar{u}(z)$, $\bar{v}(z)$, $\bar{w}(z)$, mean temperature $\bar{T}(z)$ (or $\bar{\theta}(z)$), etc. For simplicity, the overbar is omitted except for the Reynolds stress components. Then, the starting equations for investigation of the PBL follow from (2.22)–(2.24) in Chapter 8, or (1.1) in Chapter 9:

$$0 = -\alpha \frac{\partial p}{\partial x} + fv - \frac{\partial \overline{u'w'}}{\partial z},$$

$$0 = -\alpha \frac{\partial p}{\partial y} - fu - \frac{\partial \overline{v'w'}}{\partial z}, \tag{1.2}$$

$$0 = \frac{\partial u}{\partial x} + \frac{\partial v}{\partial y} + \frac{\partial w}{\partial z},$$

where $\alpha = 1/\rho$. If we replace $-\alpha \nabla p$ by the equivalent geostrophic wind $\mathbf{u}_g$ and the Reynolds stresses are expressed on the basis of semi-empirical formulae

$$-\overline{u'w'} = \frac{\tau_x}{\rho} = K \frac{\partial u}{\partial z} = X,$$
$$-\overline{v'w'} = \frac{\tau_y}{\rho} = K \frac{\partial v}{\partial z} = Y, \tag{1.3}$$

then the equations of motion (1.2) take the form

$$\frac{\partial}{\partial z} K \frac{\partial u}{\partial z} + fv = fv_g, \qquad \frac{\partial}{\partial z} K \frac{\partial v}{\partial z} - fu = -fu_g, \tag{1.4}$$

or in vector notation

$$\frac{\partial}{\partial z} K \frac{\partial \mathbf{u}}{\partial z} - f\mathbf{k} \times \mathbf{u} = -f\mathbf{k} \times \mathbf{u}_g. \tag{1.5}$$

There is one more way for a compact representation of Equations (1.4). We introduce the complex functions

$$M = u + iv, \qquad M_g = u_g + iv_g, \quad i = \sqrt{-1}. \tag{1.6}$$

Then

$$\frac{\partial}{\partial z} K \frac{\partial M}{\partial z} - ifM = -ifM_g. \tag{1.7}$$

Sometimes it is more convenient to write system (1.4) for the stresses X, Y. We differentiate (1.4) with respect to z and use (1.3). As a result we obtain

$$\frac{\partial^2 X}{\partial z^2} + \frac{f}{K} Y = \frac{f}{K} \frac{\partial v_g}{\partial z},$$
$$\frac{\partial^2 Y}{\partial z^2} - \frac{f}{K} X = -\frac{f}{K} \frac{\partial u_g}{\partial z}. \tag{1.8}$$

This system can also be written in vector or complex form.

A traditional problem in planetary boundary-layer dynamics is the determination of the wind field on a given pressure field. Since in (1.4)–(1.8) the pressure is transformed into an equivalent geostrophic velocity, the function $\mathbf{u}_g(x, y, z)$ should be considered as given.

If the atmosphere is generally barotropic, then $\mathbf{u}_g$ does not depend on z, i.e., $\mathbf{u}_g = \mathbf{u}_g(x, y)$ and will represent the geostrophic wind in the free atmosphere. In this case, we speak about *barotropic* PBL. Otherwise, if $\mathbf{u}_g$ depends on z, the PBL is *baroclinic* and to solve system (1.5) the dependence of $\mathbf{u}_g$ on z has to be specified. Examples of both cases will be discussed later.

In such a formulation both Equations (1.4) contain three unknown functions: u, v, K. A large number of so-called K-models have been proposed in which the eddy exchange coefficient is postulated a priori as a function of z. This is the simplest closure scheme for the system of Equations (1.4).

This system is of fourth order with respect to z and, consequently, needs four boundary conditions. Two of them are undoubtful

$$\mathbf{u} \to \mathbf{u}_g \quad \text{at } z \to \infty. \tag{1.9}$$

The conditions on the lower boundary can be specified in different ways. One possibility is to require

$$\mathbf{u}(h_0) = \mathbf{u}_{h_0} \quad \text{or} \quad \mathbf{u}(0) = 0 \tag{1.10}$$

where h_0 can be the upper level of the surface layer ($h_0 \equiv h$) or the height at which the anemometer is located. Obviously, the simplest case is $h_0 = 0$ and $\mathbf{u}(0) = 0$.

If we assume that at $z = 0$ $\mathbf{u} \| \boldsymbol{\tau}$ (the stress), then a boundary condition would look like

$$\frac{v}{u} \to \frac{\partial v}{\partial z} \Big/ \frac{\partial u}{\partial z}, \quad z \to 0 \tag{1.11}$$

or in vector notation

$$\rho K \frac{\partial \mathbf{u}}{\partial z} = \tilde{\beta} \mathbf{u}, \quad z \to 0, \tag{1.11'}$$

where $\tilde{\beta}$ is a coefficient of proportionality. Since at the very surface $\mathbf{u}(0) = 0$, the reference point for z should be at some level $h_0 < h$. In addition to (1.11) as a second boundary condition, the angle α_0 between $\mathbf{u}_0$ and $\mathbf{u}_g$ is considered as known.

Finally, boundary conditions consistent with the concept of the surface layer are also used

$$K \left.\frac{\partial v}{\partial z}\right|_{z=0} = 0, \qquad K \left.\frac{\partial u}{\partial z}\right|_{z=0} = \frac{\tau_0}{\rho} = v_*^2. \tag{1.12}$$

The lower boundary conditions may also be formulated at the level of the surface roughness z_0.

b. Ekman Model

As we know, $K = K(z)$, at least in the surface layer. This dependence also continues in the whole boundary layer. However, the simplest assumption concerning $K(z)$, first made by Ekman in 1905, is $K(z) = K = \mathrm{const}_z$. By the way, Ekman originally considered the influence of rotation and turbulence on ocean currents, while the present problem was first solved by Akerblom three years later.

At $K = \mathrm{const}$ Equation (1.7) takes the form

$$KM'' - ifM = -ifM_g \tag{1.13}$$

where $(') = d/dz$, with boundary conditions

$$M(\infty) = M_g, \qquad M(0) = 0. \tag{1.14}$$

In a barotropic PBL, $M_g(z) = M_g = \mathrm{const}_z$. Elementary calculations yield the solution

$$M(z) = M_g[1 - \exp(-a(1+i)z)], \quad a^2 = \frac{f}{2K}. \tag{1.15}$$

Hence,

$$\begin{aligned} &u(z) = (1 - F_1)u_g - F_2 v_g, \qquad v(z) = F_2 u_g + (1 - F_1)v_g, \\ &F_1(z) = \exp[-az] \cos az, \qquad F_2(z) = \exp[-az] \sin az, \quad \mathbf{u}_g = \mathbf{u}_g(x, y), \end{aligned} \tag{1.16}$$

or in vector notation

$$\mathbf{u} = (1 - F_1)\mathbf{u}_g + F_2\, \mathbf{k} \times \mathbf{u}_g \tag{1.17}$$

where $\mathbf{k} \times \mathbf{u}_g = \mathbf{s}_g = (-v_g, u_g, 0)$. In the case of straight-line isobars and coordinate axis $0x \parallel \mathbf{u}_g$, i.e., $v_g = 0$, solution (1.16) simplifies to

$$u(z) = (1 - F_1)u_g, \qquad v(z) = F_2 u_g. \tag{1.18}$$

Hence, for the angle between $\mathbf{u}_g$ and $\mathbf{u}$ we find

$$\begin{aligned} &\tan \alpha(z) = \frac{v}{u} = \frac{F_2}{1 - F_1}, \\ &\tan \alpha_0 = 1, \end{aligned} \tag{1.19}$$

i.e., $\alpha_0 = \pi/4$, and for the thickness of the PBL, according to (1.1), we obtain

$$h_g = \frac{\pi}{a} = \pi\left(\frac{2K}{f}\right)^{1/2} = \pi\left(\frac{K}{\omega \sin \varphi}\right)^{1/2}. \tag{1.20}$$

As can be seen from Figure 10.3, $h_g \sim 1.5$ km at $K \sim 10^3$ m^2 s^{-1} and $\varphi \sim 45°$. Besides, h_g increases with increasing K and decreasing φ, which is in qualitative agreement with the observations. The predicted decreasing of $\alpha(z)$ with z (right notation of the wind-velocity vector $\mathbf{u}$) is also in such agreement. However, the surface value $\alpha_0 = 45°$ is about twice the size of the observed one. Moreover, α_0 should depend on the meteorological conditions and the roughness z_0 which is not predicted by the Ekman model. The reason for this discrepancy is the main assumption for the constancy of K with z. If α_0 is given and (1.11) is used instead of (1.10) as Taylor did in 1915, then

$$M(z) = u_g\left[1 + \sqrt{2}\,\sin\alpha_0\,\exp\left(-az + i\left(-az + \alpha_0 + \frac{3\pi}{4}\right)\right)\right]. \tag{1.21}$$

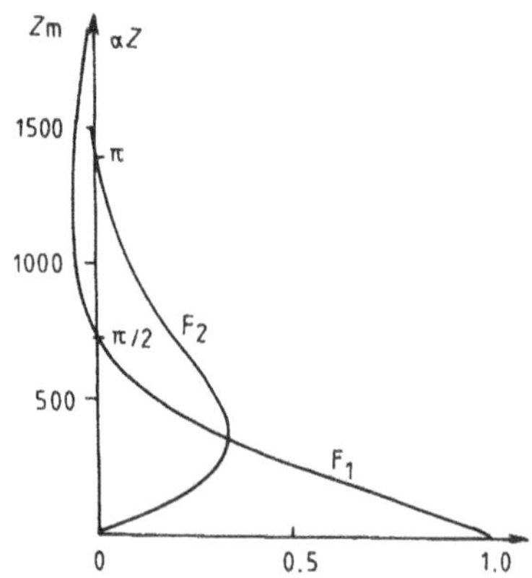

Fig. 10.3. The functions $F_1(z)$ and $F_2(z)$ from (1.16) ($K = 10$ m^2 s^{-1}).

The Ekman spiral in its ideal form, plotted in Figure 10.4, can never be observed in nature. One of the reasons is the fast increase of the eddy exchange coefficient K in the surface layer. Another reason is the instability of the Ekman PBL (instability of solution (1.17)). As a result, a secondary circulation with a scale of order h_g may develop in the

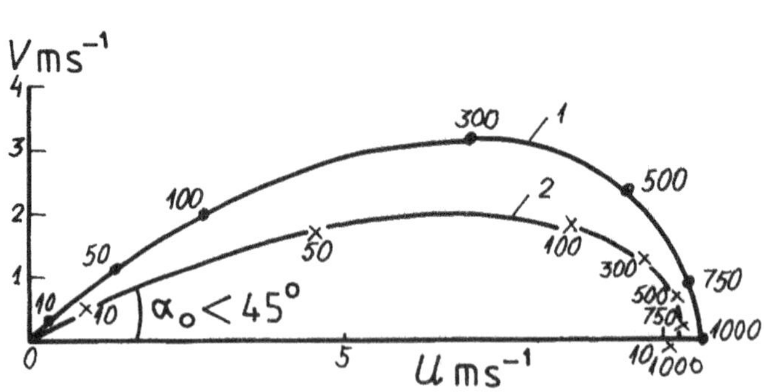

Fig. 10.4. The theoretical Ekman spiral (after [35]): (1) Constant eddy viscosity coefficient K = 8.2 m^2 s^{-1} in the entire PBL; (2) Step-like model (2.2) with h = 100 m, K_1 = 1 m^2 s^{-1}, K_2 = 9 m^2 s^{-1}. In both cases φ = 60°, u_g = 10 m s^{-1}. The figures on the curves indicate the height above the ground in meters.

PBL which transports momentum and other substances in a vertical direction but cannot be parameterized by the formulae of semi-empirical theory. The empirical hodographs (spirals) of the wind differ from the theoretical ones by their irregular form and a number of quantitative characteristics (Figure 10.5). However, the integral cross-isobaric mass transport in the whole PBL in the direction of the lower pressure is reasonably well predicted by the Ekman theory.

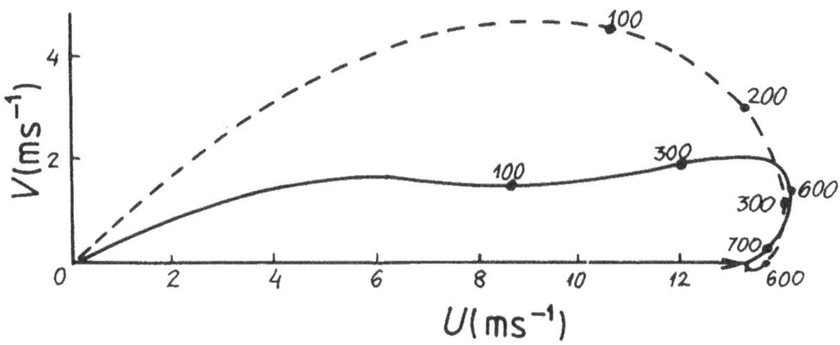

Fig. 10.5. An example of an empirical wind hodograph (solid line), compared with a theoretical one dashed line) (after [27]).

2. *K*-MODELS OF THE PBL

a. Barotropic PBL

We assume that $\mathbf{u}_g(z) = \text{const}_z$ so that in the equation of motion (1.7)

$$\frac{\partial}{\partial z} K \frac{\partial M}{\partial z} - ifM = -ifM_g \tag{2.1}$$

the only function which has to be prescribed a priori is $K(z)$. Since this can be done in an unlimited number of different ways, some speculations on this question are possible. Some of the most successful models of this type are discussed below.

1. *Step-like model:*

$$K = \begin{cases} K_1, & z \leqslant h \\ K_2, & z > h \end{cases} \tag{2.2}$$

If $M_1(z)$ and $M_2(z)$ are the general solutions of (2.1) for the corresponding layers, then the boundary conditions imposed on them are

$$\begin{gathered} M_1(0) = 0, \qquad M_2(\infty) = M_g, \qquad M_1(h) = M_2(h), \\ \tau_1(h) = \tau_2(h), \quad \left(\tau_i = \rho_i K_i \frac{\mathrm{d}\mathbf{u}_i}{\mathrm{d}z}, \quad i = 1, 2\right). \end{gathered} \tag{2.3}$$

Thus, regardless of the discontinuity of $K(z)$ at $z = h$, a continuity of the wind velocity and shear stress at the level $z = h$ is required.

Since Equation (2.1) has constant coefficients, its solution will be a combination of such exponential functions as (1.15). Details of the solution are given in [35]. Unlike the Ekman solution at $\rho_1 \approx \rho_2$

$$\tan \alpha_0 = \psi(K_1, K_2, h) \leqslant 1,$$

where the equality sign takes place at $K_1 = K_2$ and $h \to \infty$. As we see, this is a three-parametric model. However, the assumption of discontinuity of $K(z)$ at $z = h$ is rather unrealistic.

2, 3. *Linear and power models.* The simplest model of this type is the linear one

$$K(z) = k_1 z, \qquad z \geqslant z_0,$$

where z_0 is the roughness parameter, proposed almost simultaneously in 1935 and 1936 by Rossby, Montgomery, Kibel, and Blinova and later used by others. As we know, in the surface layer ($z \leqslant h$) under neutral stratification $K(z) = \kappa v_* z$, i.e., $k_1 = \kappa v_*$. Due to the fact that $|\partial \mathbf{u}/\partial z| \to 0$ at $z \to \infty$, the extrapolation of the linear dependence above $z = h$ does not considerably influence the wind profile $\mathbf{u}(z)$ in the lower part of the PBL. The solution is expressed in Bessel functions with a zero number and at $z \to z_0$ asymptotically produces the logarithmic law.

An improvement, but also a complication of the one-layer linear model, is its two-layer version*

$$K(z) = \begin{cases} k_1 z, & z \leqslant h \\ k_1 h, & z \geqslant h \end{cases} \tag{2.4}$$

under boundary conditions (2.3).

The power models generalize the linear ones:

$$K(z) = k_n z^n \quad \text{at } z \geqslant z_0,$$

or

$$K(z) = \begin{cases} k_n z^n, & z \leqslant h, \\ k_n h^n, & z \geqslant h. \end{cases} \tag{2.5}$$

Varying n, stability conditions in the PBL can be simulated. Again the solution of (2.1), i.e.,

$$k_n \frac{\partial}{\partial z} z^n \frac{\partial M}{\partial z} - ifM = -ifM_g \tag{2.6}$$

is expressed in Bessel functions, the number of which is a function of the power n. If

$$n = \frac{4m}{1 + 2m}, \qquad m = 0, 1, 2, \ldots \tag{2.7}$$

---

* In Russian literature, known as the Shwetz–Yudin (1940) model.

then they degenerate into elementary functions. For instance, at $m = 1$ ($n = \frac{4}{3}$), Panchev (1955, 1956) obtained

$$\begin{aligned} u &= u_g [1 - \tilde{z}^{-1} \exp[-q(\tilde{z} - 1)] \cos q(\tilde{z} - 1)], \\ v &= u_g \tilde{z}^{-1} \exp[-q(\tilde{z} - 1)] \sin q(\tilde{z} - 1), \end{aligned} \tag{2.8}$$

where $\tilde{z} = (z/z_0)^{1/3}$, $q = 3s\, z_0^{1/3}$, $s = (f/2k_n)^{1/2}$. Hence, it follows that

$$\tan \alpha_0 = \frac{q}{q+1} < 1,$$

i.e., $\alpha_0 < \pi/4$.

The solution with $m = \infty$, i.e., $n = 2$, also has a comparatively simple analytical form.

4. *Exponential model.* Equations (2.4) and (2.5) are two-layer models and the derivative $\partial K/\partial z$ is a discontinuous function at $z = h$, which is unrealistic. A better approximation for $K(z)$ is the exponential one

$$K(z) = K_\infty (1 - \exp[-\delta z]) \quad \text{or} \quad K_0 \exp[\delta z]. \tag{2.9}$$

Obviously, the second formula implies that $K \to \infty$ at $z \to \infty$. In both cases (2.9), analytical solutions of (2.1) are possible but only in special functions.

5. *Linearly-exponential model:*

$$K(z) = k_1 z \exp[-k_2 z]. \tag{2.10}$$

Unlike the previous models, after having a maximum ($K_m = k_1/ek_2$) at $z = 1/k_2$, then $K(z) \to 0$ at $z \to \infty$. Such K profiles are sometimes observed in stable PBL. The solving of (2.1), however, becomes more and more difficult.

Here we have listed only few of the most frequently used a priori approximations of $K(z)$ (Figure 10.6). A great number of other formulae can be found in the literature.

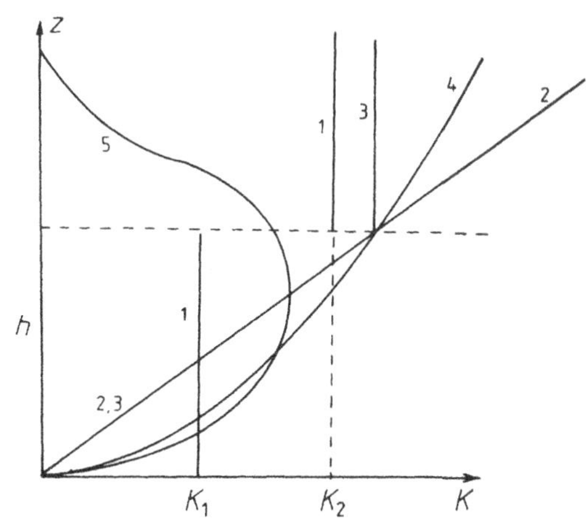

Fig. 10.6. Some a priori models for $K(z)$: (1) step-like model; (2,3) linear models; (4) exponential model; (5) linear-exponential model.

In any particular case, physical argument and experimental data, together with mathematical considerations, are needed in order to decrease the uncertainty introduced by the choice of $K(z)$. The significance of this question increases in baroclinic PBL where, in addition, $M_g(z)$ must be specified a priori too.

b. Baroclinic PBL

To concentrate our attention on the influence of baroclinicity, we assume that in Equation (2.1) $K(z) = \text{const}_z$ but $M_g = M_g(z)$. According to Section 4 in Chapter 3, two possible approximations of $\mathbf{u}_g(z)$ would be considered

$$\mathbf{u}_g(z) = \begin{cases} \mathbf{u}_g^\circ + \mathbf{u}_T' z \\ \mathbf{u}_g^\infty - \mathbf{u}_T \exp[-mz] \end{cases} \tag{2.11}$$

where $\mathbf{u}_g^\infty = \mathbf{u}_g^\circ + \mathbf{u}_T$. Thus,

$$M_g(z) = \begin{cases} M_g^\circ + M_T' z \\ M_g^\infty - M_T \exp[-mz] . \end{cases} \tag{2.12}$$

Then the equation

$$KM'' - ifM = -ifM_g(z) \tag{2.13}$$

or its vector analogue

$$K\mathbf{u}'' - f\mathbf{k} \times \mathbf{u} = -f\mathbf{k} \times \mathbf{u}_g(z) \tag{2.13'}$$

where $('') = \partial^2/\partial z^2$, has a given right-hand side and can easily be solved.

For instance, if the linear approximation is used and a new function $N = M - M_g$ is introduced, the problem to be solved is

$$KN'' - ifN = 0, \qquad N(0) = -M_g^\circ, \quad N(\infty) = 0. \tag{2.14}$$

Hence,

$$N = -M_g^\circ \exp[-az(i+1)], \quad a^2 = \frac{f}{2K}.$$

Let us assume that $\mathbf{u}_g^\circ = (u_g^\circ, 0)$, i.e., $0x \parallel \mathbf{u}_g^\circ$. Then

$$\begin{aligned} u &= u_g^\circ(1 - \exp[-az] \cos az) + \lambda_x z, \\ v &= u_g^\circ \exp[-az] \sin az + \lambda_y z, \end{aligned} \tag{2.15}$$

where $(\lambda_x, \lambda_y) = \mathbf{u}_T'$.

This simple solution predicts, though qualitatively, important features of the wind field in baroclinic PBL. For the wind directions at $z = \infty$ and $z = 0$, we find

$$\begin{aligned} \tan \alpha_\infty &= \left.\frac{v}{u}\right|_{z=\infty} = \frac{\lambda_y}{\lambda_x}, \\ \tan \alpha_0 &= \left.\frac{v}{u}\right|_{z=0} = \frac{au_g^\circ + \lambda_y}{au_g^\circ + \lambda_x}. \end{aligned} \tag{2.16}$$

Hence,

$$\tan(\alpha_\infty - \alpha_0) = \frac{au_g^\circ(\lambda_y - \lambda_x)}{\lambda_x^2 + \lambda_y^2 + au_g^\circ(\lambda_x + \lambda_y)} \tag{2.17}$$

where $\Delta\alpha = \alpha_\infty - \alpha_0$ is the total rotation of the wind vector in the atmospheric layer. But according to (5.25) in Chapter 3,

$$\lambda_x = -\left(\frac{g}{fT_0}\right)\frac{\partial T}{\partial y}, \qquad \lambda_y = \left(\frac{g}{fT_0}\right)\frac{\partial T}{\partial x}, \tag{2.18}$$

so that λ_x and λ_y are not restricted in sign and magnitude. In particular, if $\lambda_x = \lambda_y$, then $\alpha_\infty = \alpha_0 = \pi/4$ and $\Delta\alpha = 0$ (no rotation of the wind with the height). At $\lambda_x = 0$

$$\alpha_\infty = \pm\frac{\pi}{2}, \qquad \tan\alpha_0 = 1 + \frac{\lambda_y}{au_g^\circ} \gtrless 1$$

depending on the sign of λ_y. This case is illustrated in Figure 10.7.

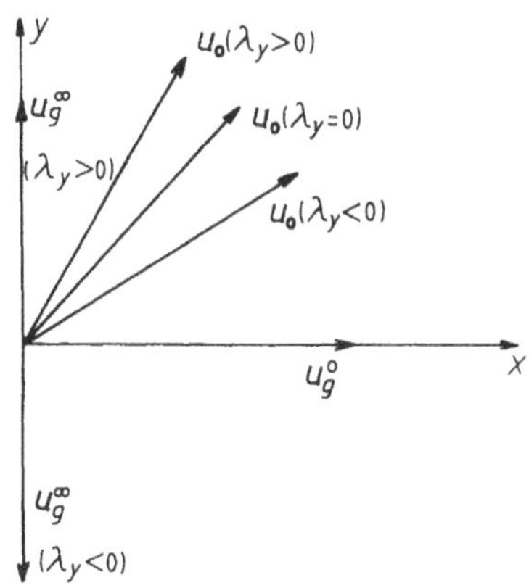

Fig. 10.7. Mutual disposition of the vector $\mathbf{u}_g^\infty$, $\mathbf{u}_g^\circ$ and $\mathbf{u}_0$ (surface wind) at $\lambda_x = 0$ and different signs of λ_y (after [35]).

It is seen that, unlike the barotropic PBL, a left rotation is also possible. Note that the sign of λ_y depends on the kind of advection (cold or warm) alongside $0x$. This conclusion is confirmed by the observations.

These model results, though only qualitative, show that the introduction of a variable-with-height geostrophic wind $M_g(z)$, i.e., baroclinicity, is of first-degree importance. Then comes the improvement due to variability of K with z.

An example of baroclinic PBL is the one in the vicinity of the seashore. The source of baroclinicity is the thermal inhomogeneity of the underlying surface.

3. NONLINEAR *l*-MODELS OF THE PBL

a. Explicit Expressions for $l(z)$

Again we assume barotropic PBL, i.e., $\mathbf{u}_g(z) = \text{const}_z$. Then Equation (1.8) reads

$$X'' + f\frac{Y}{K} = 0, \qquad Y'' - f\frac{X}{K} = 0 \tag{3.1}$$

where $X = Ku'$, $Y = Kv'$, $(') = \partial/\partial z$. On the other hand, extending expression (1.21) in Chapter 9, for the PBL, we can write

$$K = l^2 \left|\frac{\partial \mathbf{u}}{\partial z}\right|^2 = \left(\frac{l^2}{K}\right)(X^2 + Y^2)^{1/2}, \tag{3.2}$$

i.e.,

$$K = l(X^2 + Y^2)^{1/4}, \tag{3.3}$$

where $l = l(z)$ is the mixing length. As a simple geometrical characteristic of turbulence (vertical scale), an a priori choice of l can be easier motivated than that of K. Having $l(z)$ specified, system (3.1)–(3.3) becomes closed but, unlike K models, it is essentially nonlinear and can be solved only by numerical methods.

The simplest assumption concerning l is

$$l = \kappa z, \tag{3.4}$$

i.e., an unlimited growth of l in the whole PBL. More realistic and widely used is the expression for $l(z)$ proposed by Blackadar (1962)

$$\frac{1}{l} = \frac{1}{\kappa z} + \frac{1}{\lambda}, \tag{3.5}$$

i.e., $l(z) = \kappa z/(1 + \kappa z/\lambda)$.

Obviously, $\lambda = l(\infty)$ and, according to the Blackadar's interpretation,

$$\lambda = 2.7 \times 10^{-4}\ |\mathbf{u}_g|/f, \tag{3.6}$$

where $f = 2\omega \sin\varphi (> 0$ in the Northern Hemisphere). Similar in structure is the expression proposed by Lettau (1962)

$$l(z) = \frac{\kappa z}{1 + \nu z^{5/4}}. \tag{3.7}$$

A number of authors have used an exponential formula –

$$l(z) = l_\infty\left(1 - \exp\left[-\frac{\kappa z}{l_\infty}\right]\right). \tag{3.8}$$

The list of similar expressions for $l(z)$ could be continued. Some of them, and particularly (3.5), have been tested in numerical models for neutral stratification and the results are good. Appropriately modified, they could also be used for stratified PBL. In this case, however, implicit expressions for $l(z)$ seem to have the advantage.

b. Implicit Expressions for $l(z)$

Generalizing Kármán's formula, Equation (2.9) in Chapter 9, Rossby (1932) proposed

$$l = -\kappa \left| \frac{\partial \mathbf{u}}{\partial z} \right| \Big/ \frac{\partial}{\partial z} \left| \frac{\partial \mathbf{u}}{\partial z} \right| \tag{3.9}$$

for the mixing length in the PBL. With this, system (3.1) and (3.2) is closed. However, (3.9) does not reflect the influence of the stratification. From this point of view, the most suitable implicit expression for $l(z)$ is that based on the energy balance equation for the PBL

$$K \left| \frac{\partial u}{\partial z} \right|^2 - \alpha_\theta \beta K \frac{\partial \theta}{\partial z} + \alpha_b \frac{\partial}{\partial z} K \frac{\partial b}{\partial z} = \epsilon, \tag{3.10}$$

which can be written in the form

$$K^{-1}(X^2 + Y^2) - \alpha_\theta \beta K \frac{\partial \theta}{\partial z} + \alpha_b \frac{\partial}{\partial z} K \frac{\partial b}{\partial z} = \epsilon. \tag{3.11}$$

Then a possible generalization of (3.9) according to [72] would be

$$l = -\tilde{\kappa} \psi \Big/ \frac{\partial \psi}{\partial z} \tag{3.12}$$

where

$$\psi = \frac{\epsilon}{K} = K^{-2}(X^2 + Y^2) - \alpha_\theta \beta \frac{\partial \theta}{\partial z} + \frac{\alpha_b}{K} \frac{\partial}{\partial z} K \frac{\partial b}{\partial z}. \tag{3.13}$$

As additional equations, Kolmogorov's relations are used:

$$K = l\sqrt{b}, \qquad \epsilon = c_1 \frac{b^{3/2}}{l}. \tag{3.14}$$

According to [72], $\tilde{\kappa} = 2\kappa c_1^{1/4}$. A combination of (3.12) and (3.14) yields a simple relation between l and b:

$$l' - \left(\frac{b'}{2b} \right) l = \kappa b', \tag{3.15}$$

where $(') = d/dz$.

Since the structure of expression (3.12) ensures the necessary dimension of l, some authors take $\psi = (\epsilon/K)^{1/2} = (c_1 b/l^2)^{1/2}$ which physically is just as admissible as (3.13), but may have some computational advantages.

To close the system, we need one more equation for the temperature. If we are not interested in details of the temperature profile $\theta(z)$ in the PBL but wish to investigate the influence of the stratification on the wind and turbulent characteristics only, we can take the temperature equation in the simple form

$$\alpha_\theta \frac{\partial}{\partial z} K \frac{\partial \theta}{\partial z} = 0 \tag{3.16}$$

which implies

$$q = -\alpha_\theta \rho c_p K \frac{\partial \theta}{\partial z} = \text{const}_z \tag{3.17}$$

in the whole PBL.

So, we have six equations – (3.1), (3.11), (3.12) and (3.14) – for the six unknown functions of X, Y, K, l, b, ϵ. The system is closed but it is strongly nonlinear and can only be treated numerically. Many contemporary mathematical models of the PBL are based on this system.

Boundary conditions for X and Y can be easily formulated. Regardless of the stratification

$$\begin{aligned} X &\to v_*^2, \qquad Y \to 0 \quad \text{at } z \to z_0, \\ X &\to 0, \qquad Y \to 0 \quad \text{at } z \to \infty. \end{aligned} \tag{3.18}$$

More difficult is the question about the boundary conditions of b. At arbitrary stratification, due to the existence of a sublayer of dynamic turbulence, in agreement with (2.13) of Chapter 9, we can write

$$b \to \frac{v_*^2}{c_1^{1/2}} \quad \text{at } z \to z_0. \tag{3.19}$$

In the case of unstable stratification, the Monin–Obukhov similarity theory predicts the existence of a sublayer of thermal turbulence, so that

$$b = b(q_*, \beta, z) = \text{const } (q_* \beta)^{2/3} z^{2/3} \quad \text{at } z \to \infty \tag{3.20}$$

Under stable and neutral stratification, however,

$$b \to 0 \quad \text{at } z \to \infty \tag{3.21}$$

since $\partial \mathbf{u}/\partial z \to 0$ at $z \to \infty$ and in both cases the mean wind shear is the only reason for the flow to be turbulent.

The equations and the boundary conditions so formulated are enough for calculating the wind and turbulence characteristics in the PBL as functions of the height z and governing parameters $c_g = |u_g|$, z_0, f, $\beta = g/\overline{T}$, q_*, which are considered to be constant and given in advance. The friction velocity v_* and the angle α between the surface wind $\mathbf{u}_0$ and the geostrophic wind $\mathbf{u}_g$, i.e., the total rotation of the wind in the PBL, are still to be determined. If the axis $0x$ is oriented alongside the $\mathbf{u}_g$, then

$$u_g = c_g \cos \alpha, \qquad v_g = c_g \sin \alpha. \tag{3.22}$$

Here the term 'surface wind' means the wind in the surface layer whose direction does not change with z if $z < h$.

4. SIMILARITY THEORY FOR THE PBL

a. Parameterization of the PBL

The dynamics of the PBL is determined by a large complex of factors, represented by an aggregate of the dimensional physical parameters. The setting of a finite number of these (as few as possible) apart as governing parameters for the purpose of the dimensional analysis is an extremely difficult matter. Here we are going to do this in the usual way for the Ekman PBL approximations: stationarity and horizontal homogeneity. Two groups of parameters (*external* and *internal*) are distinguished.

External for the PBL parameters, governing the turbulent regime, are the following:

(1) The geostrophic wind $c_g = |\mathbf{u}_g|$ or the pressure gradient force $|\nabla p/\rho|$, which is the same. This is the 'moving force' for the PBL which is determined by the large-scale (synoptic) processes and in the problems of boundary-layer dynamics it can really be considered as external and prescribed. Aside from the strongly baroclinic areas (for instance atmospheric fronts) $\mathbf{u}_g$ is constant with respect to z.

(2) The Coriolis parameter $f = 2\omega \sin \varphi$ ($f > 0$ in the Northern Hemisphere). In surface-layer dynamics, the influence of f is negligible but in the PBL, due to the larger scale of motions, it must be taken into account.

(3) The roughness parameter z_0 of the Earth's surface. The air-surface interaction depends essentially on z_0.

(4) The stratification. If $\theta_0 = \theta(z_0)$ and $\theta_g = \theta(h_g)$ is the potential temperature at the top and bottom of the PBL, then obviously $\Delta\theta = \theta_g - \theta_0$ will be an integral stratification parameter of the entire PBL ($\Delta\theta > 0$ at stable, $\Delta\theta < 0$ at unstable, and $\Delta\theta = 0$ at the neutral stratification). Together with $\Delta\theta$ the convection parameter $\beta = g/\overline{\theta} \approx g/\overline{T}$ should be taken into account.

We now make a fundamental assumption that the turbulent regime in the temperature-stratified PBL is governed by the five above-listed dimensional parameters: c_g, f, z_0, $\Delta\theta$ and β. Three of them have independent dimensions. Consequently, only two nondimensional combinations (numbers) can be constructed:

$$\mathrm{Ro} = \frac{c_g}{f z_0}, \qquad \mathrm{S} = \beta \frac{\Delta\theta}{f c_g}, \tag{4.1}$$

the first one being called the *surface Rossby number*. Moreover three scales also exist: for length – $\lambda_g = c_g/f$, for velocity – c_g, and for temperature – $\Delta\theta$. Then, in agreement with our fundamental assumption, one can assert that the statistical characteristics of the velocity fields and temperature in the PBL, which were not given dimensions by the previous scales, can depend on the external parameters only by means of Ro and S. Therefore, for such *internal* parameters of the PBL as the friction velocity v_*, the angle α_0 between $\mathbf{u}_0$ and $\mathbf{u}_g$, and the scale $\theta_* = -q_*/\kappa v_*$ (see (3.1) in Chapter 9) we can write

$$\frac{v_*}{c_g} = \eta_u(\mathrm{Ro}, \mathrm{S}), \qquad \alpha_0 = \eta_\alpha(\mathrm{Ro}, \mathrm{S}), \qquad \frac{\theta_*}{\Delta\theta} = \eta_\theta(\mathrm{Ro}, \mathrm{S}). \tag{4.2}$$

In accordance with Equation (2.8) in Chapter 9, $v_*/c_g = C_{dg}$ is called *a geostrophic drag coefficient*. Obviously, $\theta_*/\Delta\theta$ is an integral characteristic of the heat transfer in the PBL. The universal functions $\eta_i(\mathrm{Ro}, \mathrm{S})$, $i = u, \alpha, \theta$ can be empirically or theoretically determined.

b. Universal Dependences

We consider the question of vertical profiles of the wind $\mathbf{u}(z)$ and temperature $\theta(z)$ from the similarity theory point-of-view, proposed by Kasansky and Monin (1960) and subsequently developed by many other authors.

In comparison with the analogous theory for the surface layer (Section 3, Chapter 9), some specific features can be mentioned. First of all, a new parameter arises – $f = 2\omega \sin \varphi$ – in addition to v_*, q_*, β. Moreover, under stationary conditions and neglected radiational heat exchange, the heat flux q_* is considered to be independent* of z, not only in the surface layer but in the whole PBL. Then, from parameters v_*, q_*, β, f two length scales can be constructed instead of one:

$$L = -\frac{v_*^3}{\kappa\beta q_*}, \qquad \lambda = \frac{\kappa v_*}{f}, \tag{4.3}$$

so that

$$\mu = \frac{\lambda}{L} = \kappa^3 \frac{\theta_*/\Delta\theta}{v_*/c_g} \mathrm{S} \tag{4.4}$$

is a stability parameter (proportional to S from (4.2)).

Generalizing formula (3.10) of Chapter 9, we can write

$$\begin{aligned} \mathbf{u}(z_2) - \mathbf{u}(z_1) &= \frac{v_*}{\kappa}\left[\psi\left(\frac{z_2}{\lambda}, \mu\right) - \psi\left(\frac{z_1}{\lambda}, \mu\right)\right], \\ \theta(z_2) - \theta(z_1) &= \theta_*\left[\psi_\theta\left(\frac{z_2}{\lambda}, \mu\right) - \psi_\theta\left(\frac{z_1}{\lambda}, \mu\right)\right], \end{aligned} \tag{4.5}$$

where $z_0 < z_1, z_2 \leqslant h_g$, $\psi = (\psi_u, \psi_v)$ and ψ_θ are the universal functions which can be determined with accuracy up to additive constants. It is convenient to choose these constants in such a way that

$$\psi_u = \psi_v = \psi_\theta = 0 \quad \text{at } z = h_g. \tag{4.6}$$

The latter condition requires, however, that the height h_g of the PBL in this problem be defined. If we assume $v_* \sim 10^1$ cm s^{-1} and $f \sim 10^{-4}$ s^{-1}, then from (4.3) $\lambda \sim 10^2$–10^3 m. Therefore, $\lambda \sim h_g \sim 10^3$ m – the so-far estimated value of h_g – and we may write

$$h_g = \gamma\lambda = \gamma \cdot \frac{\kappa v_*}{f} = \gamma \cdot \kappa\lambda_g \frac{v_*}{c_g}, \tag{4.7}$$

where $\lambda_g = c_g/f$, $\gamma \sim 1$. Then

$$\begin{aligned} u(h_g) &= u_g = c_g \cos \alpha_0, \qquad \theta(h_g) = \theta_g, \\ v(h_g) &= v_g = c_g \sin \alpha_0, \end{aligned} \tag{4.8}$$

and expressions (4.5) take the form

$$\begin{aligned} \mathbf{u}(z) - \mathbf{u}_g &= \frac{v_*}{\kappa}\,\psi\left(\frac{z}{\lambda}, \mu\right), \\ \theta(z) - \theta_g &= \theta_*\psi_\theta\left(\frac{z}{\lambda}, \mu\right). \end{aligned} \tag{4.9}$$

* The meteorological variables indeed respond to the flux divergence and not to the flux itself.

These formulae are valid at $z < h_g$ and also in the surface layer ($z < h$). However, at $z \ll h$, regardless of the stratification, a sublayer of dynamical turbulence exists in which the logarithmic laws hold

$$u(z) = \frac{v_*}{\kappa} \ln \frac{z}{z_0}, \quad v(z) = 0, \quad \theta(z) = \theta_0 + \theta_* \ln \frac{z}{z_0}. \tag{4.10}$$

The two groups of formulae (4.9) and (4.10) should be consistent with each other. Hence, the next relationships follow:

$$\begin{aligned}
&\ln \frac{\lambda}{z_0} - \frac{\kappa}{v_*} c_g \cos \alpha_0 = \psi_u\left(\frac{z}{\lambda}, \mu\right) - \ln \frac{z}{\lambda}, \\
&-\frac{\kappa}{v_*} c_g \sin \alpha_0 = \psi_v\left(\frac{z}{\lambda}, \mu\right) \cdot \operatorname{sign} f, \\
&\ln \frac{\lambda}{z_0} - \frac{\Delta\theta}{\theta_*} = \psi_\theta\left(\frac{z}{\lambda}, \mu\right) - \ln \frac{z}{\lambda},
\end{aligned} \tag{4.11}$$

where sign f accounts for the opposite rotation of the wind in both hemispheres of the Earth. The left-hand sides of these equations, however, do not depend on z. Consequently, in the sublayer under consideration ($z \ll h$), the right-hand sides must also not depend on z. Hence, it follows that

$$\begin{aligned}
&\lim_{\zeta \to 0} \left[\psi_u(\zeta, \mu) - \ln \zeta\right] = B(\mu) + \ln \kappa, \\
&\lim_{\zeta \to 0} \psi_v(\zeta, \mu) = A(\mu), \\
&\lim_{\zeta \to 0} \left[\psi_\theta(\zeta, \mu) - \ln \zeta\right] = C(\mu) + \ln \kappa,
\end{aligned} \tag{4.11$'$}$$

where $\zeta = z/\lambda$, the term $\ln \kappa$ is added for convenience and $A(\mu)$, $B(\mu)$, $C(\mu)$ are dimensionless universal functions of the stability parameter μ. Finally, (4.11′) and (4.11) yield the so-called *resistance* laws:

$$\begin{aligned}
&\ln(\mathrm{Ro} \cdot C_{dg}) = B(\mu) + \left[\left(\frac{\kappa}{C_{dg}}\right)^2 - A^2(\mu)\right]^{1/2} \\
&\sin|\alpha_0| = \left(\frac{C_{dg}}{\kappa}\right) A(\mu) \\
&\frac{\theta_*}{\Delta\theta} = \left[\ln(\mathrm{Ro} \cdot C_{dg}) - C(\mu)\right]^{-1}
\end{aligned} \tag{4.12}$$

where the approximation $\alpha_\theta = 1\,(K = K^\theta)$ has been used (as well as in (4.10)).

The resistance laws represent a substantial step forward in discovering the general dependences (4.2), since with (4.12) the problem is reduced to the empirical or theoretical determination of the three functions $A(\mu)$, $B(\mu)$, $C(\mu)$ of one argument μ, which is much easier than in the case of two or more arguments. At neutral stratification, $\mu = 0$ and the three numbers should be determined only: $A(0)$, $B(0)$, $C(0)$.

c. Experimental Data and Significance of the Problem

During the past few decades much experimental data has been obtained on the structure of stationary and horizontally-homogeneous atmospheric boundary layers. The similarity hypothesis was generally confirmed and the universal functions $A(\mu)$, $B(\mu)$, $C(\mu)$ were determined by many authors using the same or independent data of observations. The results have been published in the form of tables, diagrams or empirical approximation formulae. As an example, the results of Wipperman [69] are presented here. Let us mention that many attempts have been made at the theoretical determination of the universal functions too.

To determine $A(\mu)$, $B(\mu)$, $C(\mu)$ empirically, one needs data on the micrometeorological characteristics v_*, α_0, θ_* on one hand, and on the external parameters c_g, $\Delta\theta$, z_0 (f and β are constants) on the other. Using data from other authors, Wipperman [69] proposed the following empirical formulae

$$A(\mu) = 1.83 + 2.77\,e^{\mu/20},$$
$$B(\mu) = 3.67 - 2.77\,e^{\mu/20}, \qquad (4.13)$$

valid for stable ($\mu > 0$) as well as for unstable ($\mu < 0$) stratification. Consequently, $A(0) = 4.6$, $B(0) = 0.9$. The graphs of these functions are shown in Figure 10.8. $C(\mu)$ looks like $B(\mu)$ but with a greater uncertainty (not shown in Figure 10.8).

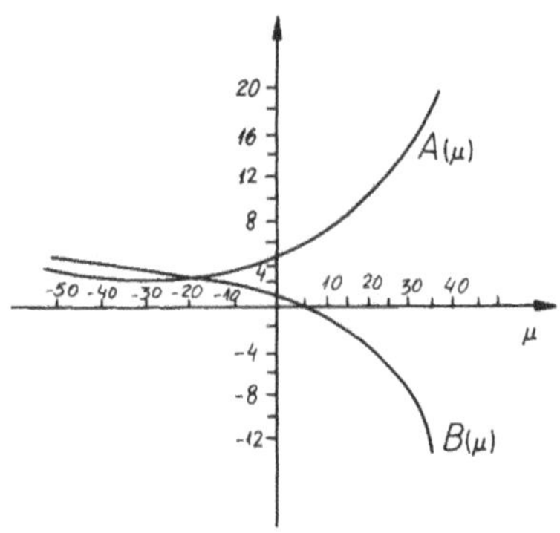

Fig. 10.8. The universal functions $A(\mu)$ and $B(\mu)$ approximated by (4.13) (after [69]).

Having the functions $A(\mu)$, $B(\mu)$, $C(\mu)$ determined, Equations (4.12) can be solved (numerically) with respect to C_{dg}, α_0 and $\theta_*/\Delta\theta$ to obtain these quantities, in agreement with (4.2), as functions of Ro and μ ($\sim S$ – see formula (4.4)!). When the results are tabulated or nomograms are built, it is possible, having synoptic data (diagnostic or prognostic) on $c_g = |\mathbf{u}_g|$ and $\Delta\theta$ for a particular region (given z_0 and f), to determine the internal parameters v_*, θ_*, α_0 and then on the basis of (4.5), or (3.10) in Chapter 9, to build the vertical profiles of the wind and temperature and to use them for solving various applied problems, i.e., when investigating turbulent diffusion of admixtures in the SL and PBL.

The practical significance of the problem for parametrization of the turbulent process in the PBL is not confined only to the above. Numerical modelling and experimentation in the field of general atmospheric and ocean circulation stimulates the development of various parametrization schemes. With a finite difference approximation of the governing system of partial differential equations, the vertical step Δz, even in models with the greatest resolution, is of the order of 1 km, which is comparable with the thickness h_g of the PBL. However, with such a vertical step, it is not possible to calculate, for example, the turbulent fluxes in the PBL, reflecting the atmosphere-underlying surface interaction, and thus playing a fundamental role in the evolution of the large-scale fields [35]; That is why the processes with scales less than the vertical step Δz have to be taken indirectly into consideration by relating them functionally to the parameters of the large-scale processes which are obtained directly from calculations in the model. This is achieved by a parametrization method that has become quite current in meteorological investigations.

5. VERTICAL MOTIONS IN THE PBL

a. General Information and Formulae

Three kinds of vertical motions exist in the atmosphere and, particularly, in the PBL – micro-scale, meso-scale and large-scale. To the first kind belong the turbulent (fluctuating) vertical motions whose velocity is usually denoted by w'. It can be precisely measured which permits the direct determination of the turbulent fluxes $\overline{u'w'}$, $\overline{w'\theta'}$, etc. In the PBL, $|w'| \sim 10^0$–10^1 cm s^{-1}.

To the second kind belong the convective vertical motions, originating under the influence of the Archimedean force. A typical example is the convection leading to a Cb-cloud formation having horizontal scales of order 10^0–10^1 km. Though the convective jets are turbulent, their instantaneous and mean velocity can also be measured instrumentally. It is of order 10^0–10^1 m s^{-1}.

Further on, we shall be interested in the third kind of vertical motions, generated by the horizontal divergence or convergence of the wind field in the PBL. These are flow-regular motions, having one and the same sign over territories with characteristic dimensions of the order 10^3 km, which is the reason why they are called large-scale vertical motions. Regardless of their slowness ($w \sim 1$ cm s^{-1}) they are responsible for St cloud formation in the PBL over large territories. This is of practical importance. Moreover, the vertical velocity $w_g = w(h_g)$ on the top of the PBL as an integral characteristic of the dynamics of this layer, transfers its influence to the dynamics of the free atmosphere, i.e., this is a kind of feedback mechanism. The information about w_g is of great importance for creating more perfect numerical schemes for weather forecasting by assimilation of the PBL influence in the lower boundary condition for prognostic equations. Unlike the first two cases, however, so far no methods have been devised for measuring this vertical velocity. Then, the only way to evaluate its magnitude is the computational one, on the basis of the equations it enters into.

We start with the equations of motion (1.5) for the PBL:

$$\mathbf{u} = \mathbf{u}_g - \frac{1}{f}\frac{\partial}{\partial z}\left[\mathbf{k} \times \left(K\frac{\partial \mathbf{u}}{\partial z}\right)\right] = \mathbf{u}_g - \frac{1}{f}\frac{\partial}{\partial z}(\mathbf{k} \times \mathbf{T}). \tag{5.1}$$

where $\mathbf{T} = (X, Y) = \tau/\rho$ is the Reynolds stress vector (1.3), and the continuity equation

$$\frac{\partial w}{\partial z} = -\nabla \cdot \mathbf{u} = -D, \tag{5.2}$$

with $w = 0$ at $z = 0$. Thus,

$$w(x,y,z) = -\int_0^z D(x, y, z')\, dz'. \tag{5.3}$$

The dependence of w on x, y comes through $\mathbf{u}_g(x, y)$ and $K(x, y, z)$. For simplicity of the notation, we shall write only $w = w(z)$.

From (5.1) and $\nabla \cdot \mathbf{u}_g = 0$ we find

$$\nabla \cdot \mathbf{u} = f^{-1}\mathbf{k} \cdot \left(\nabla \times \frac{\partial \mathbf{T}}{\partial z}\right), \tag{5.4}$$

$$\zeta = \mathbf{k} \cdot (\nabla \times \mathbf{u}) = \zeta_g - f^{-1}\frac{\partial}{\partial z}(\nabla \cdot \mathbf{T}). \tag{5.5}$$

It is seen that the wind divergence and vorticity in the PBL differ from its values in the free atmosphere (0 and ζ_g) because of height changes of the stress $\mathbf{T}$. Equations (5.3) and (5.4) yield

$$w(z) = -f^{-1}\mathbf{k} \cdot (\nabla \times \mathbf{T})\Big|_0^z. \tag{5.6}$$

But $|\mathbf{T}| \to 0$ at $z \to \infty$ so that

$$w(x, y, \infty) = f^{-1}\mathbf{k} \cdot (\nabla \times \mathbf{T})_{z=0} \tag{5.7}$$

which could be identified with $w(x, y, h_g) = w_g$.

Alternative expressions, more convenient in some respects, can be obtained by using the first half of Equation (5.1). Then, instead of (5.4) and (5.5), we have

$$\nabla \cdot \mathbf{u} = f^{-1}\frac{\partial}{\partial z}\left[K\frac{\partial \zeta}{\partial z} + \mathbf{k} \cdot \left(\nabla K \times \frac{\partial \mathbf{u}}{\partial z}\right)\right], \tag{5.8}$$

$$\zeta = \zeta_g - f^{-1}\frac{\partial}{\partial z}\left[K\frac{\partial(\nabla \cdot \mathbf{u})}{\partial z} + \nabla K \cdot \frac{\partial \mathbf{u}}{\partial z}\right], \tag{5.9}$$

where $\nabla = (\partial/\partial z, \partial/\partial y, 0)$. Similar to (5.6), we derive

$$w(z) = -f^{-1}\left[K\frac{\partial \zeta}{\partial z} + \mathbf{k} \cdot \left(\nabla K \times \frac{\partial \mathbf{u}}{\partial z}\right)\right]_0^z. \tag{5.10}$$

At $\nabla K = 0$, i.e., $K = K(z)$, these formulae simplify.

b. Ekman PBL

Let $\nabla K = 0$. Then for $\mathbf{u}$ we use the classical solution (1.17)

$$\mathbf{u} = (1 - F_1)\mathbf{u}_g + F_2 k \times \mathbf{u}_g, \tag{5.11}$$

$$F_1(z) = \exp[-az] \cos az, \qquad F_2(z) = \exp[-az] \sin az, \tag{5.12}$$

and $\mathbf{u}_g = \mathbf{u}_g(x, y)$. Hence, we obtain

$$D(z) = \nabla \cdot \mathbf{u}(z) = F_2(z) \nabla \cdot \mathbf{s}_g = -F_2(z)\zeta_g, \tag{5.13}$$

where

$$\mathbf{s}_g = \mathbf{k} \times \mathbf{u}_g = (-v_g, u_g, 0), \qquad \zeta_g = \mathbf{k} \cdot (\nabla \times \mathbf{u}_g) = -\nabla \cdot \mathbf{s}_g = \frac{\nabla^2 p}{\rho f}. \tag{5.14}$$

We substitute (5.13) into (5.3) and integrate

$$w(z) = \frac{\zeta_g}{2a} [1 - \exp[-az] (\sin az + \cos az)]. \tag{5.15}$$

Hence, at $z = \infty$ and $z = h_g = \pi/a$ (the top of the PBL) we find

$$w(\infty) = \frac{\zeta_g}{2a}, \qquad w(h_g) = w_g = w(\infty)(1 + e^{-\pi}). \tag{5.16–17}$$

The difference between $w(\infty)$ and w_g is less than 5%. Assuming that this is also the case in more sophisticated models, in future we shall let $w_g \approx w(\infty)$. Thus, for the Ekman PBL

$$w_g = \frac{\zeta_g}{2a} = \zeta_g \left(\frac{K}{2f}\right)^{1/2} = \frac{1}{2\pi} h_g \zeta_g. \tag{5.18}$$

Since $a > 0$ ($h_g > 0$), then the sign of w_g will coincide with that of ζ_g, i.e., of $\nabla^2 p$ – the Laplacian of the surface pressure (barotropic atmosphere). In a cyclone (anticyclone) $\nabla^2 p > 0$ ($\nabla^2 p < 0$) and we have $w_g > 0$ ($w_g < 0$). However, one must not identify the *frictional vertical velocity* w_g with the real one w at $z \geqslant h_g$, since w is also determined by dynamical factors (the nonlinear terms in the equations). As a result of interaction between the viscous (turbulent) and dynamic effects, it may happen that downward vertical motions in cyclonic areas and upward motions in anticyclonic areas would be observed.

The vertical motions in the PBL can also be interpreted as an element of the so-called *secondary circulation*. According to Figure 10.3, $F_2(z) \geqslant 0$ at $z < h_g = \pi/a$, so that the divergence D and the geostrophic vorticity ζ_g have opposite signs. For instance, in cyclonic geostrophic vortex $\zeta_g > 0$ and consequently (see formula (5.13) $D(z) < 0$ at $z < h_g$. Moreover, at $z = h_g/4 = \pi/4a$, D has a maximum $D_m = -\zeta_g/3$. Thus, corresponding to a large-scale cyclonic geostrophic vortex in the free atmosphere is the frictional convergence of wind and upward motion in the PBL (Figure 10.9).

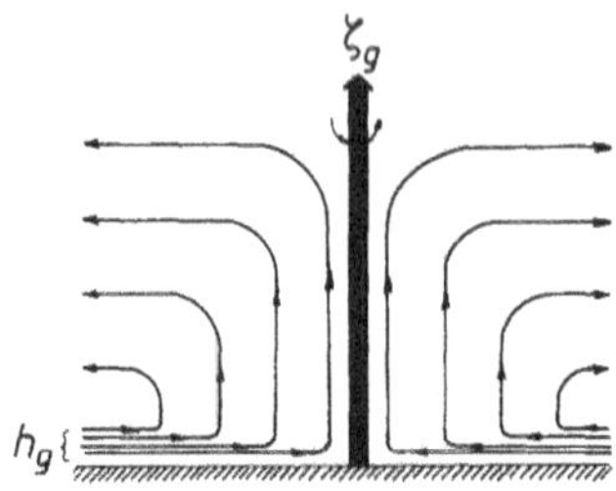

Fig. 10.9. Schematic representation of the secondary circulation in the PBL (see details in the text) (after [27]).

Let us now suppose that $\nabla K \neq 0$. In the general case $K = K(x, y, z)$. If we introduce an averaged-for-the-whole PBL eddy viscosity coefficient

$$K(x, y) = \frac{1}{h_g} \int_0^{h_g} K(x, y, z)\, dz, \tag{5.19}$$

we can use the simple Ekman solution (5.11), (5.12) in which

$$a(x, y) = \left[\frac{f}{2K(x, y)}\right]^{1/2} \tag{5.20}$$

and consequently

$$h_g(x, y) = \frac{\pi}{a(x, y)}. \tag{5.21}$$

Syrakov (1981) proposed a theory leading to an interesting expression for the averaged K from (5.19):

$$\begin{aligned} K &= 2m(\mathrm{Ro})\left(\frac{c_g^2}{f}\right)\Phi^2(\mathrm{Ro}, \mathrm{S}), \\ \Phi &= [m^{1/2}\mathrm{S} + (m\mathrm{S}^2 + 1)^{1/2}]^{-1}, \quad \mathrm{S} = \beta\frac{\Delta\theta}{fc_g}, \end{aligned} \tag{5.22}$$

where $m(\mathrm{Ro})$ is a decreasing function of Ro. Under stable conditions, $\mathrm{S} > 0$ and $\Phi < 1$ while in the opposite case, $\mathrm{S} < 0$ and $\Phi > 1$. In the intermediate case of neutral stratification $\mathrm{S} = 0$ and $\Phi = 1$, so that

$$K = 2m(0)\frac{c_g^2}{f}. \tag{5.22'}$$

The latter expression was proposed earlier in [16] as purely empirical with $2m(0) = 6.25 \times 10^{-6}$. Having definition (4.1) of Ro and S in mind, we see that the (x, y)-dependence of K is clearly determined by the horizontal inhomogeneities of u_g, $\Delta\theta$ and, eventually, z_0. Let us remember that we are considering a flat underlying surface. Through (5.20) and (5.21), the PBL thickness may also be expressed by (5.22).

The vertical velocity in case of $\nabla K \neq 0$ should be calculated through the use of (5.10). Some comparisons have shown [16] that in (5.10) the ∇K-term may be of the order of the K-term. The problem needs further investigation.

c. Vorticity Generation in the PBL

For the Ekman model, (5.9) and (5.11) yield

$$\zeta(x, y, z) = [1 - F_1(z)]\,\zeta_g(x, y). \tag{5.23}$$

It is seen from Figure 10.3 that $1 - F_1 > 0$. Thus, the vorticity ζ in the PBL has the sign of ζ_g but is less in magnitude due to the friction. At $z = 0$ $\zeta = 0$.

However, the eddy viscosity in the PBL not only influences the stationary value of the vorticity but also the vorticity generation process. The latter effect can be studied on the basis of nonsimplified momentum equations

$$\frac{\partial \mathbf{u}}{\partial t} + (\mathbf{u} \cdot \nabla)\mathbf{u} + w\frac{\partial \mathbf{u}}{\partial z} = f\mathbf{k} \times (\mathbf{u}_g - \mathbf{u}) + \frac{\partial}{\partial z} K \frac{\partial \mathbf{u}}{\partial z}. \tag{5.24}$$

Hence, an equation for $\zeta = \mathbf{k} \cdot (\nabla \times \mathbf{u}) = \partial v/\partial x - \partial u/\partial y$ can be derived. We are interested in the influence of the last term in (5.24). Consequently,

$$\left(\frac{\partial \zeta}{\partial t}\right)_{fr} = \frac{\partial}{\partial z}\left[K\frac{\partial \zeta}{\partial z} + \mathbf{k} \cdot \left(\nabla K \times \frac{\partial \mathbf{u}}{\partial z}\right)\right]. \tag{5.25}$$

Let $\nabla K = 0$. Making use of (5.10), (5.12) and (5.14), we obtain

$$\left(\frac{\partial \zeta}{\partial t}\right)_{fr} = -F_2(z)\zeta_g f = fD(z). \tag{5.25'}$$

In the case of a cyclonic geostrophic vorticity ($\zeta_g > 0$), (5.23) yields $\zeta > 0$. Then (5.25′) predicts $\partial\zeta/\partial t < 0$, i.e., the turbulent friction in the PBL tends to decrease the initial cyclonic vorticity with time. If $\zeta_g < 0$, then, according to (5.23) and (5.25′), $\zeta < 0$ and again $\partial|\zeta|/\partial t < 0$, i.e., a tendency exists for a reduction of the initial anticyclonic vorticity with time. Thus, the turbulence in the PBL always weakens the field of the vorticity, imposed by the synoptic situation in the free atmosphere. Moreover, $\partial\zeta/\partial t$ has a maximum at $z = h_g/4$, since $D = D_{max} = -\zeta_g/3$ at $z = h_g/4$ in the Ekman model.

So far we have been considering the response of the PBL to a given field of vorticity in the free atmosphere ($z > h_g$). However, an inverse influence of the PBL on the dynamics of the free atmosphere also exists, by means of the generated frictional vertical velocity w_g at the level $z = h_g$.

To study this effect, we start with the vorticity equation for a barotropic free atmosphere

$$\frac{d\zeta_a}{dt} = -\zeta_a D \approx f\frac{\partial w}{\partial z}, \tag{5.26}$$

where $\zeta_a = f + \zeta$. Following [27], we shall neglect the dependency of f on y and shall integrate (5.26) between $z = h_g$ and $z = h_T$ – the height of the tropopause, assuming that $w_T = w(h_T) = 0$:

$$\int_{h_g}^{h_T} \frac{\mathrm{d}\zeta}{\mathrm{d}t}\,\mathrm{d}z = -fw_g.$$

Making one more approximation $\zeta \approx \zeta_g$, we obtain

$$\frac{\mathrm{d}\zeta_g}{\mathrm{d}t} = -\frac{fw_g}{h_T - h_g}.$$

But $h_g \sim 1$ km and $h_T \sim 10$ km so that $h_T - h_g \approx h_T$. On the other hand, (5.18) holds. Consequently

$$\frac{\mathrm{d}\zeta_g}{\mathrm{d}t} = -\frac{\zeta_g}{\tau_g}, \quad \tau_g = \left(\frac{2\pi}{f}\right)\frac{h_T}{h_g}. \tag{5.27}$$

Hence,

$$\zeta_g(t) = \zeta_g(0)\exp\left[-\frac{t}{\tau_g}\right], \tag{5.28}$$

i.e., the vorticity of an air parcel will decay exponentially with time following the motion, because of the dynamic interaction with the friction PBL by means of w_g. At $f \sim 10^{-4}$ s^{-1}, $h_g \sim 1$ km, $h_T \sim 10$ km for the relaxation time we find $\tau_g \approx 174$ h, i.e., nearly a week. Therefore, at $t = (2–3)\tau_g$ practically $\zeta_g \approx 0$ if there is no mechanism compensating the decay.

The analysis in this section can be repeated with more realistic models for the PBL than the classical Ekman model. One can think that, qualitatively, the results will not differ from those obtained here. The introduction of baroclinicity and mountains in the problem considered here, however, is much more important and difficult than of a z-dependent eddy viscosity coefficient. Little progress has been made in this direction so far.

6. SOME SPECIAL QUESTIONS OF PBL THEORY

a. The PBL Above Mountains

Let us assume that the underlying surface is not flat and that rather low and smooth synoptic scale mountains exist. Then, as was mentioned earlier in Section 1, the concept of PBL with an upper boundary representing some continuous surface $h_g(x, y)$, determined by the geometry of the mountain $H(x, y)$ itself (but not necessarily following $H(x, y)$) and by the regularities of the turbulent regime in this layer (Figure 10.10) can still be kept.

In mathematical models, the real topography is always appropriately smoothed by some averaging procedure and $H(x, y)$ is represented by numbers in discrete points or on a map by isolines.

Air flow interacts with mountains. In the case of an individual obstacle and an inviscid

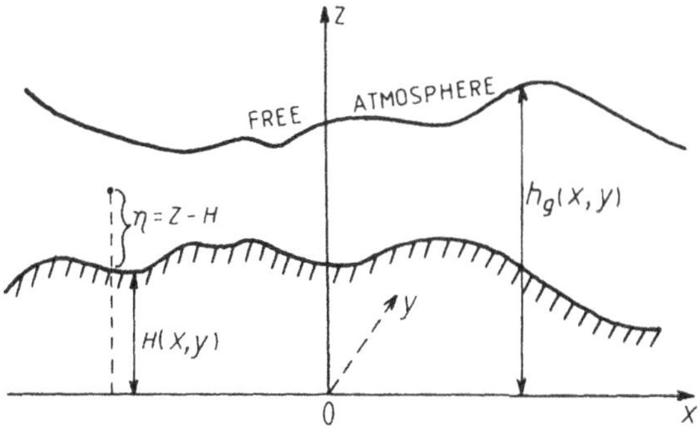

Fig. 10.10. PBL over smooth mountains.

flow, some effects were studied in Chapter 5 (topographic Rossby waves, orographic waves, etc.). However, the existence of a turbulent boundary layer and the necessity of accounting for its influence on the dynamics of the free atmosphere and in the prognostic numerical models requires the joint influence of turbulence and topography, to be studied. The fundamental PBL characteristics of interest to us are, as before, the vertical velocity $w(z)$, vorticity $\zeta(z)$ and divergence $D(z)$. For simplicity, Ekman-type PBL above mountains will be considered. Thus, the results obtained will have a qualitative character only.

We start from the equations

$$\frac{\partial}{\partial z} K \frac{\partial \mathbf{u}}{\partial z} - f\mathbf{k} \times (\mathbf{u} - \mathbf{u}_g) = \mathbf{N},$$

$$\frac{\partial w}{\partial z} = -(\nabla \cdot u) = -D, \tag{6.1}$$

$$\mathbf{N} = \frac{\partial \mathbf{u}}{\partial t} + (\mathbf{u} \cdot \nabla)\mathbf{u} + w \frac{\partial \mathbf{u}}{\partial z}.$$

In principle, a stationary regime ($\partial \mathbf{u}/\partial t = 0$) is possible in this problem. The assumption of 'rather low and smooth' mountains permits us to linearize the problem, as a first approximation, by letting $\mathbf{N} = 0$. As far as K is concerned, it is natural to consider that $K = K(x, y, z)$. A physically acceptable and convenient approximation was proposed by Godev (1970):

$$K = K(\eta), \quad \eta = z - H(x, y). \tag{6.2}$$

Since the flow in the PBL is turbulent, then the lower boundary condition to Equation (6.1) will be

$$u = v = w = 0, \quad \text{at } z = H(x, y). \tag{6.3}$$

Besides

$$\mathbf{u} \to \mathbf{u}_g, \qquad z \to \infty. \tag{6.4}$$

We also assume that $\mathbf{u}_g = \mathbf{u}_g(x, y)$, i.e., barotropic atmosphere. Therefore, the influence of the mountain 'enters' into our model by means of (6.2) and boundary condition (6.3). Even this very simple model yields interesting qualitative conclusions concerning the influence of the mountains on the distribution of w, ζ and D in the PBL.

The mathematical structure of the solution of Equations (6.1) under conditions (6.3) and (6.4) does not depend on a particular form of the function $K(\eta)$. We can write

$$u + iv = (u_g + iv_g)(P + iQ),$$

where

$$P = P(\eta), \qquad Q = Q(\eta), \quad \eta = z - H(x, y).$$

Then

$$\begin{aligned} u(x, y, z) &= u_g(x, y)\,P(\eta) - v_g(x, y)\,Q(\eta) \\ v(x, y, z) &= u_g(x, y)\,Q(\eta) + v_g(x, y)\,P(\eta) \end{aligned} \tag{6.5}$$

$$P(0) = Q(0) = Q(\infty) = 0, \qquad P(\infty) = 1. \tag{6.6}$$

Bearing in mind that $\eta_x = -H_x$, $\eta_y = -H_y$, we find from (6.5)

$$\zeta(\eta) = -Q'(\eta)\,(\nabla H \cdot \mathbf{u}_g) + P'(\eta)\,(\nabla H \cdot \mathbf{s}_g) + P(\eta)\zeta_g \tag{6.7}$$

$$D(\eta) = -P'(\eta)\,(\nabla H \cdot \mathbf{u}_g) - Q'(\eta)\,(\nabla H \cdot \mathbf{s}_g) - Q(\eta)\zeta_g \tag{6.8}$$

where

$$\mathbf{s}_g = \mathbf{k} \times \mathbf{u}_g = (-v_g, u_g, 0).$$

But $(') = \partial/\partial\eta = \partial/\partial z$ so that after the integration of (6.8) in accordance with (6.1) and (6.6), we obtain

$$w(\eta) = -\int_0^\eta D(\eta)\,d\eta = P(\eta)\,(\nabla H \cdot \mathbf{u}_g) +$$

$$+\, Q(\eta)\,(\nabla H \cdot \mathbf{s}_g) + \zeta_g \int_0^\eta Q(\eta')\,d\eta'. \tag{6.9}$$

But

$$\nabla H \cdot \mathbf{u}_g = u_g H_x + v_g H_y, \qquad (\nabla H \cdot \mathbf{s}_g) = u_g H_y - v_g H_x.$$

After some rearrangement of the first two terms in (6.9), and using (6.5), we obtain

$$w(\eta) = \mathbf{u}(\eta) \cdot \nabla H + R(\eta)\zeta_g, \tag{6.10}$$

where $R(\eta)$ stands for the integral in (6.9). Hence,

$$w(\infty) = \mathbf{u}_g \cdot \nabla H + R_\infty.\zeta_g = w_{or} + w_{fr}, \tag{6.11}$$

where w_{or} and w_{fr} are, correspondingly, the orographic and frictional vertical velocities and (6.4) has been used to derive (6.11) from (6.10).

The two-term formula of type (6.11) is widely used as a lower boundary condition in models for numerical weather forecasting. It has been constructed in three steps: first, assuming an inviscid flow over the mountains with velocity $\mathbf{u}_g$ (no boundary layer exists) so that $w_{or} = \mathbf{u}_g \cdot \nabla H$ and second, assuming that the surface is flat (no mountains), but the flow is turbulent and on the basis of some PBL model w_{fr} is calculated at the level h_g (at $z = \infty$, for simplicity). Finally, the two terms are summed up to obtain $w = w_{or} + w_{fr}$. The above derivation of (6.11) shows that this is possible under very special conditions.

To analyze (6.7) and (6.8), let us assume further that $K(\eta) = K = \text{const}$. Then

$$P(\eta) = 1 - \exp[-a\eta] \cos a\eta = 1 - F_1(\eta),$$

$$Q(\eta) = \exp[-a\eta] \sin a\eta = F_2(\eta)$$

so that $P'(0) = Q'(0) = a$ and

$$\begin{aligned} \zeta(0) &= -a[(\nabla H \cdot \mathbf{u}_g) - (\nabla H \cdot \mathbf{s}_g)], \\ D(0) &= -a[(\nabla H \cdot \mathbf{u}_g + (\nabla H \cdot \mathbf{s}_g)], \end{aligned} \tag{6.12}$$

where $a = (f/2k)^{1/2}$. At $\nabla H = 0$, i.e., a flat ground surface, $\zeta(0) = D(0) = 0$. Consequently, the inclined terrain forms ($\nabla H \neq 0$) generate nonzero orographic vorticity and divergence (6.12) and orographic vertical velocity on the top of the PBL

$$w_{or}(\infty) = \mathbf{u}_g \cdot \nabla H. \tag{6.13}$$

As an example let us consider a meridionally infinitely long mountain with a Gaussian profile along the $0x$ axis (Figure 10.11)

$$H(x, y) = H(x) = H_m \exp[-x^2/L^2], \tag{6.14}$$

where L is the characteristic width of the mountain. Then $H_y = 0$, $H_x = -2xH/L^2$. We let $u_g = c_g \cos\alpha$, $v_g = c_g \sin\alpha$ and obtain

$$\left.\begin{matrix} \zeta(0) \\ D(0) \end{matrix}\right\} = \frac{2a}{L^2} c_g x H(x) (\cos\alpha \pm \sin\alpha) \tag{6.15}$$

$$w_{or}(\infty) = -\left(\frac{2c_g xH}{L^2}\right) \cos\alpha. \tag{6.16}$$

Figure 10.11 and its table contain the results of the analysis.

In the case of geostrophic flow normal to the obstacle ($\alpha = 0$), at $x < 0$ anticyclonic surface vorticity ($\zeta(0) < 0$), surface convergence ($D(0) < 0$) and positive vertical velocity ($w_{or}(\infty) > 0$) on the top of the PBL are generated. On the lee side of the mountain ($x > 0$), the signs are opposite. These effects are generally known from observations. When the air is forced to flow obliquely over the mountain, then $\alpha \neq 0$ and the second terms in (6.12) or (6.15) contribute to the surface values of ζ and D. This effect is missing in (6.13) due to the very crude assumption that $K = \text{const}$ with respect to x, y, z.

Actually, to be convinced that this is the reason, let us assume that (Godev, 1978; Panchev and Atanasov, 1978, 1979)

$$K = K(x, y) = K[H(x, y)] \tag{6.17}$$

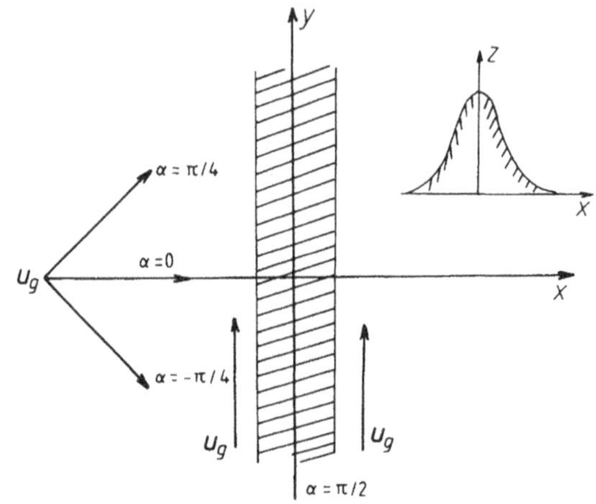

	α = 0		+ π/4		− π/4		π/2	
	$x<0$	$x>0$	$x<0$	$x>0$	$x<0$	$x>0$	$x<0$	$x>0$
$\zeta(0)$	<0	>0	<0	>0	$=0$	$=0$	<0	>0
$D(0)$	<0	>0	$=0$	$=0$	<0	>0	>0	<0
$W_{or}(\infty)$	>0	<0	>0	<0	>0	<0	$=0$	$=0$

Fig. 10.11. Orographic vorticity, divergence and vertical velocity in case of the obstacle given by (6.14).

and calculate again $\zeta(0)$, $D(0)$ and $w_{or}(\infty)$. Obviously, the Ekman solution (1.16) should be used but with $a(x, y) = [f/2K(H)]^{1/2}$. As a result, expressions (6.12) remain unchanged but, instead of (6.13), one obtains

$$w_{\mathrm{or}}(\infty) = (1 + \delta)(\nabla H \cdot u_g) - \delta(\nabla H \cdot s_g) \tag{6.18}$$

where

$$\delta = \left(\frac{1}{4aK}\right)\frac{\partial K}{\partial H}. \tag{6.19}$$

For our particular example (6.14), formula (6.16) is replaced by

$$w_{\mathrm{or}}(\infty) = c_g[(1 + \delta)\cos\alpha + \delta\sin\alpha]\frac{\partial H}{\partial x}. \tag{6.20}$$

We may think that K in (6.17) is a vertically-averaged eddy viscosity coefficient in the PBL above the mountains which, however, remains horizontally nonhomogeneous because of the complex geometry of the mountains. Physically, it seems quite reasonable to assume that $\partial K/\partial H > 0$, thus $\delta > 0$.

Let us have the case $\alpha = \pi/2$, i.e., a geostrophic wind blowing parallel to the mountain ridge (the Oy-axis, Figure 10.11). Then $w_{\text{or}}(\infty) = c_g \delta H_x$ and $w_{\text{or}} >, < 0$ at $x <, > 0$ which is in agreement with the signs of $\zeta(0)$ and $D(0)$ in this case. Moreover, there exists a critical angle $\alpha_c \neq \pi/2$, determined by $\tan \alpha_c = -(1 + \delta^{-1})$, for which $w_{\text{or}}(\infty) = 0$, (i.e., both terms in (6.20) compensate each other). Consequently, the third line of the table (Figure 10.11), should be appropriately changed.

A further generalization of (6.17) may be the assumption that $K = K(H, \eta = z - H)$, for instance, $K = k_1(H)\eta^{4/3}$. Making use of solution (2.8), it is easy to derive the expressions (compare with (6.12)!):

$$\begin{aligned} \zeta(0) &= -a^* [A(\nabla H \cdot u_g) - (\nabla H \cdot s_g)], \\ D(0) &= -a^* [(\nabla H \cdot u_g) + A(\nabla H \cdot s_g)], \end{aligned} \tag{6.21}$$

where $a^*(H) > 0$ and $A(H) > 0$ are some functions of $H(x, y)$ through $k_1(H)$.

From the above examples and discussions it follows that the spatial variations of the eddy viscosity coefficient above synoptic scale mountains are important for calculating such wind characteristics like $w(z)$, $\zeta(z)$ and $D(z)$. But the mountains are not only sources of frictional effects elevated in the atmosphere, but are also sources of cooling and warming, i.e., of horizontal thermal inhomogeneities (baroclinicity) in the PBL. The investigation of the joint action of the orography, turbulence and baroclinicity is a very difficult task. Moreover, it is generally nonlinear.

b. Local Circulations in the PBL

So far we have considered motions in the PBL initiated by large-scale horizontal inhomogeneities in the pressure field. Mathematically, this was expressed by prescribing the geostrophic wind at the h_g level. There exist, however, motions which initiate and develop in the PBL, and which also have small horizontal scales. They are called *local circulations*. Two examples will be briefly considered here.

(i) *Slope wind.* A sketch of this phenomenon is shown in Figure 10.12. For simplicity, the ground surface is idealized as an infinite, thermally homogeneous, sloped plane. At

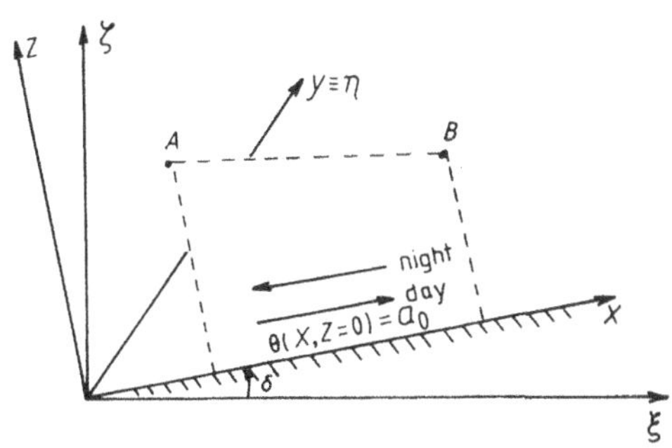

Fig. 10.12. The definition of slope wind.

two points in the atmosphere, A and B, located at a horizontal plane, the temperature during the day is generally $t_B > t_A$, while at night it is $t_B < t_A$. The temperature difference gives rise to a local pressure difference and, hence, to a motion as shown in Figure 10.12. As a rule, it is turbulent.

A major contribution to the theoretical analysis of the problem was made by Prandtl (1942) who derived the profiles of wind and temperature for a slope with a fixed surface temperature $\theta(x, y, 0) = \text{const} = a_0$ and a constant angle of inclination δ. The problem, as formulated by Prandtl, is one-dimensional. Wind and temperature variations are allowed only in a direction perpendicular to the slope. In an $x0y$ coordinate system (Figure 10.12), the Prandtl model is governed by the equations [57]

$$Ku'' + \beta_s \sin\delta \cdot \theta = 0, \qquad K\theta'' - \Gamma_s \sin\delta \cdot u = 0, \tag{6.22}$$

where $K = \text{const}$ is the eddy viscosity coefficient, $\Gamma_s = d\theta_s/dz$, $\beta_s = g/\overline{\theta}_s$, $(') = d/dz$, u and θ are velocity and temperature perturbations imposed on the motionless ($\mathbf{v}_s = 0$) reference state $\theta_s(z)$, $p_s(z)$, $\rho_s(z)$ in hydrostatic equilibrium (subscript 's'). So that we may obtain (6.22), the Boussinesq equations for a shallow convection have been linearized with respect to this state.

We have to solve (6.22), subject to the obvious boundary conditions

$$\begin{aligned} &u = 0, \qquad \theta = a_0 \quad \text{at } z = 0 \\ &u, \theta < \infty, \quad z \to \infty. \end{aligned} \tag{6.23}$$

In the case of stable conditions $\Gamma_s > 0$, and we may introduce the complex function

$$F = u + i\theta\left(\frac{\beta_s}{\Gamma_s}\right)^{1/2}, \quad i = \sqrt{-1}. \tag{6.24}$$

Then system (6.22) reduces to one equation

$$F'' - 2in^2F = 0, \qquad F(0) = ia_0\left(\frac{\beta_s}{\Gamma_s}\right)^{1/2}, \quad F(\infty) < \infty, \tag{6.25}$$

where $2n^2 = (\beta_s\Gamma_s)^{1/2} \sin\delta/K$. Hence,

$$F(z) = F(0) \exp[-n(i + 1)z],$$

so that

$$\begin{aligned} u(z) &= a_0\left(\frac{\beta_s}{\Gamma_s}\right)^{1/2} \exp[-nz] \sin nz, \\ \theta(z) &= a_0 \exp[-nz] \cos nz. \end{aligned} \tag{6.26}$$

This is Prandtl's classical solution. The profiles of u and θ are plotted in Figure 10.13. They were shown to be in reasonable agreement with observed average profiles (Defant, 1949). Despite this, Prandtl's theory is not satisfactory, since many of the assumptions made by him are not realistic. For instance, the real slopes never have an infinite extent nor are they uniformly heated by different reasons; the reference state may not be motionless and the phenomenon as a whole stationary, etc. To overcome some of these

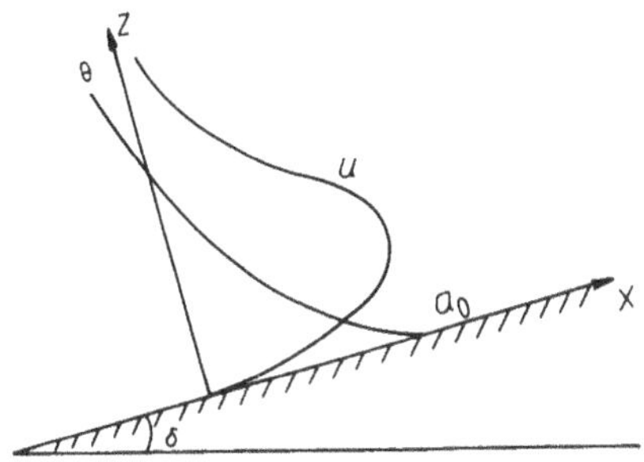

Fig. 10.13. Normal to the sloped surface profiles of the wind and temperature corresponding to the classical solution (6.26).

shortcomings of the classical theory, many authors have built complex (nonlinear) numerical models where most of these questions can be dealt with. We have to mention, however, that a reasonably successful numerical simulation does not automatically increase our understanding of the phenomenon. That is why a direct generalization of Prandtl's problem by relaxing only a few of the assumptions made by him and keeping the possibility for an analytical solution are of special interest. Recently, Egger (1981) has pursued this course, allowing the surface temperatue to vary with x only ($\theta(x, 0) = a_0(x)$) and has used K-theory with a constant coefficient. This makes the problem two-dimensional but still analytically tractable. In contrast to the one-dimensional theory, a deep penetration of the slope-wind circulation in the atmosphere was predicted. Similarly, the condition for constancy of K can be relaxed, etc.

(ii) *Sea breeze*. There are many common features in the physical nature and theoretical modelling of this and previous phenomena. The sea breeze is caused by a temperature difference between sea and land and is confined to the near-shore area – a few tenths of a kilometer on both sides. During the day the near-surface breeze blows from the sea across the land and during the night, in the opposite direction. At several hundred meters above the ground the situation is inverse (antibreeze). The phenomenon is observable in the case of a motionless reference state – there is no large-scale (synoptic) wind.

Obviously, the 'moving force' for the breeze wind is the pressure gradient caused by the thermal inhomogeneities of the surface (sea-land). In analytical models, it can be approximated successfully by the following expression:

$$\alpha_s \frac{\partial p}{\partial y} = C(1 - az)\exp[-bz] \sin \omega t, \qquad \frac{\partial p}{\partial x} = 0, \quad \left(\alpha_s = \frac{1}{\rho_s}\right), \tag{6.27}$$

where the $0x$ axis has been chosen parallel to the shore, p is the perturbation pressure, and $\omega = 2\pi/T$, ($T = 24$ h). Evidently, at level $z_1 = 1/a$ $\partial p/\partial y$ changes its sign, while at $z_2 = z_1 + 1/b$ it has an extremum. The typical values of the constants in (6.27) are $z_1 = 500$ m, $z_2 = 1500$ m, $C = 2 \times 10^{-3}$ m s^{-2} which corresponds to $|\nabla p| = 1$ mb/40 km.

Having the 'moving force' prescribed by (6.27) the equations of motion

$$\frac{\partial \mathbf{u}}{\partial t} = -\frac{1}{\rho_s}\nabla p - f\mathbf{k} \times \mathbf{u} + \frac{\partial}{\partial z} K \frac{\partial \mathbf{u}}{\partial z} \tag{6.28}$$

can be solved analytically, subject to boundary conditions

$$\mathbf{u}(0, t) = \mathbf{u}(\infty, t) = 0$$

to determine the velocity field. The additional assumption that K = const may be made.
The sea-breeze phenomenon is also a subject of many numerical models.

c. Nonstationary PBL

This problem has two aspects: a response of the PBL to a periodically-changing pressure gradient force and a diurnal variation of the wind in the PBL. Briefly, we shall pay attention to both of them, giving an idea for formulating and solving this kind of problem.

For the first problem we assume K = const and start from the equation of motion in a complex form (1.7)

$$\frac{\partial M}{\partial t} = if(M_g - M) + K\,\frac{\partial^2 M}{\partial z^2}, \tag{6.29}$$

where $M = u + iv$, $M_g = u_g + iv_g$, $\mathbf{u}_g = (1/\rho f)\mathbf{k} \times \nabla p$. Above the PBL ($z > h_g$), i.e., in the free atmosphere $K = 0$ and the velocity $\widetilde{M} = \widetilde{u} + i\widetilde{v}$ satisfies the equation

$$\frac{\partial \widetilde{M}}{\partial t} + if\widetilde{M} = ifM_g. \tag{6.30}$$

We prescribe a simple periodic wave in the field of wind velocity in the free atmosphere: $\widetilde{M} = \widetilde{A}\exp[i\nu t]$. This may be a synoptic wave with a period $T = 2\pi/\nu$ = 3–4 days and a considerably large amplitude. We are interested in the response of the PBL to such perturbation.

From (6.30) we find

$$ifM_g = i(\nu + f)\widetilde{A}\exp[i\nu t].$$

Inserting this in (6.29), we obtain the equation

$$\frac{\partial M}{\partial t} + ifM - K\,\frac{\partial^2 M}{\partial z^2} = i(\nu + f)\widetilde{A}\exp[i\nu t]. \tag{6.31}$$

We assume a solution of the form

$$M(z, t) = A(z)\exp[i\nu t].$$

Then (6.31) yields

$$KA'' - i(\nu + f)A = -i(\nu + f)\widetilde{A}, \tag{6.23}$$

where $('') = d^2/dz^2$, subject to boundary conditions $A(0) = 0$ $A(\infty) = \widetilde{A}$. Since ν, f and, consequently, $\nu + f$ can have arbitrary signs (+, –), then the general solution of (6.23) can be written as

$$A(z) = \widetilde{A}[1 - F_1(z) + i\,\mathrm{sign}\,(\nu + f)\,F_2(z)] \tag{6.33}$$

where sign $x = +1, -1$ at $x >, < 0$. $F_1(z)$ and $F_2(z)$ are taken from the Ekman solution (1.16) but now

$$a = \left(\frac{|\nu + f|}{2K}\right)^{1/2}. \tag{6.34}$$

Hence, for the thickness of the PBL we obtain

$$h_g = \frac{\pi}{a} = \pi\left(\frac{2K}{|\nu + f|}\right)^{1/2}. \tag{6.35}$$

The solution obtained has an interesting feature – $a \to 0$ and $h_g \to \infty$ when $|\nu + f| \to 0$ and, consequently, $w_g \to \infty$ since $w_g \sim h_g$ (formula (5.18)) but the solution (6.33) itself remains continuous at $f = -\nu$. The geographical latitude at which $f = -\nu$ is called critical. Since $\nu >, < 0$ such latitudes could exist in both hemispheres. Moreover, the wave disturbances in the real atmosphere are not at a single frequency only but often distributed over some frequency band with a maximum at the given frequency ν_m. Consequently, one can speak about a band of critical latitudes too. Then $w_g = w_{max}$ at $\nu = \nu_m$. According to [13], the location of the so-called intertropical convergent zone, as well as some other peculiarities of the PBL in the equatorial atmosphere, can be related to the critical latitudes.

The diurnal variations of the wind in the surface and planetary boundary layer is a well-known phenomenon in meteorology. However, it cannot be explained by the diurnal variations of the atmospheric pressure (pressure gradient force) – their amplitudes are too small. Quantitative agreement between the theoretical models and the observations can be achieved if the diurnal variations of the turbulence intensity are supposed to be the main reason. Accordingly, for the eddy viscosity K we can write a simple approximation

$$K(z, t) = \bar{K}(z)(1 + A \cos \omega t)$$

and try to solve the equation

$$\frac{\partial Q}{\partial t} = -ifQ + \frac{\partial}{\partial z} K(z, t) \frac{\partial Q}{\partial z},$$

where $Q = M_g - M$, $M_g = \text{const}$, $M = u + iv$. By letting $Q(z, t) = F(z, t) \exp[-ift]$ we obtain

$$\frac{\partial F}{\partial t} = \frac{\partial}{\partial z} K(z, t) \frac{\partial F}{\partial z}.$$

For a further solution, the function $\bar{K}(z)$ must be prescribed. No doubt, the simplest assumption is $K = \text{const}$. This approach to the problem, however, is rather formal. The diurnal variations of the wind should be studied jointly with the corresponding variations of the temperature, turbulence intensity and other meteorological elements having prescribed the diurnal variations of the fundamental physical reason – the radiative balance. The nonlinear numerical models for nonstationary PBL suggest such a possibility.

PROBLEMS

1. Solve Equations (1.4) with $K = \mathrm{const}_z$, $u_g = v_g = 0$, under the boundary conditions

$$K\frac{\partial u}{\partial z} = \frac{\tau_0}{\rho_W}, \qquad K\frac{\partial v}{\partial z} = \qquad \text{at } z = 0; \quad (u, v) \to 0 \quad \text{at } z \to \infty$$

and show that at $z = 0$

$$(u_0^2 + v_0^2)^{1/2} = \frac{\tau_0}{\rho_W}(fK)^{1/2}$$

(Ekman's problem for wind-induced ocean currents), where ρ_W is the water density, the $0z$ axis is directed downward, and τ_0 is the surface value of the Reynolds stress for the air.

2. Prove that for the Taylor solution (1.21)

$$|M_0| = u_g(\cos\alpha_0 - \sin\theta_0), \qquad M_0 = M(0).$$

3. Solve the equation (2.6) for $n = \frac{4}{3}$ and $n = 2$ and analyze the solution.
4. Using the Ekman model (1.18) calculate

$$\mathrm{Tr} = K\left[\left(\frac{\partial u}{\partial z}\right)^2 + \left(\frac{\partial v}{\partial z}\right)^2\right], \qquad E_k = \tfrac{1}{2}(u^2 + v^2)$$

and also their integrated quantities between $z = 0$ and $h_g = \pi/a$, $z = 0$ and ∞ and compare the results.

5. Starting from the equation (1.4) derive the formula

$$K(z) = \frac{f}{u\,\mathrm{d}v/\mathrm{d}z - v\,\mathrm{d}u/\mathrm{d}z}\int_{z_0}^{z}(u^2 + v^2 - uu_g - vv_g)\,\mathrm{d}z$$

allowing determination of $K(z)$ having data on $\mathbf{u}(z)$ and $\mathbf{u_g}$.

6. Find a periodic solution of the equation (6.26) if $ifM_g = \hat{A}\exp[i(\mu z - \nu t)]$ and analyze the solution.

CHAPTER 11

The General Circulation of the Atmosphere

1. CHARACTERISTIC PECULIARITIES AND STRUCTURE OF GENERAL ATMOSPHERIC CIRCULATION (GAC)

a. Factors Determining GAC

The GAC concept occupies an important place in modern meteorology, although there is no synonymous definition for it. Most often, GAC is understood to be a set of different synoptic situations which are constantly, or nearly constantly, found in the atmosphere (action centres, stream currents, monsoon circulation, intertropical convergence zone, etc.). In other cases, GAC is understood as an averaged-on-time state of the atmosphere in which the most essential peculiarities of a local character are preserved. The close relationship between both interpretations is evident and, quite often, the global condition (first of all the motion) of the atmosphere at a given instant is denoted by the term GAC. Finally, GAC is also understood to mean the combination of some large-scale statistical characteristics of the atmosphere.

The most important factors, determining GAC, are the following:

1. *The Space-time 'dimensions' of the atmosphere*. The atmospheric processes, having a determining significance for the weather, develop in the lower, let us say, 30 km air layer. At the same time, the horizontal scale of these processes are of the order of thousands of kilometers. That is why the atmosphere can be looked at as a thin air envelope around the Earth. This peculiarity, as we know, determines the order of the vertical and horizontal velocities of motion in the GAC system and, therefore, is very important. We have already used this for simplifying dynamic equations (Chapter 3).

2. *The influx of Solar radiation*. Solar radiation is the main source of energy for atmospheric motions. As is known, it is unevenly distributed over the Earth – the lower geographical latitudes receive considerably more heat than the higher ones. Moreover, the increase of received heat in a Pole-Equation direction is greatest in the middle latitudes and minimum at the Equator and the Poles. The radiational balance behaves similarly. This means that the meridional temperature gradients would also be maximum in the middle latitudes and minimum at the Poles and the Equator. This turns out to be one of the most important factors in determining GAC.

3. *The Earth's rotation*. The role of the Earth's rotation and the Coriolis force, respectively, grows with an increase of the scale of motion and becomes a determinant for the motions of the GAC system. What is more, the Rossby effect should also be taken into consideration. This also causes meridional changes in the most important characteristics

of the atmospheric motions. Besides, the Earth's rotation acts also as a stabilizing factor, thus decreasing the instability of the atmospheric motions. Along with this, the opportunity for potential to kinetic energy conversion is decreased. That is why the release of potential energy in the atmosphere turns out to be impossible in some cases. The stationary (blocked) frontal surface is a typical example in this respect.

4. *The nonhomogeneity of the Earth's surface*. Solar radiation is the main energy source of atmospheric motions but its direct influence upon the atmosphere is insignificant. The Earth's surface plays the role of 'a mediator', thus transforming the absorbed radiation into heat. This process significantly depends on its nonhomogeneities. They are usually manifested in two ways.

The existence of continents and oceans which are heated to a different extent, also leads to a nonhomogeneous heating of the air over them. Unlike the second factor, which is governed by astronomical reasons, the horizontal nonhomogeneities of the Earth's surface originate as meridional, as well as zonal thermal nonhomogeneities having well-expressed annual trends.

Also of considerable importance are the nonhomogeneities of the very continental surface and, in the first place, its large-scale roughness. The latter cause a number of dynamical processes which have been little studied. This is especially true of the influence of mountains and mountain 'clusters' on scales commensurable with those of the structural elements of the GAC itself.

5. *The friction with the Earth's surface*. Naturally, it is mostly the lower layers of the atmosphere (up to 1–1.5 km) are exposed to the influence of this factor. But it can also be felt at higher levels. It is different over the oceans and the continents and this fact is also of particular importance for the formation of GAC.

The study of GAC is of theoretical and practical importance. The air currents in the GAC system transport heat, moisture and other admixtures in the atmosphere from one geographical region into another. And so, under their influence the climate of these regions is formed. Without knowing the laws of GAC it is impossible to solve the problem of long-range weather forecasting.

b. Structural Elements of GAC

Investigations into GAC proceed in two directions: empirical and theoretical. The former one, based on years of observational data, gives an opportunity for revealing a number of typical GAC features. In brief, we shall recall the most important of them:

(i) In the troposphere, for both hemispheres, on average, the horizontal temperature gradient is directed from the lower to higher geographical latitudes. The influx of solar radiation supports this distribution in spite of the advection and eddy transport of heat in a meridional direction. In the lower stratosphere, the direction of the gradient is just the opposite – from the Poles towards the Equator.

(ii) In accordance with the barometric formula, the same peculiarities are also observed at the pressure gradient.

(iii) Subtropical zones of high pressure exist on both sides of the Equator, at about ±30° latitude. They do not encircle the Earth globe continuously but disintegrate into five closed regions (subtropical oceanic anticyclones), preserving themselves without considerable changes throughout the year. Two of them are to be found in our (Northern) hemisphere: the North Atlantic (Azores) and the North Pacific (Honolulu) anticyclones.

In the Northern Hemisphere, except for these constant baric centres, there are also Siberian and Canadian maxima and Icelandic and Aleutian minima. All of them are called *action centres* in the atmosphere.

The above-mentioned geographical pressure distribution in the free atmosphere (decreasing from the Equator toward the Poles) generates westerlies in both hemispheres and zonal transport of air masses. Because of this, the angular velocity of rotation of the atmosphere is greater than that of the Earth itself.

The zonal transport is the most characteristic feature of the general circulation in the free atmosphere at middle latitudes. Its velocity depends on the height z and the latitudes φ (Figure 11.1).

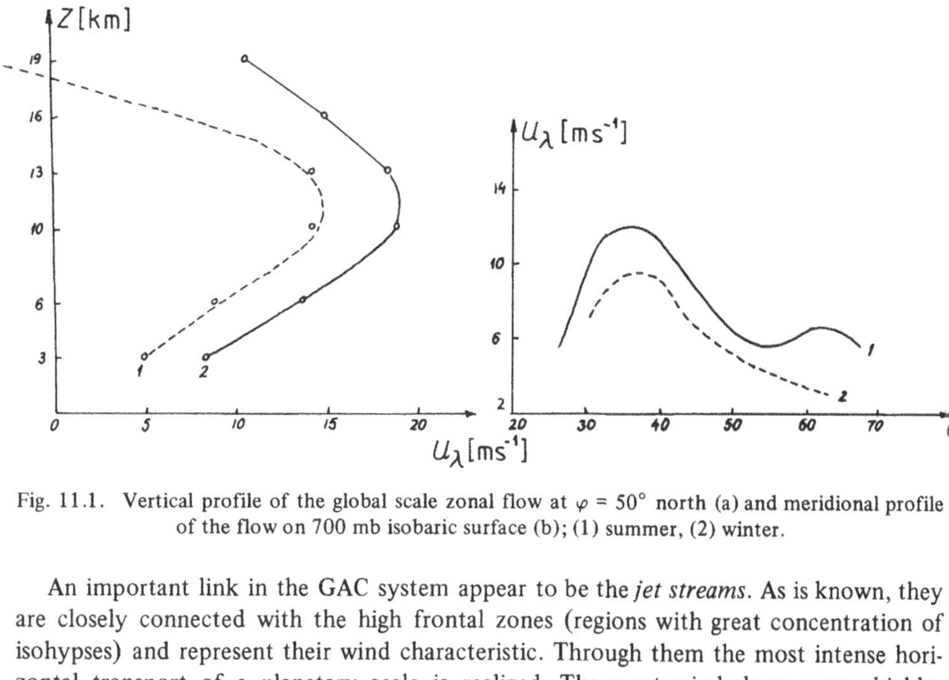

Fig. 11.1. Vertical profile of the global scale zonal flow at φ = 50° north (a) and meridional profile of the flow on 700 mb isobaric surface (b); (1) summer, (2) winter.

An important link in the GAC system appear to be the *jet streams*. As is known, they are closely connected with the high frontal zones (regions with great concentration of isohypses) and represent their wind characteristic. Through them the most intense horizontal transport of a planetary scale is realized. The great wind shear causes highly-developed turbulence in the jet streams and, consequently, an intense vertical eddy exchange. The jet streams are closely related to such important GAC elements as cyclones and anticyclones. Too many problems about jet streams are, however, still obscure.

Another important GAC element is the so-called intertropical convergence zone (ITCZ). This is a convergence region of the tropospheric trade-winds of both hemispheres. It is relatively persistent and well-defined over the Pacific and Atlantic oceans between about 5 and 10°N, but occasionally, it does appears in the Pacific between 5 and 10°S, i.e., it is centered away from the Equator. The usual confirmation of its existence is the narrow band of powerful Cb clouds over the oceans (*cloud clusters*) which have been discovered in recent years on satellite photographs and have been extensively studied. The ITCZ with its heavy convective cloudiness is a determining factor in the dynamics of the equatorial atmosphere and the neighbouring parts of the middle latitude atmosphere.

There the release of a tremendous amount of energy as a latent heat of evaporation takes place. The divergence $D = \nabla \cdot \mathbf{u}$ of the horizontal motions there is much larger than in the regions outside the ITCZ, i.e., it is no longer a small difference of great quantities, as in the middle latitudes, which allows a reliable estimation using observation data made on board exploration ships at discrete points in the ocean. Such data were collected, for instance, during TROPEX-74.

Having estimated D on the basis of the continuity equation $\partial w_p/\partial p = -\nabla \cdot \mathbf{u}$, one can also estimate the vertical velocity. Fal'kovich's recent book [13] is dedicated to the ITCZ phenomenon.

Finally, monsoon circulation, with the Indian monsoon as a typical representative, is also a very important element of GAC.

c. Theoretical Description of GAC

There exists, at present, no satisfactory theoretical description of GAC. Any serious attempt to develop such a description should take into account existing empirical data relating to real circulation. The greatest difficulties are to be found in mathematically formulating the laws governing GAC and, first of all, in composing the thermodynamic energy equation. The reason is that all three forms of heat exchange (radiative, turbulent and latent) are equally important and should all be considered. This has lead to the creation of an original semiempirical branch in which the temperature of the entire atmosphere is prescribed while the pressure and velocity fields are to be found. The greatest success in the early theories was achieved by the Soviet theoreticians Kochin [31] and Blinova (1943).

Before briefly stating their classic results, we should mention that the theoretical meditations and thoughts on the nature and essence of GAC have a long history. Indisputably, amongst the earliest were those of the 18th-century Englishman, George Hadley. Studying the tradewinds, he suggested the idea that they are part of an equatorially symmetric thermal-convective circulation, maintained by the temperature-difference between the Poles and the Equator. The heated equatorial air rises and moves poleward where it cools, sinks and moves towards the Equator. This circulation type is known as Hadley circulation. Obviously, it consists of two cells: one for each hemisphere, and they also bear Hadley's name. In other words, the GAC is interpreted only on the principles of a thermodynamic machine (heat engine) with heat and cold sources (Equator and Poles). The Hadley circulation, though possible in principle, is not observed as a primary mode of the real GAC. The reason is that under the conditions existing in the Earth's atmosphere, this circulation, when it appears, would be baroclinically unstable and, therefore, would disintegrate. This is in fact observed in the trades. It is clear that Hadley's notions had been purely qualitative and speculative.

The first successful hydrodynamic GAC model was proposed by Kochin in 1933. His main contribution in laying the foundations of GAC quantitative theory is that he applied the boundary-layer theory and succeeded in simplifying the general equations of motion in spherical coordinates to such a degree that their analytical solution became possible in a number of cases. He also paid great attention in his mathematical model to the problem related to the inclusion of turbulence. He considered the motions in the system of baric formations (cyclones, anticyclones, fronts) as disturbances with respect to the larger-scale motions and the corresponding transport of substances – to be parameterized as macroturbulent transport (horizontal and vertical). As far as the vertical dimensions of the above-mentioned elements of macroturbulence are of the order of 10 km, the influence of the eddy frictional force would extend almost into the entire troposphere. The baroclinicity of the real atmosphere, resulting in the height dependence of the pressure gradient force, would increase this effect.

Taking into account the vertical component of the eddy frictional force only, Kochin reached the following simplified system of equations for the case of *stationary zonal circulation* (the meteorological elements do not depend on time t and geographic longitude λ):

$$K \frac{\partial^2 v_\theta}{\partial r^2} + 2\omega \cos\theta \cdot v_\lambda = \frac{1}{\rho r} \frac{\partial p}{\partial \theta}, \tag{1.1}$$

$$K \frac{\partial^2 v_\lambda}{\partial r^2} - 2\omega \cos\theta \cdot v_\theta = 0, \tag{1.2}$$

$$\frac{\partial p}{\partial r} + \rho g = 0, \tag{1.3}$$

$$\frac{1}{r^2} \frac{\partial}{\partial r} (\rho r^2 v_r) + \frac{1}{r \sin\theta} \frac{\partial}{\partial \theta} (\rho v_\theta \sin\theta) = 0, \tag{1.4}$$

where, as before, v_λ, v_θ and v_r are the zonal, meridional and vertical velocity components, $\theta = (\pi/2) - \varphi$, r is the distance from the Earth's centre, and K is the vertical eddy viscosity coefficient assumed to be constant at the order 10^3 m^2 s^{-1}. In his calculations, Kochin takes $K = 2 \times 10^3$ m^2 s^{-1}.

Given (1) the temperature distribution $T(r, \theta)$ in the entire atmosphere; (2) surface pressure distribution $p(r_0, \theta) = p_0(\theta)$, where r_0 is the Earth's radius; and (3) the value of K, system (1.1)–(1.4) allows the determination of the remainder of the GAC characteristics.

Actually, from quasistatic equation (1.3) we find

$$p(r, \theta) = p_0(\theta) \exp\left[-\int_{r_0}^{r} \frac{g\, dr}{RT(r, \theta)}\right]. \tag{1.5}$$

Then

$$\rho(r, \theta) = \frac{p(r, \theta)}{RT(r, \theta)}. \tag{1.6}$$

Introducing $M = v_\lambda + i v_\theta$ we reduce (1.1) and (1.2) to one equation

$$K \frac{\partial^2 M}{\partial r^2} + 2i\omega \cos\theta \cdot M = \frac{i}{\rho r} \frac{\partial p}{\partial \theta} \tag{1.7}$$

with boundary conditions $v_\lambda = v_\theta = 0$ at $r = r_0$ and $v_\lambda, v_\theta < \infty$ at $r \to \infty$. The right-hand side of (1.7) is known from (1.5) and obviously represents an arbitrary complex function of r and θ. That is why generally Equation (1.7) should be integrated numerically.

Having found v_λ, v_θ and ρ, then from continuity Equation (1.4) one can also find the vertical velocity

$$v_r = -\frac{1}{\rho r^2 \sin\theta} \int_{r_0}^{r} \frac{\partial}{\partial\theta} (\rho r v_\theta \sin\theta)\, dr. \tag{1.8}$$

Numerical calculations performed by Kochin show that v_λ, $v_\theta \sim 10$ m s^{-1}, while $v_r \sim 10^{-1}$–10^0 cm s^{-1} and even at a height of about 6 km, the influence of the eddy friction is essential – when calculated from (1.7), the velocities at $K = 2 \times 10^3$ cm^2 s^{-1} differ from the geostrophic

$$v_\lambda(\theta) = \frac{\partial p/\partial\theta}{2\omega\rho r \cos\theta} \tag{1.9}$$

by several tenths per cent for $15° < \varphi < 75°$.

Of course, the real circulation is neither zonal nor stationary and its characteristics could be expressed in the form (at fixed r)

$$f(\theta, \lambda, t) = \bar{f}(\theta) + f'(\theta, \lambda) + f''(\theta, \lambda, t), \tag{1.10}$$

where $\bar{f}(\theta)$ are the zonal parts of the quantities v_λ, v_θ, p, ρ and T, $f'(\theta, \lambda)$ are the non-zonal parts and $f''(\theta, \lambda, t)$ are the nonstationary parts. Then $\bar{v}_\lambda(\theta)$ will be given by (1.9), while $\bar{v}_\theta(\theta) = 0$.

However, the analysis of the observational data on the distribution of the zonal geostrophic velocity (1.9) along the meridian shows that in the middle geographic latitudes $\bar{v}_\lambda(\theta)$ can be given in the form

$$\bar{v}_\lambda(\theta) = \alpha_a r \sin\theta \quad \text{or} \quad \alpha_a = \frac{\bar{v}_\lambda(\theta)}{r \sin\theta} \tag{1.11}$$

where α_a is the circulation index introduced in Section 4 of Chapter 2. Generally, α_a is a variable quantity but in most theoretical investigations the assumption α_a = const has proved satisfactory.

Having $\bar{v}_\lambda(\theta)$ prescribed by (1.11), then Equation (1.9) yields

$$\bar{p}(r, \theta) = \bar{p}(r, 0) + \alpha_a r^2 \bar{\rho}\omega \sin^2\theta. \tag{1.12}$$

Since $\bar{T} = \bar{p}/\bar{\rho}R$, then

$$\bar{T}(r, \theta) = \bar{T}(r, 0) + C(r) \sin^2\theta, \tag{1.13}$$

where

$$\bar{T}(r, 0) = \frac{\bar{p}(r, 0)}{\bar{\rho}(r, 0)R}, \qquad C = \frac{\alpha_a \omega r^2}{R}.$$

2. ANALYTICAL AND NUMERICAL MODELS OF GAC

a. Blinova's Model

In the course of three decades, E. N. Blinova created a class of analytical GAC models on a spherical Earth. The first work on this cycle dates back to 1943. It has preserved its scientific and methodological importance and we shall treat it in more detail.

The GAC elements are represented in the form of (1.10) and it is assumed that $f'' \equiv 0$ (i.e., stationarity) and $|f'| \ll |\bar{f}|$. The turbulent effects and convective derivatives $w\, \partial f/\partial z$ are omitted. The temperature field $T(z, \theta, \lambda)$ is considered as given and the velocity field is solenoidal (nondivergent). Then continuity Equation (3.27) in Chapter 2, takes the form

$$\frac{\partial v_\lambda}{\partial \lambda} + \frac{\partial}{\partial \theta}(v_\theta \sin \theta) = 0. \tag{2.1}$$

Consequently, the stream function $\psi(\theta, \lambda)$ can be introduced

$$v_\theta = -\frac{1}{r_0 \sin \theta}\frac{\partial \psi}{\partial \lambda}, \qquad v_\lambda = \frac{1}{r_0}\frac{\partial \psi}{\partial \theta}. \tag{2.2}$$

The starting equations are (3.24)–(3.26) from Chapter 2, in spherical coordinates and stationary case:

$$\frac{1}{r_0}\frac{\partial E}{\partial \theta} = -\frac{1}{\rho r_0}\frac{\partial p}{\partial \theta} + (2\omega \cos \theta + \zeta) v_\lambda,$$

$$\frac{1}{r_0 \sin \theta}\frac{\partial E}{\partial \lambda} = -\frac{1}{\rho r_0 \sin \theta}\frac{\partial p}{\partial \lambda} - (2\omega \cos \theta + \zeta) v_\theta, \tag{2.3}$$

$$E = \tfrac{1}{2}(v_\lambda^2 + v_\theta^2), \qquad \zeta = \frac{1}{r_0 \sin \theta}\left(\frac{\partial}{\partial \theta} v_\lambda \sin \theta - \frac{\partial}{\partial \lambda} v_\theta\right).$$

Evidently

$$\zeta = \frac{1}{r_0^2 \sin \theta}\left[\frac{\partial}{\partial \theta}\left(\sin \theta \frac{\partial \psi}{\partial \theta}\right) + \frac{\partial}{\partial \lambda}\left(\frac{1}{\sin \theta}\frac{\partial \psi}{\partial \lambda}\right)\right] = \frac{1}{r_0^2}\nabla^2 \psi. \tag{2.4}$$

The vorticity equation can be easily derived from (2.3). Eliminating ρ by means of the equation of state and introducing the stream function ψ, it takes the form

$$\frac{1}{r_0^2 \sin \theta}\left(\frac{\partial \psi}{\partial \theta}\frac{\partial \nabla^2 \psi}{\partial \lambda} - \frac{\partial \psi}{\partial \lambda}\frac{\partial \nabla^2 \psi}{\partial \theta}\right) + \frac{2\omega}{r_0^2}\frac{\partial \psi}{\partial \lambda}$$

$$= \frac{1}{\rho T r_0^2 \sin \theta}\left[\frac{\partial p}{\partial \theta}\frac{\partial T}{\partial \lambda} - \frac{\partial T}{\partial \lambda}\frac{\partial p}{\partial \theta}\right]. \tag{2.5}$$

Linearizing this equation with respect to the nonzonal disturbances ψ', ρ', T', we obtain

$$\frac{1}{r_0^2 \sin \theta}\left(\frac{\partial \bar{\psi}}{\partial \theta}\frac{\partial \nabla^2 \psi'}{\partial \lambda} - \frac{\partial \psi'}{\partial \lambda}\frac{\partial \nabla^2 \bar{\psi}}{\partial \theta}\right) + \frac{2\omega}{r_0^2}\frac{\partial \psi'}{\partial \lambda}$$

$$= \frac{1}{\bar{\rho}\bar{T} r_0^2 \sin \theta}\left(\frac{\partial \bar{p}}{\partial \theta}\frac{\partial T'}{\partial \lambda} - \frac{\partial p'}{\partial \lambda}\frac{\partial \bar{T}}{\partial \theta}\right). \tag{2.6}$$

But according to (1.11) and (2.2)–(2.4), $\bar{v}_\lambda = \alpha_a r_0 \sin \theta$, $\bar{v}_\theta = 0$ $\bar{\zeta} = 2\alpha_a \cos \theta = \nabla^2 \bar{\psi}/r_0^2$. Moreover, from Blinova's estimations, the second term in the right-hand side of (2.6) can be neglected. Thus, we obtain the final equation for ψ'

$$\frac{\partial}{\partial \lambda}\nabla^2 \psi' + 2\left(1 + \frac{\omega}{\alpha_a}\right)\frac{\partial \psi'}{\partial \lambda} = \frac{2\omega r_0^2 \cos \theta}{\bar{T}}\frac{\partial T'}{\partial \lambda} \tag{2.7}$$

which is nonhomogeneous since the temperature $T'(\theta, \lambda)$ is considered as a known function. Blinova assumes

$$T'(\theta, \lambda) = \sum_{n=1}^{\infty} \sum_{m=1}^{n} (X_n^m \cos m\lambda + Y_n^m \sin m\lambda) P_n^m (\cos \theta)$$

and seeks a solution of the same kind for Equation (2.7)

$$\psi'(\theta, \lambda) = \sum_{n=1}^{\infty} \sum_{m=1}^{n} (A_n^m \cos m\lambda + B_n^m \sin m\lambda) P_n^m (\cos \theta),$$

where P_n^m $(\cos \theta)$ are spherical functions. The insertion of both series into (2.7) allows the unknown coefficients A and B to be expressed in terms of X and Y and the remaining parameters of the problem entering Equation (2.7). Thus, a solution for $\psi'(\theta, \lambda)$ is found. Then, by simple differentiation according to (2.2), v_θ and v_λ can be determined.

Having determined the wind field (v_θ, v_λ), the pressure field can be found from the second equation in (2.3), and written in the form

$$\frac{1}{r_0} v_\theta \frac{\partial}{\partial \theta} v_\lambda + \frac{v_\lambda}{r_0 \sin \theta} \frac{\partial}{\partial \lambda} v_\lambda + \frac{1}{r_0} v_\theta v_\lambda \cotan \theta$$

$$= - \frac{1}{\rho r_0 \sin \theta} \frac{\partial p}{\partial \lambda} - 2\omega v_\theta \cos \theta.$$

The linearization and introduction of a stream function reduces the equation to

$$\frac{\partial p'}{\partial \lambda} = -\alpha_a \bar{\rho} \sin \theta \frac{\partial^2 \psi'}{\partial \lambda \, \partial \theta} + 2\bar{\rho}(\omega + \alpha_a) \cos \theta \frac{\partial \psi'}{\partial \lambda}. \tag{2.8}$$

A simple integration yields $p'(\theta, \lambda, r)$. Then $p(\theta, \lambda, r) = \bar{p}(\theta, r) + p'(\theta, \lambda, r)$, where $\bar{p}(\theta, r)$ is the zonal pressure at the 'middle' ('nondivergent') level $r = r_0 + z$ given by (1.12). Finally, from the barometric formula

$$p(\theta, \lambda, z) = p_0(\theta, \lambda) \exp\left[-\frac{g}{R} \int_0^z \frac{\mathrm{d}z}{T(\theta, \lambda, z)}\right] \tag{2.9}$$

the surface pressure $p_0 = p(\theta, \lambda, 0)$ can be determined. With this, the solution ends.

The significance of Blinova's work was not only in the fact that the spectral method was applied for the first time for solving a definite problem from the GAC theory. It also turned out to be very promising and, consequently, found wide application. The calculation made on the basis of the obtained solution, showed that the action centres (see Section 1) observed in the atmosphere could be reproduced. This was a great achievement at that time and a further development of Kochin's theory.

In her later investigations, Blinova rejected the assumption of prescribing the temperature field. Some new equations were added (particularly those allowing the determination of the temperature) and a closed system of equations governing the GAC in linear and nonlinear diagnostic and prognostic variants was constructed. Their common shortcoming is that macroturbulent effects are either totally disregarded or rather crudely approximated by means of K theories. These models are described in more detail in [3] and [30].

b. Monin's Model

The irregular (almost random) space-time variations of the GAC characteristics permit one to speak about *macroturbulence* and to apply the statistical methods of description which are familiar in turbulence theory. Thus, if we represent whatever variable GAC characteristic in the form

$$f(\theta, \lambda, t) = \bar{f}(\theta, t) + f'(\theta, \lambda, t), \tag{2.10}$$

where

$$\bar{f}(\theta, t) = \frac{1}{2\pi}\int_0^{2\pi} f(\theta, \lambda, t)\,d\lambda, \tag{2.11}$$

then f' can be interpreted as a macrofluctuation. If one assumes that the equations of motion

$$\begin{aligned} \frac{d}{dt} v_\lambda + \frac{1}{r_0} v_\lambda v_\theta \cot\theta &= -\frac{\partial\tilde{p}/\partial\lambda}{r_0 \sin\theta} - 2\omega v_\theta \cos\theta, \\ \frac{d}{dt} v_\theta - \frac{1}{r_0} v_\lambda^2 \cot\theta &= -\frac{1}{r_0}\frac{\partial\tilde{p}}{\partial\theta} + 2\omega v_\lambda \cos\theta, \end{aligned} \tag{2.12}$$

where

$$\frac{\partial}{\partial\lambda} v_\lambda + \frac{\partial}{\partial\theta}(v_\theta \sin\theta) = 0,$$

$$\tilde{p} = \frac{p}{\rho}, \qquad \frac{d}{dt} = \frac{\partial}{\partial t} + \frac{1}{r_0} v_\theta \frac{\partial}{\partial\theta} + \frac{v_\lambda}{r_0 \sin\theta}\frac{\partial}{\partial\lambda},$$

are satisfied by the total quantities $f = \bar{f} + f'$, a question can be put forward about the derivation from (2.12) of Reynolds equations of the mean values $\bar{f}$ or such of the Reynolds stresses and the kinetic energy of macroturbulent fluctuations in the GAC system. Such an approach has been proposed by Monin (1958).

It is supposed that $\bar{v}_\theta = 0$ and the statistical characteristics of f' do not depend on λ (homogeneity on λ). Averaging (2.12) in the sense of (2.11), we obtain the Reynolds equations for the case

$$\begin{aligned} \frac{\partial}{\partial t}\bar{v}_\lambda &= -\frac{1}{r_0 \sin^2\theta}\frac{\partial}{\partial\theta}(\overline{v_\lambda' v_\theta'} \sin^2\theta), \\ \frac{1}{r_0}\frac{\partial\bar{\tilde{p}}}{\partial\theta} &= -\frac{1}{r_0}\frac{\partial}{\partial\theta}\overline{v_\theta'^2} + \frac{1}{r_0}(\overline{v_\lambda'^2} - \overline{v_\theta'^2})\cot\theta + \bar{v}_\lambda\left(2\omega + \frac{\bar{v}_\lambda}{r_0 \sin\theta}\right)\cos\theta. \end{aligned} \tag{2.13}$$

As is seen, due to the nonlinearity of the initial equations (2.12), the evolution of the mean zonal velocity $\bar{v}_\lambda$ is determined solely by the quantity $\overline{v_\lambda' v_\theta'}$ while in the second

Equation (2.13), the variances appear. System (2.13), however, is not closed. Assuming a two-dimensional isotropy ($\overline{v_\theta'^2} = \overline{v_\lambda'^2}$), an approximate symmetry of the probability distributions for random quantities v_λ' and v_θ' (i.e., $\overline{v_\lambda'^3} \approx 0, \overline{v_\theta'^3} \approx 0$ and the same for the mixed third moments), and neglecting the cross-correlations between $\tilde{p}'$ and v_λ', v_θ', Monin derived two new equations on the basis of (2.12) and (2.13)

$$\frac{\partial}{\partial t}\overline{v_\lambda' v_\theta'} = -\overline{E}' \sin\theta \frac{\partial \alpha_a}{\partial \theta}, \qquad \frac{\partial \overline{E}'}{\partial t} = -\overline{v_\lambda' v_\theta'} \sin\theta \frac{\partial \alpha_a}{\partial \theta}, \tag{2.14}$$

where $\overline{E}' = \frac{1}{2}(\overline{v_\lambda'^2} + \overline{v_\theta'^2})$ is the kinetic energy of the fluctuating component of the motion. Together with the first equation from (2.13), written for $\alpha_a = \bar{v}_\lambda / r_0 \sin\theta$

$$\frac{\partial \alpha_a}{\partial t} = -\frac{1}{r_0^2 \sin^3\theta} \frac{\partial}{\partial \theta}(\overline{v_\lambda' v_\theta'} \sin^2\theta), \tag{2.15}$$

they form a closed system of three equations about the unknowns α_a, $\overline{E}'$, $\overline{v_\lambda' v_\theta'}$. After their determination from the second equation (2.13) which, due to the above simplifications, can be written in the form

$$\frac{1}{r_0}\frac{\partial \bar{p}}{\partial \theta} = -\frac{1}{r_0}\frac{\partial \overline{E}'}{\partial \theta} + (\alpha_a + 2\omega)\alpha_a r_0 \sin\theta \cos\theta, \tag{2.16}$$

$\bar{p}(\theta, t)$ can be found.

The system of Equations (2.14) and (2.15) is prognostic and nonlinear. They can be solved numerically. However, as was shown by its author, a single equation can easily be obtained.

Actually, as it follows from (2.14)

$$\frac{\partial}{\partial t}[(\overline{E}')^2 - (\overline{v_\lambda' v_\theta'})^2] = 0 \quad \text{or} \quad (\overline{E}')^2 - (\overline{v_\lambda' v_\theta'})^2 = F^2(\theta) \tag{2.17}$$

i.e., the system admits an invariant. Moreover, $F^2(\theta) \geqslant 0$ since at $\overline{v_\lambda'^2} \approx \overline{v_\theta'^2} \approx \overline{E}'$ we have

$$\frac{|\overline{v_\lambda' v_\theta'}|}{\overline{E}'} \approx \frac{|\overline{v_\lambda' v_\theta'}|}{(\overline{v_\lambda'^2} \cdot \overline{v_\theta'^2})}^{1/2}.$$

On the other hand, (2.15) allows us to write down

$$\alpha_a = \frac{\omega}{\sin^3\theta}\frac{\partial \Psi}{\partial \theta}, \qquad \overline{v_\lambda' v_\theta'} = -\frac{\omega r_0^2}{\sin^2\theta}\frac{\partial \Psi}{\partial t} \tag{2.18}$$

where $\Psi(\theta, t)$ is a dimensionless function (an analogue of the stream function in the coordinate plane θ, t). Then for $\Psi(\theta, t)$ a nonlinear hyperbolic equation follows

$$\frac{\partial^2 \Psi}{\partial t^2} = \left[\frac{F^2(\theta)}{r_0^4} + \left(\frac{\omega}{\sin^2\theta}\frac{\partial \Psi}{\partial t}\right)^2\right]^{1/2} \sin^3\theta \frac{\partial}{\partial \theta}\left(\frac{\partial \Psi/\partial \theta}{\sin^3\theta}\right). \tag{2.19}$$

It describes the nonlinear propagation of the waves along the meridian with a velocity equal to $(\overline{E'})^{1/2}$. Obviously, $F(\theta)$ should be determined in advance, according to (2.17), using the initial data about $\overline{E'}$ and $\overline{v'_\lambda v'_\theta}$ and the solution carried out numerically. Equation (2.19) has not yet been treated with the analytical method for $F(\theta)$ = const or some other simple expression.

A similar approach to the problem of GAC was developed by Blinova (1958, 1962) as well, but on the basis of the vorticity conservation equation

$$\frac{d}{dt}(\zeta + 2\omega\cos\theta) = 0,$$

which is the same as

$$\frac{\partial\zeta}{\partial t} + \frac{1}{r_0} v_\theta \frac{\partial}{\partial\theta}(\zeta + 2\omega\cos\theta) + \frac{v_\lambda}{r_0 \sin\theta}\frac{\partial\zeta}{\partial\lambda} = 0. \tag{2.20}$$

Taking into consideration (2.2) and (2.4), we transform this equation into the following

$$\frac{\partial}{\partial t}\nabla^2\psi + \frac{1}{r_0^2\sin\theta} J(\psi, \nabla^2\psi) + 2\omega\frac{\partial\psi}{\partial\lambda} = 0, \tag{2.21}$$

where $J(A, B) = A_\theta B_\lambda - A_\lambda B_\theta$ (the indices denote the derivatives). Letting $\psi = \overline{\psi} + \psi'$, Equation (2.21) yields directly the Reynolds-type equation in terms of $\overline{\psi}$:

$$\frac{\partial}{\partial t}\nabla^2\overline{\psi} + \frac{1}{r_0^2\sin\theta} J(\overline{\psi}, \nabla^2\overline{\psi}) + 2\omega\frac{\partial\overline{\psi}}{\partial\lambda} = -\frac{1}{r_0^2\sin\theta}\overline{J(\psi', \nabla^2\psi')}. \tag{2.22}$$

Further on, as in the previous model, new equations for the Reynolds stresses can be derived and, at the price of many approximations, the system of equations can be closed. Another approach is to parametrize the 'surplus' unknowns in (2.22) by the methods of the semiempirical theory of turbulence, e.g., by introducing a macro-eddy viscosity coefficient K.

c. Numerical Models and Experiments

Although attractive from a mathematical point of view, the GAC analytical models have limited abilities. The progress in understanding the physics and dynamics of the atmospheric processes and motions of different scales allows a much more complete setting of the GAC problem with respect to the initial equations and boundary conditions. This, however, inevitably makes it analytically unsolvable. However, the development of numerical methods for solving partial differential equations of the type encountered in geophysical hydrodynamic problems and the creation of very fast computers with large memories has stimulated the appearance of numerical models and the carrying out of extremely interesting numerical experiments.

The first model of this kind was constructed by Phillips (1956) and was a two-layer model having two basic (750 mb and 250 mb) and three subsidiary (0, 500 and 1000 mb) levels (Figure 11.2).

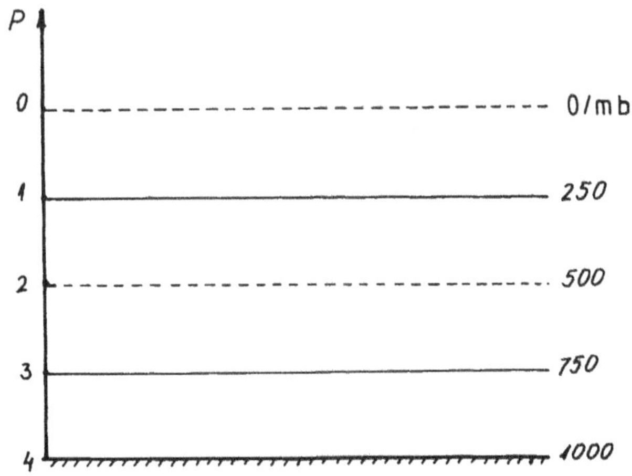

Fig. 11.2. Basic and intermediate levels in the Phillips (1956) two-layer GAC model.

Initial equations in the model are those of the motion

$$\frac{\partial \mathbf{u}}{\partial t} + (\mathbf{u} \cdot \nabla)\mathbf{u} + f\mathbf{k} \times \mathbf{u} = -\nabla\Phi + K_H \nabla^2 \mathbf{u} + g\,\frac{\partial \tau}{\partial p}, \tag{2.23}$$

of the hydrostatics and continuity

$$\frac{\partial \Phi}{\partial p} = -\frac{RT}{p}, \qquad \nabla \cdot \mathbf{u} + \frac{\partial w_p}{\partial p} = 0, \tag{2.24}$$

and the thermodynamics energy Equation (3.11) in Chapter 3

$$\frac{\partial T}{\partial t} + \mathbf{u} \cdot \nabla T - m_T w_p = \sum \frac{Q_i}{c_p}, \tag{2.25}$$

where ΣQ_i is the residual heat influx,

$$w_p = \frac{\mathrm{d}p}{\mathrm{d}t}, \qquad \tau = -\rho^2 gK\,\frac{\partial \mathbf{u}}{\partial p}$$

and K, K_H are vertical and horizontal eddy viscosity coefficients. In this and the subsequent models, $K \sim 10$ m^2 s^{-1} and $K_H \sim 10^5$–10^6 m^2 s^{-1} are the most frequently used values.

Each model is based on possibly the most complete system of equations in one or other coordinate. The construction of the initial system, however, is only a part of the work involved in creating the model. Initial and boundary conditions are needed as well, and an algorithm which is suitable for the numerical integration of this system. The latter requirement is extremely important since every algorithm does not transform the continual initial system into an equivalent system in finite differences. When the initial system is constructed and the algorithm for the numerical solution chosen, then the varying of the initial conditions and the 'outer' parameters of the problem allows the carrying out of the so-called *numerical experiments* within the framework of the model. For instance, in searching for stationary solutions or such with an annual period of smoothed (climatic) initial conditions, we shall obtain the regular seasonal changes of the large-scale atmospheric processes. This will be an experiment on GAC. The same system of equations, solved using concrete initial data, will be an experiment on long-range weather forecasting.

The very first numerical models and experiments demonstrated the great potential of this new method of investigation. Thus, the Phillips model reproduces the three-cell structure of GAC in both hemispheres, the evolution of the fronts, and the origin of the jet streams. For a period of about a month, from a state of rest the model atmosphere reaches a quasi-stationary fluctuating regime with a wind velocity $|\mathbf{u}| \sim 10$ m s^{-1} and pressure variations $\delta p \sim 10$ mb, i.e., the same as those actually observed. In addition, to this, some of the later models predict 'thinner' effects such as selective baroclinic instability – the fastest increase of planetary waves with a zonal wave number $m = 5$ or 6, global two-week GAC variations, etc.

In all GAC models, including the present-day ones, one of the main problems is the assimilation of the planetary boundary layer and, consequently, the interaction between the atmosphere and the underlying surface (land and ocean) despite their small vertical resolution. Now this is achieved by the parametrization of the corresponding effects and, for that purpose, some special parametrization schemes are being developed. Most often, for instance, the turbulent fluxes of impulse, heat and water vapour at the ground are prescribed by

$$\begin{aligned}\tau_s &= -\rho_h C_d \,|\mathbf{u}_h|\, \mathbf{u}_h,\\ Q_s &= \rho_h c_p C_d \,|\mathbf{u}_h|\, (T_s - T_h),\\ R_s &= \rho_h C_d \,|\mathbf{u}_h|\, (r_s - r_h),\end{aligned} \tag{2.26}$$

where $C_d \sim 10^{-3}$ is the drag coefficient (empirically determined), h is the height of the first level above the ground, and r is the mixing ratio, a characteristic of the moisture.

The recent numerical GAC models describe the atmosphere-ocean system as a whole. Numerical experiments with such models allow us to answer a number of questions from climate theory concerning its past, present and future. If in nature it is not possible to carry out an experiment in order to justify how the atmosphere would respond if, for instance, the influx of solar radiation considerably changes, if the ice cover melts or if the land and ocean distribution on the Earth globe is accomplished otherwise, by using the model it all becomes absolutely possible. These experiments revealed the extremely important role which the large-scale interaction between atmosphere and ocean plays in the formation of our planet's climate. It turns out that an explanation of past climatic variations can be given by 'purely Earth-like' reasons without necessarily involving hypotheses about cosmic influences.

Another result, obtained by means of numerical experiments on long-range forecasting, is no less important. It concerns the problem of the predictability of the large-scale atmospheric processes. Statistical analysis of numerical predictions has shown that satisfactory results are obtained for a period of up to two weeks. For instance, the correlation coefficient between the prognostic and actual variability of the geopotential on the 1000,

500 and 50 mb levels drops below 0.5 for a period $t \geqslant 14$ days. For example, after such an interval of time the mean-square value of the prognostic error becomes greater than the actual mean-square variability. This result supports the hypothesis about the existence of a limit (maximum period) of predictability, not greater than a month, according to recent evaluations. It depends considerably on the correct estimation of the permanently-acting nonadiabatic factors. It is clear, in principle, that the maximum period of predictability can be lengthened if a prognosis of the individual peculiarities of the synoptic processes leads to a prognosis of their statistical characteristics on the basis of Equations (2.15), (2.16), (2.22) and others similar to them which could be derived from the initial ones. Of course, in a proper statistical form, the initial conditions should also be given.

3. GAC AS QUASI-TWO-DIMENSIONAL TURBULENCE

a. Empirical Data

As we have already seen, irregularity as a property is also inherent to the motions of the GAC system, so that whichever characteristic, $f(\mathbf{x}, t)$ can be decomposed into two parts

$$f(x, t) = \bar{f}(\mathbf{x}, t) + f'(\mathbf{x}, t) \tag{3.1}$$

where f' is a random function of the spatial coordinates $\mathbf{x}$ and time t (random field), while the overbar denotes averaging (climatic, zonal – on the parallels or some other kind, specially defined). Thus, $f = \bar{f} + f'$ is a random function (scalar or vector) as well. The simplest statistical characteristic of $f(\mathbf{x}, t)$ is its mean value $\bar{f}(\mathbf{x}, t)$, already treated as a nonrandom function of $\mathbf{x}$ and t. With exception of (2.14), all the prognostic equations considered so far have concerned $\bar{f}$. However, the information which we get studying the mean field $\bar{f}$ only is not enough. As in the theory, so in the practical applications too, it is necessary to know some other statistical characteristics of the field (f) and, in the first place, the correlation, structure and spectral functions (see Section 1 of Chapter 8 and, for more information, [15, 5, 53]).

For simplicity, we shall confine ourselves to the 'flat earth' approximation. We shall also recall that the synoptic scale motions are quasi-two-dimensional – the vertical scale in the coordinate system (x, y, z) is sharply 'shortened'. That is why a full isotropy of the random fields f, which takes place in the theory of three-dimensional microturbulence, cannot exist here. In return, the hypothesis about two-dimensional isotropy (in the horizontal plane $x0y$ only) of the fields of the main meteorological elements: wind velocity, temperature, pressure (geopotential) and others, appears to be a good-enough approximation, but only for the free atmosphere (for instance at 500 mb level). More precisely, these fields are locally homogeneous and isotropic in the horizontal plane and stationary in time with the structure function

$$D_{ff}(r, \tau) = \overline{[f'(x + r, t + \tau) - f'(x, t)]^2} \tag{3.2}$$

as a fundamental statistical characteristic. At $r = 0$ or $\tau = 0$ the temporal or spatial structure function $D_{ff}(\tau)$ and $D_{ff}(r)$ can be obtained, respectively. Theoretically, r and τ vary from 0 to ∞. In reality, however, the arguments r and τ are limited from below and above.

Thus, for instance, $10 < r < 10^4$ km due to the requirement for two-dimensional isotropy and the impossibility of having motions with a scale greater than the Earth's dimensions.

We accept saying that the structure, correlation and spectral functions (D_{ff}, B_{ff}, E_{ff}) at $r \sim 10^2$–10^3 km and $\tau \sim 10$–10^2 h characterize the statistical macrostructure of a particular element f. These functions can be determined empirically by using the standard aerological data. The results are represented in tables, graphs or by empirical approximation formulae which smooth the discrete values and ensure existence of nonnegative Fourier transformation (i.e., $E_{ff} \geqslant 0$). Most of the published data (see, for instance, [5]) concern the structure functions of geopotential $D_{\phi\phi}(r)$, temperature $D_{TT}(r)$, and wind velocity $D_{cc}(r)$ or the corresponding correlation functions $B_{ff}(r)$. Thus, for example, the empirical curves for the normalized correlation function $R_{\phi\phi}(r) = B_{\phi\phi}(r)/B_{\phi\phi}(0)$ are well approximated by some of the following analytical expressions

$$R_{\phi\phi}(r) = (1 + a_1 r)\exp[-a_1 r],\ \exp[-a_2 r^2] \tag{3.3}$$

or

$$R_{\phi\phi}(r) = \exp[-a_3 r]\cos a_3 r,\ \exp[-a_4 r^{3/2}]\ J_0(a_5 r), \tag{3.4}$$

where $J_0(x)$ is the Bessel function, the distance between the points r is in 10^3 km and the constants a_i have empirical values: $a_1 = 0.98$, $a_2 = 0.6$, $a_3 = 0.3$, $a_4 = 0.195$, $a_5 = 1.065$.

For $r < 10^3$ km, all formulae approximate the empirical data equally well. However, for $r > 10^3$ km expressions (3.3) always remain positive. Most general physical considerations for mass conservation of the atmosphere require $R_{\phi\phi}(r)$ (or $R_{pp}(r)$, p – pressure) to have negative values for large r as (3.4) do. The empirical values of $R_{\phi\phi}(r)$ are determined up to r = 3000–4000 km. Similar results have also been obtained for the temperature.

Let us now consider the wind velocity $\mathbf{u}(\mathbf{x}, t)$ as an example of a random vector field. It is called statistically homogeneous and isotropic on the plane $x0y$ if the longitudinal $u_l(\mathbf{x}, t)$ and lateral $u_n(\mathbf{x}, t)$ components of $\mathbf{u}$ considered as random scalar fields are homogeneous and isotropic in a statistical sense and also mutually noncorrelated ($\overline{u_l u_n} = 0$). Hence, it follows that the correlation (structure) functions B_{ll}, B_{nn}, $B_{cc} = B_{ll} + B_{nn}$ (D_{ll}, D_{nn}, $D_{cc} = D_{ll} + D_{nn}$) will depend only on $r = |\mathbf{x}'' - \mathbf{x}'|$ but as can be seen on Figure 11.3, the correlation functions B_{uu} and B_{vv} of the components u, v (most often these are zonal and meridional components) depend not only on r but also on the angle θ, i.e., (u, v) are not isotropic random fields. It can be proved [53] that

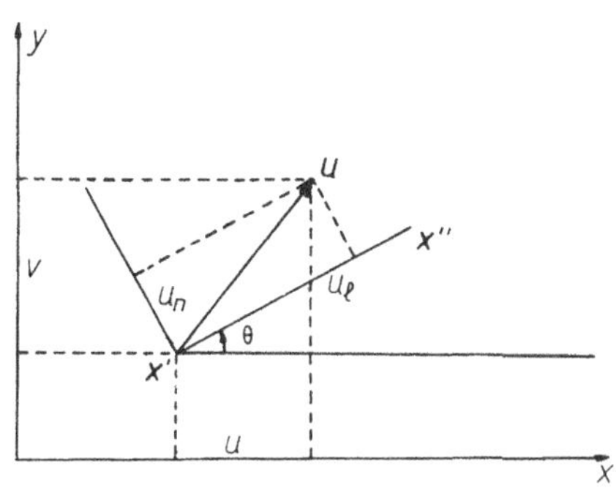

Fig. 11.3. A two-dimensional random vector field.

$$B_{uu}(r, \theta) = B_{ll}(r) \cos^2 \theta + B_{nn}(r) \sin^2 \theta,$$
$$B_{vv}(r, \theta) = B_{ll}(r) \sin^2 \theta + B_{nn}(r) \cos^2 \theta, \tag{3.5}$$
$$B_{uv}(r, \theta) = [B_{ll}(r) - B_{nn}(r)] \sin \theta \cos \theta,$$

so that

$$B_{cc}(r) = \overline{\mathbf{u}' \cdot \mathbf{u}''} = \overline{u'u''} + \overline{v'v''} = B_{ll}(r) + B_{nn}(r), \tag{3.6}$$

i.e., the total or vector-correlation function depends only on r.

The aerological data for the wind velocity **u** at different levels in the atmosphere are more uncertain than those for the geopotential or temperature. Having such data about **u**, any of the above-mentioned correlation (B_{ff}) or structure (D_{ff}) functions can be computed. Yet the first empirical results in the early Fifties showed that for $10 < r < 10^3$ km

$$D_{cc}(r) = a_c r^n, \quad n \approx 1. \tag{3.7}$$

The recent empirical results also imply a 'linear section' in the general dependence of D_{cc} (consequently of B_{ll} and B_{nn} too) on r. This problem has a history of its own about which a few words will be said below.

Here we shall mention that at synoptic scale motions the independently-determined statistical characteristics of the wind **u** and geopotential Φ (pressure p) should be consistent with each other. Actually for such motions, the individual realizations of both fields are related by the quasi-geostrophic equation

$$\mathbf{u} = f^{-1}\mathbf{k} \times \nabla\Phi. \tag{3.8}$$

Assuming f = const, it follows from (3.8) that

$$B_{cc}(r) = \overline{\mathbf{u}' \cdot \mathbf{u}''} = f^{-2}\overline{(\mathbf{k} \times \nabla'\Phi') \cdot (\mathbf{k} \times \nabla''\Phi'')}$$
$$= -f^{-2}\nabla_r^2 B_{\phi\phi}(r), \tag{3.9}$$

where $\mathbf{u}' = \mathbf{u}(\mathbf{x}')$, $\mathbf{u}'' = \mathbf{u}(\mathbf{x}'')$, $\nabla_r^2 = \mathrm{d}^2/\mathrm{d}r^2 + \mathrm{d}/r\,\mathrm{d}r$ is the Laplace operator in case of circular symmetry (isotropy) on the plane, $\mathbf{r} = \mathbf{x}'' - \mathbf{x}'$, $r = |\mathbf{r}|$. Therefore, (3.9) is a statistical equivalent of (3.8). It has a simpler form as a connection between the spectral densities (see formula (1.16) of Chapter 8)

$$E_{cc}(k) = f^{-2}k^2 E_{\phi\phi}(k). \tag{3.10}$$

In this way a formula for $B_{cc}(r)$ corresponds to each one of the formulae (3.3) and (3.4). For example, the first expression (3.3) inserted into (3.9) yields

$$B_{cc}(r) = (a_1 \sigma_\phi/2)^2\,(2 - a_1 r)\exp[-a_1 r]. \tag{3.11}$$

Hence, an interesting relation follows

$$f\sigma_c = \sqrt{2}a_1\sigma_\phi. \tag{3.12}$$

It allows the estimation of the characteristic scale of correlation $L_1 = 1/a_1 = \sqrt{2}\sigma_\phi/f\sigma_c$ having empirical data about the variances σ_c and σ_ϕ. At $a_1 r \ll 1$, i.e., at $r \ll 10^3$ km, (3.11) has a linear asymptote which is in agreement with (3.7).

Having determined the correlation function $B_{ff}(r)$, the corresponding spectral density $E_{ff}(k)$ can be found using the inverse transformation

$$E_{ff}(k) = \int_0^\infty B_{ff}(r) J_0(kr)\,dr. \tag{3.13}$$

For instance, (3.3) and (3.13) yield

$$E_{\phi\phi}(k) = \frac{3\sigma_\phi^2}{a_1^2} k \Big/ \left(1 + \frac{k^2}{a_1^2}\right)^{5/2} \quad \text{or} \quad \frac{\sigma_\phi^2}{2a_2^2} \exp\left(-\frac{k^2}{4a_2^2}\right). \tag{3.14}$$

Obviously, $E_{\phi\phi}(k)$ never becomes negative.

Correlation, structure and spectral functions are equally used in the investigations with preference given to one or another of them, depending on the case.

On constructing statistical GAC schemes over a hemisphere or a sphere, spherical coordinates are compulsorily used and then some definitions alter. For instance, about the correlation function we have

$$B_{ff}(\lambda_1, \theta_1; \lambda_2, \theta_2) = \overline{f'(\lambda_1, \theta_1) f'(\lambda_2, \theta_2)}, \tag{3.15}$$

where λ, θ are spherical co-ordinates, $f' = f - \bar{f}$. In the case of statistical homogeneity along the parallels, B_{ff} will depend on $\lambda = \lambda_2 - \lambda_1$ only. So far, the empirical data about the correlation functions of type (3.15), as well as about the space-time statistical characteristics, e.g., (3.15), are very scarce.

A considerable part of the published empirical data on the statistical macrostructure of the meteorological elements refer to the frequency and space spectra and comprise a narrower or wider range of frequences ω or wavenumbers k. Here we shall adduce only two examples.

The kinetic energy frequency spectrum of the ground wind composed by Van der Hoven (1957) is shown in Figure 11.4. The existence of a wide and deep minimum (gap)

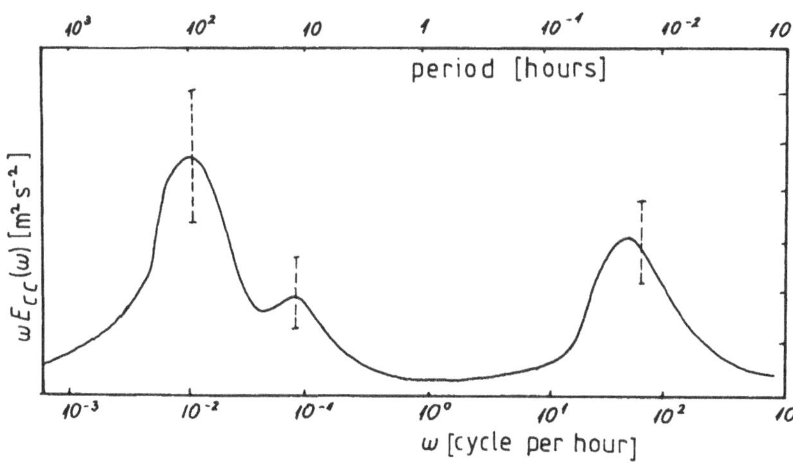

Fig. 11.4. Kinetic energy spectrum curve of the surface layer wind, composed by Van der Hoven (1957).

corresponding to oscillations with periods of between 0.1 and 10 h (meso-meteorological minimum) that separates two well-formed maxima (micro-meteorological maximum on the right and synoptic maximum on the left) is worth mentioning. The existence of such a spectral gap makes the statistical characteristics of the micro- and meso-scale turbulent motions slightly sensitive to the time period of averaging when it belongs to the range of the minimum. Here we are mainly interested in the synoptic part of the spectrum, corresponding to the quasi-two-dimensional synoptic wind variations. It has been the subject of detailed investigations in recent years. What is more, the statistics are based both on actual data from observations and upon ensembles of prognostic values for the wind, based on one or other numerical GAC models. Part of the so-obtained results, analysed by Leith (1971), are shown in Figure 11.5 in a double-logarithmic plot, where k is the

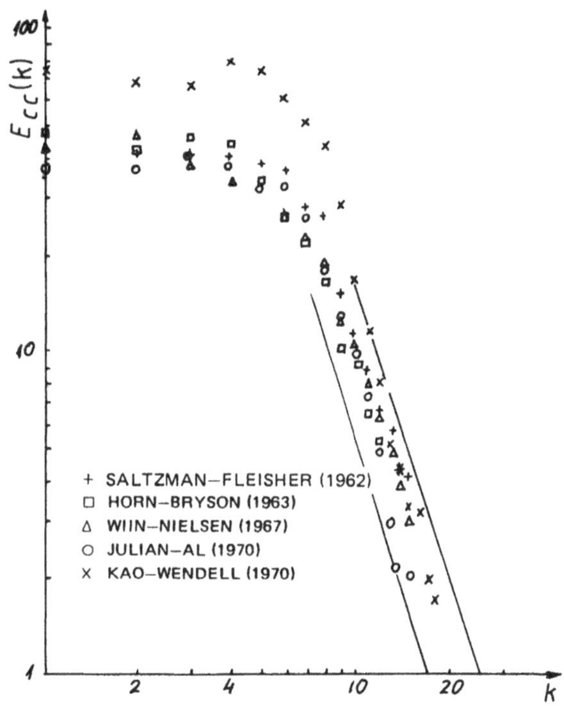

Fig. 11.5. Kinetic energy spectrum $E_{cc}(10^{-4}\ \text{rad}^2\ \text{day}^{-2})$ of the large-scale quasi-two-dimensional eddy motions in the atmosphere (after Leith, 1971).

zonal wavenumber ($k = 1$ corresponds to a wavelength $2\pi r_0 \cos\varphi = L_\varphi$, i.e., the length of the circumference at latitude φ, $k = 2$ – to a wavelength $0.5\ L_2$, etc.). It is worth mentioning that around wavenumber $k = 10$ (or waves on the parallel $\varphi = 45^\circ$ with a length of about 2700 km), there exists a quite large spectral range ($8 \leqslant k \leqslant 16$) with -3 slope of the energy spectrum:

$$E_{cc}(k) = q_c k^{-3}. \tag{3.16}$$

This dependence has been deduced by many authors for different seasons, geographical regions, and isobaric levels. One of the most interesting problems of GAC theory, that of the geostrophic (quasi-two-dimensional) turbulence, is connected with the physical interpretation of (3.16). A brief exposition is given in the next section.

b. Theory of Atmospheric Macroturbulence

(1) The notion *macrotubulence* was introduced by the Austrian meteorologist Defant in 1921. Later this was given a clear physical meaning within the framework of the GAC problem and particularly after the development of Kolmogorov's theory of locally-isotropic turbulence and its application to the atmosphere. At a given scale $L(\sim 10^2-10^3$ km), conditionally, all smaller-scale motions can be considered as turbulence and their effect described statistically, while the larger-scale motions can be described individually by the equations of atmospheric fluid dynamics and thermodynamics in which, as we have already seen, 'turbulent terms' also are present.

The role of L can be played by the distance ('step') between the grid points in the numerical schemes for solving the equations. Then a concept for subgrid turbulence is introduced. In the case of GAC, this conditional scale (L) is suggested by Nature itself – the dimensions of the biggest elements are of the order of 3000–4000 km. Then, the cyclones and anticyclones appear as the most dominant elements (eddies) of the atmospheric macroturbulence. A systematic effect of the motions with scales smaller than the fixed one (L) is the horizontal transport of heat, momentum and other properties, as well as a reduction in the predictability of motions with scales larger than L.

Defant also applied the methods of the semi-empirical theory to the horizontal atmospheric macroturbulence, mainly for calculation of the meridional heat transport. Really, although the advective mechanism, for instance by means of the tradewind circulation, makes some contribution to the meridional heat exchange, due to the fact that the Coriolis force presents the establishment of continuous meridional circulation, this exchange, particularly in the middle geographical latitudes, is mainly realized by macroturbulence. The heat flux $q = \rho c_p \overline{v'_\theta T'}$ is expressed by

$$q = -\rho c_p K_H^T \frac{\partial \overline{T}}{r_0\, \mathrm{d}\varphi}, \tag{3.17}$$

where $K_H^T > 0$ $(\sim 10^6-10^7$ m^2 s$^{-1})$ is the coefficient of horizontal macroturbulent heat exchange. At $\varphi \sim 40°$N $|\partial\overline{T}/r_0\ \partial\varphi| \sim 5°$C/1000 km and $q \sim 30$ cal cm^{-2} min^{-1} which exceeds the solar constant nearly 15 times. Since $K_H^T > 0$ and $\partial\overline{T}/\partial\varphi < 0$ then $q > 0$, i.e., the heat flux is directed from the Equator to the Poles.

As was mentioned in Section 3 of Chapter 8, such a kind of momentum flux parametrization by means of large-scale eddy viscosity coefficient K_H in the case of GAC is not always possible. This is because the quantity $Tr = -(\overline{v'_\theta v'_\lambda}/r_0 \sin\theta)\ \partial\overline{v}_\lambda/\partial\theta$, characterizing the rate of exchange of the kinetic energy between the mean zonal motion and the macroturbulence, according to the empirical results [22], for some regions turns out to be positive, while for others, it is negative. For instance, in the middle latitudes atmosphere $Tr > 0$ on average, i.e., a 'dissipation' of the mean motion energy occurs – it partly turns into the energy of the macroturbulence (of the cyclonic activity) which yields to a strenghtening of the intermeridional exchange. If this process is parametrized by introducing K_H, then it would have a positive value ($K_H > 0$). However, in the subtropical zone of the maximum zonal velocity $\overline{v}_k$, very often $Tr < 0$, i.e., the mean flow partially obtains its kinetic energy at the expense of the fluctuating motion which, for its part, is created by the local temperature contrasts. Here we should have $K_H < 0$ (negative eddy viscosity) which is physically unacceptable.

The quoted results are average. They do not exclude the possibility that the energy exchange between the mean and the fluctuating motion takes place the other way round under particular conditions. All this makes the theoretical determination of the structure and spectral functions $D_{ff}(r)$ and $E_{ff}(k)$ extremely difficult.

(2) The result quoted earlier (3.7), i.e., $D_{cc}(r) = a_c r$ at $r \sim 10^2$ km shows that the '$\frac{2}{3}$ law' of Kolmogorov, confirmed so definitely by observations of the atmospheric (three-dimensional) microturbulence, here ceases to be valid, at least for such distances. Assuming the linear expression $D_{cc} \sim r$ as an empirical fact, an attempt can be made for its physical interpretation, at least for the case when $Tr > 0$.

In this case, a cascade process of kinetic energy transport from larger to smaller inhomogeneities in the velocity field (conditionally 'eddies') at a rate $\epsilon = Tr$ (stationarity assumed) can be postulated. Of course, a part of the kinetic energy of the quasi-two-dimensional turbulent 'eddies' can pass directly into the energy of three-dimensional microturbulence in any part of the spectrum (not in the short-wave part of the synoptic range only), where the horizontal scales are commensurable with the vertical ones ($\sim$10 km). In addition to this, the magnitudes of the mean and fluctuating velocities are of one and the same order, the latter one thus being subject to the influence of the Earth's rotation and curvature too. So, the structure characteristics D_{cc} and E_{cc}, except on r and k, may also depend on a number of dimensional parameters. For instance,

$$\begin{aligned} D_{cc}(r) &= F_1(r, \epsilon, f, \beta, \ldots), \\ E_{cc}(k) &= F_2(k, \epsilon, f, \beta, \ldots). \end{aligned} \tag{3.18}$$

where $f = 2\omega \sin \varphi$, $\beta = df/dy$. If β and f are neglected, then from (3.18), Kolmogorov's $\frac{2}{3}$ and $-\frac{5}{3}$ laws immediately follow

$$D_{cc}(r) = \alpha_r (\epsilon r)^{2/3}, \qquad E_{cc}(k) = \alpha_k \epsilon^{2/3} k^{-5/3}. \tag{3.19}$$

Many authors argue that Equations (3.19) satisfactorily approximate the empirical data on the atmospheric macroturbulence. One very strong argument proves to be the consistency of (3.19) with the empirical law of Richardson about the coefficient of macroturbulent diffusion in the atmosphere $K \sim l^{4/3}$, derived for $l \lesssim 10^6$ m scale. But as we have already seen, there are also data, contradictory to this concept, e.g., (3.7).

Retaining in (3.18) the explicitly-indicated arguments, we can write it in a more general form [53]

$$\begin{aligned} D_{cc}(r) &= \epsilon f^{-1} D(r/L_f, L_f/L_\beta), \\ E_{cc}(k) &= \epsilon^{3/2} f^{-5/2} E(kL_f, L_f/L_\beta), \end{aligned} \tag{3.18'}$$

where

$$L_f = \epsilon^{1/2} f^{-3/2}, \qquad L_\beta = \epsilon^{1/5} \beta^{-3/5} \tag{3.20}$$

are two linear scales, characterizing the influence of the Earth's rotation and curvature on the macrostructure of the wind field. With Brunt's averaged-for-the-whole-atmosphere value of $\epsilon = 5\ \mathrm{cm^2\ s^{-3}}$ and f and β taken for $\varphi \sim 45°$ we obtain

$$L_f \sim 10\ \mathrm{km}, \qquad L_\beta \sim 10^3\ \mathrm{km}.$$

Making (3.18′) consistent with (3.7), we obtain

$$\begin{aligned} D_{cc}(r) &= d(\mu)\,(\epsilon f)^{1/2} r = a_c r, \\ E_{cc}(k) &= e(\mu)\,(\epsilon f)^{1/2} k^{-2} = b_c k^{-2}, \end{aligned} \tag{3.21}$$

where $\mu = L_f/L_\beta$, d and e are dimensionless functions of μ, which are not independent as far as $D_{cc}(r)$ and $E_{cc}(k)$ are Fourier transformations to each other.

As we have seen, at motions in the GAC system there are also cases when $Tr < 0$. Besides, the empirical law (3.16) is in evidence too, not agreeing with (3.21). These facts are united in the theory of the so-called *geostrophic turbulence* (Charney, 1971), based on the previously-developed theory of two-dimensional turbulence (Batchelor, 1965; Kraichnan, 1967, etc.).

As was shown in Section 2 of Chapter 7, two integral invariants exist in a two-dimensional inviscid flow – the kinetic energy and enstrophy (one-half of the mean square vorticity) assigned to a unit mass are preserved. Moreover, according to Fjortoft's theorem, the kinetic energy flux across the wavenumber spectrum is directed from smaller to larger scales, while the cascade of enstrophy is realized from larger to smaller scales. Batchelor and Kraichnan showed that in a two-dimensional flow these restrictions lead to the existence of two inertial intervals (Figure 11.6).

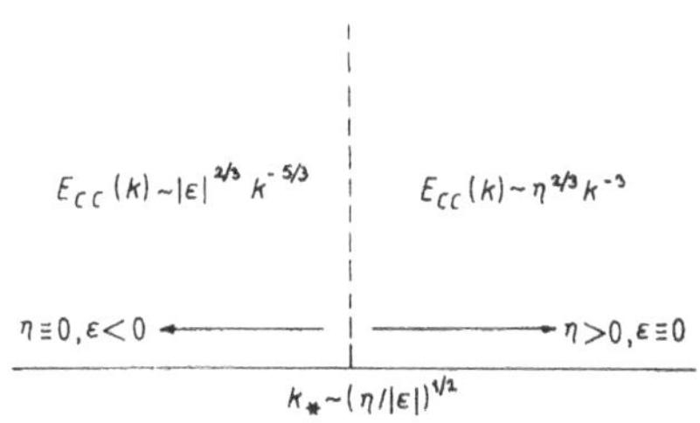

Fig. 11.6. Inertial subranges in a two-dimensional turbulent flow.

If kinetic energy and vorticity are generated in the vicinity of wavenumber k_*, then the enstrophy cascades towards $k \to \infty$ *only* with a constant (across the spectrum) rate $\eta > 0$, while the energy cascades towards $k \to 0$ *only* with a constant rate $\epsilon < 0$. The inertial intervals block the energy and enstrophy transfer correspondingly. Then simple dimensional arguments yield the following power laws for the energy spectrum

$$E_{cc}(k) = \alpha_1 \eta^{2/3} k^{-3}, \qquad k \geqslant k_* \tag{3.22}$$

$$E_{cc}(k) = \alpha_2 |\epsilon|^{2/3} k^{-5/3}, \quad k \leqslant k_* \tag{3.23}$$

where α_1 and α_2 are universal constants. The full analogy among (3.23) and (3.19) should not be a reason for the different physical notions, on whose basis they have been derived, to be forgotten.

On the other hand, the consistency of (3.22) with the empirical law (3.16), as well as the quasi-two-dimensional character of the atmospheric macroturbulence, give a reason to think that (3.16) is an empirical confirmation of (3.22) and is a direct consequence of the specific geometry of the flow. In spite of all this, we cannot neglect the fact that the atmosphere has a finite thickness which is enough to determine a number of specific phenomena, missing in purely two-dimensional flow. It would be enough to mention the baroclinic properties of the atmosphere and the related mean wind shear, vertical motions, etc. These peculiarities have been taken into consideration in Charney's (1971) theory for quasi-geostrophic turbulence. This is a random, nonlinear atmospheric motion in approximate geostrophic and hydrostatic balance. That is why the Rossby number (Ro $= U/fL$), the Ekman number ($E = \nu/fZ^2$) and the Kibel number ($Ki = 1/fT$) have to be small (Z – vertical scale). In such a flow (stratified and baroclinic), the potential vorticity and the kinetic-plus-available potential (i.e., the total) energy are preserved. The first quantity, however, characterizes the third (the vertical) dimension of the atmosphere. Introducing vertically stretched coordinate $dz^* = (\omega_{BV}/f)\, dz$, where ω_{BV} is the Brunt–Vaisala frequency, Charney postulates a three-dimensional local homogeneity and isotropy in the space (x, y, z^*) and cascade transfer of potential enstrophy ($\frac{1}{2}$ of the mean-square potential vorticity) and total energy towards $k \to \infty$ and $k \to 0$ correspondingly with rates η and ϵ. Dimensional analysis yields again (3.22) but now $k^2 = k_1^2 + k_2^2 + k_3^{*2}$ where k_3^* is a vertical wave-number in the frame (x, y, z^*). In such a way, it follows from this theory that not only the kinetic energy but the potential energy too (i.e., the temperature) would obey the -3 spectral law. Charney's theory suggests a greater certainty that empirical expression (3.16) is not accidental but reflects the truly-essential characteristics of the atmospheric macroturbulence. As long as the '-3 power law' gains new confirmations from observations and numerical experiment, the existence of the '$\frac{5}{3}$ law' at $k \leqslant k_*$ has not yet been substantiated. It is believed that in the atmosphere $k_* = 4$–8, which is a region with maximum baroclinic instability and most intense conversion of available potential energy into kinetic energy, at which vorticity is also generated.

(3) The problem related to the value of the power in the empirical or theoretical formulae of the kind $E_{cc}(k) \sim k^{-m}$ became extremely important when it turned out that it is connected with the predictability range of the wind field in a definite spectral region.

Actually, as was mentioned earlier, motions with scales smaller than the distances between the observation points cannot be fixed and used in the initial conditions of numerical forecasting. The larger-scale motions, however, are precisely observable. Consequently, the initial data would possess some 'uncertainty' concentrated in the short-wave part of the synoptic spectrum, which by means of the nonlinear interactions described by the corresponding terms in the prognostic equations, will propagate towards its long-wave part.

Lorenz (1969) assumes that the time necessary for a complete uncertainty at wavenumber $2k$ to infect wavenumber k and to turn it completely uncertain as well, is proportional to the timescale of a turbulent turn-over at scale $l \sim k^{-1}$. The latter might be defined as

$$\tau(k) = k^{-1}\,[E_{cc}(k)k]^{-1/2} = k^{-1/2} E_{cc}^{-1/2}(k). \tag{3.24}$$

Then the predictability time necessary for propagating the uncertainty N octaves from, say, $2^N k$ to k is given by

$$\tau_N = \sum_{n=0}^{N-1} \tau(2^n k) = \sum_{n=0}^{N-1} (2^n k)^{-3/2} [E_{cc}(2^n k)]^{-1/2}. \tag{3.25}$$

if $E_{cc}(k) = e_c k^{-5/3}$, then $\tau(k) = e_c^{-1/2} k^{-2/3}$ and

$$\tau_N = [1 + 2^{-2/3} + 2^{-4/3} + \cdots + 2^{-(2/3)(N-1)}]\tau(k). \tag{3.26}$$

Remarkably, this series has a finite sum for $N \to \infty$, i.e.,

$$\tau_\infty = \frac{\tau(k)}{1 - 2^{-2/3}} \approx 2.7\tau(k). \tag{3.27}$$

If, however, $E_{cc}(k) = q_c k^{-3}$, then

$$\tau_N = (1 + 1 + \cdots)\tau = N\tau = \text{const}_k. \tag{3.28}$$

At $N \to \infty$, the sum diverges to $\tau_N \to \infty$. Consequently, the predictability range depends essentially on the form of the energy spectrum $E_{cc}(k)$. Generally speaking, 'more predictable' are the spectral components of the motion for which $E_{cc}(k) \sim k^{-3}$, since, theoretically for them according to (3.28), $\tau_\infty = \infty$, while at $E_{cc}(k) \sim k^{-5/3}$, $\tau_\infty \approx 3\tau$. In other words, the predictability period for the larger scales of motion, obeying the $-\frac{5}{3}$ spectral law, is only about three times the turn-over time for those scales, essentially regardless of how well the initial state is known, while in the case of the -3 power law for kinetic energy spectral density, long predictions would be possible with sufficiently good initial data. In meteorological practice, perhaps 7 or 8 octaves of information are likely to be available as initial data, so that τ_N from (3.28) will be only 2–3 times larger than τ_N from (3.26), i.e., the difference is only a factor of 3 for even the planetary scale. Nevertheless, the conclusions made on the basis of (3.28) are physically more substantiated and reveal more optimistic perspectives for reaching the theoretical period of predictability within the frame-work of a particular numerical model.

4. LAGRANGIAN DESCRIPTION OF THE ATMOSPHERIC MACROTURBULENCE AND DIFFUSION

a. Theoretical Results

In Section 8 of Chapter 2, a number of problems with initial conditions concerning the deterministic description of the motion of air masses ('particles', 'parcels') in a flat or spherical Earth approximation have been discussed. Although theoretically possible, such an approach is hardly of practical benefit. A reason for this is the extremely complex form of the air particle trajectories in the real atmosphere, i.e., practically, the random character of the Lagrangian characteristics of the particles: the radius-vector of their position $\mathbf{x}(t)$, the velocity along the trajectory $\mathbf{u}_L(t)$ and so on (see Section 1 of Chapter 1). This makes us apply statistical methods of analysis.

Let us, for simplicity, consider macroturbulence which is planar isotropic and stationary in time. Obviously

$$\mathbf{x}(t) = \int_0^t \mathbf{u}_L(t')\,\mathrm{d}t' \tag{4.1}$$

where $\mathbf{x}(0) = 0$ is assumed. The same relationship holds between the deviations of $\mathbf{x}$ and $\mathbf{u}_L$ from their mean values. That is why we denote the deviations by $\mathbf{x}$ and $\mathbf{u}_L$. Due to the assumed isotropy, all directions on the plane are equal and we can consider one of the two scalar equalities (4.1), say

$$x(t) = \int_0^t u_L(t')\,\mathrm{d}t' \tag{4.1'}$$

Since $u_L(t)$ is assumed to be a stationary random function, then its integral $x(t)$ will not be stationary [53]. A given particle would drift away from its initial position and, a measure this would be the variance

$$\sigma_x^2(t) = \overline{x^2(t)} = \overline{\left[\int_0^t u_L(t')\,\mathrm{d}t'\right]^2}. \tag{4.2}$$

Hence,

$$\sigma_x^2(t) = \int_0^t \int_0^t \overline{u_L(t')u_L(t'')}\,\mathrm{d}t'\,\mathrm{d}t'' = \overline{u_L^2} \int_0^t \int_0^t R_L(t'' - t')\,\mathrm{d}t'\,\mathrm{d}t'', \tag{4.3}$$

where

$$R_L(\tau) = \frac{\overline{u_L(t')u_L(t'')}}{\overline{u_L^2}}, \quad \tau = t'' - t' \tag{4.4}$$

is the Lagrangian auto-correlation coefficient, $\overline{u_L^2} = \sigma_u^2$ is the corresponding variance, and the overbar denotes time averaging along the particle's trajectory. In case of many realizations, statistical averaging (over an ensemble) is also included. Due to the stationarity, $R_L(\tau)$ is an even function. Then a single integration of (4.3) yields

$$\sigma_x^2(t) = 2\overline{u_L^2} \int_0^t (t - \tau) R_L(\tau)\,\mathrm{d}\tau, \tag{4.5}$$

which is known as Taylor's formula. It relates two important statistical Lagrangian characteristics – $\sigma_x^2(t)$ and $R_L(\tau)$. If one of them is known, the other can be determined from (4.5) or from the reverse relationship

$$R_L(\tau) = \frac{1}{2\overline{u_L^2}} \left[\frac{\mathrm{d}^2}{\mathrm{d}t^2}\,\sigma_x^2(t)\right]_{t=\tau}. \tag{4.6}$$

Another measure for the particles' turbulent diffusion with respect to the source (the degree to which they have been spread in the flow), is the diffusion coefficient

$$K(t) = \frac{1}{2}\frac{\mathrm{d}}{\mathrm{d}t}\sigma_x^2(t) = \overline{u_L^2}\int_0^t R_L(\tau)\,\mathrm{d}\tau. \tag{4.7}$$

Let us consider two asymptotic cases:

(1) $t \to \infty$. Then Equations (4.7) and (4.5) yield

$$K(t) = \overline{u_L^2}\,T_L = K_\infty, \qquad \sigma_x^2(t) = 2K_\infty t, \tag{4.8}$$

where

$$T_L = \int_0^\infty R_L(\tau)\,\mathrm{d}\tau \tag{4.9}$$

is the so-called Lagrangian integral scale of correlation. It is seen that the classical theory of turbulent diffusion with a constant diffusion coefficient K is valid for a large diffusion time when $K(t) \approx K_\infty$ and $\sigma_x^2 \to \infty$.

(2) $t \to 0$. Then we can put $R_L(\tau) \approx 1$ in (4.5) and (4.7) so that

$$\sigma_x^2(t) = \overline{u_L^2}\,t^2, \qquad K(t) = \overline{u_L^2}\,t = \sigma_u\sigma_x(t). \tag{4.10}$$

To determine $\sigma_x(t)$ and $K(t)$ for the intermediate values of t, one must know $R_L(\tau)$ for arbitrary τ. Some empirical data imply that

$$R_L(\tau) = \exp\left(-\frac{|\tau|}{T_L}\right) \tag{4.11}$$

would be a good approximation formula. Substitution into (4.5) and (4.7) and integration yields

$$\begin{aligned}\sigma_x^2(t) &= 2\overline{u_L^2}\,T_L\left(\frac{t}{T_L} - 1 + \exp\left[-\frac{t}{T_L}\right]\right),\\ K(t) &= \overline{u_L^2}\,T_L\left(1 - \exp\left[-\frac{t}{T_L}\right]\right).\end{aligned} \tag{4.12}$$

Let us now study the relative diffusion of a pair of particles having coordinates $\mathbf{x}_1$ and $\mathbf{x}_2$. Then, the variance of the vector $\mathbf{r} = \mathbf{x}_2 - \mathbf{x}_1$

$$\sigma_r^2(t) = \overline{[r(t)]^2} \tag{4.13}$$

will be a measure for the relative separation of the particles. Obviously, responsible for the magnitude of σ_r will be turbulent inhomogeneities with scales $\lambda \sim \sigma_r$, i.e., σ_r is a local characteristic. Thus, on the basis of the theory of the atmospheric quasi-two-dimensional macroturbulence (Section 3), the intermediate asymptotes of $\sigma_r(t)$ can be found, depending on which inertial interval (energy or enstrophy) $\sigma_r(t)$ belongs to. Considering, for example, the coefficient of relative diffusion

$$K_r = \frac{1}{2}\frac{\mathrm{d}}{\mathrm{d}t}\sigma_r^2(t),$$

we can write

$$K_r = \begin{cases} F_1(\eta, \sigma_r), \\ F_2(\epsilon, \sigma_r). \end{cases} \tag{4.14}$$

Hence, it follows that

$$K_r = \begin{cases} a_1 \eta^{1/3} \sigma_r^2, \\ a_2 \epsilon^{1/3} \sigma_r^{4/3}, \end{cases} \tag{4.15}$$

and also

$$\sigma_r^2(t) = \sigma_r^2(0) \begin{cases} e^{t/T_1}, \\ (1 + t/T_2)^3, \end{cases} \tag{4.16}$$

where $2a_1\eta^{2/3} = 1/T_1$, $\frac{4}{3}a_2\epsilon^{1/3}/\sigma_r^{2/3}(0) = 1/T_2$. Substituting (4.16) into (4.15), one can find K_r as a function of the diffusion time t.

The relative macrodiffusion of particles in the atmosphere as a phenomenon was studied for the first time in 1926 by Richardson who found empirically that $K_r \sim \sigma_r^{4/3}$ for values of σ_r up to 10^3 km. Later on, an explanation of this dependence was given by Obukhov on the basis of Kolmogorov's theory of locally-isotropic three-dimensional turbulence. Since the energy inertial $-\frac{5}{3}$ interval (3.23) has not yet received empirical support, then the $\frac{4}{3}$ diffusion law (4.15) in the case of atmospheric macro-turbulence, if it really holds, should be considered as physically unexplained. And yet the difference between $\frac{4}{3}$ and 2 (the powers in (4.15)) is not so large that a firm preference to one or the other should be given according to the empirical data.

b. Empirical Data

For the empirical determination of the above-introduced diffusion characteristics of a single or pair of particles, observational data are needed concerning the trajectories, which are schematically shown in Figure 11.7. Such data can be obtained in two ways.

The first one consists in releasing a large number of constant-volume balloons at some isobaric level, say 200 or 500 mb that serve as indicators of the atmospheric motions at

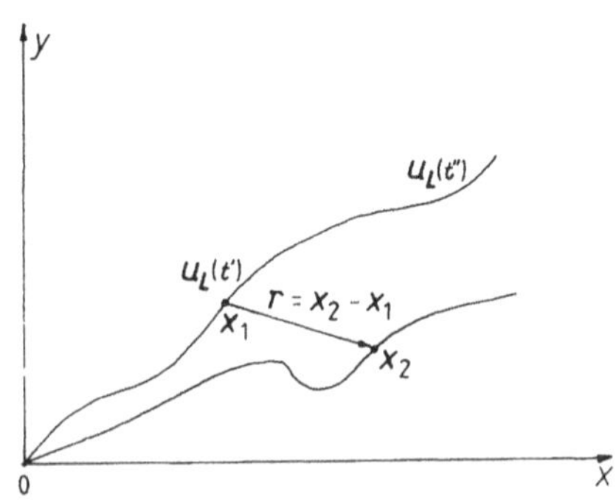

Fig. 11.7. Particles' trajectories and the Lagrangian characteristics of their relative motion.

these levels. The balloons are located periodically by means of special navigation devices from the ground or a satellite. Then various Lagrangian statistics can be extracted from the observational data. Unique results of such observations have been published by Morel and Larcheveque (1974) based on the so-called EOLE experiment with 480 balloons distributed over the Southern Hemisphere at the 200 mb level. It was found that the eddy dispersion process is homogeneous, isotropic and stationary up to scales of the order of 1000 km. This agrees well with the prediction of the two-dimensional turbulence model of atmospheric motions. Thus, the mean square relative separation $\sigma_r^2(t)$ increases exponentially with time up to 6 days (Figure 11.8), which is in agreement with (4.16), whereas $\sigma_r(0) = 80$ km, $T_1 = 1.35$ days, and more slowly like $t^{1/2}$ later. Accordingly, the diffusivity (Figure 11.9), obeys the first law of (4.15). Its saturated value at $t \to \infty$ is estimated to be $K_\infty \approx 1.6 \times 10^6$ m^2 s^{-1}.

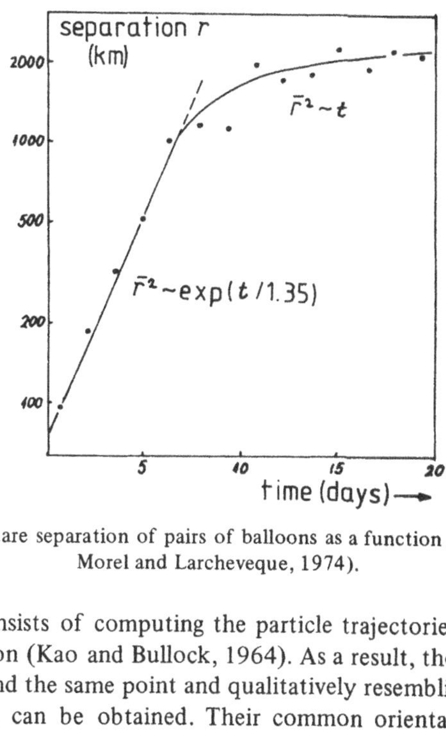

Fig. 11.8. Root mean square separation of pairs of balloons as a function of time after launch (after Morel and Larcheveque, 1974).

The second way consists of computing the particle trajectories using actual maps of the large-scale circulation (Kao and Bullock, 1964). As a result, thousands of trajectories, coming out from one and the same point and qualitatively resembling the two trajectories shown in Figure 11.7, can be obtained. Their common orientation, naturally, would follow the mean flow at the level, but the form of each one would show the random character of the law of motion $\mathbf{x}(t)$; the ends of the trajectories for one and the same time after the initial instant would constitute an ensemble of random points in the plane xOy. An example is shown in Figure 11.10. The trajectories are called geostrophic because only the horizontal motion is considered.

Based on these data, the Lagrangian and Eulerian time-lag correlation coefficients $R_L(\tau)$ and $R_E(\tau)$ of the zonal (u) and meridional (v) components have been computed by Kao and Bullock (1964) (see also the supplement to [53] written by Kao). Two examples are shown in Figures 11.11 and 11.12. Their analysis yields the following conclusions:

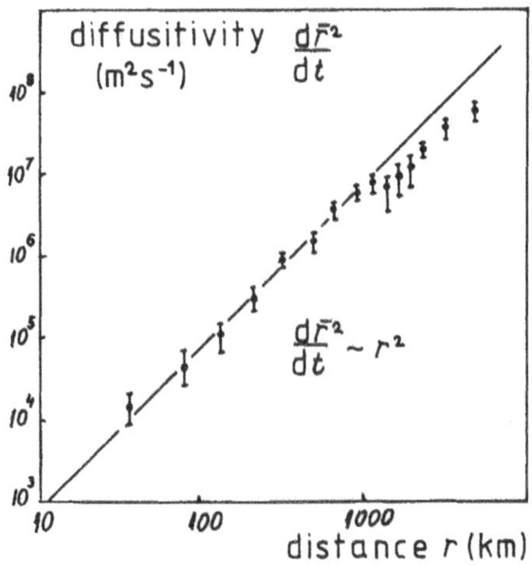

Fig. 11.9. The large-scale relative diffusivity as a function of the distance (after Morel and Larcheveque, 1974).

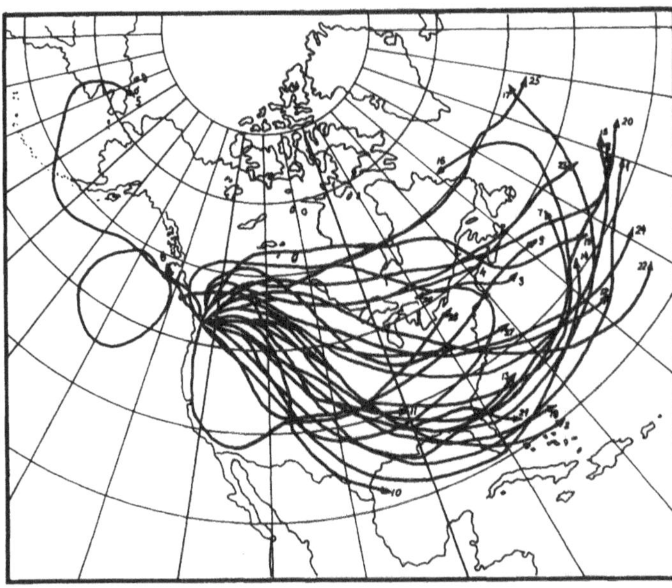

Fig. 11.10. Daily initiated geostrophic trajectories on a 500 mb isobaric surface computed for the period 1–28 February 1958. Each trajectory has been followed to 72 h (after Kao and Bullock, 1964).

(i) $R_{LU}(\tau)$ and $R_{EU}(\tau)$ monotonously approach zero, while $R_{LV}(\tau)$ and $R_{EV}(\tau)$ first take negative values and then approach zero. Their form (decaying oscillations) implies a periodic process responsible for the formation of the meridional velocities. It is not difficult for us to guess that these are the planetary waves.

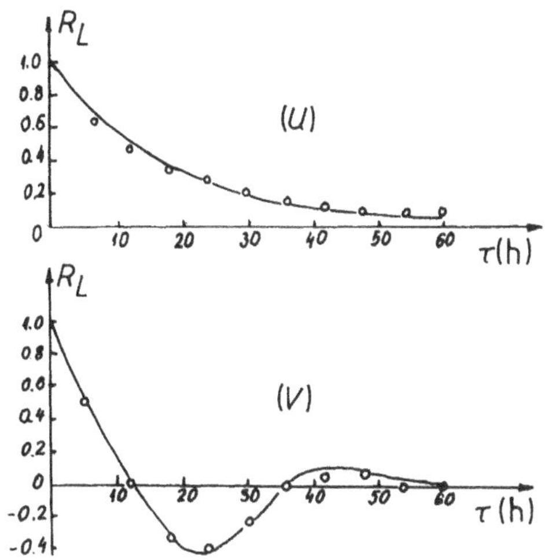

Fig. 11.11. Lagrangian auto-correlation functions $R_L(\tau)$ for the zonal (u) and meridional (v) wind components (after [53] – supplement by S. K. Kao).

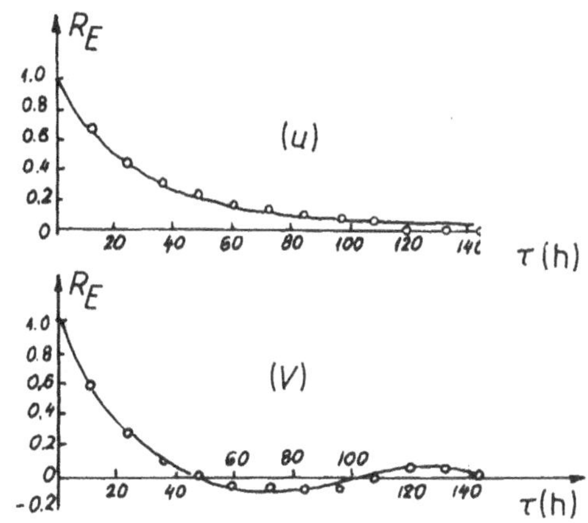

Fig. 11.12. Eulerian auto-correlation functions $R_E(\tau)$ for the zonal (u) and meridional (v) wind components (after [53] – supplement by S. K. Kao).

(ii) The Lagrangian and Eulerian correlation coefficients are similar to each other, which mathematically means that

$$R_L(\eta) = R_E(\tau), \quad \eta = \beta\tau. \tag{4.17}$$

It has been found that in the middle latitudes, the dimensionless constant $\beta < 1$. Obviously, the existence of relationships (4.17) is of extreme importance since they give us an opportunity of using the standard data of observation to calculate $R_L(\tau)$, by calculating first $R_E(\tau)$, if only β is known. As is seen from the graphs, a good approximation for $R_{LU}(\tau)$ and $R_{EU}(\tau)$ would be the exponential formula (4.11).

The practical significance of the problem for the atmospheric macroturbulence and diffusion in connection with the global transport of pollutants in the atmosphere and also its difficulty in the theoretical aspects are undoubted. This is one of the topical problems related to modern dynamic meteorology.

PROBLEMS

1. Show that from the general expression for the correlation tensor of the homogeneous and isotropic vector field $v_\alpha(x)$

$$B_{ij}(r) = \overline{u_i(x)u_j \cdot (x+r)} = [B_{ll}(r) - B_{nn}(r)] \frac{1}{r^2} r_i r_j + B_{nn}(r)\, \delta_{ij}$$

   formulae (3.5) follow.
2. Prove that formula (3.9) follows from (3.8) and that (3.10) is a spectral equivalent to (3.9).
   Note: see [53].
3. Calculate τ_N using (3.25) with $E_{cc}(k)$ taken from (3.21).
4. Starting from (4.3), derive (4.5),
5. Calculate T_L, $K(t)$ and $\sigma_x^2(t)$ if

$$R_L(\tau) = (1 + c|\tau|)^{-n}, \quad n > 0.$$

References

I. TEXTBOOKS AND MONOGRAPHS

1. Andreev, V. M. and Panchev, S.: 1975, *Dynamics of Atmospheric Thermals*, Gidrometeoizdat, Leningrad (in Russian).
2. Beer, T.: 1975, *Atmospheric Waves*, Adam Hilger, London.
3. Belov, P. N.: 1975, *Numerical Methods for Weather Prediction*, Gidrometeoizdat, Leningrad (in Russian).
4. Brown, R. A.: 1974, *Analytical Methods in the Planetary Boundary Layer Modelling*, Adam Hilger, London.
5. Czelnai, R., Gandin, L. S., and Zachariew, W. I. (eds.): 1976, *Statistische Structur der Meteorologischen Felder*, Budapest.
6. Chapman, S. and Lindzen, R. S.: 1970, *Atmospheric Tides*, D. Reidel, Dordrecht.
7. Csanady, G. T.: 1973, *Turbulent Diffusion in the Environment*, D. Reidel, Dordrecht.
8. Dikii, L. A.: 1969, *Theory of Oscillations of the Earth's Atmosphere*, Gidrometeoizdat, Leningrad (in Russian).
9. Dikii, L. A.: 1976, *Hydrodynamic Stability and the Atmospheric Dynamics*, Gidrometeoizdat, Leningrad (in Russian).
10. Dobrishman, E. M.: 1980, *Dynamics of the Equatorial Atmosphere*, Gidrometeoizdat, Leningrad (in Russian).
11. Dutton, J. A.: 1976, *The Ceaseless Wind*, McGraw-Hill, N.Y.
12. Du-Chjen, E., and Bao-Chjen, Chju: 1961, *Some Most Important Problems of the General Atmospheric Circulation*, Gidrometeoizdat, Leningrad (in Russian, translated from Chinese).
13. Falkovich, A. I.: 1979, *Dynamics and Energetics of the Intertropical Convergent Zone*, Gidrometeoizdat, Leningrad (in Russian).
14. Fein, J. S.: 1978, *Boundary Layers in Homogeneous and Stratified Rotating Fluids*, Univ. of Florida Press, Gainesville.
15. Gandin, L. S. and Kagan, R. L.: 1968, *Statistical Methods for Interpretation of Meteorological Data*, Gidrometeoizdat, Leningrad (in Russian).
16. Gandin, L. S. and Dubov, A. S.: 1968, *Numerical Methods for Short Range Weather Prediction*, Gidrometeoizdat, Leningrad (in Russian).
17. Gaponov-Grehov, A. V. (ed.): 1979, *Nonlinear Waves*, Nauka, Moscow (in Russian).
18. Gill, A. E.: 1982, *Atmosphere-Ocean Dynamics*, International Geophysical Series, vol. 30, Academic Press, N.Y., London.
19. Gledzer, E. B., Dolzhansky, F. V., and Obukhov, A. M.: 1981, *The Systems of Hydrodynamic Type and Its Application*, Nauka, Moscow (in Russian).
20. Godev, N. G.: 1976, *Synoptic Meteorology*, Nauka i Izkustvo, Sofia (in Bulgarian).
21. Gossard, E. E. and Hooke, W. H.: 1975, *Waves in the Atmosphere*, Elsevier, Amsterdam.
22. Gruza, G. V.: 1961, *Macroturbulence in the General Atmospheric Circulation*, Gidrometeoizdat, Leningrad (in Russian).
23. Haltiner, J. R.: 1971, *Numerical Weather Prediction*, John Wiley, New York.
24. Haltiner, G. J. and Martin, F. L.: 1957, *Dynamical and Physical Meteorology*, McGraw-Hill, N.Y.
25. Haugen, D. A.: 1973, *Workshop on Micrometeorology*, AMS, Boston.

26. Hinze, J. O.: 1975, *Turbulence*, McGraw-Hill, N.Y.
27. Holton, J. R.: 1972, *An Introduction to Dynamic Meteorology*, Academic Press, N.Y.
28. Holton, J. R.: 1975, *The Dynamic Meteorology of the Stratosphere and Mesosphere*, AMS, Boston.
29. Houghton, J. T. 1977, *The Physics of Atmospheres*, Cambridge University Press.
30. Kibel, I. A.: 1957, *Introduction to the Hydrodynamic Methods for Short Range Weather Prediction*, Gidrometeoizdat, Leningrad (in Russian).
31. Kochin, N. E.: 1949, *Collected Papers*, Acad. Sci. USSR, Moscow (in Russian).
32. Kostukov, V. V.: 1982, *Objective Analysis and Initialization of Meteorological Fields*, Gidrometeoizdat, Leningrad (in Russian).
33. Krastanov, L., Panchev, S., and Andreev, V.: 1978, *General Meteorology*, Nauka i Izkustvo, Sofia (in Bulgarian).
34. Kuznetsov, D. S.: 1951, *Hydrodynamics*, Gidrometeoizdat, Leningrad (in Russian).
35. Laikhtman, D. L. (ed.): 1976, *Dynamic Meteorology*, Gidrometeoizdat, Leningrad (in Russian).
36. Laikhtman, D. L.: 1970, *Physics of the Atmospheric Boundary Layer*, Gidrometeoizdat, Leningrad (in Russian).
37. Lebedev, V. L., Aizatulin, T. A., and Hailov, K. M.: 1974, *The Ocean as Dynamical System*, Gidrometeoizdat, Leningrad (in Russian).
38. Leibovich, S. and Seebass, A. R. (eds.): 1974, *Nonlinear Waves*, Cornell University Press, Ithaca and London.
39. *Lectures on Numerical Short Range Weather Prediction*, WMO Regional Training Seminar, Moscow, 17 Nov. – 14 Dec. 1965, Gidrometeoizdat, Leningrad (English version) 1969.
40. Lorenz, E. N.: 1967, *The Nature and Theory of the General Circulation of the Atmosphere*, WMO.
41. Lumley, J. L. and Panofsky, H. A.: 1964, *The Structure of Atmospheric Turbulence*, Interscience, N.Y., London, Sydney.
42. Marchuk, G. I.: 1967, *Numerical Methods in Weather Prediction*, Gidrometeoizdat, Leningrad (in Russian).
43. Marchuk, G. I.: 1974, *Numerical Solution of Atmospheric and Oceanic Dynamic Problems*, Gidrometeoizdat, Leningrad (in Russian).
44. Matveev, L. T.: 1965, *Fundamentals of General Meteorology*, Gidrometeoizdat, Leningrad (in Russian).
45. Mesinger, F.: 1978, *Dinamicka Meteorologija*, Belgrade (in Serbo-Croatian).
46. Monin, A. S. (ed.): 1967, *Dynamics of Large-Scale Atmospheric Processes*, Proc. IAMAP Intern. Symposium, 23–30 June, 1965, Moscow.
47. Monin, A. S.: 1969, *Weather Forecasting as a Problem of Physics*, Nauka, Moscow (in Russian); 1972, English translation, MIT Press, Cambridge, Mass.
48. Monin, A. S. and Yaglom, A. M.: 1965 – vol. 1, 1967 – vol. 2, *Statistical Fluid Mechanics*, Nauka, Moscow (in Russian); 1971 and 1975, English translation, MIT Press, Cambridge, Mass.
49. Morel, P. (ed.): 1973, *Dynamic Meteorology*, D. Reidel, Dordrecht.
50. Mussaelyan, Sh. A.: 1961, *Mountain Waves in the Atmosphere*, Gidrometeoizdat, Leningrad (in Russian).
51. Ozmidov, R. V.: 1968, *Horizontal Turbulence and Turbulence Exchange in the Ocean*, Nauka, Moscow (in Russian).
52. Panchev, S.: 1968, *Dynamic Meteorology*, Nauka i Izkustvo, Sofia (in Bulgarian).
53. Panchev, S.: 1971, *Random Functions and Turbulence*, Pergamon Press, Oxford.
54. Panchev, S.: 1976, *Mathematical Spectral Analysis*, Nauka i Izkustvo, Sofia (in Bulgarian).
55. Pedlosky, J.: 1979, *Geophysical Fluid Dynamics*, Springer, N.Y.
56. Petterssen, S.: 1956, *Weather Analysis and Forecasting*, McGraw-Hill, N.Y.
57. Prandtl, L.: 1942, *Strömungslehre*. Vieweg & Sohn, Braunschweig.
58. Proceedings of the WMO/IUGG Symposium on Numerical Weather Prediction in Tokyo, Nov. – Dec. 1968, Japan Meteorol. Agency, Tokyo, 1969.
59. Scorer, R. S.: 1978, *Environmental Aerodynamics*, John Wiley, N.Y.
60. Starr, V. P.: 1968, *Physics of Negative Viscosity Phenomena*, McGraw-Hill, N.Y.

61. Syono, S. (ed.): 1961, *Proceedings of the International Symposium on Numerical Weather Prediction in Tokyo*, Met. Soc. of Japan.
62. Tatarsky, V. I.: 1967, *Propagation of Waves in Turbulent Atmosphere*, Nauka, Moscow (in Russian).
63. Thennekes, H. and Lumley, J. L.: 1972, *A First Course in Turbulence*, MIT Press, Cambridge, Mass.
64. Thompson, Ph.D.: 1961, *Numerical Weather Analysis and Prediction*, Macmillan, N.Y.
65. Turner, J. S.: 1973, *Buoyancy Effects in Fluids*, Cambridge University Press.
66. Van Meighem, J,: 1973, *Atmospheric Energetics*, Clarendon Press, Oxford.
67. Vapniar, D. U.: 1976, *Planetary Waves and Currents in the Equatorial Ocean*, Naukova Dumka, Kiev (in Russian).
68. Whitham, G. B.: 1974, *Linear and Nonlinear Waves*, John Wiley, N.Y., London.
69. Wipperman, F.: 1973, *The Planetary Boundary Layer of the Atmosphere*, Deutscher Wetterdienst, Offenbach.
70. Yaglom, A. M. and Tatarsky, V. I. (eds.): 1967, *Atmospheric Turbulence and Radio Wave Propagation*, Nauka, Moscow.
71. Yudin, M. I.: 1963, *New Methods and Problems in the Short Range Weather Prediction*, Gidrometeoizdat, Leningrad (in Russian).
72. Zilitinkevich S. S.: 1970, *Dynamics of Atmospheric Boundary Layer*, Gidrometeoizdat, Leningrad (in Russian).

II. PAPERS FROM SCIENTIFIC JOURNALS

Chapter 2

Kasahara, A.: 1974, *Monthly Weather Rev.* **102**, 509–522.

Chapter 3

Chakalov, R. and Panchev, S.: 1977, *Bulg. Geophys. J.* **3**, No. 3, 21–28.
Chakalov, R. and Panchev, S.: 1978, *Bulg. Geophys. J.* **4**, No. 3, 17–28.
Dobrishman, E. M.: 1964, *Meteorologia i Gidrologia* (*Meteorology and Hydrology, USSR*) No. 5, 10–19 (in Russian).
Ertel, H.: 1942, *Meteorol. Z.* **59**, 277–281 (in German).
Herbert, F.: 1978, *Arch. Meteorol. Geophys. Bioclim. Ser A* **27**, 87–93.
Ingel, L. H. and Lepikash, E. R.: 1981, *Meteorologia i. Gidrologia*, 47–54.
Lorenz, E. N.: 1963, *J. Atmos. Sci.* **20**, 130–141.
Monin, A. S.: 1972, *Fizika Atmosferi i Okeana* (*Physics of the Atmosphere and Ocean*) **8**, 1035–1041 (in Russian).
Nguyen, Suan Zui and Yudin, M. I.: 1972, *Fizika Atmosferi i Okeana* **8**, 1029–1035. (in Russian).
Obukhov, A. M.: 1962, *Dokl. Akad. Nauk.* (*Bulletin of the Academy of Sciences of the USSR*) **145**, 1239–1242.
Obukhov, A. M.: 1964, *Meteorologia i. Gidrologia*, No. 2, 3–9.
Panchev, S.: 1983, *Bulg. Geophys. J.* **9**, No. 2, 3–13.
Platzman, G. W.: 1964, *Tellus*, **4**.
Zaltzman, B.: 1962, *J. Atmos. Sci.* **19**, 329–341.
Shirer, H. N. and Dutton, J. A.: 1979, *J. Atmos. Sci.* **36**, 1705–1721.
Wu, R. and Blumen, W.: 1982, *J. Atmos. Sci.* **39**, 1774–1782.

Chapter 4

Kasahara, A. 1974, *Monthly Weather Rev.* **102**, 509–522.
Lorenz, E. N.: 1955, *Tellus* **7**, 157–167.
Lorenz, E. N.: 1960, *Tellus* **12**, 346–373.
Plumb, R. A.: 1982, *J. Atmos. Sci.* **40**, 1669–1688.

Chapter 5

Abdullah, A. S.: 1949, *J. Meteorol.* **6**, 86–97.
Abdullah, A. S.: 1955, *Bull. Amer. Meteorol. Soc.* **36**, 515–518.
Abdullah, A. S.: 1956, *J. Meteorol.* **13**, 381–387.
Blandford, R.: 1966, *Deep Sea Res.* **13**, 941–961.
Christie, D. R., Muirhead, K. J., and Hales, A. L.: 1978, *J. Atmos. Sci.* **35**, 805–525.
Dobrishman, E. M.: 1977, *Meteorologia i Gidrologia*, No. 1, 92–103 (in Russian).
Eady, E. J.: 1949, *Tellus* **1**, 33–52.
Kuo, H. L.: 1951, *Tellus*, **3**, 268–284.
Matsuno, T.: 1966, *J. Meteorol. Soc. Jpn.* **11**, 44, 25–43.
Redekopp, L. G. and Weidman, P. D.: 1978, *J. Atmos. Sci.* **35**, 790–804.
Wiin-Nielsen, A. C.: 1975, *European Center for Medium-Range Weather Forecasts. Seminar*, Part 1. Reading, 1–2 Sept., 139–202.

Chapter 6

Blumen, W.: 1972, *Rev. Geophys. Space Sci.* **10**, 485–528.
Dikii, L. A.: 1969, *Fizika Atmosferi i Okeana* **5**, 188–192 (in Russian).
Yaglom, A. M.: 1953, *Izv. Acad. Sci. USSR, Ser. Geophys.*, 346–369 (in Russian).
Wiin-Nielsen, A. C.: 1976, *Beit. Phys. Atmos.* **49**, 245–271.

Chapter 7

Fjortoft, R.: 1953, *Tellus* **5**, 225–230.
Kibel, I. A.: 1958, *Dokl. Akad. Nauk.* **118**, 687–690 (in Russian).
Kibel, I. A.: 1960, *Dokl. Akad. Nauk.* **132**, 319–322 (in Russian).
Kibel, I. A.: 1962, *Dokl. Akad. Nauk.* **143**, 1336–1339 (in Russian).
Lorenz, E. N.: 1960, *Tellus* **12**, 364–373.
Lorenz, E. N.: 1965, *Tellus* **17**, 321–333.
Lorenz, E. N.: 1959, *Tellus* **21**, 289–307.
Lorenz, E. N.: 1981, *European Center for Medium-Range Weather Forecasts Seminar*, Reading, 14–18 Sept., 1–20.
Lorenz, E. N. 1982, *J. Meteorol. Soc. Jpn. Ser II* **60**, 255–267.
Novikov, E. A.: 1959, *Izv. Acad. Sci. USSR, Ser. Geophys.*, 1721–1723 (in Russian).
Sasaki, Y.: 1958, *J. Meteorol. Soc. Jpn.* **36**, 77–88.
Scorer, R.: 1967, *Proc. Symp. Mountain Meteorology, Colorado, USA.*
Thompson, Ph.D.: 1957, *Tellus* **9**, 275–295.

Chapter 9

Bolgiano, R.: 1959, *J. Geophys. Res.* **64**, 2226–2229.
Obukhov, A. M.: 1959, *Dokl. Akad. Nauk.* **125**, 1246–1248 (in Russian).

Chapter 10

Blackadar, A. K.: 1962, *J. Geophys. Res.* **67**, 3095–3102.
Bodin, Sv.: 1978: *J. Rech. Atmos.* **12**, 71–95.
Defant, F.: 1949, *Arch. Meteorol. Geophys. Bioclim.* **A1**, 421–450.
Egger, J.: 1981, *Beit. Phys. Atmos.* **54**, 465–482.
Godev, N.: 1970, *Arch. Meteorol. Geophys. Bioclim. Ser. A* **19**, 29–46, 299–310.
Godev, N.: 1975, *Bulg. Geophys. J.* **1**, No. 2, 43–56 (in Bulgarian).
Godev, N.: 1978, *Bulg. Geophys. J.* **4**, No. 1, 12–20 (in Bulgarian).
Kasansky, A. B. and Monin, A. S.: 1960: *Izv. Acad. Sci. USSR Ser. Geophys.*, 165–168 (in Russian).
Lettau, H. H.: 1962, *Beit. Phys. Atmos.* **35**, 195–212.
Panchev, S.: 1955, *Hydrologia i Meteorologia* (*Hydrology and Meteorology, Bulgaria*), No. 5, 23–32 (in Bulgarian).
Panchev, S.: 1956, *Hydrologia i Meteorologia*, No. 1, 44–56 (in Bulgarian).
Panchev, S.: 1977, *Bulg. Geophys. J.* **3**, No. 1, 22–29 (in Bulgarian).
Panchev, S. and Atanassov, D.: 1978, *Hydrologia i Meteorologia*, No. 5, 3–10 (in Bulgarian).

Panchev, S. and Atanassov, D.: 1979, *Bulg. Geophys. J.* **5**, No. 1, 10–18.
Panchev, S. and Stanev, E.: 1977, *Meteorologia i Gidrologia*, No. 4, 29–34 (in Russian).
Rossby, C. G.: 1932, *MIT Meteorol. Paper* **1**, No. 4.
Syrakov, E.: 1981. *IX Conference on Karpatian Meteorology*, Sofia, 13–16 Nov. 1979, 289–294.

Chapter 11

Batchelor, G. K.: 1979, *Phys. Fluids*, suppl. II, 223–239.
Blinova, E. N.: 1943, *Dokl. Acad. Nauk.* **39**, 284–287 (in Russian).
Blinova, E. N.: 1958, *Dokl. Acad. Nauk.* **123**, 440–442 (in Russian).
Blinova, E. N.: 1962, *Dokl. Acad. Nauk.* **147**, 1355–1358 (in Russian).
Boltenkov, V. P.: 1966, *Trudi Glavnoi Geofizicheskoi Observatorii* (*Proc. of the Main Geophysical Observatory, USSR*), No. 191, 58–71 (in Russian).
Charney, J. G.: 1971, *J. Atmos. Res.* **28**, 1087–1095.
Gavrilin, B. L. 1965, *Fizika Atmosferi i Okeana* **1**, 1229–1257 (in Russian).
Kao, S. K. and Bullock, W. S.: 1964, *Quart. J. Royal Meteorol. Soc.* **90**, 166–174.
Kasahara, A. and Washington, W. M.: 1971, *J. Atmos. Sci.* **28**, 657–701.
Kraichnan, R.: 1967, *Phys. Fluids* **10**, 1417–1423.
Leith, C. E.: 1971, *J. Atmos. Sci.* **28**, 145–161.
Lorenz, E. N.: 1969, *Tellus* **21**, 289–307.
Monin, A. S.: 1958, *Izv. Acad. Sci. USSR, Ser. Geophys.*, 1250–1253 (in Russian).
Morel, P. and Larcheveque, M.: 1974, *J. Atmos. Sci.* **31**, 2189–2196.
Phillips, N. A.: 1956, *Quart. J. Royal Meteorol. Soc.* **82**, 123 pp.
Smagorinsky, J.: 1963, *Monthly Weather Rev.* **91**, 99–165.
Van der Hoven, J.: 1957, *J. Meteorol.* **14**, 160–164.

APPENDIX

Retrospective view of Dynamic Meteorology; Perspectives

BRIEF HISTORY

The goal of dynamic meteorology is the theoretical study of atmospheric motions and then applying the results to weather prediction. In the nineteenth century, theoretical meteorologists were occupied with similar investigations. Their efforts were concentrated on the construction of a set of equations of motion, combined with those of atmospheric thermodynamics, and obtaining some simple analytical solutions. However, the paper of V. Bjerknes, 'The Problem of Weather Prediction from the Point of View of Mathematics and Mechanics' [1], published in 1904, can be accepted as the beginning of contemporary dynamic meteorology. In that paper, for the first time, the problem is formulated as an initial value problem for the hydrodynamic equations of a baroclinic fluid. This problem received further development in the works of J. Bjerknes and many other representatives of the famous Bergen School, founded by V. Bjerknes. As is well known, its name is connected with the development of the concept of atmospheric fronts (boundary surfaces between different air masses) and the formation of cyclones as a result of the instability of wave motions on frontal surfaces. These concepts were considerably developed further in the theoretical aspect by two eminent representatives of the young Soviet School of dynamic meteorology, A. A. Friedman and N. E. Kochin and became the basis of the contemporary synoptic method for short-range weather prediction.

A crucial event in the history of dynamic meteorology, which initiated a new scientific discipline, was the appearance of Richardson's work *Weather Prediction by Numerical Process* [11] in 1922. The history of this little-known Richardson work is curious. He started his investigations in 1913 and continued them on the front line during World War I as a British army officer. All his computations were lost during the war and only some years afterwards were they gradually recovered and eventually published in 1922.

Richardson himself accomplished all computations using his own method. It took approximately a year to prepare a 24-hour forecast. He reckoned that in order to keep pace with time, that is, to prepare a 24-hour forecast some 64 000 calculators were needed. However, such a forecast would have no practical value for it could only be completed after the event. A useful forecast should be accomplished earlier which means hiring more calculators. That, however, didn't bother Richardson as much as the fact that, in spite of the correctness of the computations, the final result was unrealistic. The only forecast he managed to prepare for 20 May 1910 for the Nurenberg-Augsburg region proved to be completely unsatisfactory. Nevertheless, he described his attempt and the results obtained in his book, which still enjoys great vogue, without being aware of the reasons for the failure. Those reasons are now clear:

– Insufficient initial data. In those years only surface routine observations were made by the existing meteorological network.

– At the dawn of numerical methods for solving partial differential equations, Richardson, being one of the founders, used an inadequate numerical scheme. Precisely, the later-discovered Courant–Friedrichs–Levy (CFL) stability criterion was not observed.

– Richardson used unnecessarily sophisticated equations, describing not only the slow synoptic processes but also a variety of fast wave motions ('parasitic noises') without any meteorological significance.

Under the influence of Richardson's unsuccessful attempt at numerical weather forecasting and the exceptional labor-consumption of the new method proposed by him, an opinion was formed that numerical weather prediction was a hopeless task. This opinion reigned in meteorology for nearly two decades.

Meanwhile, in 1930, the Soviet aerologist P. A. Molchanov invented the radiosonde – that indispensible instrument, used for mass observations of principal meteorological elements (wind, pressure, temperature and humidity) to altitudes of up to 30 or more kilometers. Numerical methods for integrating differential equations were also developed to a considerably high degree. In particular, the above-mentioned CFL stability criterion was established, thus, creating arguments for reducing and minimizing the influence of the first two reasons.

A way of overcoming the third reason, which has much a more physical than mathematical aspect, was shown by I. A. Kibel [9] in his classical work of 1940. He proposed a fundamental principle for the simplification of weather equations through asymptotic 'quasi-geostrophic expansion' (expansion on a small parameter) which filters 'parasitic noises' from the equations and retains the meteorologically-significant motions. Further development and refinement of this principle, as is known, led to the construction of the modern hydrodynamic theory for short-range weather prediction.

The year 1940 is also remarkable for the appearance of another classical work that left deep traces in the history of the dynamics of the atmosphere (and ocean) and is related to the name of another colossus in this science – the Swede C. G. A. Rossby [12]. In this independent work and in a collective one from the previous year, the data from observations were analyzed and a theory was created of the phenomenon now called planetary (long) Rossby waves, which has great meteorological significance. Essentially, a filtering problem is solved but only in one particular case. The name of Rossby is also connected with one of the rather complete formulations of the theorem of potential vorticity conservation and to the earlier problem of mutual adjustment process (in an oceanographic aspect).

The next link in this series of landmark dates includes the 1947–1950 period when the contemporary formulation of the problem of quasi-geostrophic approximation in weather equations, namely in the vorticity equation, was given. This important theoretical achievement is related, first of all, to the names of the American theoretical meteorologist J. Charney [3–6] and the Russian A. M. Obukhov [10] who published their studies during that period. Thus, the artificial boundary condition at the level of the tropopause, imposed by Kibel, which did not follow from hydrodynamic laws and impeded the revelation of the great potentialities of his method was removed.

It was a fortunate coincidence that during this period the possibility emerged of applying the first electronic computer, ENIAC (Electronic Numerical Integrator and Computer), built in 1945. Furthermore, during the elaboration of the computer, its principal constructor John van Neumann had participated in solving the problem of numerical weather prediction. In collaboration with Charney and Fjortoft in 1950, von Neumann [6] published the first scientific paper on the computer realization of the barotropic quasi-geostrophic prediction scheme. Thus, the onset of a new branch in the meteorological prediction practice was established – the computer-aided prediction of weather elements.

Over the past 30 years many generations of computers have been constructed, but meteorological problems have always been among those demanding the greatest requirements for storage and speed, thus stimulating their development. This holds first of all for the problems of climate theory and long-range weather prediction whose physical foundations are suggested in the classical work of E. N. Blinova [2].

Among this constellation of names of distinguished scientists who left bright traces in the history of dynamic meteorology, the name and work of the Swedish oceanologist V. W. Ekman [8], the founder of the theory of the drift and gradient currents in the ocean, remains a little at one side on first sight. The history of his classical work from 1905 is little known and rather curious.

The famous Norwegian seafarer and scientist F. Nansen was the first to note the rotation of the velocity vector of ocean currents with depth during his measurements in the Atlantic. He guessed that one of the reasons was certainly the Coriolis force but asked his compatriot V. Bjerknes to treat this phenomenon mathematically. However, the equally well-known Bjerknes assigned this task to his young collaborator V. W. Ekman with whom Nansen discussed his discovery and his suggestions about the reasons for the observed rotation. "Right in the evening of the same day", Bjerknes writes later in his memoirs, "Ekman brought the solution of the problem in the form of the now well-known Ekman spiral." Therefore, it was not principally difficult to "put on feet" Ekman's solution, i.e., to relate it to the wind rotation with height in the so-called atmospheric planetary boundary layer (PBL). Knowing the immense role of the interaction between the atmosphere and underlying surface (particularly the ocean), realized through boundary layers, we can now estimate the importance of the scientific heritage of Ekman and his followers and contemporaries (Sverdrup, Hesselberg, etc.). The notion of the 'Ekman Boundary Layer' is now one of the most frequently encountered notions in atmospheric and ocean dynamics.

Many living representatives of this vigorously developing scientific discipline have contributed and still contribute substantially to its development. Practice, the supreme judge of theory, will assign to these efforts an appropriate recognition.

PERSPECTIVES

Each scientific discipline, including dynamic meteorology, has periods of vigorous development and periods of crises. As a rule, crises are due to two radically different reasons:

– The majority of investigators carry out experiments (observations) and process the results, but very few of them give these results enough consideration relating to analysis and generalization.

– Many of them are occupied with theory (calculating something) while only a few process the observational data in order to make a comparison with theory.

If dynamic meteorology is now considered to be undergoing a somewhat critical period, this is probably due to the second reason in spite of the growing information flux about the processes in the atmosphere and the ocean obtained by classical and modern observational means (satellites, etc.), and the available powerful computer equipment for its processing. Besides, the genuine theories – representing internally consistent systems of outlooks on the physical causes for one or another phenomenon – set in a proper mathematical framework, an innumerable variety of pseudo-theories exists whose authors, misunderstanding the tasks of the theory because they are incapable of doing anything else, just perform mathematical exercises which attempt to give a 'theoretical character' to their work. They do not leave any positive trace. On the contrary, such work impedes progress.

We are tempted to cite here the aphorism borrowed from [13]: "Each theory (the physico-mathematical too) is so much better elaborated as less mathematics is needed for the interpretation (and not getting!) of its results."

Finally, combining the speculative character of many of the existing 'theories' with the uniqueness of the concrete atmospheric processes and phenomena and of their observations, the difficulties of this science become clear. That is why we are not surprised by the statements of "good or perfect correspondence of the theory to the empirical data" seen in the conclusions of many papers.

In spite of all this, dynamic meteorology has not exhausted its object of study nor the methods available. Following [7], we can identify four groups of problems:

1. The study of the interaction of motions with different spatial scales. The corresponding spectrum in the atmosphere occupies 11 orders of magnitude: from 1 mm to 10^5 km. Strictly speaking, a motion of a certain scale interacts with all the remaining motions because of the nonlinearity of the equations. Therefore, research into the motions in separate independent spectral regions should proceed to a more comprehensive study of the interaction between them.

2. Predictability of atmospheric motions. The problem relates to the sensitivity of the solutions to small differences in the initial or boundary conditions or to the approximation of one or other terms in the equations. The behaviour of the real atmosphere suggests the importance of this problem. At constant astronomic and geophysical factors (solar energy influx, the Earth's rotation rate, ocean-continent distributions, topography, etc.), in spite of some common features, a great variety in the realization of atmospheric motions is observed. Why? The problem has yet to be treated with the necessary thoroughness.

3. Climate theory. While the previous problem was related to the predictability of individual meteorologically-significant forms of motion, here we have in view those global phenomena, determined by the above-mentioned constantly operating factors. They determine the climate of our planet and their causal relationship in the past could give us the key to the superlong-range forecast for the future. That is why the studies of climate theory are in the front line of dynamic meteorology.

4. Inclusion of moisture convection in experimental (numerical) and prognostic models of the atmosphere and accounting for the role of clouds and precipitation on the remaining processes and phenomena of various scales. This problem is still far from its solution.

Other classifications of the foregoing problems and more detailed specifications are also possible. In all cases, however, it is reasonable to believe in the forecast of scientific achievement in the next century, according to which such problems as controlled thermonuclear reactions, the conquest of the solar system, and some others which now appear as a nearer or further perspective will be solved first and only then will come the turn of the problem of weather prediction for an arbitrary (practically!) range in the future.

The problem for a more reliable and for a longer period weather forecast is extremely difficult and it will be a problem of the twenty-first century too!

REFERENCES

1. Bjerknes, V.: 1904, 'Das Problem der Wettervorhersage, Betrachtet vom Standpunkt der Mechanik und der Physik', *Meteor. Z.* **21**, 1–7.
2. Blinova, E. N.: 1943, 'Hydrodynamical Theory of the Pressure Waves and Atmospheric Centers of Action', *Dokl. Acad. Nauk.* **39**, 284–287 (in Russian).
3. Charney, J.: 1947, 'On the Scale of Atmospheric Motions', *Geophys. Publ. Norske Vids. Akad., Oslo* **17**, No. 2, 7 pp.
4. Charney, J.: 1947, 'The Dynamics of Long Waves in a Baroclinic Westerly Current', *J. Meteorol.* **4**, 135–163.
5. Charney, J.: 1949, 'On a Physical Basis for Numerical Prediction of Large-Scale Motions in the Atmosphere', *J. Meteorol.* **6**, 371–385.
6. Charney, J., Fjørtoft, R., and von Neumann, J.: 1950, 'Numerical Integration of the Barotropic Vorticity Equation', *Tellus*, 237–254.
7. Dutton, J. A.: 1976, *The Ceaseless Wind*, McGraw-Hill, N.Y.
8. Ekman, V. W.: 1905, 'On the Influence of the Earth's Rotation on Ocean Currents', *Arch. Math. Astron.* **2**, 1–52.
9. Kibel, I. A.: 1940, 'Application to Meteorology of the Fluid Mechanics Equations – Baroclinic Cases', *Izv. Acad. Sci. USSR, Ser. Georg. Geophys.*, 627–638 (in Russian).
10. Obukhov, A. M.: 1949, 'On the Problem of Geostrophyc Wind', *Izv. Acad. Sci. USSR, Ser. Geogr. Geophys.*, 281–306 (in Russian).
11. Richardson, L. F.: 1922, *Weather Prediction by Numerical Process*, Cambridge Univ. Press, London (reprinted: Dover, 1965), 236 pp.
12. Rossby, C. G.: 1940, 'Planetary Flow Patterns in the Atmosphere', *Quart. J. Royal Meteorol. Soc.* **66**, 68–97.
13. Shtokman, V. B.: 1970, *Collected Papers on Ocean Physics*, Gidrometeoizdat, Leningrad (in Russian).

Biographical Data

1. O. Reynolds (1842–1912)
2. A. A. Friedman (1888–1925)
3. N. E. Kochin (1901–1944)
4. V. Bjerknes (1862–1951)
5. L. F. Richardson (1881–1953)
6. L. Prandtl (1875–1953)
7. V. W. Ekman (1874–1954)
8. C.-G. Rossby (1898–1957)
9. J. von Neumann (1903–1957)
10. H. U. Sverdrup (1888–1957)
11. T. von Kármán (1881–1963)
12. V. Väisälä (1889–1969)
13. I. A. Kibel (1904–1970)
14. G. I. Taylor (1866–1975)
15. V. P. Starr (1908–1976)
16. E. N. Blinova (1906–1981)
17. J. Charney (1917–1981)

1. *O. Reynolds* (1842–1912). English physicist and founder of the turbulent motion theory. He proposed that a turbulent flow could be regarded as a superposition of an averaged (laminar-like) motion and a fluctuating one and derived the well-known equations for the mean motion, which now bear his name. Reynolds was the author of the first works on the stability of laminar flows and transition to turbulence.

2. *A. A. Friedman* (1888–1925). An eminent Soviet natural scientist and mathematician. His cosmological model of the expanding Universe based on the solution of Einstein's equations is well known. Friedman's contributions to the development of fluid dynamics and theoretical meteorology are equally important. With L. V. Keller he cofounded the contemporary statistical theory of turbulence. In 1924 they proposed various correlation functions of a most general type to be considered as the fundamental statistical characteristics of turbulence and derived the corresponding dynamic equations for them. These are known as the Friedmann–Keller equations.

3. *N. E. Kochin* (1901–1944). Soviet mathematician, hydrodynamicist and meteorologist. Cofounder with A. A Friedman of the Soviet school of dynamic meteorology, later representatives of which were I. A. Kibel and E. N. Blinova. He first proposed a hydrodynamical theory of general atmospheric circulation. Kochin's publications in dynamic meteorology are collected in one volume [31].

4. *V. Bjerknes* (1862–1951). Norwegian scientist, founder of the famous Bergen school in meteorology. He first proposed a mathematical formulation of the weather prediction problem on the basis of the general equations of atmospheric thermo- and fluid dynamics.

5. *L. F. Richardson* (1881–1953). An English scientist with encyclopedic knowledge, who made considerable contributions to the theory of differential equations and dynamic meteorology. He first realized Bjerknes' idea for weather prediction by applying numerical methods developed by him [11]. Richardson also made considerable contributions to the theory of turbulence.

6. *L. Prandtl* (1875–1953). German scientist who, in 1940, first derived the equations of fluid motion in boundary layers. Cofounder with T. von Kàrmàn of the semi-empirical theory of turbulence which was later applied with some necessary generalizations to the case of atmospheric turbulence, particularly in the boundary layer.

7. *V. W. Ekman* (1874–1954). Swedish oceanographer and originator of the theory of drift and gradient currents. In connection with these works, his name has entered into scientific terminology: a homogeneous boundary later which appears in a rotating fluid under the influence of the Coriolis and frictional forces, is called the Ekman layer. In both the atmosphere and ocean these are turbulent layers.

8. *C.-G. Rossby* (1898–1957). Swedish meteorologist and oceanographer, author of fundamental studies and papers on what is now called geophysical fluid dynamics. Rossby was also a great organizer of science and a teacher. His collaborators have been such prominent scientists as B. Bolin (Sweden), A. Eliassen (Norway), J. Charney, N. Phillips and H. Stommel (U.S.A.), and many others. (Photo by permission of G. W. Platzman.)

9. *J. von Neumann* (1903–1957). American mathematician and physicist of Hungarian origin, author of brilliant papers on mathematics and quantum mechanics, and founder of the automation theory on the basis of which the first electronic computer, ENIAC, was built in 1945 at Princeton University. He took an active part in the realization of the first numerical weather forecasts by ENIAC. (Photo by permission of G. W. Platzman.)

10. *H. U. Swerdrup* (1888–1957). Norwegian geophysicist who also worked in the U.S.A., author of fundamental papers on the hydrodynamical theory of marine currents. Swerdrup was one of the first to point out the importance of the air-sea interaction process.

11. *T. von Kàrmàn* (1881–1963). A prominent American mechanist and fluid dynamist of Hungarian origin, cofounder of the semiempirical and statistical theory of turbulence. (Photo by permission of the Royal Aeronautical Society.)

12. *V. Väisälä* (1889–1969). Mathematician and physicist in education, later Professor of Meteorology at Helsinki University (Finland). In 1930 Väisälä became interested in meteorological instrumentation. A year later after the Soviet aerologist Moltshanov, he constructed his own radiosonde, launched on 30 December 1931. In 1936 Väisälä founded a company for the commercial manufacturing of radiosondes, which is now a world-known company for meteorological instruments. (Photo by permission of Väisälä Co., Helsinki.)

13. *I. A. Kibel* (1904–1970). Soviet scientist-mathematician, fluid mechanist and meteorologist, author of the first successful solution to the problem for weather prediction on the basis of thermo- and fluid dynamic equations of the atmosphere (1940). Later on he made considerable contributions towards solving the problems of atmospheric dynamics (geostrophic adjustment, mesometeorology, etc.). After Richardson in 1922, Kibel (1957) was the author of the second book on the theory of weather forecasting. (Photo by permission of 'Nauka', Moscow.)

14. *G. I. Taylor* (1886–1975). English scientist and fluid mechanist, author of fundamental works on the semiempirical and statistical theory of turbulence. In 1935 he introduced the concept for isotropic turbulence, which was later developed by A. N. Kolmogoroff in his 1941 theory of locally homogeneous and isotropic turbulence. (Photo by permission of Prof. G. K. Batchelor.)

15. *V. P. Starr* (1908–1976). American scientist-meteorologist, who contributed much to the understanding of the 'physics of negative viscosity phenomena' encountered in Nature (earth and planetary atmospheres) and described in his unique book of the same title [60]. (Photo by permission of G. W. Platzman.)

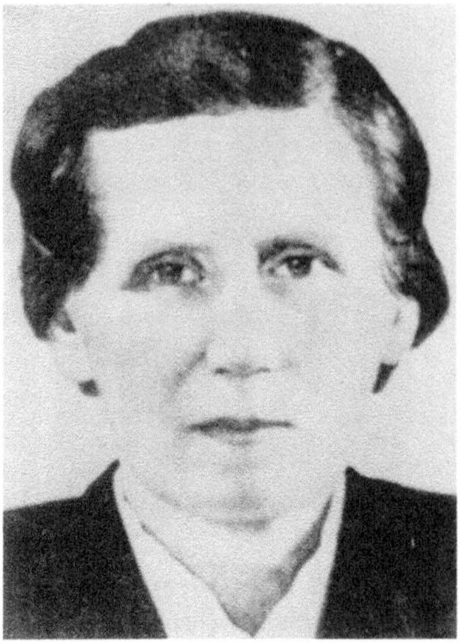

16. *E. N. Blinova* (1906–1981). Soviet meteorologist-theoretician. She is considered as the cofounder, during the 1940s, with Rossby and others of the contemporary theory of general atmospheric circulation. Kibel and Blinova were a unique married couple in the history of dynamic meteorology. (Photo by permission of 'Nauka', Moscow.)

17. *J. Charney* (1917–1981). One of the most prominent American meteorology theoreticians. Charney related his name to the formulation of the quasi-geostrophic equations of motion in the late forties and with a successful machine realization of the first scheme for numerical weather forecasting in 1950. Among his later achievements of great importance is the theory of the so-called quasi-geostrophic turbulence (1971). (Photo by permission of N. A. Phillips.)

Index

The manufacturer's authorised representative in the EU is Springer
Nature Customer Service Centre GmbH, Europaplatz 3, 69115 Heidelberg,
Germany. If you have any concerns regarding our products, please
contact ProductSafety@springernature.com

Printed and bound by CPI Group (UK) Ltd, Croydon, CR0 4YY
22/09/2026
02220985-0002